Introduction to Biological and Small Molecule Drug Research and Development

Introduction to Biological and Small Molecule Drug Research and Development

Theory and Case Studies

Edited By

Robin Ganellin

Stanley Roberts

Roy Jefferis

AMSTERDAM • BOSTON • HEIDELBERG • LONDON
NEW YORK • OXFORD • PARIS • SAN DIEGO
SAN FRANCISCO • SINGAPORE • SYDNEY • TOKYO

Academic Press is an imprint of Elsevier

Academic Press is an imprint of Elsevier
225, Wyman Street, Waltham, MA 02451, USA
The Boulevard, Langford Lane, Kidlington, Oxford OX5 1GB, UK
Radarweg 29, PO Box 211, 1000 AE Amsterdam, The Netherlands

Notice
No responsibility is assumed by the publisher for any injury and/or damage to persons or property as a matter
of products liability, negligence or otherwise, or from any use or operation of any methods, products, instructions
or ideas contained in the material herein. Because of rapid advances in the medical sciences, in particular,
independent verification of diagnoses and drug dosages should be made

British Library Cataloguing in Publication Data
A catalogue record for this book is available from the British Library

Library of Congress Cataloging-in-Publication Data
A catalog record for this book is available from the Library of Congress

ISBN: 978-0-12-397176-0

For information on all Academic Press publications
visit our website at www.store.elsevier.com

Working together
to grow libraries in
developing countries

www.elsevier.com • www.bookaid.org

Contents

Biographies xiii
Preface xix

CHAPTER 1 **Introduction to enzymes, receptors and the action of small molecule drugs** 1
    1.1. Section I: Background Information 2
        1.1.1. Communication between cells: the roles of receptors and enzymes 2
        1.1.2. Neurotransmitters, receptors and the nervous system 9
        1.1.3. Introduction to enzymes and enzyme inhibitors 15
        1.1.4. Other types of bioactive molecules 15
        1.1.5. Factors influencing drug action 16
        1.1.6. The impact of the sequencing of the human genome 19
    1.2. Section II: More About Enzymes 19
        1.2.1. Configuration of enzymes 19
        1.2.2. Enzyme specificity, classification and nomenclature 21
        1.2.3. Characteristics of enzyme catalysis 21
        1.2.4. Enzyme reaction rates 25
        1.2.5. Enzyme substrates as drugs 27
        1.2.6. Enzyme inhibition and enzyme inhibitors as drugs 28
        1.2.7. Enzyme regulation 32
    1.3. Section III: More About Receptors 35
        1.3.1. Bioassay and the measurement of drug effects 35
        1.3.2. Quantifying drug–receptor interactions 37
        1.3.3. Radioligand-binding studies – a direct measure of occupancy 39
        1.3.4. Receptor structure 40
        1.3.5. Relating occupancy to response 44
        1.3.6. Competitive antagonism and the Schild equation 48
        1.3.7. Desensitization and the control of receptor number 49
        1.3.8. Partial agonists, agonist efficacy and inverse agonism 51
    1.4. Section IV 53
        1.4.1. Conclusions: uncertainties in drug design and development 53
        Further Reading 55

CHAPTER 2 **Protein structure and function** 57
    2.1. Introduction 58
    2.2. General Aspects of Protein Structure 58
        2.2.1. Primary, secondary, tertiary and quaternary structure 63
        2.2.2. Protein folds and structural bioinformatics 69
    2.3. Enzymes 71
        2.3.1. Enzyme inhibitors 71
        2.3.2. Examples of enzyme inhibition 72

2.4. Protein Receptors...74
2.4.1. Specific examples of protein receptors 74
2.5. Structural Proteins as Drug Targets..75
2.6. Proteins as Drugs..76
2.7. Concluding Remarks 77
References...78

CHAPTER 3 **The small molecule drug discovery process – from target
selection to candidate selection** **81**
3.1. Introduction...82
3.2. Where Do Leads Come From?...84
3.2.1. Definitions...84
3.2.2. High-throughput screening 85
3.2.3. Structure-based drug design and virtual screening...............86
3.2.4. Natural products 89
3.2.5. Fragment-based drug discovery 91
3.2.6. Fast-follower approaches..94
3.3. Lead Generation: Active-to-Hit..95
3.4. Lead Generation: Hit-to-Lead 96
3.4.1. Lipophilicity 96
3.4.2. Solubility...97
3.4.3. DMPK considerations...98
3.4.4. Transporter proteins 104
3.4.5. Plasma protein binding and whole blood potency...............105
3.4.6. Inhibition and induction of CYP enzymes...........................107
3.4.7. Time-dependent CYP inhibition 108
3.4.8. Ligand lipophilicity efficiency – identifying and working at the
'leading edge' 109
3.4.9. Hit and lead criteria..109
3.4.10. Case study – IKK-2 inhibitors 111
3.5. LO – Establishing the Screening Cascade and Candidate Biological
Target Profile...113
3.5.1. Initial safety considerations and potential liabilities 114
3.5.2. Therapeutic ratio/margin of safety......................................115
3.5.3. Common toxicities...115
3.5.4. Cardiovascular side-effects – hERG 115
3.5.5. Genotoxicity caused by Ames-positive anilines...................119
3.5.6. Phospholipidosis 119
3.5.7. Pharmacokinetic–pharmacodynamic relationship and dose-to-man
prediction 120
3.5.8. LO case study – the discovery of aprepitant 121
References and Notes 123

CHAPTER 4 **Protein therapeutics (introduction to biopharmaceuticals)** **127**
4.1. Introduction..128
4.1.1. Brief history of protein therapeutics 129
4.1.2. Overview: Protein therapeutics compared to small molecule drugs.........131

4.2. Types of Protein Therapeutics I: Regulatory and Enzymatic Activity 133
    4.2.1. Insulin 133
    4.2.2. Interferons.. 134
    4.2.3. Epoetins 135
    4.2.4. Granulocyte colony-stimulating factor (G-CSF)/granulocyte-
            macrophage colony-stimulating factor (GM-CSF) 136
    4.2.5. Growth factors 137
    4.2.6. Coagulation and fibrinolytic regulation 138
    4.2.7. Therapeutic enzymes 138
    4.2.8. Peptide therapeutics 140
4.3. Types of Protein Therapeutics II: Targeting Activity 141
    4.3.1. Monoclonal antibodies 142
    4.3.2. Monoclonal antibody therapeutic discovery and engineering................ 146
    4.3.3. Fc engineering 147
    4.3.4. Fc fusion proteins (FcFPs) 147
    4.3.5. Engineered antibody fragment therapeutics...................................... 149
    4.3.6. Non-antibody binding proteins (NABP) 151
4.4. Challenges of Protein Therapeutics 153
    4.4.1. Immunogenicity... 153
    4.4.2. Developability and manufacturing 153
    4.4.3. Route of administration 155
4.5. Future Directions for Protein Therapeutics 156
4.6. Biosimilar Protein Therapeutics.. 157
4.7. Summary and Conclusions 158
    References.. 158

CHAPTER 5 **Similarities and differences in the discovery and use of
               biopharmaceuticals and small-molecule chemotherapeutics 161**
    5.1. Introduction... 163
    5.2. How do SMDs Differ from Biomolecular Drugs?................................. 169
         5.2.1. Structure 169
         5.2.2. Distribution 174
         5.2.3. Metabolism 174
         5.2.4. Serum half-life 174
         5.2.5. Typical dosing regimen 174
         5.2.6. Species reactivity 174
         5.2.7. Antigenicity and hypersensitivity... 175
         5.2.8. Clearance mechanisms... 175
         5.2.9. Drug–drug interactions 175
         5.2.10. Pharmacology 176
    5.3. Historical Changes to the FDA Approach to Handle the Biotech Boom—
         the Differing Nomenclature for Small Molecules vs Biologics
         Entering the Clinic... 176
    5.4. Comparisons of Clinical Metrics—Biologics vs Small Molecules 179
         5.4.1. Overall clinical success rates of biologics vs small molecules 180
         5.4.2. Stage-related success rates for small molecules vs biologics.................... 180

        5.4.3. Stage-related clinical cycle times for small molecules vs biologics 180
        5.4.4. Comparative cost of R&D for biologics vs small molecules 183
        5.4.5. The challenges in comparing small-molecule and biomolecular drug
                metrics—the influence of biomolecular scaffold 185
  5.5. Are Peptide Drugs Small Molecules or Biologics? 186
  5.6. The Manufacture and Supply of SMDs vs Biomolecular Drugs 191
  5.7. The Pricing of SMDs vs Biomolecular Drugs 191
  5.8. Comparing Small-Molecule, Peptide and Biomolecular Drugs
        in the Market 192
        5.8.1. GPIIb/IIIa antagonists 192
        5.8.2. Her2 inhibitors 194
  5.9. Biosimilar Biomolecules vs Generic Small Molecules 195
  5.10. Discovery and Preclinical Stages for SMDs vs Biomolecular Drugs—Where
        the Technologies Differ the Most 196
        5.10.1. Target discovery 197
        5.10.2. Lead discovery 197
        5.10.3. Lead optimization 197
        5.10.4. Preclinical evaluation 198
        5.10.5. Clinical phases 198
  5.11. Small-Molecule and Biologics Approvals by Therapy Areas 198
  5.12. Managing Small-Molecule & Biomolecular Drug R&D in the Same
        Company 198
  5.13. Conclusion 200
        References 200

CHAPTER 6   Therapies for type 2 diabetes: modulating the incretin
            pathway using small molecule peptidase inhibitors or peptide
            mimetics 205
  6.1. Introduction 206
  6.2. Pharmacotherapy of Type 2 Diabetes 210
  6.3. The Rationale for Incretin-Based Therapies for Type 2 Diabetes 211
  6.4. Discovery and Pharmacokinetics of the Incretin-Based Therapies 214
  6.5. Clinical Efficacy of the Incretin-Based Therapies 216
        6.5.1. Glucose control 216
        6.5.2. Weight loss 218
  6.6. Evidence for Disease Modification 218
  6.7. Impact on Cardiovascular Risk 219
  6.8. Clinical Safety and Tolerability of Incretin-Based Therapies 220
        6.8.1. Nausea 220
        6.8.2. Anti-drug antibodies (see Chapter 3) 220
        6.8.3. Injection site reactions 220
        6.8.4. Miscellaneous risks 220
  6.9. Conclusions 221
        References 222

CHAPTER 7   The structure and business of biopharmaceutical companies including the management of risks and resources 225
    7.1. Introduction.. 227
        7.1.1. The business of biopharmaceutical research and development................ 227
        7.1.2. The science of biopharmaceutical R&D.................................... 230
    7.2. The Organization of Biopharmaceutical R&D 231
        7.2.1. Biopharmaceutical R&D is drug discovery and drug development.......... 231
        7.2.2. The logical organization of biopharmaceutical R&D 231
        7.2.3. Stage-gate organization – the project pipeline 232
        7.2.4. Stage-related goals... 234
        7.2.5. Attrition.. 235
        7.2.6. Risk – how it influences POS, cost, value and commitment................... 237
        7.2.7. Resource – who does what, when and where.............................. 238
        7.2.8. In-house vs. in-license 240
        7.2.9. In-house vs. out-license 241
        7.2.10. Big vs. small 242
        7.2.11. The necessity of standard operating procedures...................... 243
    7.3. Project Management in Biopharmaceutical R&D 244
    7.4. Portfolio Management in Biopharmaceutical R&D 244
    7.5. Cost Reduction Experiments in the Business of Biopharmaceutical Discovery and Development 245
        7.5.1. Process improvement 245
        7.5.2. Reducing resource – vertical disintegration............................ 245
        7.5.3. Increasing success through changes in the organizational model............ 249
    7.6. Conclusion 251
        References... 251

CHAPTER 8   Discovery and development of the anticancer agent gefitinib, an inhibitor of the epidermal growth factor receptor tyrosine kinase 255
    8.1. Introduction and Biological Background 256
    8.2. Biological and Chemical Approach to EGFR Tyrosine Kinase Inhibitors 258
        8.2.1. Isoflavones 258
        8.2.2. Anthelmintics... 259
        8.2.3. Anilinoquinazolines.. 259
    8.3. Structure–Activity Studies 262
    8.4. *In vivo* Studies 267
    8.5. Early Development 271
    8.6. Development and Clinical Studies 273
    8.7. Next-Generation Approaches 277
        8.7.1. Dual-ErbB inhibitors 277
        8.7.2. First-generation irreversible inhibitors.................................. 277
        8.7.3. Second-generation irreversible inhibitors 279
        8.7.4. EGFR monoclonal antibodies 279
        References... 279

CHAPTER 9   **Targeting HER2 by monoclonal antibodies for cancer therapy... 283**

    9.1. Introduction.. 284

    9.2. Structure and Function of HER2 285

    9.3. Trastuzumab (Herceptin™) 286

        9.3.1. Trastuzumab manufacturing and quality control 287

        9.3.2. Trastuzumab's multiple mechanisms of action....................... 288

        9.3.3. Nonclinical pharmacology 289

        9.3.4. Clinical data and therapeutic efficacy of trastuzumab 290

    9.4. Pertuzumab (Perjeta™) 297

        9.4.1. Mechanisms of action and preclinical activity
              of pertuzumab.. 297

        9.4.2. Clinical efficacy of pertuzumab 299

    9.5. Trastuzumab Emtansine (T-DM1; Kadcyla™) 300

        9.5.1. Characteristics of trastuzumab emtansine and its mechanisms
              of action.. 301

        9.5.2. Clinical efficacy.. 302

    9.6. Conclusion 303

         References 303

CHAPTER 10   **Recombinant human erythropoietin and its analogues.............. 307**

    10.1. Introduction 308

    10.2. Discovery of Erythropoietin (Epo) 309

    10.3. Physiology of Epo... 310

        10.3.1. Sites and control of production.. 310

        10.3.2. Structure of Epo 310

        10.3.3. *EPO* expression 313

        10.3.4. Action of Epo.. 313

        10.3.5. Assay of Epo 316

    10.4. Manufacture of Erythropoiesis-Stimulating Agents (ESAs)....... 316

        10.4.1. Innovator rhEpo preparations.. 316

        10.4.2. Biosimilar epoetins 318

        10.4.3. Second-generation ESAs ('biobetter')................................... 319

    10.5. Clinical Application of ESAs 320

        10.5.1. Indications for ESA therapy 320

        10.5.2. Safety.. 321

        10.5.3. Non-haematopoietic actions of Epo 322

    10.6. Perspectives 323

         References 325

CHAPTER 11   **Lysosomal storage disorders: current treatments and future directions 327**

    11.1. Introduction 328

    11.2. Clinical Aspects of LSDs... 329

11.3. Therapies for LSDs 330
  11.3.1. Enzyme replacement therapies 330
  11.3.2. Improving Efficacy by Enzyme Modifications
           to Improve Cellular Uptake and Tissue Targeting 335
  11.3.3. Cross-correction through Stem Cell Therapies 336
  11.3.4. Enzyme stabilization (chaperone therapy) 337
  11.3.5. Substrate reduction therapy 337
11.4. Conclusions 338
      References 338

CHAPTER 12 Hormone replacement therapy **343**
12.1. Background and Overview 344
12.2. History of Oestrogen Therapy 344
  12.2.1. Premarin® 344
  12.2.2. Alternatives to Premarin® 345
12.3. HRT and Cancer 346
12.4. HRT and Osteoporosis 348
12.5. HRT and Heart Disease 348
12.6. The Women's Health Initiative 348
12.7. Improved Protocols for HRT 352
12.8. Conclusion 352
      References 353

CHAPTER 13 Design of the anti-HIV protease inhibitor darunavir **355**
13.1. Introduction 356
13.2. The Target Enzyme: HIV-1 Aspartic Acid Protease 357
13.3. Advent of Protease Inhibitors, HAART, and Structural Insights from
      Saquinavir 358
13.4. Cyclic Ethers to Mimic Peptide Bonds: Inspiration from Natural Products 360
13.5. Design of Conceptually New Cyclic Ether-derived Ligands and
      Corresponding Protease Inhibitors 361
  13.5.1. Exploration of ligands from cyclic ethers to cyclic sulfones 364
  13.5.2. Design and development of bicyclic bis-tetrahydrofuran (bis-THF)
           ligand 366
13.6. Design Strategy to Combat Drug Resistance by Targeting Protein Backbone:
      'Backbone Binding Concept' 367
13.7. Design of PIs Promoting Strong Backbone Interactions from S2 to S2'
      Subsites 369
13.8. The 'Backbone Binding Concept' and Its Relevance to Combat Drug
      Resistance 370
13.9. Selection of Darunavir as a Promising Drug Candidate 372
  13.9.1. Thermodynamic and kinetic effects behind darunavir's high
           binding affinity for the protease 374
  13.9.2. Darunavir inhibits HIV-1 protease dimerization: a unique dual
           mode of action 375

13.10. Convenient Syntheses of the *bis*-THF Ligand 376
13.11. Clinical Use of Darunavir in the Management of HIV-1 Infection................ 377
13.12. Design of Potent Inhibitors Targeting the Protease Backbone 377
13.13. Conclusion.. 378
References.. 379

CHAPTER 14   The case of anti-TNF agents.................................... 385
14.1. The Story of anti-TNF: From Septic Shock to Inflammatory Diseases 386
14.2. From a Clinically Validated Target to the Design of anti-TNF
Biopharmaceuticals 387
14.2.1. Development of anti-TNF-α mAbs.. 387
14.2.2. Development of soluble receptors 391
14.3. Marketing Approvals and Post-Marketing Clinical Experience 392
14.3.1. Efficacy (biomarkers and sequential treatment)................................. 392
14.3.2. Tolerance 393
14.4. Structure–Clinical Activities Relationships Drawn from the Clinical
Experience 393
14.4.1. Immunogenicity 394
14.4.2. mAbs, granulomas, and membrane TNF.................................... 394
14.5. Conclusions 395
References.. 395

CHAPTER 15   Discovery of the cholesterol absorption inhibitor, ezetimibe..... 399
15.1. Introduction 400
15.2. Past Approaches to Lowering Levels of Circulating Cholesterol...................... 401
15.2.1. Bile acid sequestrants.. 401
15.2.2. Hypocholesterolaemic drugs.. 402
15.2.3. Inhibition of cholesterol absorption by fibrates: inhibitors
of ACAT 403
15.2.4. Statins: inhibitors of HMG-CoA reductase 404
15.2.5. Inhibition of cholesterol absorption by saponins 405
15.3. Further Work on ACAT Inhibitors.. 406
15.4. Inhibition of a Novel Mechanism for Cholesterol Uptake and the Discovery
of SCH 48461.. 408
15.5. Design of Ezetimibe (SCH 58235)... 408
15.6. Ezetimibe in Human Studies 412
15.7. Identification of a New Mechanism for Cholesterol Uptake 413
15.8. Conclusion.. 414
References.. 414

Index.. 417

Colour Plate (Refer end of the book)

Biographies

Stanley M. Roberts was appointed honorary visiting professor in the School of Chemistry at the University of Manchester in 2004. Previously he held posts in other universities in the UK (Salford 1972–80; Exeter 1986–95; Liverpool 1995–2004) and in industry (Head of Chemical Research, Glaxo, Greenford, UK), before moving to Manchester University to become founder and inaugural director of the Biotech. Centre CoEBio3.

Dr A. J. Gibb is currently reader in pharmacology at University College London, UK. He graduated in biochemistry and pharmacology and completed his PhD at the University of Strathclyde. He was appointed lecturer in pharmacology at UCL in 1990 and teaches pharmacology to students in medicine, biomedical sciences, natural sciences, medicinal chemistry and pharmacology. He is a past editor of the Journal of Physiology and of the British Journal of Pharmacology. He currently leads the General and Advanced Receptor Theory Workshop for the British Pharmacological Society Diploma in Pharmacology. He has published over 50 research papers, reviews and contributions to books. His current main research interest is in the pharmacology and function of NMDA receptors.

Prof. Jennifer Littlechild is professor of biological chemistry and director of the Henry Wellcome Centre for Biocatalysis at Exeter. She carried out her PhD in the Biophysics Laboratory, Kings College, London University, UK followed by a postdoctoral fellowship at the biochemistry department at Princeton university, USA. In 1975, she became a group leader at the Max Planck Institute for Molecular Genetics in Berlin, Germany. In 1980, she returned to the UK to Bristol University and in 1991 to Exeter. Her current research grants are from UK research councils, BBSRC, EPSRC and the EU and large and SME industries.

Her research studies involve the structural and mechanistic characterisation of the C–C bond forming enzymes transketolase and aldolase, vanadium haloperoxidases, Baeyer–Villiger monooxygenases, aminoacylases, novel esterases and lipases, gamma lactamases, alcohol dehydrogenases, dehalogenases, transaminases and other enzymes from thermophilic bacteria and archaea. Many of these enzymes are used in combination with conventional chemical synthesis for the production of new optically pure drugs of interest to pharmaceutical companies. She has published over 120 publications in refereed high impact journals and presented her research work internationally.

Dr Michael Stocks is currently associate professor in medicinal chemistry in the school of pharmacy at the University of Nottingham. He received his PhD from Southampton University (1991) under the supervision of Prof. Philip Kocienski and has worked for Fison's Pharmaceuticals, Astra Pharmaceuticals and AstraZeneca. He has wide experience and successful delivery of projects in GPCR, enzyme and ion channel drug discovery in both the lead optimisation and generation phases.

Jill M. Carton, director, Biologics Research, Biotechnology Center of Excellence, Janssen R&D. Dr Jill Carton received her PhD in molecular and cellular biology from Fordham University, New York. While at Fordham, Jill identified and characterized novel interferon-regulated genes and the role the gene products play in interferon's antiviral and antiproliferative activities. Following graduate school, Jill took a postdoctoral fellowship at Johnson & Johnson, Pharmaceutical Research Institute. In 2000, Jill transferred into a research scientist position at Centocor, Inc. working on the discovery and development of monoclonal antibody therapeutics. Since 2000, Jill has worked in Biologics Research with particular interest in early discovery projects through therapeutic project transition from discovery to development. Currently, Jill heads the Molecular and Protein Biosciences group at Janssen R&D supporting the biologics discovery portfolio.

William R. Strohl, vice president, Biologics Research, Biotechnology Center of Excellence, Janssen R&D. Dr William Strohl received his PhD in microbiology from Louisiana State University, and worked as a researcher at GBF, Braunschweig, Germany. From 1980 to 1997, he rose from assistant to full professor in the Department of Microbiology and Program of Biochemistry at The Ohio State University, researching natural product biosynthesis. In 1997, Dr Strohl moved to Merck to lead Natural Products Microbiology, and then from 2001 to 2008, Dr Strohl was a leader in Merck monoclonal antibody discovery. In April 2008, Dr Strohl joined Centocor to lead antibody drug discovery, and then was promoted to vice president, Biologics Research, Biotechnology COE, Janssen R&D. Dr Strohl has over 120 publications and has recently written the book "*Therapeutic Antibody Engineering: Current and Future Advances Driving the Strongest Growth Area in the Pharma Industry*" (Woodhead Publishing, October 2012).

James Samanen is a pharmaceutical consultant with a track record of 30+ years leadership and innovation in drug discovery project and portfolio management, and portfolio data analysis at GlaxoSmithKline. While at GSK, James helped in developing and implementing worldwide Portfolio Management for GlaxoSmithKline Discovery and Genetics Research businesses encompassing 2000 staff and $400M budget—involving negotiation and implementation of processes for selection and prioritization of research programmes, developing IT systems, mirroring past to present performance with fit-for-purpose benchmarking analysis, modeling future performance, change management and team training. His work enabled an increase in productivity of drug-like leads in GSK by >50% in 2 years by cross-functional team working, prioritization and 40% portfolio downsizing. He also enabled strategy changes reducing cycle times by 50%. As a group leader in medicinal chemistry he discovered and championed lotrafiban (antithrombotic) to Phase III, and two preclinical candidate antithrombotics. He has been a project leader in diverse areas of drug discovery and authored 25 patent applications and over 95 publications. He established James Samanen Consulting in December 2008.

Matthew P. Coghlan is currently a project director leading Cardiovascular and Metabolic Disease Research and Early Development Projects at MedImmune (Cambridge, UK); the biologics division of AstraZeneca. Previously Dr Coghlan was tasked with overseeing the growth of the biologics portfolio in the cardiovascular and gastrointestinal therapy area at MedImmune/AstraZeneca (2008–2010). Prior to entering biologics R&D in early 2008, Dr Coghlan led *in vitro* biology teams and small molecule research projects in the metabolic disease therapy area at SmithKline Beecham and AstraZeneca (1998–2007). Dr Coghlan has published over 25 scientific papers and reviews. Dr Coghlan has a 1st class honours degree in biochemistry from Imperial College, University of London, and a PhD from the University of Cambridge. Dr Coghlan conducted postdoctoral research at Harvard Medical School as a recipient of NATO Royal Society Fellowship.

David Fairman is a pharmacodynamic/pharmacokinetic modeller working within the clinical pharmacology and drug metabolism and pharmacokinetics (DMPK) department at MedImmune (Cambridge, UK), the biologics division of AstraZeneca. His current portfolio of preclinical and clinical biologics projects spans across the oncology and cardiovascular and gastrointestinal therapy areas with a focus on diabetes. Immediately prior to this position he led the preclinical modelling team within the allergy and respiratory areas within Pfizer (Sandwich, UK) and was responsible for the delivery of multiple small molecule candidates for evaluation in human clinical studies. Earlier in his career he led DMPK and wider research teams within the allergy and respiratory and cardiovascular areas. David Fairman has a BSc in pharmacology from the open university, an MSc in modelling and simulation from the University of Manchester and has authored or co-authored over 10 scientific papers.

Dr Andy Barker is currently an independent scientific consultant. He trained as an organic chemist at Nottingham University with Prof. Gerry Pattenden and has held positions at Beecham Pharmaceuticals, Hoechst and AstraZeneca working in drug discovery groups as a medicinal chemist, project leader and manager of the chemistry department. He has experience in a variety of therapeutic areas and has been the member or leader of project teams, which have put 12 drugs into development, two of which have reached the market. He was awarded the American Chemical Society 'Hero of Chemistry' award in 2011.

Dr David Andrews trained as a pharmacist at Nottingham University and then as a PhD organic chemist under the supervision of Drs Ian Galpin and Rick Cosstick at Liverpool University. From 1990 to 2003, he held positions at GlaxoSmithKline as a medicinal chemist, project leader and department director. Since 2003, he has worked as an associate director in the oncology iMed chemistry department of AstraZeneca. He has experience of working in CNS, inflammation, antiviral and oncology discovery and development chemistry. He currently heads AstraZeneca's oncology lead generation chemistry group.

Dr M. Hasmann is a preclinical science leader and project team leader at Roche's Pharma Research and Early Development (pRED) unit in Penzberg, Germany (since 2002). Previously, he was Head of Cell Culture Laboratories at Klinge Pharma GmbH (Fujisawa group; 1987–2001). His main background is microbiology (Technical University Munich, Germany) and cell biology (Max Planck Institute for Biochemistry, Martinsried, Germany). He has essentially contributed to several early and late stage oncology research and development programmes including small molecules and therapeutic antibodies.

Prof. W. Jelkmann is currently Chairman of the Institute of Physiology at the University of Luebeck, Germany. After studying medicine in Hannover (1967–73), he worked at university institutes in Regensburg, New Orleans and Bonn. His research interests include hemopoietic growth factors and cytokines with emphasis on erythropoietin, thrombopoietin and vascular endothelial growth factor. He has authored over 150 original publications, 130 review articles/book chapters and (co-)edited five books.

Charles W. Richard III is currently the Chief Medical Officer of Oxyrane, a biotechnology company devoted to discovering and developing next-generation enzyme replacement products for lysosomal storage disorders. Previously he held positions as Head of translational medicine at Shire Human Genetic Therapies (2006–2012); vice president and Head of the genomics department at Wyeth Research (1997–2005), and assistant professor of psychiatry and human genetics at the University of Pittsburgh (1992–1995). Dr Richard holds an MD and PhD from Ohio State University and a BS in biology from Stanford University, and has published over 40 scientific papers and reviews.

Arun K. Ghosh is currently the Ian P. Rothwell Distinguished Professor of Chemistry and Medicinal chemistry at Purdue University. He received his BS in Chemistry and MS in Chemistry from the University of Calcutta and Indian Institute of Technology, Kanpur, respectively. He obtained his PhD in organic chemistry (1985) from University of Pittsburgh and pursued postdoctoral research at Harvard university (1985–1988). He was a research fellow at Merck Research Laboratories and in 1994, he joined University of Illinois, Chicago as an assistant professor, eventually becoming professor of chemistry in 1998. In 2005, he moved to Purdue University, with a joint appointment in chemistry and medicinal chemistry. His research interests include diverse areas of organic, bioorganic, and medicinal chemistry with particular emphasis on organic synthesis and protein–structure-based drug design. He has published over 250 scientific papers and edited a book on aspartic acid protease as a therapeutic target.

Bruno D. Chapsal graduated with a M.Sc. in chemistry & chemical engineering from the Lyon's school of chemistry (CPE Lyon), France. He then obtained his Ph.D. from Stony-Brook University, NY, where he worked on synthetic methodologies and catalysis under the guidance of Professor Iwao Ojima. In 2006, he joined the group of Professor Arun K. Ghosh at Purdue University. His post-doctoral research focused on the structure-based design of novel HIV protease inhibitors with broad-spectrum activities against drug-resistant viruses. He is currently a Senior Research Scientist at AMRI's in-sourcing operations at Eli Lilly and Company, Indianapolis.

Denis Mulleman, MD, PhD, is professor of rheumatology at the University of Tours (France). He is member of the CNRS research unit 7292 (team 'Antibodies, Fc receptors and clinical responses'). His main research interest is to characterize the concentration–response relationship of anti-TNF monoclonal antibodies and Fc-containing fusion proteins, in order to help clinicians in personalizing drug administration.

Marc Ohresser is a research associate in Hervé Watier's group at the Université François-Rabelais, Tours. His interests focus on therapeutic antibodies and their interactions with Fc receptors. He is also a member of the CNRS unit 7292 and has management responsibilities within the Laboratory of Excellence (LabEx) 'MAbImprove' programme, in particular as scientific leader of the MAbMapping exercise that maintains 'technology watch' and 'patent mapping' in the field of therapeutic antibodies.

Hervé Watier, MD, PhD, is currently Professor of Immunology at the Faculty of Medicine, Tours, France, and Head of the Laboratory of Immunology at the University Hospital of Tours. He discovered, in 2001, that patients homozygous for a polymorphism in a gene encoding an IgG Fc receptor better responded to different therapeutic antibodies, which is now considered as a milestone in the recent history of therapeutic antibodies development. Since 2009, he is Director of a French CNRS Research

Network bringing together a hundred of public or private French research teams, engaged in antibody research. Since 2011, he coordinates the laboratory of excellence (LabEx) 'MAbImprove', a consortium of 14 research teams from Tours and Montpellier, whose 10 year-programme is the optimization of therapeutic antibodies development.

Robin Ganellin led the chemistry for the discovery of histamine H_2 receptors at SmithKline & French Labs (UK), is a coinventor of cimetidine (Tagamet^{TM}) and became Vice President for Research. He has authored over 260 scientific publications, coedited eight books and has received many international awards for his work in Medicinal Chemistry. In 1986, he was elected a Fellow of the Royal Society and became Professor of Medicinal Chemistry at University College London. He has been inducted into the USA National Inventors Hall of Fame and the ACS Division of Medicinal Chemistry Hall of Fame.

Preface

The history of medicines has been documented in detail but even at a broad and superficial level, different eras emerge quite clearly. First of all, for almost 2000 years, European medicine was dominated by Galen's theory of the humours. This held that health depended on the correct balance of the four 'humours', or 'principal fluids' namely black bile, yellow bile, phlegm and blood produced in the body. The correct balance could be maintained, it was thought, by diet, blood-letting and medicines. The medicines were natural products used as infusions and mixtures based primarily on folklore; the early applications of morphine and quinine quickly come to mind. Indeed, some traditional therapies are still very popular and the practice has become part of the field of 'complementary medicine'.

However, as it became possible to define detailed mechanisms of infection and the causes of disease, it was then feasible to design biological assays which could lead scientists to new therapies. Still, natural products provided a good proportion of the new medicines (for example, penicillins and steroids) although, increasingly, these natural materials were modified to provide optimized drug substances. In the latter half of the twentieth century, the naturally occurring small-molecule neurotransmitters were being modified to provide blockbuster drugs in areas such as heart disease, asthma and the treatment of stomach ulcers.

Medicinal chemistry came to the forefront at this time; chemists manufactured molecules on a small scale to test in the increasingly rapid (fast-throughput) assays. Novel, small molecules were preferred to allow patent protection and easy production at the multi-kilogram/tonne scale, respectively. Those entities passing the stringent safety requirements emerged as the drugs for the clinic. Still, today, small-molecule chemotherapeutics are sought after to provide greater protection against cancer and heart disease *inter alia* as well as debilitating conditions such as arthritis and dementia.

The contemporaneous development of protein chemistry allowed for the identification, isolation, purification and use of large biomolecules as drugs, for example, porcine insulin. With the advent of genetic engineering, the commercial production of biomolecules *ex vivo* became possible, for example, the production of human insulin in the bacterium *Escherichia coli* and erythropoietin in mammalian cells. A further seminal advance was the introduction of hybridoma technology that allows for the generation of human monoclonal antibodies of predefined specificity on an industrial scale.

This overall ability more efficiently to prepare selected proteins has given rise, particularly between 1991 and 2010, to a complementary set of medicines, commonly known as biopharmaceutics or biopharmaceuticals. Currently only *ca* 15% of medicines come into this bracket, but the way things are progressing, specifically the number of biopharmaceuticals presently in clinical trials and the proportion of biopharmaceuticals in the market place, this percentage could double by 2025.

It has been a fascinating challenge for us to present the background to these two types of manufactured medicine for a reader who is becoming interested in medicinal science. We start with a review of biological targets and the strategies to search for small molecules that might be therapies for a particular disease. Having established that base, the structures and properties of biopharmaceuticals are described before the two approaches are compared from scientific and commercial perspectives.

The case studies that are provided in the later chapters serve to illustrate the general concepts with specific examples. Where possible, large and small molecules targeting related therapies are grouped together. Overall, the coverage encompasses major therapeutic sectors as well as niche and orphan diseases. The clinical experience with the new drugs is outlined in some detail in most cases.

The Editors are extremely fortunate that the many world-leading experts agreed to contribute to this work. We are most grateful for their expositions, particularly in the chapters which describe complex issues, wherein great effort has been made to employ comprehensible language, keeping jargon to a minimum.

We hope that the book will provide a good basis to allow the reader to start to compare and contrast the science of small-molecule chemotherapies with that of the peptide/protein biopharmaceuticals.

C. Robin Ganellin

Roy Jefferis

Stanley M. Roberts

Introduction to enzymes, receptors and the action of small molecule drugs

1

Stanley M. Roberts*, Alasdair J. Gibb[†]

* *School of Chemistry, Manchester University, Manchester M1 7ND, UK,*
† *Research Department of Neuroscience, Physiology and Pharmacology, Division of Biosciences, University College London, London WC1E 6BT, UK*

CHAPTER OUTLINE

1.1 Section I: Background Information 2
    1.1.1 Communication between cells: the roles of receptors and enzymes 2
    1.1.2 Neurotransmitters, receptors and the nervous system 9
    1.1.3 Introduction to enzymes and enzyme inhibitors 15
    1.1.4 Other types of bioactive molecules 15
    1.1.5 Factors influencing drug action 16
    1.1.6 The impact of the sequencing of the human genome 19
1.2 Section II: More About Enzymes 19
    1.2.1 Configuration of enzymes 19
    1.2.2 Enzyme specificity, classification and nomenclature 21
    1.2.3 Characteristics of enzyme catalysis 21
    1.2.4 Enzyme reaction rates 25
    1.2.5 Enzyme substrates as drugs 27
    1.2.6 Enzyme inhibition and enzyme inhibitors as drugs 28
        1.2.6.1 Irreversible inhibitors 28
        1.2.6.2 Competitive inhibitors 28
        1.2.6.3 Noncompetitive inhibitors 32
    1.2.7 Enzyme regulation 32
1.3 Section III: More About Receptors 35
    1.3.1 Bioassay and the measurement of drug effects..35
        1.3.1.1 Bioassay 35
    1.3.2 Quantifying drug–receptor interactions 37
    1.3.3 Radioligand-binding studies – a direct measure of occupancy 39
    1.3.4 Receptor structure...40
        1.3.4.1 Nicotinic AChR structure: a ligand-gated ion channel 42
        1.3.4.2 β-adrenoceptor structure: a G-protein-coupled receptor 43
    1.3.5 Relating occupancy to response 44
        1.3.4.1 Ligand-gated ion channels 44
        1.3.4.2 Receptor mechanisms that involve second messengers 46

Introduction to Drug Research and Development. http://dx.doi.org/10.1016/B978-0-12-397176-0.00001-7

1.3.6 Competitive antagonism and the Schild equation 48
    1.3.6.1 Drug blockade of open ion channels: a noncompetitive antagonism.......................... 49
1.3.7 Desensitization and the control of receptor number 49
1.3.8 Partial agonists, agonist efficacy and inverse agonism...51
1.4 Section IV... 53
1.4.1 Conclusions: uncertainties in drug design and development 53
Further Reading 55

ABSTRACT

Increasingly, pharmaceutical research and development is based on a detailed understanding of molecular interactions in diseased and healthy states of the human body. Over the past 50 years, most drug research has concentrated on the effects of small molecules on naturally occurring entities called enzymes and receptors. Hence, this chapter commences with an overview of the interactions of low-molecular-weight compounds (some natural [e.g. neurotransmitters] and some non-natural [e.g. drugs that inhibit certain enzymes]) with these natural macromolecules. This high-level introduction is followed by a more detailed inspection of the structures of some typical enzymes and receptors, emphasizing the complex shapes and subtle intermolecular interactions of these high-molecular-weight proteins. In addition, the importance of understanding the 'on–off' interaction between a small molecule and the target protein is illustrated by introducing the rate equations which dictate the kinetics of these episodes. The concluding section provides the first insight into the problems that have to be faced and overcome in moving from the point of having a compound with the desired effect on an enzyme or receptor *in vitro* to the position of introducing a useful drug to the marketplace.

Keywords/Abbreviations: Nerve cell; Neurotransmitter; Acetylcholine (receptor) (ACh(R)); Noradrenaline (NorA); Serotonin/5-hydroxytryptamine (5-HT); Dopamine (DA); Agonist/antagonist; Cyclic adenosine-$3',5'$-monophosphate (cyclic-AMP); Cyclic guanosine -$3',5'$-monophosphate (cyclic GMP); Adenosine/guanosine triphosphate (A/GTP); Catechol O-methyl transferase (COMT); Selective serotonin reuptake inhibitor (SSRI); Nicotinic/muscarinic AChR; Prodrug; Enzyme denaturation; Enzyme active/binding/catalytic sites; Co-enzyme; Enzyme cofactor; Michaelis–Menten equation; Lineweaver-Burk plot; L-DOPA; Allosteric enzyme inhibitor; Guanosine diphosphate (GDP); Hill-Langmuir equation; Scratchard plot; Lysergic acid diethylamide (LSD); G-protein coupled receptor (GPCR); Transmembrane (TM) domain; Gamma-aminobutyric acid (GABA); N-methyl-D-aspartate (NMDA); Sino-atrial (SA) node cell; Schild equation/plot; Efficacy.

1.1 SECTION I: BACKGROUND INFORMATION

This chapter is adapted from the first three chapters of the book 'Medicinal Chemistry: the Role of Organic Chemistry in Drug Research' (eds. C. R. Ganellin and S. M.Roberts) Academic Press, London, 1992. While the basic principles remain the same, the text has been updated and modified to reflect the different focus of this book.

1.1.1 Communication between cells: the roles of receptors and enzymes

Some forms of life are composed of a single independent cell (the protozoa), while mammals are multicellular organisms. In between these two extremes there are life forms of varying complexity. All

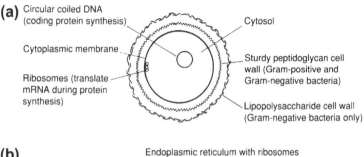

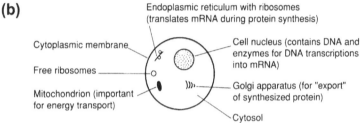

FIGURE 1.1

Simplistic representations of a prokaryotic bacterial cell (a) and a eukaryotic (possessing a nucleus) human cell (b) Not all substructures are shown.

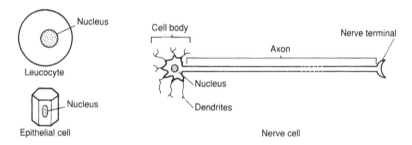

FIGURE 1.2

Shapes and sizes of mammalian cells.

these organisms possess cell(s) to compartmentalize various chemical reactions in order to use available materials for energy and the maintenance of life's processes.

Cells of different life forms have different characteristics (Figure 1.1) and, indeed, different cells from the same organism can be distinguished readily. For example, mammalian cells come in all shapes and sizes: compare the spheroidal leucocyte (the white blood cell), the flat epithelial cells found lining the mouth and the nerve cell (Figure 1.2).

The cells are organized such that chemical transformations can be accomplished efficiently, the rate of these transformations being controlled by Nature's catalysts – enzymes. Enzymes are high-molecular-weight compounds which catalyse anabolic (synthesis) and catabolic (degradation) reactions. The trivial name of the enzyme often gives a guide to its role (see Eqn 1.1–1.3); a more comprehensive list of enzyme activities is contained in Section 1.2.

$$RCO_2R^1 \xrightarrow[H_2O]{\text{Esterase enzyme}} RCO_2H + R^1OH \tag{1.1}$$

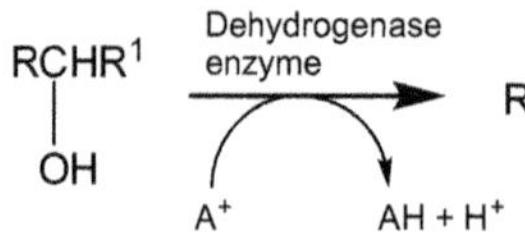

$$\underset{\underset{OH}{|}}{RCHR^1} \xrightarrow{\text{Dehydrogenase enzyme}} RCOR^1 \tag{1.2}$$

A$^+$ AH + H$^+$

$$\underset{\underset{O-P-O}{|\;\;\;\;|}}{RCH\text{-}CH_2\text{-}CHR^1} \xrightarrow[H_2O]{\text{Phosphodiesterase enzyme}} \underset{\underset{O^-}{|}}{\underset{OPO_2H\;\;OH}{|\;\;\;\;\;\;|}}{RCH\text{-}CH_2\text{-}CHR^1} \tag{1.3}$$

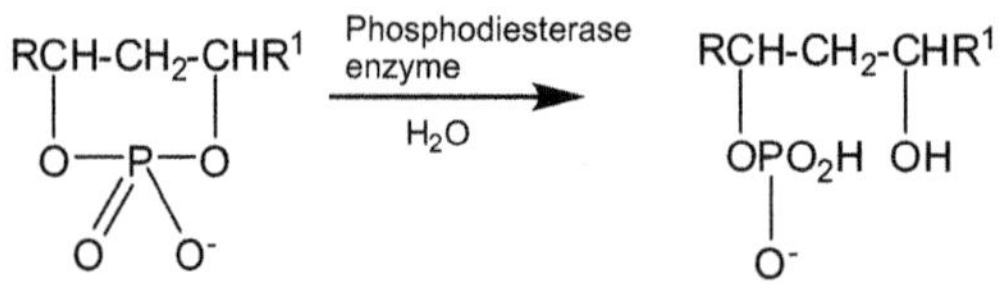

In order to coordinate their activities, the different cells in multicellular organisms need to communicate and this correspondence is accomplished mainly by small chemical molecules. For example, on receiving the appropriate signal, nerve terminals may release substances such as acetylcholine (ACh) (**1**), noradrenaline[1] (NorA) (**2**), serotonin (**3**) (otherwise known as 5-hydroxytryptamine or 5-HT) or dopamine (DA) (**4**), and these substances, known as neurotransmitters, can interact with the appropriate receptors.

$$CH_3COOCH_2CH_2\overset{+}{N}(CH_3)_3$$

(**1**)

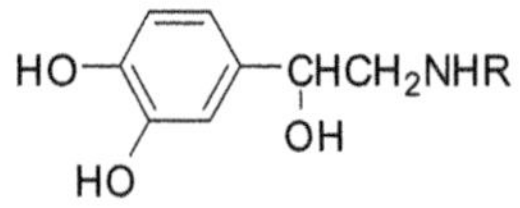

HO—⟨benzene⟩—CHCH$_2$NHR
          |
         OH
HO

(**2**) R = H
(**8**) R = CH$_3$

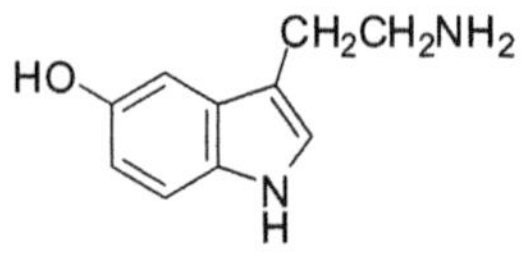

HO—⟨indole⟩ CH$_2$CH$_2$NH$_2$

(**3**)

HO—⟨benzene⟩—CH$_2$CH$_2$NH$_2$
HO

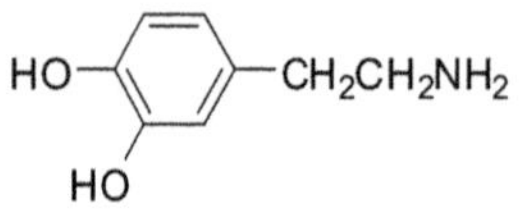

(**4**)

[1] Also known as *norepinephrine*.

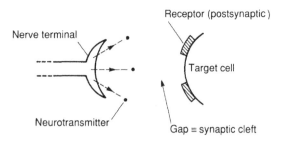

FIGURE 1.3

A neuroeffector junction (synapse).

The receptors can lie, for example, on the surface of the cells opposite the nerve terminal (Figure 1.3). The interaction of a neurotransmitter (agonist[2]) with its receptor usually effects a change in conformation of the macromolecular receptor, leading to a change in enzyme activity within the cell (Figure 1.4), and/or movement of ions into or out of the cell (Figure 1.5).

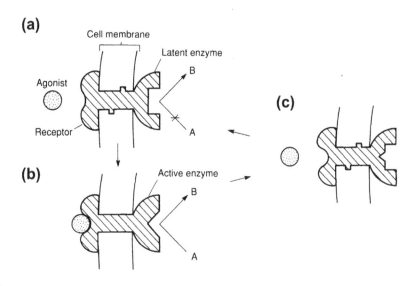

FIGURE 1.4

Activation of an enzyme by occupation of a receptor by an agonist. (a) Receptor free, enzyme inactive. (b) Receptor occupied, enzyme triggered into action (allosteric activation of enzyme). (c) Agonist leaves receptor surface and enzyme quickly returns to inactive form.

---

[2] Agonist is a name coined by Gaddum in 1937 and is now used to describe all physiological mediators and drugs which mimic their action by activating the same cellular reactions.

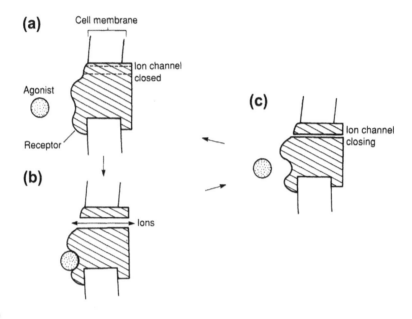

FIGURE 1.5

Opening of an ion channel by the occupation of a receptor by an agonist. (a) Receptor free, channel closed. (b) Receptor occupied, channel opened, ion migration rapidly takes place down the electrochemical gradient. (c) Channel closes, neurotransmitter diffuses away. Difference in metal ion (e.g. Na^+) concentration across the membrane is re-established by exergonic metal–ion pump.

One specific example of the process diagrammatically illustrated in Figure 1.4 is given by NorA, which will act on a receptor (for detail see Section 1.3.4.2 describing the G-protein-coupled receptor (GPCR)) basically resulting in the activation of the intracellular enzyme adenylate cyclase to produce cyclic adenosine $3',5'$-monophosphate (**6**) (cyclic AMP) from adenosine triphosphate (ATP) (**5**) as shown in Eqn (1.4).

Cyclic AMP initiates a cascade of other enzyme activations leading to the observed biological response. Cyclic AMP is inactivated by a phosphodiesterase enzyme (Eqn (1.4)).

An example of the process represented in Figure 1.5 is the interaction of ACh with its receptor which causes opening of the receptor ion channel (i.e. it is a ligand-gated ion channel – for more details see Section 1.3.4.1) resulting in the movement of sodium ions into the cell. Figures 1.4 and 1.5 illustrate the general idea that receptor occupation may result in an effect (e.g. change in enzyme activity) that can last for the lifetime of the occupation or the effect (e.g. ion movement) may be triggered and will not re-occur (for example due to receptor 'desensitization' – see also Sections 1.3.2 and 1.3.7) until disengagement of the agonist, re-priming of the system and re-engagement of the agonist and the receptor.

Note that if a neurotransmitter remains in the synaptic cleft, disengagement/re-engagement will continue and the receptor will be activated repeatedly; this will carry on until the chemical diffuses away from the site. To allow a faster return to the resting state after the neurotransmitter has ceased to be released from the nerve terminal, an enzyme may be present which will convert the neurotransmitter into an inactive substance or the neurotransmitter or hormone may be removed from the extracellular space by active transport ('uptake') into nearby cells. For example, ACh is deactivated by an esterase (Eqn (1.5)), while NorA is rendered inactive by methylation of one of the phenolic groups (Eqn (1.6)) through the enzyme catechol O-methyltransferase.

$$CH_3COOCH_2CH_2\overset{+}{N}(CH_3)_3 \xrightarrow[\text{H}_2\text{O}]{\text{Acetylcholinesterase}} CH_3CO_2H + HOCH_2CH_2\overset{+}{N}(CH_3)_3 \qquad (1.5)$$

(1.6)

Most neurotransmitters including 5-HT and NorA are removed by uptake into the presynaptic terminal glial cells; indeed, the majority of the NorA released from a nerve terminal is removed from the synapse by this method. Drugs have been developed to slow down the process of uptake of a particular neurotransmitter into the presynaptic terminal. Selective serotonin reuptake inhibitors such as fluoxetine (Prozac) (**7**) have been developed as treatments for depression.

(**7**)

These examples illustrate that enzymes are not always contained within cells. Equally, not all chemical messengers are released from nerve terminals to act on adjacent terminals before being degraded. For example, adrenaline (**8**), NorA (**2**) and various steroids are released into the circulation from endocrine glands. Local hormones or autocoids such as histamine (a key component in the inflammatory response) are released from cells and travel through extracellular fluid to act on nearby cells. The three types of intercellular communication processes are shown in Figure 1.6.

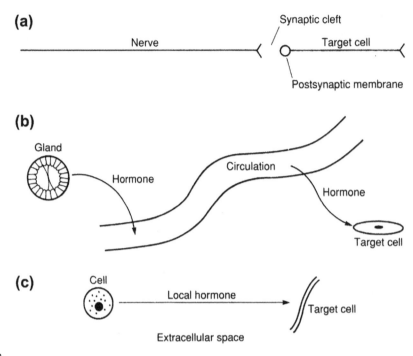

FIGURE 1.6

Intercellular communication processes. (a) Nerve releases a neurotransmitter substance, e.g. ACh, which diffuses across a synapse to act on a postsynaptic membrane. (b) Endocrine gland releases a hormone, e.g. steroid, which is distributed throughout the body by the circulatory system. (c) Local hormone (autocoid) is released by cells and diffuses through the extracellular space to act locally, e.g. histamine released from mast cell or cells in the stomach.

It is important to understand that enzymes and receptors are both composed of amino acids condensed into high-molecular-weight polypeptide chains and can be associated with ions and small molecules; however, the likeness ends here. One of the major differences is that enzymes catalyse bond-making and bond-breaking reactions, while the receptors release the agonist unchanged.

In the normal healthy state, all cells are communicating, synthesizing and degrading molecules and changing ion concentrations for the overall well-being of the organism. As a result of disease, damage or degeneration, cellular activities may become impaired and the correct dynamic equilibrium must then be re-instated by means of a suitable drug. (Sometimes it may also be desirable to alter the normal physiology (in anaesthesia for example) by administration of a drug substance.)

It may be desirable to amplify the effect of a neurotransmitter. This can be accomplished by the following:

1. Increasing the concentration of the natural neurotransmitter by (a) direct supplementation through introduction of the substance into the body, (b) inhibition of enzymes that degrade the transmitter; or (c) inhibition of reuptake.
2. Using a more-potent and/or less readily metabolized surrogate of the natural substance (an unnatural agonist).

Alternatively, it may be prudent to decrease the effect of a particular neurotransmitter or hormone at a given receptor. This can be done using an antagonist substance, i.e. an unnatural compound which will bind strongly to a receptor without eliciting a response and which will prevent access of the natural agonist to the receptor.

An antagonist diminishes or abolishes the effects of the corresponding agonist. Two types of antagonists are known: a competitive antagonist competes with the agonist for a binding site at the active site of the receptor (also known as the orthosteric site), while a noncompetitive antagonist binds at a different site (an allosteric site) from that of the agonist. In the former situation, the effect of the antagonist decreases in the presence of increasing concentrations of agonist, while in the latter situation, the effect of the antagonist is often independent of agonist concentration.

Similarly, certain enzyme substrates and specific or highly selective enzyme inhibitors can prove to be useful drug substances. The inhibition of human immunodeficiency virus (HIV) protease is one example (see Chapter 13), complementing drugs such as azidothymidine which block another HIV enzyme called reverse transcriptase.

Before amplification of these points, the central control of cell communication and a more detailed consideration of certain neurotransmitters and receptors are warranted.

1.1.2 Neurotransmitters, receptors and the nervous system

In complex organisms such as man, there are a considerable number of receptors and a multitude of different enzymes. Actions are coordinated by the central nervous system (CNS) (the brain and the spinal cord) and some actions are the result of sensory input (sight, sounds, touch, hearing, etc.). Output from the CNS is directed towards the autonomic nervous system (the sympathetic and parasympathetic systems) and nerves associated with voluntary motor functions as illustrated in Figure 1.7. Voluntary motor function deals with the controlled movement of muscles (skeletal muscle) and the associated limbs; some of the organs controlled by the parasympathetic and sympathetic nerves are listed in Figure 1.8 and the effects on selected organs from the two systems are listed in Table 1.1. In short, the sympathetic and parasympathetic systems operate in a complementary fashion. In response to a particular external influence, enhanced stimulation of the sympathetic nervous system occurs and leads to preparation for 'fight or flight' (Figure 1.9). In the relaxed state (Figure 1.10), stimulation of the parasympathetic nervous system predominates and deals with secretion and voidance of materials from the body.

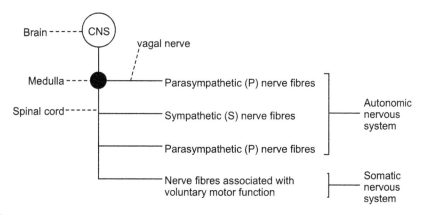

FIGURE 1.7

The peripheral nervous system.

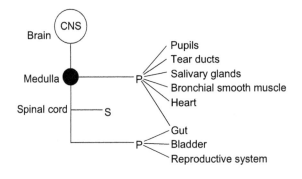

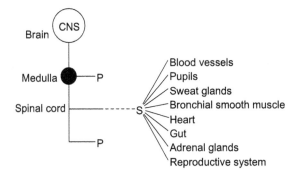

FIGURE 1.8

Organs controlled by sympathetic (S) and parasympathetic (P) nerves.

Table 1.1 Response to Activation of Sympathetic and Parasympathetic Nervous System

	Response to Activation	
Organ	**Sympathetic**	**Parasympathetic**
Pupil	Dilation	Constriction
Bronchi	Dilation	Constriction
Heart	↑ Acceleration	↓ Slowing
Digestive tract	↓ Slowing	↑ Increase
Sphincters of gut	Constriction	Relaxation
Bladder	Relaxation	Contraction Emptying
Blood vessels	Constriction	
Glands of alimentary canal (salivary, gut, pancreas)		↑ Increase in activity

FIGURE 1.9

Stimulation of the sympathetic nervous system due to *fright* leads to preparation of the system for *flight* or *fight* increase in heart rate, dilation of bronchi, dilation of pupils, constriction of peripheral blood vessels (pallor), etc.

Source: From B.L.A.T. Booklet 'Action of Drugs', Centre for Health and Medical Education, London.

FIGURE 1.10

In the relaxed state, stimulation of the parasympathetic nervous system is predominant and leads to (*inter alia*) slowing of the heart and increase in the activity of the gastrointestinal tract.

Source: From B.L.A.T. Booklet 'Action of Drugs', Centre for Health and Medical Education, London.

Some important neurojunctions and the associated neurotransmitters in the nervous system are shown in Figure 1.11.

Note that the acetylcholine receptors (AChRs) are divided into two categories, the classification being based on the actions of two drugs of plant origin.[3] The receptors which are activated by

---

[3] The subdivision of the various receptors is mainly based on agonist and/or antagonist actions of substances not normally found in the mammalian system; it is known that the amino acid sequence of receptor subtypes is similar, but not the same.

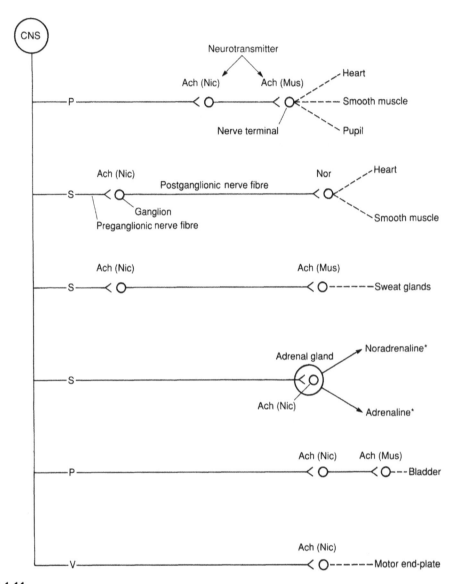

FIGURE 1.11

Neurotransmitters in the peripheral nervous system. Asterisk indicates release into the circulation to act on distant receptors in the periphery. S, sympathetic nervous system; P, parasympathetic nervous system; V, voluntary motor function; ACh, acetylcholine; NorA noradrenaline; Mus, muscarinic receptor; Nic, nicotinic receptor.

Table 1.2 Situation of Some Acetylcholine-Controlled Neuroeffector Junctions

Organ	Type of Receptor	Control by Voluntary Motor (VM), Parasympathetic (P), or Sympathetic (S) Nervous System	Effect of Agonist
Heart	Muscarinic	P	Slowing rate of contraction, decreasing force of contraction
Eye	Muscarinic	P	Constriction of pupil
Sweat glands	Muscarinic	S	Activation
Gastrointestinal smooth muscle	Muscarinic	P	Increase in tone, increase in contractions and peristalsis
Adrenal gland	Nicotinic	S	Release of adrenaline and noradrenaline
Bladder	Muscarinic	P	Contraction
Skeletal muscle	Nicotinic	VM	Muscle contraction
CNS	Muscarinic and Nicotinic	-	Various

muscarine are termed muscarinic receptors and those activated by binding nicotine are called nicotinic receptors. Thus nicotine and muscarine mimic the action of ACh at two different distinct receptors. The effects on various end organs of stimulation of the appropriate AChRs are listed in Table 1.2. Inhibition of the action of ACh at the neuromuscular junction leads to muscle relaxation.

NorA and adrenaline stimulate adrenoceptors. When NorA is released from a presynaptic nerve terminal, it crosses the synaptic cleft and initiates a response in the postsynaptic tissue by combining with one of two types of receptor called α-adrenoceptors and β-adrenoceptors. The type of receptor found postsynaptically to noradrenergic nerves in the sympathetic system depends on the type of tissue; classification of adrenoceptors in different tissues is again based on the ability of the agonists to initiate responses and antagonists to prevent responses. A further subclassification of β-adrenoceptors has been made: in man the majority of β-adrenoceptors in the heart are called β_1-adrenoceptors and these are distinct from other β-adrenoceptors (dubbed β_2-adrenoceptors) found elsewhere in the periphery (outside the CNS). Many tissues have a mixed population of β_1- and β_2-receptors. NorA and adrenaline have different effects on α- and β-receptors; NorA is potent at stimulating α-receptors but is less potent at activating β-receptors, while adrenaline elicits activity from both α-and β-receptors at about the same level.

Some important sites of α- and β-adrenergic receptors are given in Table 1.3.

Note that activation of α-receptors generally results in a stimulant response (except in the gut), while activation of β-receptors leads to an inhibitory response, namely, relaxation of muscle (except in the heart). Blocking of α- and β-receptors causes, *inter alia*, relaxation of peripheral blood vessels and slowing of the heart, respectively, and can have beneficial effects in the treatment of angina and hypertension, while β_2-stimulants can alleviate mild to moderate asthmatic attacks by relaxation of

Table 1.3 Important Adrenaline and Noradrenaline Receptors

Organ	Major Adrenoceptor Present	Effect of Stimulation
Eye	α	Pupillary dilation
Blood vessels (periphery including skin)	α	Constriction
Gastrointestinal tract	$\alpha + \beta$	Relaxation
Heart	β_1	Tachycardia (speeding of the heart) Increased force of contraction
Small intestine	β_1	Relaxation
Blood vessels (skeletal muscle)	β_2	Relaxation (vasodilation)
Bronchioles (lung)	β_2	Relaxation (bronchodilation)

bronchiolar muscle and widening of airways. Thus the antiasthma drug salbutamol (Ventolin) (**9**) is a β_2 agonist while Propranolol (**10**) was one of the first 'beta-blockers' useful for the treatment of hypertension.

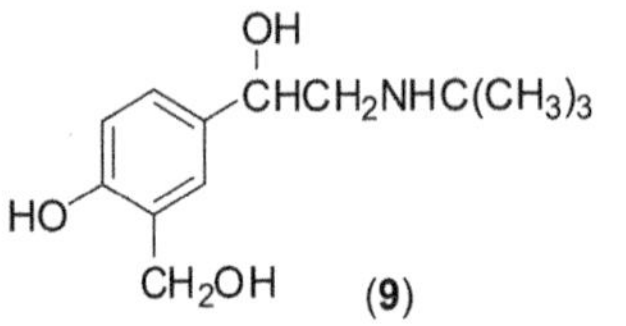

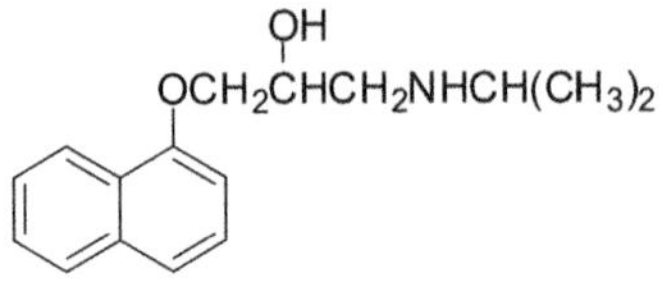

Some peptides act as neurotransmitters. For example, receptors for the enkephalins (**11**) have been demonstrated within the CNS. It is believed that morphine (**12**) and the other opiates exert their analgesic action by interaction with these receptors. Thus, morphine (**12**), heroin (**13**) and codeine (**14**) can be considered to be agonists at the enkephalin receptor; other compounds such as the painkilling drug buprenorphine are partial agonists, i.e. have mixed agonist and antagonist properties. Opiate receptor blockers (antagonists), e.g. naloxone, are also known.

H - Tyr - Gly - Gly - Phe - Met - OH   Methionine enkephalin

H - Tyr - Gly - Gly - Phe - Leu - OH   Leucine enkephalin

(**11**)

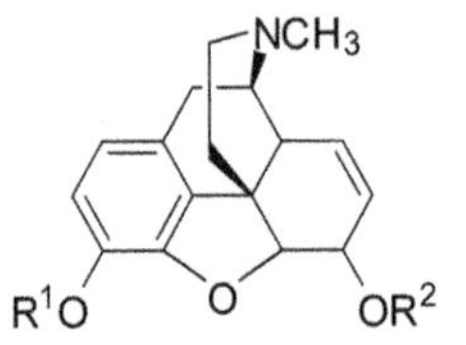

(**12**) $R^1 = R^2 = H$
(**13**) $R^1 = R^2 = COCH_3$
(**14**) $R^1 = CH_3 ; R^2 = H$

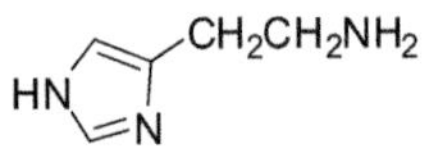

(**15**)

Histamine (**15**) is a neurotransmitter which also occurs in many tissues in the body. It is stored in mast cells and platelets and is released from these sites in response to stimuli such as allergic reactions and injury. Four main types of histamine receptors, termed H_1, H_2, H_3 and H_4, have been identified; they differ in sensitivity to various unnatural agonists and antagonists. Bronchiolar smooth muscle has H_1 receptors; activation by histamine causes contraction of the muscle and bronchoconstriction. The actions of histamine on vascular smooth muscle are complex, species dependent and mediated by both H_1 and H_2 receptors, while gastric acid secretion from parietal cells is stimulated by mucosally released histamine acting at H_2 receptors. H_2 receptor blockers, such as cimetidine and ranitidine are useful in the treatment of conditions in which there is excess acid secretion in the stomach, especially in duodenal ulceration. The H_3 receptors appear to be autoreceptors which inhibit further histamine synthesis or release; they occur especially in nervous tissue, while H_4 receptors are involved in immunologically based responses.

It is noteworthy that all the previously mentioned transmitters, i.e. ACh, NorA, 5-HT, enkephalins and dopamine, as well as histamine and adrenaline, have receptors in the CNS. Interaction with these receptors will elicit a response and if interaction of a potential drug substance with peripheral receptors (e.g. at the neuromuscular junction, on bronchiolar smooth muscle, in parietal cells, in heart muscle, on blood vessels etc.) is beneficial, then it is often necessary to ensure that the compound does not cross the 'blood–brain barrier' to cause unwanted side effects through interaction with receptors in the CNS. Of course, if a drug has its effect by interacting with receptors in the brain (e.g. an antidepressant), the compound must have the appropriate physical properties (high lipophilicity, low polarity) in order to cross the blood–brain barrier.

1.1.3 Introduction to enzymes and enzyme inhibitors

The active centres of enzymes are similar in many ways to the agonist-binding sites of receptors, and enzyme inhibitors are identical in principle to receptor blockers. Simple inhibitors derange the active centre by engaging it directly (isosteric inhibition) or by inducing a conformational change affecting the active site through binding to a distant site (allosteric inhibition) (see Section 1.2.7). Unnatural substrates for an enzyme which are slowly processed are also effective inhibitors of the physiological enzyme action provided that they have a substantially greater affinity than the natural substrate for the enzyme centre. Substantial inhibition of a key enzyme-controlled process in an organism will generally lead to the demise of the organism. If the enzyme in question is peculiar to a bacterium or fungus that has invaded the mammalian host, then inhibition will eradicate the pathogen and leave the host unharmed. For example, penicillins and cephalosporins act as highly selective antibacterials because they inhibit an enzyme that is found only in the cell wall of bacteria.

1.1.4 Other types of bioactive molecules

Not all drugs act on discrete receptors or at active sites of particular enzymes. Others, such as some general anaesthetics, act partly by insertion in lipid membranes, thereby modifying ion transport across the membrane. Yet others, such as antacids or some diuretics, produce their effects by means of their physicochemical properties. Many steroid drugs and hormones must first pass into cells, becoming associated with specific cytosolic proteins which facilitate their transport into the cell nucleus where gene expression, and ultimately protein synthesis, is modified. Facilitated transport is also crucial in getting polar building blocks (e.g. nucleosides) across the hydrophobic blood–brain barrier.

1.1.5 Factors influencing drug action

The aim of administering a drug is to get it to the right place, in the right concentration and for the right period of time. Except for topical treatments (e.g. application of an anti-inflammatory agent to the skin), humans are usually dosed by a route which is remote from the intended site of action. Thus there will generally be a certain latent period before the action of the drug is initiated. This latent period will depend upon the route of administration, the formulation of the compound and the mode of distribution. The duration and intensity of action will, in turn, depend on the relative rates of arrival at, and removal from, the site of action. These rates depend primarily on the distribution, metabolism and excretion of the drug. The overall chronology of events between drug administration and elimination is summarized in Figure 1.12.

Some drugs (prodrugs) will not exert pharmacological activity until they have undergone biotransformation within the body. For example, the penicillin ester pivampicillin (**16**) is well absorbed when taken by mouth unlike the parent acid; the ester must be de-esterified by a blood-borne esterase enzyme before antibacterial activity is exhibited.

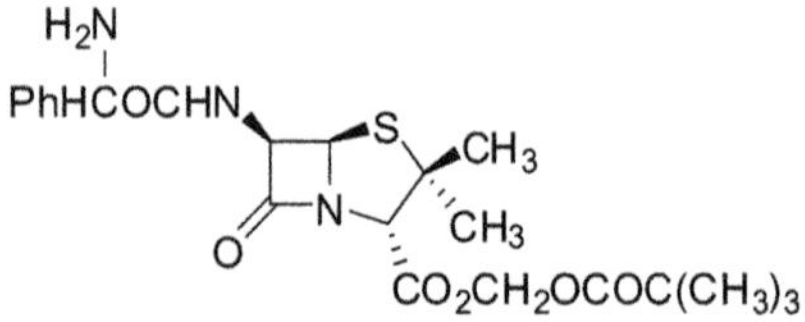

(**16**)

Having succeeded in getting a drug into the biophase, binding to the appropriate site of an enzyme or receptor must take place. The binding of the small drug molecule to the macromolecule involves many complementary forces (electrostatic forces, hydrogen bonding, hydrophobic bonding, and van der Waals forces).

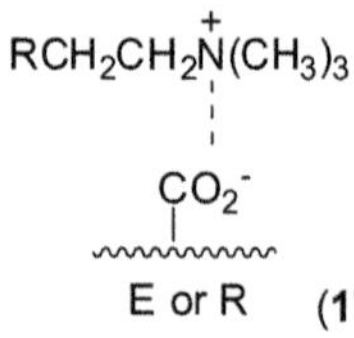

E or R   (**17**)

Electrostatic attraction, for example, between the quaternary ammonium group of ACh analogues and a carboxylate residue in the macromolecule (**17**), is a powerful binding and stabilizing influence. Ion–dipole and dipole–dipole interactions are of less importance. Hydrophobic bonding is also very important and leads to good binding mainly due to the increase in entropy of the system ($\Delta G = \Delta H - T\Delta S$) through displacement of water molecules from 'uncomfortable' quasicrystalline arrangements adjacent to the hydrocarbon surfaces to a less-ordered more-favourable situation (Figure 1.13). On the other hand, the gain in free energy through hydrogen bonding between enzyme or receptor and a drug molecule is not as significant as may be imagined at first sight.

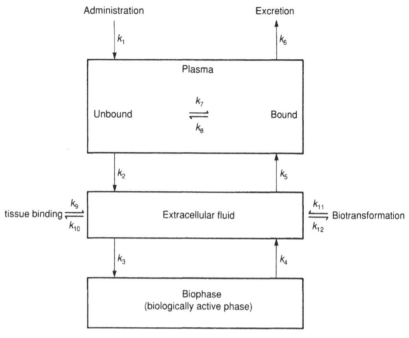

FIGURE 1.12

Summary of the events between drug administration and elimination.

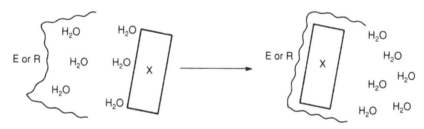

FIGURE 1.13

Approach of hydrophobic faces of a drug molecule (X) and a receptor (R) or enzyme (E).

For example, the phenol moiety forming the part structure of the drug molecule D-C_6H_4OH can be nicely accommodated at the active site of an enzyme (E) or receptor (R) as shown in Figure 1.14(a); strong hydrogen bonds are formed and good binding takes place. However, the hydrogen bonds between the drug and the enzyme or receptor can only be formed at the expense of the hydrogen bonds between the separate entities and associated water molecules (Figure 1.14(b)). Hence, the net gain in energy through hydrogen bonding between the drug and macromolecule is marginal. Note that

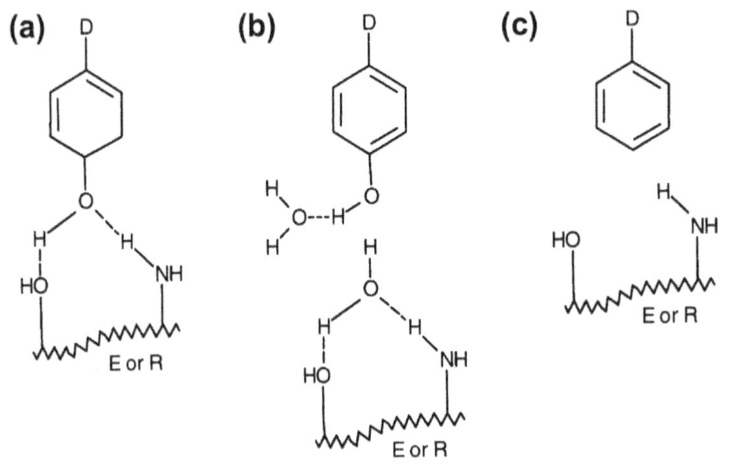

FIGURE 1.14

(a–c) Changes in hydrogen bonding that take place on approach of a drug to the surface of an enzyme (E) or receptor (R).

a closely related drug substance D-C$_6$H$_5$ lacking the phenolic hydroxy group would displace the water molecule(s) from the surface of the macromolecule without forming the compensating hydrogen bonds (Figure 1.14(c)); in this case binding would be less secure.

The pharmacokinetic profile of a drug will vary from species to species and be influenced by many factors such as the route of administration, formulation, age, sex, disease, diet, environmental influences and the presence of other drugs.

Clearly, the design of a drug molecule is hampered by the lack of detailed knowledge regarding pharmacological receptors, accessibility to target enzymes or other intended sites of action, and other undesirable and toxic effects. In addition, the development of a useful therapeutic agent is fraught with difficulties as the ideal structure to fit an enzyme or a receptor may not be suitable to allow that particular molecule to run the gauntlet between administration and elimination so as to be present near the enzyme/receptor at a concentration which would elicit the required physiological effect. It is fair to say that, currently, it is the inability to predict toxicity profiles that slows down the pace of pharmaceutical research.

It can be appreciated that a drug molecule has somewhat greater difficulty in travelling to its intended site of action than, for example, a natural neurotransmitter which is released close to the appropriate receptor. The localized release and rapid disposal of these small molecules means that a complex organism can operate with the minimum number of different neurotransmitters; when an active substance such as histamine or adrenaline is released into a large part of the system, many different effects are seen owing to the stimulation of receptors at many sites. Drugs can be regarded as being more closely analogous to hormones and local hormones (autocoids) in that they are widely distributed in the body. Generally, these molecules are sterically more bulky and possess more functional groups; the extra features within the molecules restrict binding to relatively few receptors or enzymes.

1.1.6 **The impact of the sequencing of the human genome**

One of the most spectacular achievements over the past few years was the sequencing of the human genome, the first draft appearing in 2000.

However, a survey in 2011 showed that about 75% of protein research still focused on the 10% of proteins that were known before the genome was mapped! For example, it has become clear that the human genome encodes for over 500 kinases (enzymes catalysing phosphorylation reactions), the majority of which have links to diseases. However, research has remained fixed on the 'top 50' kinases that were already established as targets in the 1990s. In a related analysis, it emerged that 11 protein kinases have been identified as important enzymes in the signalling pathways underlying breast cancer. Surprisingly, one of these enzymes, CDC2, received more attention (as judged by the output of research papers) than the other 10 kinases combined, even though the relative importance of the different kinases is not so marked.

Similarly, for nuclear receptors (transcription factors that bind small signalling molecules such as steroids and hormones), three-quarters of research activity in 2009 focused on the six receptors that were most studied in the 1990s. The other 42 nuclear receptors have not received anywhere near the same level of attention.

Obviously, there is a colossal amount of work still to be accomplished with regard to the understanding of biological processes and from that knowledge, the design of new drug substances.

1.2 SECTION II: MORE ABOUT ENZYMES

1.2.1 **Configuration of enzymes**

Although some metabolic reactions can occur spontaneously, by far the majority require *enzymes* as catalysts. Each cell requires over 500 different enzymes to enable it to carry out all its functions, although the types of enzymes required will depend on the nature of the cell. Some enzymes are membrane bound, some are found only in particular subcellular organelles such as mitochondria and others are cytoplasmic. Over 2000 enzymes have been described, the vast majority of which are proteins (although many proteins, such as haemoglobin and insulin, are not enzymes). Many enzymes also require additional nonprotein components (known as *cofactors*) in order to be catalytically active; the role of cofactors will be discussed later.

As described in detail in Chapter 2, enzymes have characteristic primary, secondary and tertiary structures. The *primary structure* is based on the sequence of amino acids linked by peptide bonds. A total of 22 amino acids are available for use in the primary sequence and each polypeptide chain may contain several hundred amino acid residues. An enzyme's *secondary and tertiary structure* gives it its three-dimensional characteristics, often giving rise to beautiful structures as revealed by X-ray crystallography (Figure 1.15). The secondary structure refers to the coiling and twisting of the polypeptide chain due to hydrogen bonding, and tertiary structure refers to the way in which the enzyme folds into its characteristic globular form and is retained in this configuration by a number of types of bonding including disulfide bonds formed between adjacent thiol groups and ionic bonds between adjacent free carboxylic acid and amino groups. Hydrogen bonding and hydrophobic bonding are also crucial in determining the overall shape of an enzyme. The digestive

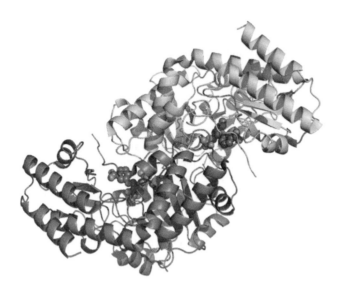

FIGURE 1.15

Structure of an enzyme from a microorganism (Sulfolobus solfatarius). (See colour plate.)

enzymes such as pepsin and trypsin consist of single polypeptide chains and have molecular weights varying from 15,000 to 35,000.[*]

Most enzymes catalysing metabolic reactions inside cells have what is called a *quaternary structure*, which means that they consist of more than one polypeptide subunit. Ultracentrifugation has shown that anything from 2 to 60 subunits can occur in a single enzyme, giving rise to molecular weights ranging from 35,000 to several hundred thousand. Most enzymes have from two to eight subunits. The enzyme may consist of aggregates of identical subunits or of more than one type of subunit. For example, the enzyme lactate dehydrogenase, important for carbohydrate metabolism in muscle, consists of four subunits of which there are two types. As a consequence, there are five different forms of this enzyme, depending on the combination of subunits. Enzymes like this which can occur in more than one molecular form are called *isoenzymes*. These have an important role in the regulation of metabolism and are also valuable clinically in diagnostic enzymology, as elevated plasma levels of a particular isoenzyme may indicate that damage has occurred in a particular organ such as the liver or heart.

Some metabolic systems involve even more complex levels of organization known as *multienzyme complexes* in which several different enzymes associate together physically to catalyse a series of sequential reactions. An example of this occurs in fatty acid synthesis in bacteria and plants in which seven different enzymes aggregate together to give a highly ordered and efficient system. *Multifunctional enzymes* are also known, in which one protein can catalyse a sequence of reactions. For example, HIV reverse transcriptase catalyses the synthesis of DNA on an RNA template, then cleavage

---

[*] These enzymes, incidentally, have an inactive precursor form called *zymogens* or proenzymes, which have to be converted to the active form by hydrolytic removal of a protective section of the peptide chain.

of the RNA strand away from the new DNA sequence and finally the synthesis of double-stranded viral DNA using the first formed strand as the template.

Because of their protein structure, enzymes are highly sensitive to their physical environment. Organic solvents, high salt concentrations, oxidation or reduction and extremes of pH or temperature will readily cause *denaturation*, precipitation due to breakdown of the tertiary and secondary structure. Loss of enzyme activity without denaturation can also occur as a result of relatively small changes in temperature and pH. Most enzymes have their optimal activity at physiological pH and temperature, that is, pH 7 and 37 °C.

1.2.2 Enzyme specificity, classification and nomenclature

One of the properties of enzymes that most distinguishes them from inorganic catalysts is their high degree of specificity. This is due to the chemical and physical characteristics of an enzyme's protein structure. Enzymes are normally specific for one particular type of reaction and one particular type of substrate only. For example, a hydrolytic enzyme such as a peptidase or a lipase will not catalyse oxidation–reduction or group transfer reactions. Likewise, the enzyme alcohol dehydrogenase, which catalyses the oxidation of ethanol to acetaldehyde, will also convert a variety of other alcohols to their corresponding aldehydes, with differing affinities for the substrates but will have no effect on nonalcoholic substrates. Most enzymes will accept a range of chemically related compounds as their substrates; relatively few enzymes show absolute specificity for a single substrate.

Enzymes also display stereochemical specificity and may be specific for particular optical or geometric isomers of a substrate. For example, D-amino acid oxidase will oxidize a variety of D-amino acids but has little or no activity with L-amino acids, and fumarase catalyses the hydration of the *trans* unsaturated dicarboxylic acid fumaric acid but not the *cis* isomer (maleic acid).

Enzymes are generally classified as belonging to one of the six categories shown in Table 1.4.

Enzymes have complex systematic names based on further subdivisions of the categories in Table 1.4. These are used for precise identification in technical literature, but trivial names are normally used in laboratories. These usually include the name of the substrate and the type of reaction (e.g. alcohol dehydrogenase). Other common trivial names simply add the suffix *-ase* to the name of the substrate (e.g. pig liver esterase), while many of the digestive enzymes are known by their historic names (e.g. amylase, lipase, pepsin and trypsin).

1.2.3 Characteristics of enzyme catalysis

Although enzymes show much greater specificity and sensitivity to temperature and pH than inorganic catalysts, they have the same function as all catalysts, accelerating the rate of reaction by lowering the activation energy required for a reaction to proceed. Enzymes also obey the following normal rules of catalysis: (1) enzymes will not catalyse thermodynamically unfavourable reactions, (2) they will not change the direction of a reaction, (3) they will not change the equilibrium of a reaction, and (4) they remain unchanged at the end of the reaction (although some enzymes undergo temporary covalent changes such as phosphorylation during the course of a reaction). Enzymes normally catalyse reactions much more efficiently than inorganic catalysts. For example, both platinum and the enzyme peroxidase (also known by its old name catalase) catalyse the decomposition of hydrogen peroxide

Table 1.4 Enzyme Classification

Enzyme Category	Examples	Types of Reaction Catalysed
Oxidoreductases		Oxidation or reduction of substrate
	Dehydrogenases	Transfer of H from substrate to cofactor
	Reductases	Addition of H to substrate
	Oxidases	Transfer of H from substrate to oxygen
Transferases		Transfer of group from one molecule to another
	Aminotransferases	Transfer of amino groups
	Transacetylases	Transfer of acetyl groups
	Phosphorylases	Transfer of phosphate groups
Hydrolases		Hydrolysis of substrate (irreversible)
	Glycosidases	Hydrolysis of glycosidic bonds
	Esterases	Hydrolysis of ester bonds
	Peptidases	Hydrolysis of peptide bonds
Lyases		Elimination and addition reactions
	Hydratases	Addition of water to double bonds (reversible)
	Decarboxylases	Removal of CO_2 from substrate (reversible)
	Aldolases	Aldol condensations (reversible)
Isomerases		Molecular rearrangement
	Racemases	D- and L-isomer interconversion
	cis-trans isomerases	Geometrical isomerization
	Mutases	Intramolecular group transfer
Ligases		Energy-dependent bond formation (generally irreversible)
	Synthetases	Condensation of two molecules
	Carboxylases	Addition of CO_2 to substrate

into water and oxygen. The effects of the different catalysts on the activation energy (in kilojoules/mole) are as follows:

No catalyst	75
Platinum	50
Peroxidase	8

One mole of peroxidase can catalyse the decomposition of over 1 million moles of H_2O_2 per minute.

Enzymes are able to lower the activation energy for a reaction involving substrate (S) through the formation of an intermediate *enzyme–substrate complex* (ES), which in turn can break down to the enzyme and the product (P) as follows (Eqn (1.7)):

$$E + S \rightarrow E - S \rightarrow E + P \qquad (1.7)$$

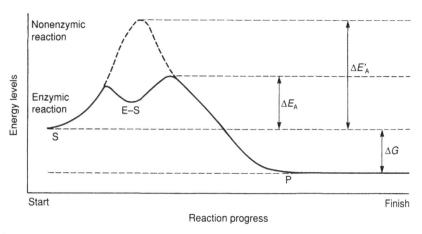

FIGURE 1.16

The effect of enzyme catalysis on activation energy. ΔE_A is the activation energy for the enzyme-catalysed reaction and $\Delta E'_A$ for the nonenzymatic reaction. ΔG is the free energy of the reaction.

S may consist of more than one substrate and P of more than one product. The effect of the formation of this complex is shown in Figure 1.16.

The formation of this enzyme–substrate complex occurs at a small region on the surface of the enzyme called the *active site* or *active centre*, usually present as a crevice or pit. Only a few amino acids (5–10) are directly involved in the formation of this complex and subsequent catalysis at the active site. These amino acid residues are not normally consecutive, as different parts of the polypeptide chain come together at the active site as a result of the characteristic folding of the protein. The amino acids not directly involved at the active site are still important, however, as they are essential for maintaining the configuration of the protein required for the active site to function. This explains why some modifications to an enzyme's structure are more crucial than others. A modification or mutation leading to a change in amino acid sequence at the active centre can lead to complete loss of enzyme activity, while a change in sequence elsewhere may have far less effect.

The amino acids most frequently found at the active site are shown in Table 1.5; these amino acids possess groups on the side chain that are known to interact with the substrate.

Not only are hydrogen bonding and electrostatic interactions are involved between the substrate and complex but also hydrophobic interactions are important as mentioned above.

It has been shown that some of the amino acids at the active centre are only involved in binding the substrate to the enzyme. This part of the active site is called the *binding site*. Other amino acids at the active site are exclusively involved in catalysis, making up the *catalytic site*. The active site thus consists of binding and catalytic sites. Both binding and catalysis are essential for conversion of substrate(s) to product(s). The active site of an enzyme is shown diagrammatically in Figure 1.17.

The presence of the binding and catalytic sites at the active site accounts for the specificity of enzyme reactions discussed earlier. Only the correct substrate (and closely related analogues) can bind to the enzyme; only the appropriate type of reaction can be catalysed by a particular enzyme.

Table 1.5 Amino Acids Frequently Involved in the Active Centre

Amino Acid	Side Group	Interaction with Substrate
Serine	$-OH$	Hydrogen bonding
Cysteine	$-SH$	Hydrogen bonding/disulfide bridging
Histidine	$-Imidazole$	Hydrogen bonding/ electrostatic
Lysine	$-NH_3^+$	Electrostatic
Arginine	$-NH_3^+$	Electrostatic
Aspartic acid	$-COO^-$	Electrostatic
Glutamic acid	$-COO^-$	Electrostatic

The old analogy of a rigid lock and key mechanism introduced in the nineteenth century by Emil Fischer has been largely superseded since the 1960s by Koshland's induced fit hypothesis, in which the substrate is believed to induce the required orientation of groups in the active site required for binding and catalysis.

Formation of the enzyme–substrate complex leads to a lowering of activation energy and subsequent catalysis due to the interaction between the substrate and the catalytic site. The binding between the substrate and amino acid groups at the active centre causes changes in bond energies and electron densities within the substrate, possibly accompanied by physical distortion and strain, resulting in a thermodynamically unstable conformation from which the reaction can readily proceed. When there is more than one substrate for a reaction, the enzyme catalyses the reaction by bringing the substrates together at the binding site in the juxtaposition required for the reaction to proceed. Provided the affinity of the product for the binding site is lower than that of the substrate, the product will dissociate from the active site, leaving the active site free for further reactions. As will be seen later, certain inhibitors have their effect by binding to the active site without further reaction and thus remain there, blocking the site.

Many enzymes require additional nonprotein substances called cofactors for catalysis to occur. These substances can be either metal ion *activators* or relatively small organic compounds called

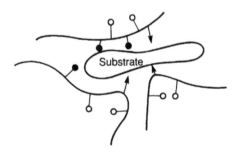

FIGURE 1.17

Active site of an enzyme: $-\bullet$, amino acids involved in binding site; $\rightarrow$, amino acids involved in catalytic site; $-\circ$, amino acids not directly involved in active site.

Table 1.6 Some Commonly Occurring Coenzymes

Coenzyme	Function	Vitamin Processor	Deficiency Disease
Nicotinamide adenine dinucleotide (NAD)	Hydrogen transfer	Nicotinamide (niacin)	Pellagra
Thiamine pyrophosphate	Decarboxylation	Thiamine (B_1)	Beriberi
Flavin mononucleotide	Hydrogen transfer	Riboflavin (B_2)	Skin lesions
Flavin adenine dinucleotide (FAD)	Hydrogen transfer	Riboflavin (B_2)	Skin lesions
Pyridoxal phosphate	Amino transfer	Pyridoxine (B_6)	Neurological disturbances
Ascorbic acid	Hydroxylation	Vitamin C	Scurvy
Cobalamin	Methylation	Vitamin B_{12}	Pernicious anaemia
Tetrahydrofolic acid (FH_4)	One-carbon transfer	Folic acid	Megaloblastic anaemia

coenzymes. These cofactors are normally attached to the enzyme by electrostatic bonds, but some coenzymes are linked covalently and are then called *prosthetic groups*.

Metal ion activators include Mg^{2+}, Ca^{2+}, Zn^{2+}, Fe^{2+}, Fe^{3+}, Cu^{2+}, Co^{2+}, Mo^+, K^+ and Na^+ and account for the trace metal elements required in the diet. Their requirement for enzyme activity also explains the inhibitory effect of chelating agents.

Coenzymes are derived from water-soluble vitamins and are involved in a variety of reactions, as shown in Table 1.6. They operate as second substrates, undergoing chemical modifications during the course of a reaction and reverting back to the original form by a further reaction. Diseases (such as scurvy) caused by the deficiency of a water-soluble vitamin are due to the resultant loss of enzyme activity.

1.2.4 Enzyme reaction rates

Enzyme activity is measured in units which indicate the rate of reaction catalysed by that enzyme expressed as micromoles of substrate transformed (or product formed) per minute. An enzyme unit is the amount of enzyme that will catalyse the transformation of 1 µmol of substrate/min under specified conditions of pH and temperature. The specific activity of an enzyme is expressed as the number of units per milligram of protein.

The rate of a biochemical reaction at a given temperature and pH depends on the enzyme concentration and the substrate concentration. Provided the substrate concentration remains in excess, the initial rate is directly proportional to the enzyme concentration.

When the enzyme concentration is kept constant and the substrate concentration varies, the effect of the substrate concentration on the rate of reaction is as shown in Figure 1.18. Initially, the reaction follows first-order kinetics with the rate proportional to substrate concentration, and eventually, zero-order kinetics is followed with the velocity reaching a limiting value V_{max}. V_{max} is the reaction rate when the enzyme is fully saturated by substrate, indicating that all the binding sites are being constantly reoccupied. V_{max} is constant for a given amount of enzyme. The substrate concentration corresponding to half the maximum velocity is known as K_m, the Michaelis constant, which is inversely proportional to the affinity of the enzyme for the substrate. K_m is constant for a particular enzyme and is independent of the amount of enzyme. A useful rule of thumb is that the substrate concentration has to be

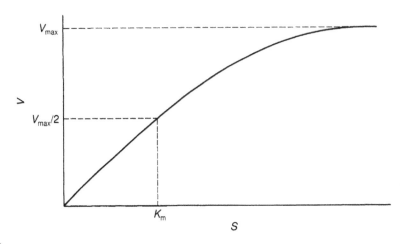

FIGURE 1.18

The effect of substrate concentration (S) on the rate of reaction (V).

about $100 \times K_m$ to achieve near-maximum velocity (~99% saturation). The relationship between reaction rate and substrate concentration is described by the Michaelis–Menten equation (Eqn (1.8)):

$$V = \frac{V_{max}S}{K_m + S} \tag{1.8}$$

This can also be expressed in the form of Eqn (1.9).

$$\frac{1}{V} = \frac{K_m}{V_{max}S} + \frac{1}{V_{max}} \tag{1.9}$$

A graph showing the reciprocal of the rate against the reciprocal of the substrate concentration, called the Lineweaver–Burk plot, will therefore give a straight line, as shown in Figure 1.19. As the

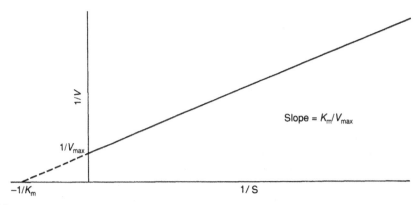

FIGURE 1.19

A Lineweaver–Burk plot.

intercepts correspond to $1/V_{max}$ and $-1/K_m$ and the slope is K_m/V_{max}, this plot provides a useful way of determining the kinetic constant V_{max} and equilibrium constant K_m from experimental data. As will be shown shortly, it is also a useful way of showing what type of inhibition may be occurring.

1.2.5 Enzyme substrates as drugs

Many of the basic building blocks (e.g. amino acids and sugars) for enzyme-controlled biosynthesis of complex natural products are provided in the diet. In contrast, there are few cases where an enzyme substrate is given to counter a disease. One excellent example is in the treatment of Parkinson disease.

Parkinson disease is a disorder characterized by rigidity of the limbs, torso and face; tremor; abnormal body posture and an inability to initiate voluntary motor activity (akinesia). It occurs mainly in elderly people and is a chronic and progressively degenerative disorder.

The disease is associated with decreased dopaminergic function in the brain, and many of the symptoms of the disease can be alleviated by oral administration of L-dopa (L-dihydroxyphenylalanine) (**18**). L-Dopa is given orally, absorbed from the gastrointestinal tract, and carried to the brain by the bloodstream. In the brain, the compound is a substrate for the enzyme dopa decarboxylase and DA (**19**) is produced, thus raising the activity of the dopaminergic neurons in the brain. DA cannot be given itself because it does not have the lipophilic properties necessary to cross the blood–brain barrier (Figure 1.20).

The enzyme dopa decarboxylase is also present in the liver and other tissues in the periphery. The action of peripheral dopa decarboxylase decreases the amount of L-dopa reaching the brain. To obtain

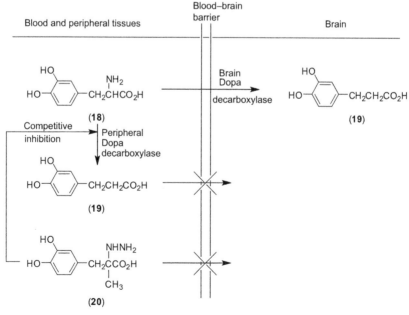

FIGURE 1.20

Structural relationship of dopa, dopamine and carbidopa.

a sufficient amount of compound at the necessary sites, large doses must be given, with a consequent increase in side effects (nausea and vomiting). The administered dose of L-dopa can be reduced by inhibiting the peripheral dopa decarboxylase enzyme using a compound such as carbidopa (**20**) which does not cross the blood–brain barrier.

1.2.6 Enzyme inhibition and enzyme inhibitors as drugs

Enzyme inhibitors are substances which bind to the enzyme with resulting loss of activity, without damaging the enzyme's protein structure. Inhibitors exert their effect by decreasing the affinity of the enzyme for the substrate, by decreasing the amount of active enzyme available for catalysis, or by a combination of these effects. Different categories of inhibitors are described below.

1.2.6.1 Irreversible inhibitors

These are compounds which bind covalently to specific groups on the protein's surface, preventing binding and catalysis of the reaction. These compounds will inhibit a wide range of enzymes. Examples include iodoacetamide (**21**) and *p*-chloromercuribenzoate (**22**) (which bind to the sulfhydryl group of cysteine residues), diisopropylfluorophosphate (**23**) (which binds to the hydroxyl groups of serine residues) and 1-fluoro-2,4-dinitrobenzene (**24**) (which binds to the amino groups in lysine residues and to the phenolic groups in tyrosine residues). Compounds such as these are mainly used for *in vitro* research studies, although some (e.g. mustards) are also used in anticancer therapy.

| (21) | (22) | (23) | (24) |

1.2.6.2 Competitive inhibitors

These compounds often structurally resemble the substrate and can thus compete with the substrate for the enzyme's binding site. Such inhibitors are highly specific for a particular enzyme. Binding of the inhibitor is reversible, so the inhibitor can be displaced from the binding site by excess substrate. The apparent affinity of the enzyme for the substrate is thus lowered in the presence of a competitive inhibitor (i.e. K_m is increased) but as catalysis is not directly affected, V_{max} can still be attained, even though a higher concentration of substrate will be required for this. The competitive inhibitor may be a naturally occurring alternative substrate for the enzyme which is also metabolized by it, although with a different degree of affinity, or may be a chemical analogue of the substrate which binds to the enzyme without being further metabolized. A number of toxins and drugs operate in the latter manner. One of the earliest examples to be recognized was sulfanilamide (**25**), which inhibits a bacterial enzyme dihydropteroate synthetase (Scheme 1.1) required for folic acid synthesis through its chemical similarity to *p*-aminobenzoic acid (**26**), a component of folic acid.

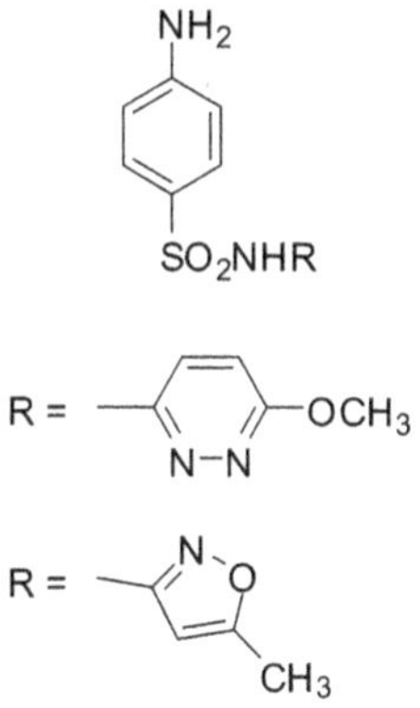

NH$_2$

SO$_2$NHR

NH$_2$

CO$_2$H

(26)

R = [ring] —OCH$_3$ Sulphamethoxypyridazine

R = [ring] CH$_3$ Sulphamethoxazole

(25)

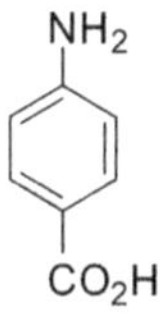

NH$_2$

H$_2$N— —NHCH$_2$— OCH$_3$ / —OCH$_3$ / OCH$_3$

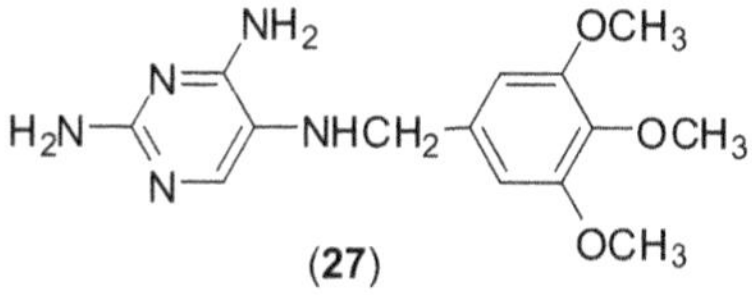

(27)

Since the growth and proliferation of many bacteria depend on the production of folates by this process (whereas in humans folic acid is used and supplied in the diet as a preformed vitamin), sulfanilamide (and related sulfur drugs) will stop the infecting bacteria in their tracks but will not affect the human host.

A different situation arises when it is required to inhibit an enzyme which is common to bacteria and humans. Consider transformation 2 in Scheme 1.1. This step is essential for the well-being of both bacteria and humans; in order to have an antibacterial agent that is nontoxic to the host, advantage must be taken of the fact that the structures of bacterial dihydrofolate reductase and mammalian dihydrofolate reductase are different. Selective inhibitors of the bacterial enzyme have been found: for example, the antibacterial drug trimethoprim (**27**) binds to dihydrofolate reductase from the bacterium *Escherichia coli* about 10^4 times more strongly than to the same enzyme derived from rat liver.

Methotrexate (**28**) is another dihydrofolate inhibitor that is in clinical use, for the treatment of the widespread and sometimes crippling disease psoriasis as well as in the treatment of some forms of cancer. The effect of reducing purine synthesis and cell division by blocking tetrahydrofolate production in psoriatic cells is beneficial; however, the drug must be used with caution since other structurally similar, and in many cases indistinguishable, dihydrofolate reductase enzymes in other tissues are affected, leading to severe toxicity on long-term treatment.

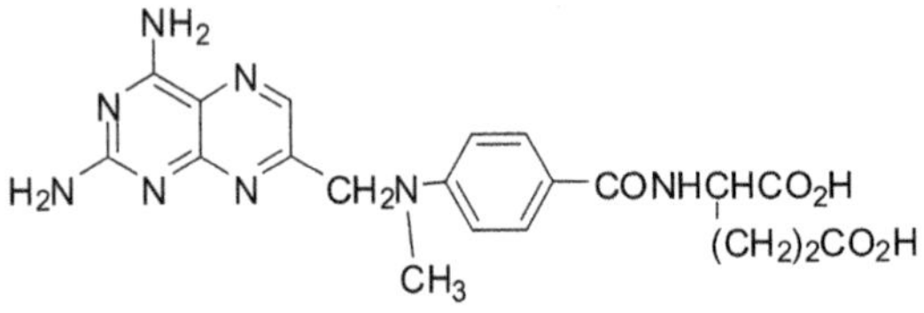

(28)

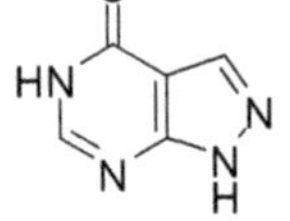

(29)

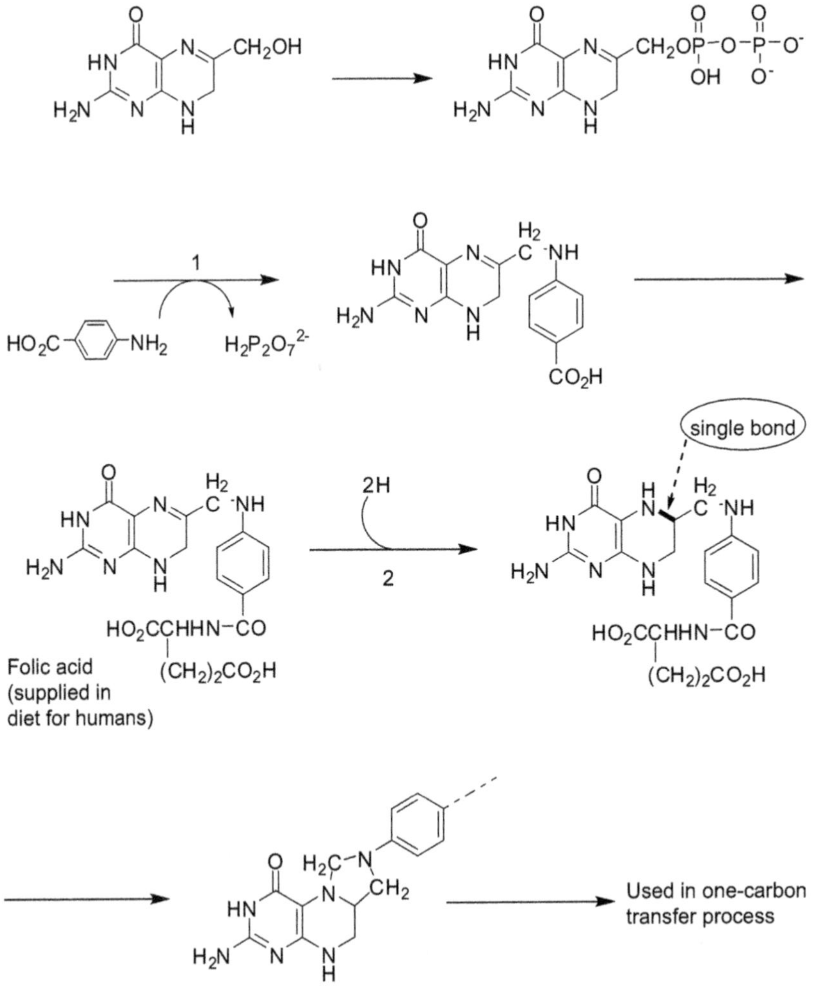

Folic acid
(supplied in
diet for humans)

single bond

2H

Used in one-carbon
transfer process

SCHEME 1.1

Biosynthesis of the folates. Transformation 1 is catalysed by dihydropteroate synthetase. This enzyme is inhibited by sulfanilamides (**25**). Transformation 2 is catalysed by dihydrofolate reductase. This bacterial enzyme is inhibited by trimethoprim (**27**).

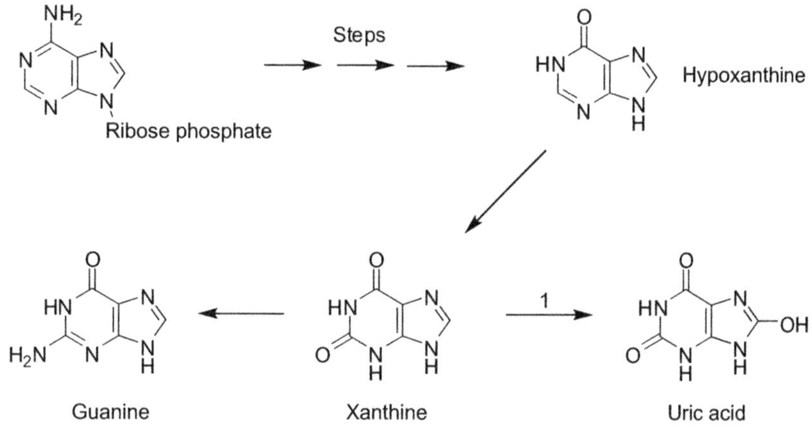

SCHEME 1.2

Biosynthesis of uric acid. Transformation 1 is catalysed by xanthine oxidase. This enzyme is inhibited by allopurinol (**29**).

Another important enzyme inhibitor is allopurinol (**29**), which is used in the treatment of gout. This condition is due to a build-up in the concentration of uric acid in joints. The enzyme controlling the production of urate from xanthine is xanthine oxidase (Scheme 1.2); on inhibition of this enzyme with allopurinol, the cascades from guanine and hypoxanthine are stopped at xanthine and this compound is rapidly excreted. Such enzymes, having a very limited function in the body, may be safely and usefully inhibited by drugs.

The above cases, as well as a cross-section of other examples of drugs acting as enzyme inhibitors, are summarized in Table 1.7.

Table 1.7 Drugs Acting as Enzyme Competitive Inhibitors

Drug	Enzyme Inhibited	Disorder Treatment
Sulfanilamides	Dihydropteroate synthetase	Bacterial infections
Trimethoprim	Bacterial dihydrofolate reductase	Bacterial infections
Penicillins and cephalosporins	Bacterial peptidoglycan transacylases	Bacterial infections
Azidothymidine (AZT)	HIV reverse transcriptase	AIDS
Allopurinol	Xanthine oxidase	Hyperuricemia (primary metabolic gout)
Methotrexate	Dihydrofolate reductase	Psoriasis, cancers
AIDS: Acquired immune deficiency syndrome.		

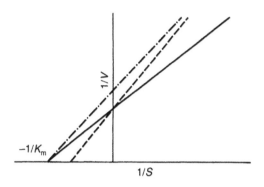

FIGURE 1.21

Effects of competitive and noncompetitive inhibition on enzyme kinetics: —, no inhibitor present; — —, competitive inhibitor present; —·—, noncompetitive inhibitor present.

1.2.6.3 Noncompetitive inhibitors

These substances, which are generally structurally unrelated to the substrate, bind reversibly to groups distant from the binding site of the enzyme and are thus less specific than competitive inhibitors. The rate of reaction is decreased because the catalytic site is affected by the presence of the inhibitor. V_{max} is thus reduced, but because the binding site is not affected, the affinity of the enzyme for the substrate and therefore the K_m remains unchanged. The effect of noncompetitive inhibition is the same as that of less enzyme being present. Competitive and noncompetitive inhibition can be distinguished from each other by the Lineweaver–Burk plot discussed earlier, as shown in Figure 1.21.

A number of poisons are harmful to cells because they are potent noncompetitive inhibitors. Some examples are listed in Table 1.8.

1.2.7 Enzyme regulation

Metabolic systems require regulation to ensure that adequate output occurs (whether this be energy production from nutrients or biosynthesis of complex molecules) while avoiding the wasteful and potentially harmful consequences of overproduction. In general, this fine control is achieved by

Table 1.8 Poisons Acting as Noncompetitive Enzyme Inhibitors

Compound	Mode of Action	Biological Effect
Organophosphate nerve gases and pesticides	Bind to serine —OH groups in cholinesterase active site	Paralysis
Mercuric salts and arsenic salts	Bind to —SH groups in many enzymes	Widespread cellular damage
Cyanide	Binds to cytochrome oxidase	Respiratory failure
Digitoxin	Binds to Na, K-ATPase	Inhibits sodium ion migration

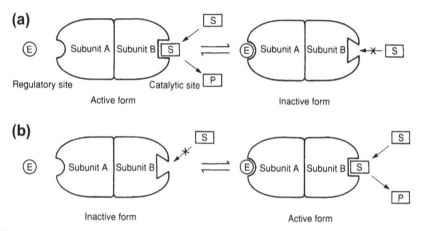

FIGURE 1.22

Allosteric regulation of enzyme activity showing (a) inhibition and (b) activation. The effector is E, the substrate S and product P.

a coordination of the regulation of enzyme synthesis and the regulation of enzyme activity. Enzyme activity is mainly controlled by the process of *allosteric regulation*, which can produce activation or, more commonly, inhibition of enzyme activity. The regulator or *effector* molecule is normally structurally unrelated to the substrate but binds specifically and reversibly to the enzyme. However, this does not occur at the substrate binding site but at a quite separate regulatory site (the name allosteric is derived from the Greek *allos steros*, meaning other space). Binding of the effector induces a conformational change which either increases or decreases the affinity of the enzyme for the substrate, depending on the nature of the effector.

Only certain enzymes, called regulatory or allosteric enzymes, are sensitive to this form of control, and all such regulatory enzymes have been shown to have quaternary structure, that is, two or more subunits. In many cases, the regulatory site is on a different subunit from the active site, although in some enzymes both sites are on the same subunit. Some enzymes are regulated by more than one allosteric effector; in such cases, there are separate regulatory sites for each effector. Allosteric regulation is illustrated in Figure 1.22.

In biosynthetic pathways, a system of feedback inhibition frequently occurs whereby the end product of the pathway allosterically inhibits the first enzyme specific to that pathway. In branched-chain pathways, as occurring in amino acid biosynthesis, the different products can separately inhibit the primary enzyme in the pathway and also the enzymes immediately after branching, as shown in Figure 1.23.

This system of regulating biosynthesis ensures that the pathways operate normally when the products are required, but can be partially or completely shut down when adequate supplies of product have been formed. Some examples of allosteric regulation are shown in Table 1.9.

As well as the change in activity achieved through allosteric regulation, some enzymes undergo changes in activity due to *covalent modification*, normally phosphorylation. Well-studied examples of this form of regulation occur in the synthesis of the polysaccharide glycogen. Glycogen

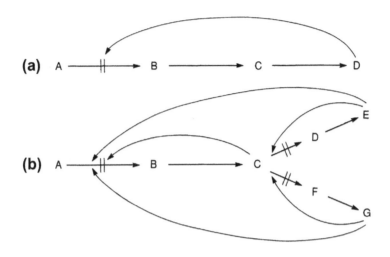

FIGURE 1.23

Allosteric regulation of biosynthetic pathways by end product feedback inhibition illustrating (a) an unbranched pathway and (b) a branched pathway.

synthetase exists in an inactive phosphorylated form and an active dephosphorylated form. The protein hormone insulin stimulates the dephosphorylation of the inactive form of the enzyme, thus increasing its activity in synthesizing glycogen. (Insulin also stimulates the uptake of the precursor glucose into the cells, further enhancing glycogen synthesis.) Glycogen breakdown (producing blood glucose in the liver and energy in muscles) requires the enzyme phosphorylase, which exists as a relatively inactive dimer or as a highly active phosphorylated dimer. Conversion to the active form requires intracellular cyclic AMP, which is produced from ATP by the membrane enzyme adenylate cyclase in response to the extracellular hormone adrenaline (Eqn (1.4)) or glucagon binding to specific receptors on the membrane. Cyclic AMP also leads to phosphorylation and hence inactivation of glycogen synthetase. The opposite effects of the hormones adrenaline and glucagon from that of insulin can thus be explained by their indirect effects on enzyme activity. Further discussion on the origin and actions of insulin, as well as drugs to control type 2 diabetes, can be found in Chapter 6.

Table 1.9 Some Examples of Allosteric Regulation

Regulatory enzyme	Effector	Effect	Biochemical pathway
Phosphofructokinase	AMP	Activation	Glycolysis
	ATP	Inhibition	
Pyruvate carboxylase	Acetyl-CoA	Activation	Gluconeogenesis
Mevalonate synthetase	Cholesterol	Inhibition	Steroid synthesis
Aspartokinase	Threonine, isoleucine, methionine, lysine	Inhibition	Amino acid synthesis

CoA, coenzyme A.

Cyclic AMP is referred to as a secondary messenger as it is produced inside cells in response to plasma hormones binding to specific receptors on the outer surface of cell membranes. Many hormones other than adrenaline and glucagon operate in this way, including NorA.

Apart from this indirect effect on enzyme activity, many hormones help to control metabolism by their effect on enzyme synthesis. An increase in synthesis and subsequently the amount of enzyme is called *induction*, and a decrease in synthesis is called *repression*. In bacterial cells, these processes frequently occur as a direct response to levels of nutrients present and do not involve hormones. For example, when bacteria are grown in media rich in carbohydrates, the enzymes involved in carbohydrate absorption and breakdown are induced in order to optimize their utilization, whereas when bacteria are grown in media rich in amino acids, enzymes required for their synthesis are repressed in order to avoid wasteful overproduction.

In animals, however, induction and repression of enzyme activity occur in response to hormones and some drugs. Most hormones cannot readily cross the cell membrane and will therefore achieve this effect by secondary messengers, but the steroid hormones and possibly the thyroid hormones do cross the membrane and exert a direct influence on protein synthesis. The steroid hormones affect *transcription*, the transfer of genetic information required for protein synthesis from DNA to messenger RNA. Other hormones affect *translation*, the production of proteins at the ribosomes. In both cases, metabolism is controlled through the regulation of enzyme synthesis. It is noteworthy that the mechanisms by which such hormones operate are still not well understood at the molecular level.

1.3 SECTION III: MORE ABOUT RECEPTORS

During the past 100 years or so, a succession of eminent pharmacologists such as Clark, Gaddum, Stephenson and Schild developed the receptor concept and laid the quantitative foundations of modern receptor pharmacology.

In principle, the term receptor could refer to any site to which drug molecules may bind. In practice, it is more useful in pharmacology to restrict the term receptor to those specific protein molecules whose functional role is to act as chemical sensors or transducers in the chemical communications systems that coordinate the activity of cells and organs within the body.

1.3.1 Bioassay and the measurement of drug effects

1.3.1.1 Bioassay

In many areas of the drug discovery process, the use of bioassay preparations has been replaced by the use of cell lines expressing cloned human receptors linked to a suitable transduction system or reporter gene that can be assayed robotically. It may be argued though, that some reliance on bioassays is still desirable in order to confirm drug action in the intact target tissue or to assess the significance of 'off-target' effects.

Historically, one of the earliest examples of the use of bioassay is the experiments of Otto Loewi, who in 1920 demonstrated the chemical nature of neurotransmission using two perfused frog hearts. Loewi showed that chemical transmitters (later found to be ACh and NorA) were released from the nerve endings in the heart on stimulation of the vagus nerve. Vagal stimulation caused either slowing (mediated by ACh) or speeding up (mediated by NorA) of the frog heart rate.

Historically, bioassay techniques were also crucial to the identification by Dale and co-workers of ACh as the neurotransmitter at the skeletal muscle neuromuscular junction. Bioassay techniques have been particularly useful in the identification of local hormones which are too labile to be studied by chemical means. This was particularly so in the identification of the prostaglandins and prostacyclins by Vane and Moncada et al. which led to the discovery that aspirin-like drugs act by inhibiting the synthesis of these local hormones.

Another example of the use of bioassay techniques is the work which led to the identification of endothelium-derived relaxant factor as nitric oxide (NO) (Figure 1.24). The relaxation of arterial

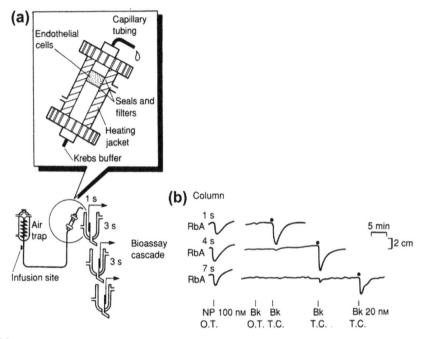

FIGURE 1.24

(a) Diagrammatic representation of the bioassay cascade and the column containing cultured endothelial cells which were used by Moncada and collaborators to determine the nature of the endothelium-derived relaxant factor (EDRF). The principle of the bioassay is that perfusate from the preparation under investigation (in this case endothelial cells) is passed over a series of test preparations in the bioassay cascade (the responses of the test preparations could be contraction, relaxation, light emission, secretion, etc.). The effects of the test solution are then compared with the effects of known concentrations of hormones suspected of being the unknown. Owing to the exquisite sensitivity of some biological preparations, they are often preferred to chemical techniques as a means of detecting physiological mediators. (b) Relaxation of contracted rabbit aortic strips (RbAs) in the bioassay cascade. There was a 1 s delay in perfusion between the column of endothelial cells and the first tissue in the cascade and a 3 s delay in perfusion between each bioassay tissue. The effect of nitroprusside over the bioassay tissues (NP, 100 nM O.T.) is compared with that of bradykinin (Bk, 20 nM) which has no effect when given O.T. but induces the release of a labile relaxing substance when given through the column (T.C.). The second and third Bk infusions T.C. are given when the effluent from the column directly superfuses the second and the third tissues in the cascade. The tissues were precontracted with a prostaglandin analogue U46619 (15 nM).

Source: From Gryglewski et al. (1986) with permission.

smooth muscle by bradykinin was studied and it was demonstrated that bradykinin did not relax strips of aorta which had been denuded of endothelium. When the superfusate from cultures of aortic endothelial cells was perfused over the aortic strips, these relaxed only when bradykinin was in the superfusate to cause release of NO from the endothelial cells. Bioassay techniques were crucial to these experiments because the half-life of NO is only about 6 s.

The receptor for NO is the cytosolic form of the enzyme guanylate cyclase which catalyses the conversion of guanosine triphosphate (GTP) to cyclic guanosine $3',5'$-monophosphate (cGMP). This forms part of the transduction mechanism mediating vasodilatation due to a range of vasodilators such as bradykinin, 5-HT or the peptide called substance P which act on receptors on the endothelial cells. After generation of NO *in vivo*, it diffuses into the neighbouring smooth muscle cell to cause relaxation. NO is also an important mediator in blood platelet activation and immunological reactions.

Studies of the effects of drugs on living systems require careful design because living systems display *biological variation*. No two people are alike, and no two preparations of biological tissue can be considered identical either. For example, it is likely that the β_2-adrenoceptors of all laboratory rats have exactly the same amino acid sequence. However, owing to possible variations in the assay being employed, nutritional state, hormonal balance, etc., different preparations will respond in a slightly different way to a β_2-adrenoceptor agonist. Moreover, variations in the experimenter's technique must be taken into account. Repetition of pharmacological experiments is therefore essential, not only to avoid being misled by variation between different preparations but also to avoid slight variation in experimental technique. For example, if control responses and test responses are taken in the same tissue, the test response can be expressed as a percentage of the control response. Where the absolute magnitude of responses in each tissue may be quite different, results can be normalized in this way and combined. If during each experiment, control responses are taken before *and* after each test response, then each preparation can be checked for changes in drug sensitivity during the experiment.

1.3.2 Quantifying drug–receptor interactions

The theory of drug–receptor interactions was developed using the Law of Mass Action. This law states that the rate of any reaction is proportional to the concentrations of the reactants. It can be used to derive a relationship that, in many cases, describes drug binding to receptors.

A drug molecule D is assumed to combine with a receptor, R, to form a drug–receptor complex, DR. By the Law of Mass Action, the rate of this reaction is proportional to the product of the concentrations of D and R. Thus (Eqn 1.10)

$$\text{Rate} \propto [\text{D}][\text{R}] \tag{1.10}$$

If we call the proportionality constant k_{+1} this means (Eqn (1.11)).

$$\text{Rate} = k_{+1}[\text{D}][\text{R}] \tag{1.11}$$

[D] represents the concentration of drug molecules in solution and [R] is the concentration of free (unoccupied) receptors. The constant k_{+1} is known as the *microscopic association rate constant* (this is a first-order rate constant with dimensions of $M^{-1}s^{-1}$). The reverse reaction (*dissociation* of drug from receptor) is described by the microscopic dissociation constant k_{-1} (dimensions s^{-1}). In short, this reaction is written as (Eqn (1.12)).

$$\text{D} + \text{R} \underset{k_{-1}}{\overset{k_{+1}}{\rightleftharpoons}} \text{DR} \tag{1.12}$$

At equilibrium the rate of the forward reaction ($k_{+1}[D][R]$) will be equal to the rate of the reverse reaction $k_{-1}[DR]$. So (Eqn (1.13))

$$k_{+1}[D][R] = k_{-1}[DR] \qquad (1.13)$$

Drugs are characterized by their *dissociation equilibrium constant* K_D which is the ratio of dissociation to association rate constants. Alternatively, drugs are sometimes characterized by their *affinity* or *association constant*, which is the reciprocal of K_D. Rearranging Eqn (1.13) gives Eqn (1.14).

$$K_D = \frac{k_{-1}}{k_{+1}} = \frac{[D][R]}{[DR]} \qquad (1.14)$$

Note that K_D has units of concentration.

It is not very useful to consider the 'concentration of receptors' since receptors are not free in solution but embedded in the cell membrane in most cases. It is more useful to consider this situation in a different way. If the total *number* of receptors, N_{tot}, reacts with a concentration, x_D of drug D to produce N_D drug–receptor complexes, then the fraction of receptors occupied by the drug will be (Eqn (1.15)).

$$p_D = \frac{N_D}{N_{tot}} \qquad (1.15)$$

which is known as the *occupancy*. The number of free receptors is therefore $N_{tot} - N_D$. Substitution into Eqn (1.14) gives Eqn (1.16).

$$K_D = \frac{k_{-1}}{k_{+1}} = \frac{x_D(N_{tot} - N_D)}{N_D} \qquad (1.16)$$

It is assumed here that because the number of drug molecules in solution is much greater than the number of receptors, x_D is not significantly changed by binding of the drug to the receptor. Rearranging Eqn (1.16) gives the occupancy (Eqn (1.17)).

$$p_D = \frac{N_D}{N_{tot}} = \frac{x_D}{K_D + x_D} \qquad (1.17)$$

Equation (1.17) is known as the *Hill–Langmuir equation*. It was derived by A.V. Hill to describe the binding of nicotine in muscle and later by Langmuir to describe the adsorption of gases on to metal surfaces. Notice that when $p_D = 0.5$, $x_D = K_D$. Thus the dissociation equilibrium constant is the concentration of drug that will produce 50% receptor occupancy.

This relationship between drug concentration and occupancy given by Eqn (1.17) generates a curve which is (part of) a rectangular hyperbola (Figure 1.25(a)). In pharmacology, drug concentration is generally plotted on a logarithmic scale which converts the rectangular hyperbola to a sigmoid shape (Figure 1.25(b)).

The Hill equation (Eqn (1.18)) is an empirical generalization of the Hill–Langmuir equation (Eqn (1.17)).

$$p_D = \frac{x_D^n}{K_D^n + x_D^n} \qquad (1.18)$$

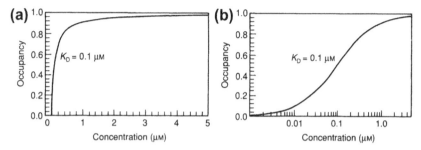

FIGURE 1.25

Characteristic shape of the rectangular hyperbola. The curves show the relationship between drug concentration and receptor occupancy for a hypothetical drug–receptor interaction with $K_D = 0.1\ \mu M$. (a) Hyperbola plotted on linear scales illustrates the rapid rise in occupancy at low drug concentration and the slow approach to saturation at high concentrations. (b) With a logarithmic concentration scale, the hyperbola takes on the familiar symmetrical sigmoid shape. For many drug-mediated effects, the relationship between concentration and response often has the sigmoid shape. Nevertheless, the relationship between occupancy and response is usually complex.

where the *Hill coefficient, n,* is a measure of the steepness of the curve ($n = 1$ in Eqn (1.17)). For graphical purposes, Eqn (1.18) is often rearranged in the form (Eqn (1.19)).

$$\log \frac{p_D}{1 - p_D} = n\log x_D - n\log K_D \qquad (1.19)$$

A log–log plot of $p_D/(1 - p_D)$ versus x_D (a *Hill plot*) will be linear, with a slope (n) of 1.0, in principle. It is often the case, though, that when a Hill plot is made from experimental data, the slope is not unity. This is usually due to the complexity of the relationship between drug binding and response measured. Despite this complexity, the slope of the Hill plot is a useful indicator of the steepness of the dose–response relationship and may be an indicator of cooperativity in drug binding to the receptor. Hill slope greater than 1.0 indicates positive cooperativity and less than 1.0 suggests a possible negative cooperativity, the presence of multiple receptor types, or the presence of an interfering process such as desensitization (see below; briefly, *desensitization* is the tendency of an agonist response to wane despite the continued presence of the agonist).

1.3.3 Radioligand-binding studies – a direct measure of occupancy

The theoretical basis for radioligand-binding studies is the same as used to derive the Hill–Langmuir equation (Eqn (1.17)). In principle, receptor occupancy can be measured using a radiolabelled drug which can be either agonist or antagonist (usually labelled with ^3H, ^{35}S or ^{125}I). At equilibrium, the number of drug molecules bound (B) is a function of the drug concentration x_D (Eqn (1.20)).

$$B = B_{max}\frac{x_D}{x_D + K_D} \qquad (1.20)$$

and B_{max}, the total number of binding sites in the tissue. (Note the similarity to the Michaelis–Menten relationship in enzyme catalysis). This relationship is often used in the linearized form (Eqn (1.21)).

$$\frac{B}{x_D} = \frac{B_{max}}{K_D} - \frac{B}{K_D} \tag{1.21}$$

which, when plotted as B/x_D versus B, is known as a *Scatchard plot*. This has a slope of $1/K_D$ and an x-axis intercept of B_{max}.

The main requirement for ligand-binding studies is to have a ligand that binds with sufficiently high affinity ($K_D < 10$ nM) and specificity to identify clearly the receptors of interest. Nonspecific binding (drug binding to anything other than the receptor of interest) is measured by performing the binding experiment in the presence of an excess of nonradioactive ligand which will hopefully prevent the radioligand binding to the receptor but not prevent the radioligand binding to the nonspecific sites. Specific binding is then calculated by subtracting the nonspecific from total binding. It is the relationship between ligand concentration and specific binding which is often plotted as a Scatchard plot to estimate K_D and B_{max}. It should be borne in mind that because of possible receptor conformational changes subsequent to ligand binding, binding studies will not necessarily give a true measure of the microscopic K_D (see Section 1.3.8). Nevertheless, radioligand-binding studies are now very widely used to study many different types of receptor. When combined with functional studies, radioligand binding becomes a very powerful technique because it is a direct measure of occupancy.

Sometimes binding studies can identify a binding site that does not immediately have a functional correlate. An interesting example of this is in the 5-HT receptor field (Bradley et al., 1986). Gaddum originally designated 5-HT receptors as D or M types based on studies in the guinea pig ileum. Then various radioligand-binding sites for 5-HT were identified in brain tissue based on relative affinities for several new drug tools. Two sites have high affinity for lysergic acid diethylamide. At one site (termed the 5-HT$_1$ receptor) 5-HT also has a high affinity, while at the other (5-HT$_2$ receptor) spiperone has a high affinity. Subsequent studies demonstrated that 5-HT$_2$ receptors correlate well with effects mediated by the D receptor.

The 5-HT$_1$-binding site was found to be heterogeneous. Initial studies based on the use of three different ligands suggested that there should be 5-HT$_{1A}$, 5-HT$_{1B}$ and 5-HT$_{1C}$ binding sites, and subsequently the use of sumatriptan identified a 5-HT$_{1D}$-binding site.

None of the 5-HT$_1$-binding sites correlated with the functional D and M receptors and so the M receptor was designated 5-HT$_3$ on the basis of results obtained with the selective agonist 2-methyl-5-HT and the antagonist ondansetron. Interestingly, while a 5-HT$_4$-binding site was identified, a binding site equivalent to the 5-HT$_3$ receptor was not identified at that time.

This fascinating story of 5-HT receptor discovery and classification continues with several 5-HT receptors having now been cloned (Table 1.10) and identified using binding and functional data. The extensive diversity of 5-HT receptors, first identified using radioligand-binding studies, has therefore now been confirmed by molecular genetic investigations.

1.3.4 Receptor structure

The idea that drugs interact with specific receptive substances, present in biological tissues, was developed long before there was any evidence regarding the physical nature of receptors. In his classical paper in 1956 (*A Modification of Receptor Theory*) Stephenson said, "What a drug combines

Table 1.10 Some Examples of Ion Channel and G-protein-coupled Receptors

Agonist	Receptor	Effector Mechanism
Acetylcholine	Nicotinic (muscle/neuronal) Muscarinic: m_1, m_3, m_5 Muscarinic: m_2, m_4	Ion channel (Na^+, K^+, Ca^{2+}) $G_{q/11}$ (couples via PLC and IP_3 to Ca^{2+} regulation) $G_{o/i}$ (inhibits adenylate cyclase, Ca^{2+} channels/activates K^+ channels)
Glutamate	NMDA Non-NMDA (kainate and AMPA) Metabotropic: $mGluR_{1,5}$ Metabotropic: $mGluR_{2,3,4,6,7,8}$	Ion channel (Na^+, K^+, Ca^{2+}) Ion channel (Na^+, K^+, some Ca^{2+} permeability) $G_{q/11}$ (couples via PLC and IP_3 to Ca^{2+} regulation) $G_{o/i}$ (inhibits adenylate cyclase, Ca^{2+} channels/activates K^+ channels)
GABA	$GABA_A$ $GABA_B$	Ion channel (Cl^-) $G_{o/i}$ (inhibits adenylate cyclase, Ca^{2+} channels/activates K^+ channels)
Glycine	Glycine	Ion channel (Cl^-)
5-HT	$5\text{-}HT_{1A}$, $5\text{-}HT_{1B}$, $5\text{-}HT_{1D}$, $5\text{-}HT_{1E}$, $5\text{-}HT_{1F}$ $5\text{-}HT_{2A}$, $5\text{-}HT_{2B}$, $5\text{-}HT_{2C}$ $5\text{-}HT_4$, $5\text{-}HT_6$ $5\text{-}HT_3$	$G_{o/i}$ (inhibits adenylate cyclase, Ca^{2+} channels/activates K^+ channels) $G_{q/11}$ (couples via PLC and IP_3 to Ca^{2+} regulation) G_s (activates adenylate cyclase) cAMP and PKA Ion channel (Na^+, K^+, Ca^{2+})
Dopamine	D_1, D_5 D_2, D_3, D_4	G_s (activates adenylate cyclase) cAMP and PKA $G_{o/i}$ (inhibits adenylate cyclase, Ca^{2+} channels/activates K^+ channels)
Noradrenaline and adrenaline	$\alpha_{1A,1B,1D}$ $\alpha_{2A,2B,2C}$ β_1, β_2, β_3	$G_{q/11}$ (couples via PLC and IP_3 to Ca^{2+} regulation) $G_{o/i}$ (inhibits adenylate cyclase, Ca^{2+} channels/activates K^+ channels) G_s (activates adenylate cyclase) cAMP and PKA
Histamine	H_1 H_2 H_3, H_4	$G_{q/11}$ (couples via PLC and IP_3 to Ca^{2+} regulation) G_s (activates adenylate cyclase) cAMP and PKA $G_{o/i}$ (inhibits adenylate cyclase, Ca^{2+} channels/activates K^+ channels)
Adenosine	A_1 A_2	$G_{o/i}$ (inhibits adenylate cyclase, Ca^{2+} channels/activates K^+ channels) G_s (activates adenylate cyclase) cAMP and PK
ATP	$P2X_{1-7}$ $P2Y_{1,2,4,6,11,14}$ $P2Y_{12,13}$	Ion channel (Na^+, K^+, Ca^{2+}) $G_{q/11}$ (couples via PLC and IP_3 to Ca^{2+} regulation) $G_{o/i}$ (inhibits adenylate cyclase, Ca^{2+} channels/activates K^+ channels)

Abbreviations: G_i, inhibitory G-protein (couples to adenylate cyclase); PLC, phospholipase C; IP_3, 1,4,5-inositol trisphosphate; NMDA, N-methyl-D-aspartate; GABA, γ-aminobutyric acid; 5-HT, 5-hydroxytryptamine (serotonin); G_s, stimulatory G-protein; cAMP, cyclic adenosine $3'$-$5'$-monophosphate; ATP, adenosine triphosphate; PKA, protein kinase A (cAMP-dependent protein kinase). It should be noted that in different cell types, some G-protein receptors may couple to different G-proteins and hence link to different second messenger systems.

with to produce its effect is a subject for speculation." By the 1970s the difficult and time-consuming procedures of protein purification had allowed the structure of a few receptors to be studied, particularly the nicotinic AChR. However, during the 1980s, applications of biochemical and molecular genetic techniques and X-ray crystallographic studies resulted in spectacular advances in the knowledge of receptor structure.

Receptors are often classified into two groups: the ligand-gated ion channels and the GPCRs (Table 1.10). (It was found that the major drugs ranitidine (Zantac, GSK), Zyprexa (Eli Lilly), Clarinex (Schering-Plough) and Zelnorm (Novartis) exert their effects through interaction with a GPCR). Indeed, these two groups constitute by far the majority of neurotransmitter and hormone receptors which are important pharmacological targets. There are, however, other receptor classes, exemplified by the receptors for insulin and epidermal growth factor which have intrinsic tyrosine kinase activity, receptors that transport ligands across the cell membrane such as the low-density lipoprotein receptor or the transferrin receptor, and the steroid receptors which influence DNA transcription.

1.3.4.1 Nicotinic AChR structure: a ligand-gated ion channel

Of the ligand-gated ion channels, by far the best studied is the nicotinic AChR (Figure 1.26). Most studies have utilized the AChR from the electric organ of the Californian ray, *Torpedo californica* which is an exceedingly rich source of these receptors. By utilizing α-bungarotoxin (a component of cobra snake venom with extremely high affinity for the nicotinic AChR), it was possible in the late 1970s and early 1980s to isolate and characterize this receptor. The *Torpedo* AChR was found to be composed of four different protein subunits in a stoichiometry $\alpha_2\beta\gamma\delta$ arranged in a pseudosymmetrical fashion around a central ion channel pore (Figure 1.26). Each subunit crosses the cell membrane four times giving four transmembrane (TM) domains which are numbered from the amino terminus of the protein. Since five subunits make up each receptor, there are 20 TM domains per receptor (Figure 1.26). The amino acid residues in TM2 of each subunit line the central ion channel and determine its conductance properties.

ACh, tubocurarine and other nicotinic receptor ligands were found to compete with α-bungarotoxin for its binding site on the α-subunit. The α-subunit was found to be unique in having two cysteine residues at amino acid positions 192 and 193. Subsequently, molecular genetic techniques were used to isolate and sequence multiple subtypes of neuronal (found in the CNS and in autonomic ganglia) nicotinic receptor subunits and all α-subunits were found to contain two cysteine residues at positions analogous to 192 and 193. These cysteine residues are located before the first TM domain in the extracellular part of the protein and the agonist-binding site is thought to be close by, perhaps in a shallow cleft between the α- and adjacent subunits.

The neuronal nicotinic receptors are different from muscle receptors in a number of ways, including the fact that they have a higher sensitivity to nicotine and are not sensitive to α-bungarotoxin. In contrast with the α-, β-, γ- and δ-subunits of electric organ and muscle AChRs, neuronal nicotinic receptor subunits are divided into α- and β-subunits. The exact stoichiometry of neuronal nicotinic receptors is unknown but is probably $\alpha_2\beta_3$. Because nine different α-subunits and five different β-subunits have been identified, the possibilities for variation in neuronal nicotinic receptor structure are substantial. A similar situation has arisen with the receptors for the inhibitory amino acid γ-aminobutyric acid (GABA) and with the excitatory amino acid glutamate receptors of the kainate, 2-amino-3-(3-hydroxy-5-methyl-isoxazol-4-yl)propanoic acid and *N*-methyl-D-aspartate type. Multiple subtypes of the

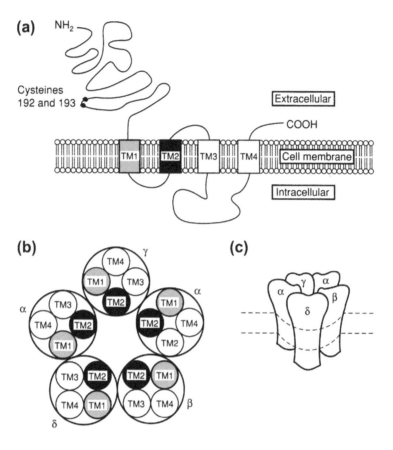

FIGURE 1.26

Schematic diagram of the structure of the nicotinic AChR. This general structure is thought to hold for all pentameric ligand-gated receptors. The receptor is composed of five subunits each with the general structure shown in (a). (a) Schematic arrangement of a nicotinic AChR α-subunit. The four transmembrane domains (TM1–TM4) are shown crossing the cell membrane. The large extracellular domain at the amino terminus of the α-subunit is where the main determinants of agonist and antagonist binding are located. (b) Arrangement of the AChR subunits when viewed from above the cell membrane. Each subunit is arranged so that the trans-membrane segment TM2 lines the central ion channel of the receptor (c) Schematic of the three-dimensional structure of the nicotinic AChR as it sits in the cell membrane.

Source: (a) and (b) adapted from Huganir and Greengard (1990) with permission.

subunits that make up the glutamate receptors and the GABA receptors (Table 1.10) have been cloned. In each case, 10 or more different subunits (based on amino acid sequence) are known.

1.3.4.2 β-adrenoceptor structure: a G-protein-coupled receptor

The β-adrenoceptor is one of the group of over 300 receptors that are coupled via G-proteins (GTP-binding proteins) to their effector system (Table 1.10). They are perhaps the largest and most diverse group of receptor proteins. Their effector systems include adenylate cyclase (stimulated by G_s

and inhibited by G_i proteins), guanylate cyclase, phospholipase C, phospholipase A_2, phosphodies-terases and Ca^{2+} and K^+ channels.

The 'prototype' for GPCRs is retinal rhodopsin. Rhodopsin is the most studied GPCR because of its abundance in halobacteria and in the retina of mammals. The ligand for this specialized receptor is retinal. Absorption of a photon of light activates rhodopsin which, via the G-protein transducin, activates a phosphodiesterase causing hydrolysis of cGMP and so indirectly causes the closure of the cGMP-activated channels in the photoreceptor membrane. Bacteriorhodopsin has been crystallized and its structure elucidated. It is a single polypeptide with seven TM-spanning regions (numbered 1–7 from the amino terminus of the protein). All GPCRs such as the β-adrenoceptors have analogous structures on the basis of amino acid sequence homology and crystallographic studies (Figure 1.27). They are single polypeptide receptors commonly with approximately 400 amino acids but ranging to over 1000 amino acids for the metabotropic glutamate receptors. Other examples of this receptor type include the muscarinic AChRs, the DA receptors, the histamine receptors and the 5-HT receptors (with the exception of the 5-HT$_3$ receptor which is a ligand-gated ion channel). In each case, the intracellular domain between TM5 and TM6 is important for G-protein binding.

These receptors form distinct structural groups: the largest group (Group A) being similar in size to rhodopsin and includes the receptors for monoamines and muscarinic AChRs. Unlike the ligand-gated channels, in these receptors, agonist and antagonist binding occurs within the TM domains involving particularly amino acids in TM3, TM5 and TM6. The receptor is thought to fold round on itself (Figure 1.27(c)) so that a pocket is formed within the TM domains. This deep binding within the receptor structure, where there may be scope for multiple interactions between agonist and receptor, perhaps accounts for the very high (nanomolar) affinity often found for agonist binding to GPCRs compared with the relatively low affinity (micromolar) of agonists for the ligand-gated receptors, where the agonist-binding site is thought to be more superficial. In contrast, peptides and glycoprotein receptors (Group B) bind to the receptor extracellular domains and the metabotropic glutamate receptors, or GABA$_B$ receptors (Group C) bind the neurotransmitter on a large extracellular N-terminal domain.

The rapid advances made possible by molecular genetic techniques and the human genome sequencing project have resulted in the cloning and sequencing of all human receptors (some examples are listed in Table 1.10). This has made structural comparisons between receptors possible and has led to a much more detailed understanding of the factors that determine why one receptor is, for example, selectively activated by muscarine and another by histamine, even though both receptors are of a very similar overall structure.

1.3.5 Relating occupancy to response

Equation (1.12) could be described as a *model* or a *mechanism* for drug binding to a receptor. However, it is rather a simple model since it gives no indication of how the occupied receptor might generate a response. Thus Eqn (1.12) may be quite sufficient to describe the action of an antagonist, but requires extension to describe agonist action adequately.

1.3.4.1 Ligand-gated ion channels

Receptors like the nicotinic AChR or the GABA$_A$ receptor have both the agonist-binding site and the transduction mechanism (ion channel) as part of a single macromolecule. Relative to receptors which are linked to second messenger systems, the ligand-gated ion channels are simpler to study.

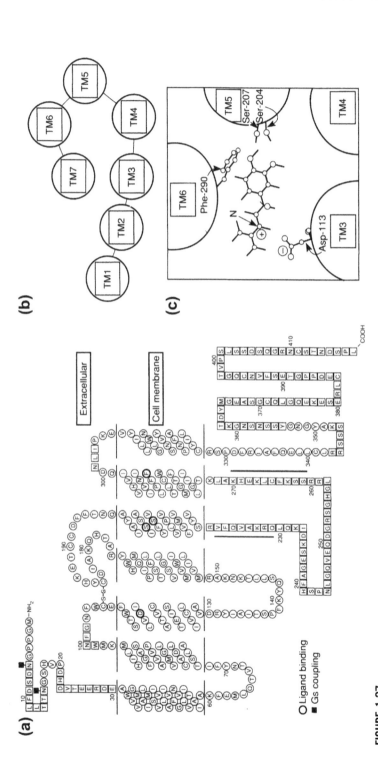

FIGURE 1.27

Structure of the β-adrenoceptor (βAR). (a) Model of the transmembrane topology of the βAR. The horizontal lines indicate the likely limits of the plasma membrane. The protein's amino terminus is extracellular and the carboxy terminus is intracellular. The receptor has seven trans-membrane helices which are numbered from TM1 nearest to the amino terminus to TM7 nearest to the carboxy terminus. The amino acid residues represented by squares can be deleted without apparently affecting folding or protein folding whereas deletion of those rep-resented by circles adversely affects folding or ligand binding. The residues shown in bold circles in transmembrane domains 3, 5 and 6 are proposed to be directly involved in interacting with βAR agonists (c). The regions of the third intracellular loop between TM5 and TM6 delineated by the solid black bar are critical for G-protein coupling. (b) Schematic of the arrangement of the seven transmembrane helices of the βAR as seen from above the cell membrane labelled TM1 to TM7 from the amino terminus. (c) Detailed view of the arrangement of the transmembrane domains around the ligand-binding site of the βAR. The positioning of the helices is in accordance with the structure of bacteriorhodopsin. In (Henderson and Unwin, 1975). The residues that are suggested by mutagenesis studies to interact with the agonist isoprenaline are indicated by arrows. The structure of isoprenaline in the proposed binding pocket is shown with the amino nitrogen arrowed. It is almost certain that several other amino acids in the receptor will interact with agonists to a greater or lesser extent.

Source: (a) and (c) adapted with permission from Strader et al. (1989).Ã

A simple extension to Eqn (1.12) can allow it to become a useful model of receptor activation. This was first used to describe AChR activation by del Castillo and Katz (Eqn (1.22)).

$$D + R \underset{k_{-1}}{\overset{k_{+1}}{\rightleftharpoons}} DR \underset{a}{\overset{b}{\rightleftharpoons}} DR^* \tag{1.22}$$

Equation (1.22) describes how the agonist drug, D, combines with receptor, R, to form the agonist–receptor complex (DR) which then undergoes a conformational change which opens (activates) the AChR ion channel (the open channel is denoted by DR*). Channel opening is described by a rate constant b and channel closing by a rate constant a. Equation (1.22) describes a mechanism for the nicotinic AChR where the receptor can exist in only three *discrete* states. Two of these states are obviously going to be different: liganded and unliganded. The open state is also structurally different because it is where the receptor ion channel is open. If, for example, receptor desensitization were to be included in this mechanism, then an additional state (a desensitized state) would need to be added. In general, for practical purposes, receptors may be thought of as existing in only a few discrete conformational states, although this is obviously a simplification, given that these are large and complex protein molecules.

The equilibrium fraction of occupied receptors in the closed state will be (Eqn (1.23)).

$$p_{DR} = \frac{x_D/K_D}{1 + x_D/K_D(1 + b/a)} \tag{1.23}$$

However, the *response* of the receptor to agonist occupancy is p_{open}, the fraction of time for which an individual channel is open (DR*) or the fraction of a population of channels which are open at equilibrium. This is expressed in Eqn (1.24).

$$p_{open} = \frac{x_D/K_D(b/a)}{1 + x_D/K_D(1 + b/a)} \tag{1.24}$$

This simple scheme provides a plausible mechanism for activation of the AChR (but see Section 1.3.8).

1.3.4.2 Receptor mechanisms that involve second messengers

With receptors that couple to second messenger systems, the transduction mechanism is often much more complex than for the ligand-gated ion channels. For this reason, it is generally not possible to postulate any specific mechanism linking receptor occupancy to response.

It is quite often the case that a long chain of events will separate occupation of the receptor from the actual response being measured. A good example of this might be the increase in heart rate caused by β-adrenoceptor agonists such as adrenaline which forms part of the 'fight or flight' response (see Section 1.1.2). The β-adrenoceptors in the heart are of the β_1-subtype. When adrenaline occupies the β_1-receptors of the heart, the G-protein (G_s) (a guanine-nucleotide-binding protein associated with the β_1-receptor) releases guanosine diphosphate (GDP) and binds one molecule of GTP. This causes dissociation of the G-protein α-subunit (α_s) which can then bind to and activate adenylate cyclase (Figure 1.28). Adenylate cyclase catalyses the conversion of ATP to cAMP. cAMP activates specific cAMP-dependent protein kinases which by phosphorylation of calcium channels in the heart sinoatrial (SA) node cells increases the calcium current into these cells during each action potential and so speeds up the rate of firing of the SA node cells. Because the SA node cell firing determines the heart rate, β-adrenoceptor agonists cause an increase in heart rate.

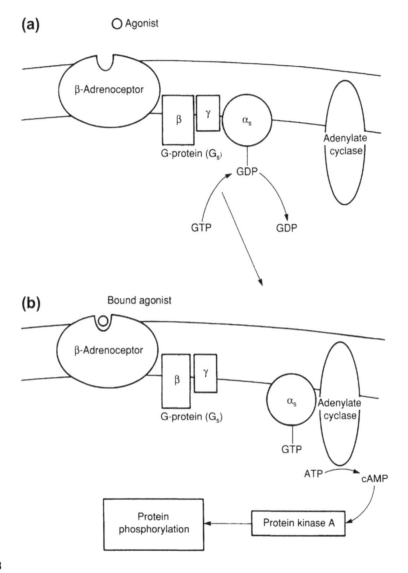

FIGURE 1.28

Schematic representation of the mechanism of activation of adenylate cyclase by β-adrenoceptor agonists. (a) Resting state of the β-adrenoceptor and associated G-protein. The β-adrenoceptor couples to the G-protein, G_s. G_s is composed of three subunits denoted α_s, β and γ. On binding an agonist molecule, the β-adrenoceptor conformation changes such that G_s loses one molecule of GDP and binds GTP (b). The α_s subunit of G_s then dissociates from the G-protein β- and γ-subunits and on collision with adenylate cyclase causes activation of the enzyme and conversion of ATP to cAMP. cAMP can then activate protein kinase A to cause phosphorylation of a variety of target proteins. Note that other receptors such as the muscarinic m_2 and m_4 receptors (Table 1.10) by coupling with G_i, can cause inhibition of adenylate cyclase, whereas by coupling with G_o can cause inhibition of Ca^{2+} or K^+ channels.

The action of the β-adrenoceptor agonist is terminated because the GTP attached to the α_s-subunit is slowly hydrolysed back to GDP. With an attached GDP, α_s can then associate with the G-protein β- and γ-subunits again, until the next collision with an activated β-receptor occurs (Figure 1.28).

Consider for a moment the following points in this process where it might be possible to measure the response of the system to a β-adrenoceptor agonist: (1) heart rate, (2) cell calcium current, (3) cAMP level, and (4) adenylate cyclase activity. With receptors that couple to second messenger systems, it cannot be assumed that in different preparations the relationship between receptor occupancy and response will be the same, even though the receptors expressed in each preparation are the same.

1.3.6 Competitive antagonism and the Schild equation

As explained earlier, drugs are often characterized as being either *agonists* or *antagonists*. Agonists have the ability to evoke a response on binding to the receptor, while antagonists block the response of the receptor to an agonist, without evoking any response themselves. ACh and nicotine are agonists at the nicotinic AChR, whereas tubocurarine (one of the main components of the South American Indians' arrow poisons) is a nicotinic antagonist used as a muscle relaxant during surgery. The β_2-adrenoceptor agonist salbutamol is used by asthmatics to dilate the airways while the β-adrenoceptor antagonist propranolol is widely used as an antihypertensive. The Schild equation provides a method of quantifying the actions of an antagonist. Many fundamental advances in pharmacology occurred as the result of the development of quantitative methods for the characterization of antagonist actions.

When an agonist drug (D) and antagonist (B) at concentrations of x_D and x_B are in equilibrium with a single population of receptors, and the binding of D and B are mutually exclusive then the fraction of receptors occupied by the agonist will be (Eqn (1.25)).

$$p_D = \frac{x_D/K_D}{(x_D/K_D) + (x_B/K_B) + 1} \tag{1.25}$$

where K_B is the dissociation equilibrium constant of the antagonist. In effect, the antagonist reduces the occupancy of the receptor by the agonist. If the agonist concentration is increased to say x_D' such that agonist occupancy is restored to that level obtained by x_D in the absence of antagonist, then the factor (r) by which the agonist concentration has to be increased is given by (Eqn (1.26)).

$$r = \frac{x_D'}{x_D} \tag{1.26}$$

where r is known as the *dose ratio*. For a competitive antagonist, Schild demonstrated that (Eqn (1.27)).

$$r = \frac{x_B}{K_B} + 1 \tag{1.27}$$

This simple result, known as the *Schild equation*, is obtained by assuming that equal agonist occupancies (whether in the presence or absence of antagonist) will always give equal responses. Equating the occupancies in the absence and presence of antagonist gives Eqn (1.28).

$$\frac{x_D}{K_D + x_D} = \frac{x_D'/K_D}{(x_D'/K_D) + (x_B'/K_B) + 1} \tag{1.28}$$

which can be rearranged (using substitution with a rearrangement of Eqn (1.26) – $x'_D = r\,x_D$) to give Eqn (1.27). The importance of the Schild equation is that the dose ratio, r, is a characteristic *only* of the antagonist (neither the agonist concentration nor the agonist affinity appear in the Schild equation). The relationship between r and antagonist concentration means that a competitive antagonist will produce a parallel rightward shift of the log (concentration)–response curves with no change in maximum response. The Schild equation is usually plotted on log–log scales giving (Eqn (1.29)).

$$\log(r-1) = \log x_B - \log K_B \tag{1.29}$$

Thus a plot of $\log(r-1)$ versus $\log x_B$ (the *Schild plot*) for a competitive antagonist will give a straight line with slope of 1 and an intercept on the x-axis of $\log K_B$. The negative of the x-axis intercept is usually called the pA_2 value. Where a slope of 1 is found for experimental data, the pA_2 value will be the negative logarithm of the dissociation equilibrium constant (K_B) for the antagonist. Verifying that the slope of a Schild plot for a particular antagonist is unity over a wide range of antagonist concentrations is by far the best way of demonstrating competitive antagonism. The Schild method is independent of agonist concentration or affinity and therefore a particular antagonist should give the same Schild plot for all agonists which act on the same population of receptors.

Where a Schild slope of greater than unity is encountered, *noncompetitive* antagonism may be indicated: this is the situation whereby raising the agonist concentration cannot completely overcome the effect of the antagonist. This situation can arise when agonist and antagonist binding are not mutually exclusive. Figure 1.29 shows an example of the use of the Schild method to determine the K_B for tubocurarine block of nicotinic AChRs at the frog neuromuscular junction. The main action of tubocurarine is a competitive block of the AChR. However, as illustrated in Figure 1.29, tubocurarine also blocks the AChR ion channel which is a noncompetitive effect.

1.3.6.1 Drug blockade of open ion channels: a noncompetitive antagonism

Apart from drug effects at receptors, the site of drug action that has received most attention has been the open ion channel. Several major classes of drugs mediate their effects by blocking specific ion channels in cell membranes. Prominent among these are the local anaesthetics such as lidocaine which block sodium channels, the dihydropyridine Ca^{2+} channel blockers such as nifedipine and the K^+ channel blockers.

By blocking axonal Na^+ channels, the local anaesthetics inhibit conduction in pain fibres and hence cause local anaesthesia. Lidocaine is also used to reduce arrhythmias of the heart where it blocks Na^+ channels in cardiac muscle causing a slowing of the cardiac action potential. The Ca^{2+} channel blockers on the other hand are used mainly in the treatment of angina and hypertension.

1.3.7 Desensitization and the control of receptor number

Desensitization can be defined as the tendency of a drug response to become smaller with repeated doses, or during a single constant application of drug. It generally occurs with agonist responses and is absent with antagonists. Desensitization is probably a general phenomenon, although it varies widely in extent and rate of appearance with different receptor systems. One example might be the tolerance that develops over a few days when repeated doses of opiate analgesics such as morphine are given to a patient: in order to obtain similar pain relief at each administration, the dose has to be stepped up. After cessation of treatment, several days may be needed to recover full opiate sensitivity. Another example

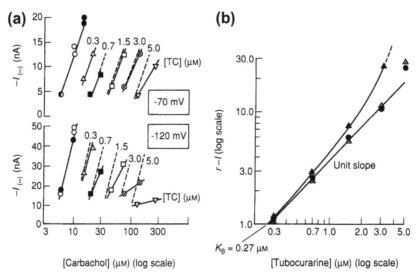

FIGURE 1.29

Use of the Schild method to determine the K_D for tubocurarine (TC) blockade of nicotinic AChRs at the frog neuromuscular junction. (a) Partial log (concentration)–response curves for net inward current ($-I_{(\infty)}$) evoked by the agonist carbachol in the presence of increasing concentrations of tubocurarine. In the top panel, the responses are recorded at a membrane potential of -70 mV and it can be seen that, except at the highest tubocurarine concentration (5 µM), the antagonist produces an approximately parallel shift in the dose–response curves as expected for competitive antagonism. The lower panel shows the same experiment but at a membrane potential of -120 mV. When the inside of the cell is made more negative with respect to the outside, the shift in dose-response curves is far from parallel. This is because the positively charged tubocurarine molecule is being attracted into the AChR ion channel by the negative membrane potential, increasing the channel-blocking action of tubocurarine: a noncompetitive effect. The dashed lines in each panel show the responses predicted for pure competitive antagonism with $K_B = 0.27$ µM (b) Dose ratios were calculated at a response level of -8 nA at -70 mV and at a response level of -24 nA at -120 mV. Schild plot of dose ratio (r) against tubocurarine concentration (x_B) according to Eqn (1.27). (Filled circles, Dose ratios from equilibrium response at -70 mV; filled triangles, Dose ratios from equilibrium response at -120 mV; open triangles, Dose ratios from peak responses at -120 mV). Because the noncompetitive open channel block by tubocurarine is slow to develop, if the peak response is measured, then mainly competitive antagonism is seen and the Schild slope is close to unity. The fact that both curves coincide at low agonist concentrations (small dose ratios), suggests that the K_B for competitive binding to the receptor is independent of membrane potential as might be expected if the agonist-binding site is on the outermost part of the receptor molecule, outside the membrane potential field.

Source: Adapted from Colquhoun et al. (1979), with permission.

occurs with responses to β-adrenoceptor agonists which desensitize over a period of a few minutes to an hour and then require a similar period for recovery. A much faster desensitization is observed with the ligand-gated receptors. The response of these receptors to a high agonist concentration can desensitize in less than 1 s, but recovery occurs over a period of seconds to minutes. The mechanisms involved in desensitization of any receptor type are not fully understood. However, it is clear that desensitization of the ligand-gated ion channel receptors involves conformational changes in the receptor-channel protein itself where the receptor enters a desensitized state. For the ion channel receptors, desensitization can be expressed in a model (Eqn (1.30)) as an extension to the simple agonist mechanism (Eqn (1.22)):

$$D + R \underset{k_{-1}}{\overset{k_{+1}}{\rightleftharpoons}} DR \underset{a}{\overset{b}{\rightleftharpoons}} DR \underset{d_-}{\overset{d_+}{\rightleftharpoons}} DR_d \tag{1.30}$$

In this mechanism, the receptor forms the desensitized state DR_d from the open state with a rate constant d_+, and recovers from desensitization with a rate constant d_-. While in the desensitized state, the receptor cannot be activated by ACh and so desensitization reduces the size of the ACh response. In fact, a more complex mechanism (the *cyclic model* of receptor desensitization) is necessary to describe nicotinic AChR desensitization. In the cyclic model, the desensitized state of the receptor can lose its agonist molecule and then revert to the resting state without going through the open state.

Desensitization of receptors which couple to second messenger systems generally involves both alterations in the transduction system (or coupling between receptor and transduction system) and changes in the number of receptors on the cell surface. In the case of the β-adrenoceptors, this involves internalization of receptors which are subsequently reinserted into the plasma membrane. With the β-adrenoceptor and probably many other GPCRs, desensitization involves phosphorylation of one of the internal domains of the receptor.

1.3.8 Partial agonists, agonist efficacy and inverse agonism

The concept of *partial agonism* was introduced by Ariens and used by Stephenson to provide a general method of dealing with the situation where the relationship between receptor occupancy and tissue response is different for different agonists. Stephenson distinguished between the ability of an agonist to *bind* to the receptor (measured by its equilibrium dissociation constant, or affinity constant) and its ability to elicit a response once bound (measured by an empirical constant termed the *efficacy* of the agonist). This concept conveniently accounts for the experimental observation that in a single tissue, two different agonists, acting on the same receptors, will often generate different maximal responses from the tissue. Even at 100% receptor occupancy, one agonist may generate a much smaller response than a different agonist. Associated with the idea of efficacy is the concept of *spare receptors*. A high-efficacy agonist may be able to elicit a maximal response from the tissue when occupying only a small fraction of receptors (hence the agonist is said to have a 'receptor reserve' or *spare* receptors), whereas an agonist of low efficacy may be unable to generate a maximal tissue response, even at 100% receptor occupancy.

The term *intrinsic activity* is sometimes used synonymously with efficacy. This term was introduced by Ariens along with the concept of partial agonism. The intrinsic activity of drugs was suggested to range from 0 to 1. However, this was a more restrictive term than efficacy as defined by Stephenson because the intrinsic activity of all full agonists was assumed to be 1, whereas efficacy may vary from zero to a large positive value.

The difference between agonists, partial agonists and antagonists becomes simply a matter of efficacy: antagonists have zero efficacy, partial agonists have low efficacy and full agonists have high efficacy. For example, different β-adrenoceptor agonists may, by occupying the receptor in slightly different ways, cause more or less G-protein activation. An interesting example of this occurs with the β-adrenoceptor agonist salbutamol (Ventolin). This agonist selectively activates the β_2-adrenoceptor subtype present in the airways relative to β_1-adrenoceptor activation in the heart. Yet there is only a small difference in the affinity of salbutamol for β_1- and β_2-receptors. Salbutamol apparently has higher efficacy at β_2-adrenoceptors leading to its development as an effective antiasthma agent.

It is a straightforward matter to include a term for efficacy in the simple receptor theory. So far we have the occupancy p_D expressed in terms of the agonist concentration x_D, and equilibrium dissociation constant K_D (Eqn (1.31)).

$$p_D = \frac{x_D}{K_D + x_D} \tag{1.31}$$

The magnitude of the tissue response was written by Stephenson as some function, f, of the efficacy ε multiplied by the occupancy, i.e. Eqn (1.32),

$$\text{Response} = f(\varepsilon p_D). \tag{1.32}$$

Notice that there is no explicit term here for the number of receptors in the tissue, although obviously for any particular occupancy, the larger the total number of receptors in the tissue, N_{tot}, the larger is the stimulus given to the tissue (and hence the response) likely to be for a given agonist concentration. This was taken into account by Furchgott who used the term *intrinsic efficacy* (ε_i say) to define the ability of the agonist to activate any individual receptor. The response of the tissue then becomes a function, f, of the intrinsic efficacy, ε_i, multiplied by the total number of receptors in the tissue, N_{tot}, times the occupancy p_D. Differences in the number of receptors in different tissues are one reason why a drug may show tissue selectivity without there being any difference between the receptors in each tissue.

In the classical approach, it is not necessary to know what the function f is. Usually, efficacy can only be measured in relative terms by comparing one agonist with another. Unfortunately the classical methods used to estimate agonist affinities and efficacies may not be accurate because the Stephenson approach neglects the fact that the efficacy of the agonist will generally affect the binding of the agonist, as discussed above in relation to ligand-binding measurements. However, in the case of the nicotinic AChR, single channel recording experiments have allowed direct estimates of both receptor affinity and intrinsic efficacy for several different agonists. This is because the AChR system is simple enough that a precise receptor activation mechanism can be postulated. For the model described in Eqn (1.22), the ratio b/a is a direct measure of efficacy (i.e. it reflects the ability of the agonist once bound to the receptor) to open the receptor ion channel.

An interesting question arises here in relation to comparing agonists, at different receptors. Antagonists, it will be recalled, are supposed to have zero efficacy and so can be classified purely on the basis of their dissociation equilibrium constant (K_D) using the Schild method. Comparing the selectivity of an *antagonist* between two different receptor types is done simply by comparing K_Ds to obtain a selectivity ratio. In contrast, the selectivity of an *agonist* for one receptor type compared with another will depend on both the K_D at each receptor and the efficacy of the agonist at each receptor. Indeed, the separation of these factors, which, in principle, is necessary to make full use of structure–activity relationships, was one of Stephenson's original aims.

When *agonist* and *antagonist* K_Ds are estimated using radioligand-binding studies, these will not necessarily give a direct measure of the microscopic equilibrium dissociation constant. Generally, antagonist K_Ds estimated in this way (where either another antagonist or agonist can be used as the displacing agent) agree well with K_Ds estimated using the Schild method. In contrast, agonist K_Ds estimated from binding studies sometimes suggest a much higher affinity than predicted from functional studies, not only because of problems with receptor desensitization (see below) but also because of the fundamental problem that the actual binding measured will generally reflect the efficacy of the agonist, as well as the K_D (e.g. for the mechanism in Eqn (1.22), the occupancy is $p_{DR} = p^*_{DR}$ which depends on b/a as well as K_D).

The del Castillo and Katz model (Eqn (1.22)) provided a useful description of agonist activation of the nicotinic AChR. However, this model allows for only a single agonist-binding site and there is now considerable functional, biochemical and structural evidence that there are two agonist-binding sites on the nicotinic AChR (see also Section 1.3.5). For example, the Hill coefficient for AChR activation at low ACh concentrations approaches 2, suggesting that two ACh molecules must bind to the receptor to produce efficient receptor activation. Biochemically, there are found to be two ACh- and α-bungarotoxin-binding sites per receptor molecule and two agonist-binding subunits are present in each receptor. This evidence suggests that the del Castillo and Katz model could be usefully extended to include a second agonist-binding reaction (Eqn (1.33)).

$$D + R \underset{k_{-1}}{\overset{2k_+}{\rightleftharpoons}} DR + D \underset{2k_{-2}}{\overset{k_{+2}}{\rightleftharpoons}} D_2R \underset{a}{\overset{b}{\rightleftharpoons}} D_2R^* \tag{1.33}$$

(The factor of 2 before k_{+1} and k_{-2} occurs because there are two agonist-binding sites and this mechanism presumes that either site can be occupied or vacated first.) The occupancy for this scheme of the doubly liganded closed state will be as expressed in Eqn (1.34).

$$p_{D_2R} = \frac{c_1 c_2}{c_1 c_2 (b/a + 1) + 2c_1 + 1} \tag{1.34}$$

where $c_1 = x_A/K_1$ and $c_2 = x_A/K_2$ and K_1 and K_2 are the microscopic dissociation equilibrium constants for the first and the second agonist-binding reactions, respectively. Finally, it follows that the occupancy of the open state (p_{open}) will be (Eqn (1.35)).

$$p_{D_2R^*} = \frac{b}{a} p_{D_2R} \tag{1.35}$$

A major advance in understanding the action of drugs at receptors came about from the recognition that some receptors (initially studied at the β-adrenoceptor) could occasionally adopt the active conformation in the absence of agonist binding. This activity, termed 'constitutive activity' was a surprise, as was the discovery that some drugs, on binding to the receptor, resulted in a decrease in receptor constitutive activity (e.g. resulting in a decrease in the generation of cAMP within the cell). These drugs are now classed as 'inverse agonists'. They will antagonize the action of the natural hormone, full or partial agonist and so are 'antagonists' in a functional sense. Inverse agonists are distinguished from traditional competitive antagonists that bind to the receptor without producing any change in receptor signalling and so the traditional competitive antagonist is also termed a 'neutral antagonist'.

1.4 SECTION IV

1.4.1 Conclusions: uncertainties in drug design and development

There has been a phenomenal increase in recent years in the amount of information about cellular mechanisms and, therefore the outlook for rational drug design ought to be very optimistic. Unfortunately, however, we still lack sufficient insight to make drug discovery a straightforward process.

One reason is that the body is under multifactorial control: there are many natural checks and balances. For any given function, there are usually several messengers and several types of receptors; there are also amplification systems, modulating systems, feedback inhibitory mechanisms, various ion fluxes and so on: if we block one pathway by drug action, another pathway is likely to take over. The consequence is that we cannot be sure at the outset that designing a drug to act on a particular receptor or enzyme will necessarily provide treatment for a medical condition, even though we may know that it is involved in the physiological controlling mechanisms. Indeed, there are now many examples of nicely designed 'drugs' which are still looking for a suitable disease! Furthermore, enzymes and receptors are ubiquitous and occur in many different tissue systems. Blocking them at a tissue site involved in a disease may be therapeutically effective but the concurrent blockade in other tissues (not involved in the disease) may be thoroughly undesirable and give rise to unwanted 'side effects'. Selectivity is very important.

For the above reasons, there is still a strong element of speculation in drug design and a considerable uncertainty in achieving success. Thus, only a small proportion of drug discoveries are destined to become useful therapeutic agents with substantial commercial implications.

As mentioned in Section 1.1.5, complications also arise during drug design because a drug has to be administered and find its way to the desired site of action, whereas the natural messenger may be generated locally or stored nearby to its required site of action. Also, after it has acted, the natural transmitter is removed by specific enzymes or reuptake mechanisms, whereas a drug is disposed of by being excreted. We have to balance the desired pharmacology with the biochemical needs to achieve drug access and elimination. In altering the chemistry of the drug to achieve adequate disposition, one may inadvertently introduce other pharmacological properties, thereby reducing the selectivity of action.

The discovery of a potential new drug as a possible therapeutic agent is followed by its development: the studies that have been carried out in the laboratory and on animals to characterize the properties of any new drug and to assess its safety, before it can be tried in the human body, are often referred to as 'preclinical R&D'. Then comes the third stage, 'clinical R&D', namely, human studies progressing from healthy volunteers through to clinical trials on patients with the relevant condition. Although one can differentiate between these three stages, there is usually considerable overlap in timing.

Safety assessment often starts with testing *in vitro* on cell cultures to examine for potential effects on cell reproduction. This is usually followed by repeated daily doses (at different dose levels) to groups of laboratory animals for 7–14 days to observe what signs of toxicity might be revealed. Eventually, this will be repeated for longer periods, e.g. daily dosing for 1 month, 3 months or 6 months and in at least two species of animal. During this period, various indicators of biochemical function will be assessed and, at the end of the period, the animals will be killed and tissue samples from all the main organs will be examined microscopically for possible damage or unusual effects. These safety studies (toxicity profiling) are enormously time consuming and involve many skilled people. A study *in vivo* to show that a potential drug is not likely to cause cancer can take more than 3 years to complete; it will involve daily dosing to groups of laboratory rats or mice for almost their lifetime (18 months–2 years) followed by extensive microscopic examination of the tissues by qualified pathologists.

The scale of the safety assessment is considerable and can require 500–750 kg of compound to be synthesized before it makes the manufacturing stage. In addition, the synthetic route used to generate the material for safety assessment must be close to that used in the intended production process. This is to ensure that the identity and amount of any chemical impurities remain essentially the same. This requires that considerable process investigation should have taken place and there is great pressure to do this rapidly because the generation of sufficient chemical at this juncture can be rate limiting. This investment in time, effort and money has to be made before it is known whether the new drug will be therapeutically effective or whether it may show an unacceptable 'side effect' problem. None of the product is sold during this time. This phase is very costly and there is a high chance that there will be no financial return on the investment, thus it is a very risky business, a theme which is elaborated in Chapter 7.

It takes many years to invent and develop a new drug, typically 10–15 years. Today's new research programmes will generally not come to fruition until the following decade. Yet, it should be said that, despite all the uncertainties and frustrations, drug discovery and development remains one of the most exciting and rewarding pursuits.

Acknowledgement

The section 'More About Enzymes' was adapted from Chapter 2 of the book 'Medicinal Chemistry: the Role of Organic Chemistry in Drug Research' (Academic Press, 1992) written by Dr Michael G. Davis. We acknowledge Dr Davis's contribution.

Further Reading

1. Sneader, W. *Drug Discovery – A History;* J. Wiley and Sons, 2005.
2. Patrick, G. L. *Introduction to Medicinal Chemistry,* 4th ed.; Oxford University Press, 2009.
3. King, F. D., Ed. *Medicinal Chemistry: Principles and Practices,* 2nd ed.; RSC, 2003
4. Thomas, G. *Fundamentals of Medicinal Chemistry;* J. Wiley and Sons, 2003.
5. Wilson, C. O., Block, J. H., Beale, J. M., Eds. *Wilson and Grisvold's Organic, Medicinal and Pharmaceutical Chemistry,* 12th ed.; Lippincott, Williams and Wilkins, 2011.
6. Smith, C. G.; O'Donnell, J. T. *The Process of New Drug Discovery and Development;* Informa Healthcare, 2006.
7. Rang, H. P. *Drug Discovery and Development;* Churchill Livingstone Elsevier, 2006.
8. Lednicer, D. *New Drug Discovery and Development;* J. Wiley and Sons, 2007.
9. Rydzewski, R. M. *Real World Drug Discovery;* Elsevier, 2008.
10. Wermuth, C. G. *Practice of Medicinal Chemistry,* 3rd ed.; Elsevier, 2008.
11. Silverman, R. B. *Drug Discovery, Design and Development;* Elsevier Academic, 2004.
12. Ng, R. *Drugs from Discovery to Approval,* 2nd ed.; J. Wiley and Son, 2009.
13. Li, J. J.; Johnson, D. S. *Modern Drug Synthesis;* J. Wiley and Sons, 2010.
14. Lemke, T. L., Williams, D. A., Roche, V. F., Zito, S. W., Eds. *Foye's Principles of Medicinal Chemistry,* 7th ed.; Lippincott, Williams and Wilkins, 2012.

Protein structure and function

2

Jennifer Ann Littlechild

Exeter Biocatalysis Centre, Biosciences, College of Life and Environmental Sciences,
Stocker Road, Exeter EX4 4QD, UK

CHAPTER OUTLINE

2.1 Introduction 58
2.2 General Aspects of Protein Structure 58
    2.2.1 Primary, secondary, tertiary and quaternary structure................................63
    2.2.2 Protein folds and structural bioinformatics 69
2.3 Enzymes 71
    2.3.1 Enzyme inhibitors..71
    2.3.2 Examples of enzyme inhibition 72
        2.3.2.1 Interaction of aspirin and COX 72
        2.3.2.2 Interactions of angiotensin cleaving enzyme with inhibitors 73
2.4 Protein Receptors.. 74
    2.4.1 Specific examples of protein receptors 74
        2.4.1.1 Ion channel receptors.. 74
        2.4.1.2 G-Protein-linked receptors 74
        2.4.1.3 Kinase-linked receptors (1-TM)... 74
2.5 Structural Proteins as Drug Targets 75
2.6 Proteins as Drugs.. 76
2.7 Concluding Remarks.. 77
References 78

ABSTRACT

This chapter covers the general features of protein structure and recent advances in structural bioinformatics. The importance of the three-dimensional structure of the protein target in order to understand its mechanism of action as an aid for drug design is illustrated by specific examples of enzyme inhibition, receptor interactions and drugs binding to structural proteins. The impact of proteomics and bioinformatics is stressed, while protein interactions with other proteins and different biological macromolecules are discussed.

Keywords/Abbreviations: D/L amino acids; Protein chain rotation; Protein folding; Alpha-helix; Beta-sheet; Ramachandran plot; Protein data bank (PDB); Bioinformatics; Composer program; Molecular operating environment (MOE); European structural genomics project (SPINE); European bioinformatics institute (EBI); Enzyme classification (EC); Cyclooxygenase (COX); Angiotensin cleaving enzyme (ACE); Nicotinic AChR; Adrenergic-G protein receptor complex; Insulin receptor.

Introduction to Drug Research and Development. http://dx.doi.org/10.1016/B978-0-12-397176-0.00002-9

2.1 INTRODUCTION

Proteins make up a large proportion of living cells. They are important (1) as enzymes to carry out all the metabolic reactions going on inside the cell; (2) as receptors and ion channels to communicate from the outside to the inside of the cell and to allow flow of ions and small molecules across the cell membrane; (3) as structural proteins making up connective tissue, muscle, bones and the cellular division machinery; (4) as the basis of our immune system in the form of antibodies and (5) being involved in macromolecular complexes with RNA, DNA and carbohydrate molecules. Many proteins in eukaryotic cells have complex glycosylation patterns on their surface which are involved in their recognition by other macromolecules.

2.2 GENERAL ASPECTS OF PROTEIN STRUCTURE

Despite their varied biological roles all proteins are made up of the 20 standard amino acids which are encoded on DNA. The proteins can undergo post-translational modification after they have been made on the ribosome. This increases the diversity of protein function without increasing the size of the genome. All amino acids have a common general structure. As the name implies the general structure involves an amino group and a carboxyl group, but the variations in the side chains (the so-called R groups) distinguish them from each other (Figure 2.1). The acid and amino groups are bonded to the same carbon atom, the α-carbon atom. The order of carbon atoms is numbered by the Greek alphabet, α, β, γ, δ, ε, in the order in which they are bonded, starting with the carbon adjacent to the carboxyl group.

The α-carbon is also bonded to a hydrogen atom and to the side-chain group, R. The identity and properties of the amino acid depend on the nature of the R group.

All proteins can be described as chiral macromolecules. This is because the α-carbon atom of each constituent amino acid is tetrahedral and, if decorated with four different substituents, will exhibit chirality (Greek handedness). Indeed, all amino acids except glycine (R = H) will have two forms related as object to mirror image, designated L and D (Figure 2.1). Amino acids in proteins are almost always in the L form. The D form occurs in bacterial cell walls and some human disease states have identified amino acid changes in proteins from the healthy L to D forms (1). Some D-amino acids have been shown to be present in the brain and could either act as biologically active small molecules or be

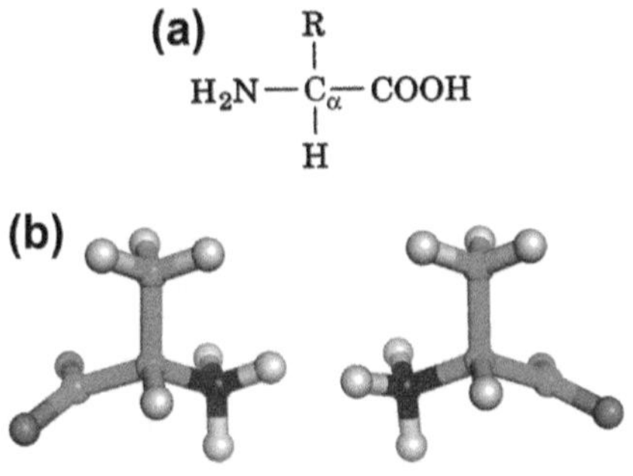

FIGURE 2.1

(a) The basic structure of an amino acid; (b) the L and D forms of the amino acid alanine. (See colour plate.)

incorporated into insoluble deposits associated with some neurological diseases such as Alzheimer's disease (*2,3*).

Different amino acids can be referred to by a three-letter code e.g. Ala – alanine and Lys – lysine. More commonly a one-letter code is used e.g. A – alanine, K – lysine. As seen from the examples given in Table 2.1 sometimes this code is obvious but in other cases it needs to be memorized.

Depending on the nature of the R group each amino acid will have different properties. Table 2.1, taken from Voet and Voet (*4*), show the amino acids as their three- and one-letter codes, as well as their structure, mass, average occurrence and their properties.

The properties and order of the amino acids in the protein will determine how it will fold and its overall properties. Glycine is the smallest amino acid and its content and position is important in both globular and structural proteins. Since its R group is hydrogen it shows no chirality. It is the most flexible amino acid. It is found in structural proteins where there is only limited space such as collagen and silk. It is also found in small loops in globular proteins and in the active sites of enzymes where there is a requirement for flexibility in the protein chain.

Proline is an unusual amino acid where the side chain comes back to the main chain to form a covalent bond. It is often called a cyclic amino acid or an imino acid. Proline is the most rigid amino acid and therefore is an important component in proteins that need to be rigid, such as the special polyproline helix structure of collagen. It is also found in specific short turns that are found in globular proteins. Its other important property is that it can exist in both the *cis* form and the *trans* form which is used to confer a specific function in many proteins. The isomerization from the *cis* to *trans* form of proline is often involved in protein folding and protein conformational changes. Apart from proline most other amino acids found in proteins are in the *trans* conformation where the R groups are consecutively located on different sides of the polypeptide chain.

Histidine is another important amino acid found in the active site of many enzymes. It is the only amino acid with a pKa of the R group of c. 6.7. This means that the extent of protonation of the side chain can alter around physiological pH. For example, the extent of the positive charge on this residue can be slightly altered by other amino acids surrounding it in the three-dimensional (3D) structure of the protein.

Cysteine and methionine are the two amino acids that contain sulfur in their R groups. The occurrence of methionine is low in proteins. It is the amino acid that is the starting point in protein synthesis and as such all proteins have a methionine residue at their N-terminal end. This residue may be cleaved from the protein after translation or it may remain at the N-terminus.

Cysteine is important for several reasons, one of which is its ability to form a covalent bond with another cysteine residue on the protein chain when these amino acids approach each other in the 3D structure. The covalently linked disulfide bond form is called cystine, as shown in Figure 2.2. This disulfide bond is the only interchain covalent bond that helps to hold the protein in its correct fold. It is found mainly in proteins which are exported from the cell and in proteins isolated from thermophilic organisms where the protein has to be stable to temperatures up to 100 °C. Transient disulfide bonds are now known to play important roles in eukaryotic systems (*5*).

Cysteine can also exist in different oxidation states (*6*). The sulfenic and sulfinic forms of this amino acid are involved in the human peroxiredoxin system which acts to remove damaging free radicals that are generated in cells and various diseased states. Natural processes such as ageing are linked to a malfunction of this free radical scavenging system (*7*).

The fundamental amino acid unit, without any charge contribution from the R group, is a zwitterion and carries no overall charge around the physiological pH (c. 7.0) (Figure 2.3).

Table 2.1 List of the Different Amino Acids and Their Properties

Name, Three-Letter Symbol, and One-Letter Symbol	Structural Formula*	Residue Mass (D)†	Average Occurrence in Proteins (%)**	pK₁ α-COOH††	pK₂ α-NH₃⁺ ††	pKR Side Chain††

Amino acids with nonpolar side chains

Glycine Gly G		57.0	7.2	2.35	9.78	
Alanine Ala A		71.1	7.8	2.35	9.87	
Valine Val V		99.1	6.6	2.29	9.74	
Leucine Leu L		113.2	9.1	2.33	9.74	
Isoleucine Ile I		113.2	5.3	2.32	9.76	
Methionine Met M		131.2	2.2	2.13	9.28	
Proline Pro P		97.1	5.2	1.95	10.64	

Amino acid	Structure				
Phenylalanine Phe F		147.2	3.9	2.20	9.31
Tryptophan Trp W		186.2	1.4	2.46	9.41

Amino acids with uncharged polar side chains

Amino acid	Structure				
Serine Ser S		87.1	6.8	2.19	9.21
Threonine Thr T		101.1	5.9	2.09	9.10
Asparagine§ Asn N		114.1	4.3	2.14	8.72
Glutamine§ Gln Q		128.1	4.3	2.17	9.13
Tyrosine Tyr Y		163.2	3.2	2.20	9.21	10.46 (phenol)
Cysteine Cys C		103.1	1.9	1.92	10.70	8.37 (sulfhydryl)

(continued on next page)

Table 2.1 List of the Different Amino Acids and Their Properties *(continued)*

Amino acids with charged polar side chains

Name, Three-Letter Symbol, and One-Letter Symbol	Structural Formula*	Residue Mass (D)[†]	Average Occurrence in Proteins (%)**	pK₁ α-COOH[††]	pK₂ α-NH₃⁺ [††]	pK_R Side Chain[††]
Lysine Lys K	$H-C-CH_2-CH_2-CH_2-CH_2-NH_3^+$ with COO⁻ and NH₃⁺	128.2	5.9	2.16	9.06	10.54 (ε-NH₃⁺)
Arginine Arg R	$H-C-CH_2-CH_2-CH_2-NH-C$ with COO⁻, NH₃⁺, NH₂, NH₂⁺	156.2	5.1	1.82	8.99	12.48 (guanidino)
Histidine His H	$H-C-CH_2$ with COO⁻, NH₃⁺, imidazole ring	137.1	2.3	1.8	9.33	6.04 (imidazole)
Aspartic acid[§] Asp D	$H-C-CH_2-C$ with COO⁻, NH₃⁺, O, O⁻	115.1	5.3	1.99	9.90	3.90 (β-COOH)
Glutamic acid[§] Glu E	$H-C-CH_2-CH_2-C$ with COO⁻, NH₃⁺, O, O⁻	129.1	6.3	2.10	9.47	4.07 (γ-COOH)

*The ionic forms shown are those predominating at pH 7.0 although residue mass is given for the neutral compound. The C_α atoms, as well as those atoms marked with an asterisk, are chiral centers with configuration as indicated according to Fischer projection formulas. The standard organic numbering system is provided for heterocycles.

[†]The residue masses are given for the neutral residues. For the molecular masses of the parent amino acids, add 18.0 D, the molecular mass of H_2O, to the residue masses. For side chain masses, subtract 56.0 D, the formula mass of a peptide group from the residue masses

**Calculated from a database of nonredundant proteins containing 300, 688 residues as compiled by Doolittle, R. F. In Fasman, G. D., Ed.; Predictions of Protein Structure and the Principles of Protein Conformation; Plenum Press, 1989.

[††]Source: Dawson, R. M. C.; Elliott, D. C.; Elliott, W. H. and Jones, K. M., Data for Biomedical Research 3rd ed.; Oxford Science Publications, 1986, pp. 1–31.

[§]The three-and one-letter symbols for asparagine or aspartic acid are Asx and B, whereas for glutamine or glutamic acid ther are Glx and Z. The one-letter symbol for an undetermined or "nonstandard" amino acid is X.

Taken From Voet and Voet (4)

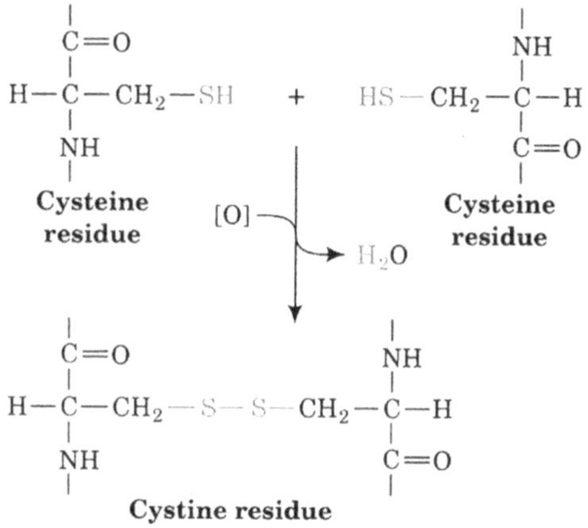

FIGURE 2.2

Conversion of two cysteine residues into a cystine unit. (For colour version of this figure, the reader is referred to the online version of this book.)

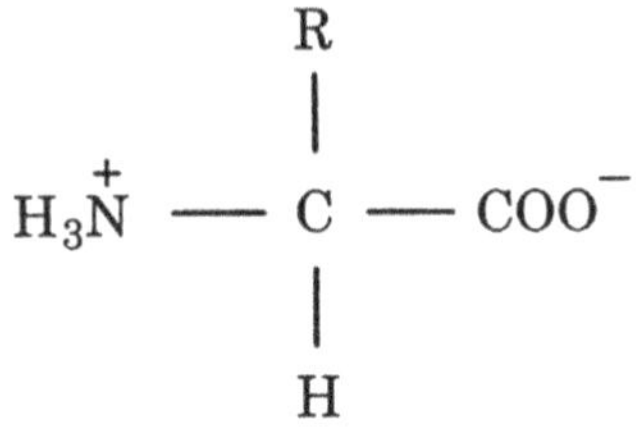

FIGURE 2.3

The fundamental amino acid unit.

If glycine or an amino acid that has no charge contribution from its side chain is titrated by increasing pH then its carboxyl group will ionize at an approximate pKa of 2.2 and its amino group at an approximate pKa at 9.8. The pI of the protein where it is electrically neutral is pH 6.1. Basic amino acids, such as arginine and lysine, carry a positive (+) charge at pH 7.0 while other amino acids, e.g. glutamic acid and aspartic acid, carry a negative charge (−) at pH 7.0, due to the functional group on the R side chain (Figure 2.4).

2.2.1 Primary, secondary, tertiary and quaternary structure

Amino acids join together to make a polypeptide by the elimination of water to form a peptide bond (Figure 2.5). This process occurs on the ribosome and the peptidyl transferase site (where this is catalysed) is now known to be a specific adenine residue held in an unusual conformation in the large subunit of the ribosomal RNA (8).

The peptide bond is planar due to its partial double-bond characteristic. This feature dictates how the protein chain can fold since the rotation of the bonds is restricted to either side of the α-carbon

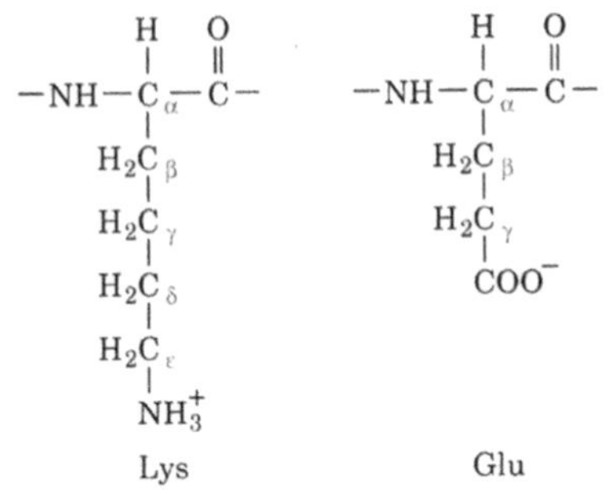

FIGURE 2.4

The basic amino acid, lysine and an acidic amino acid, glutamic acid showing the numbering of the carbon atoms. (For colour version of this figure, the reader is referred to the online version of this book.)

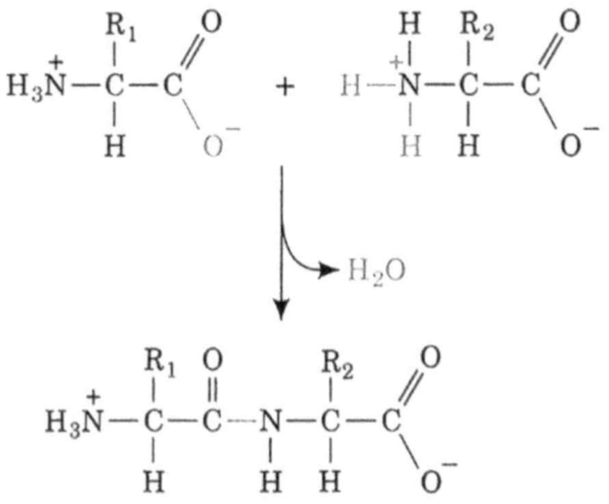

FIGURE 2.5

Formation of the peptide bond by the elimination of water. (For colour version of this figure, the reader is referred to the online version of this book.)

atom. Two angles ψ and φ are the allowed rotations of the chain that define the values for the secondary structure of all proteins (Figure 2.6). A polypeptide chain is always formed from the N-terminus to the C-terminus as defined by the direction of synthesis on the ribosome.

The secondary structure of proteins is represented by two main folds of the polypeptide chain. First is the α-helix (predicted by Linus Pauling in 1951), formed from a single polypeptide chain arranged in a right-handed helix which has 3.6 amino acid residues for each turn of the helix. This gives a pitch to the helix (distance of the helix rises along its axis/turn) of 5.4 Å (1 Å = 0.1 nm) (Figure 2.7). It is stabilized by H-bonding such that the peptide C=O group is bonded to the N–H group of the fourth amino acid along the chain. This configuration results in an H-bond at nearly optimal length (2.8 Å). The R groups of the different amino acids all point outwards from the helix. The length of the helices is on average about 11 amino acids long in globular proteins but can be much longer in fibrous proteins such as human hair.

The second main type of secondary structure, the β-sheet, was proposed by Pauling and Corey in 1951 and is made up of at least two polypeptide chains. The chains can be organized in a parallel or an

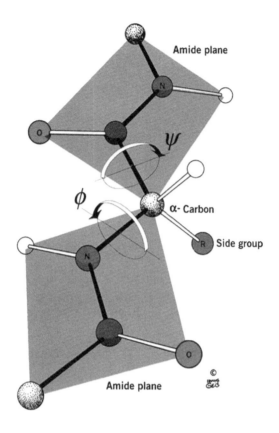

FIGURE 2.6

Figure showing the planar peptide bond. (See colour plate.)

anti-parallel fashion and can be said to have a rippled or pleated appearance. Each type of sheet is stabilized by H-bonds where the C=O group of one chain is bonded to the N–H group of the other chain (Figure 2.8). In globular proteins these are, on average, 15 residues long but can be much longer in fibrous proteins such as silk.

The allowed values of angles ψ and φ for these different secondary structures are defined by the Ramachandran plot (Figure 2.9). Some amino acids can be outliers on this plot especially if they are held in an unfavourable conformation for purposes of catalysis. The quality of protein structures is assessed by analysis of this plot using the program PROCHECK (9).

Another secondary structure that has to be formed in globular proteins is the turn between the different α-helices and β-sheets in the protein structure. These can fall into different classes such as the type 1 and type 2 turns that have an H-bond between the C=O of the first residue of the turn and the N–H of the last residue and are referred to as β-bends.

The secondary structural elements of proteins seem to like to be arranged in specific 'motifs' such as (1) α-helix, β-sheet, α-helix; (2) β-meander and (3) α-helical bundles. These arrangements are also referred to as the super-secondary structures.

(a)   **(b)**

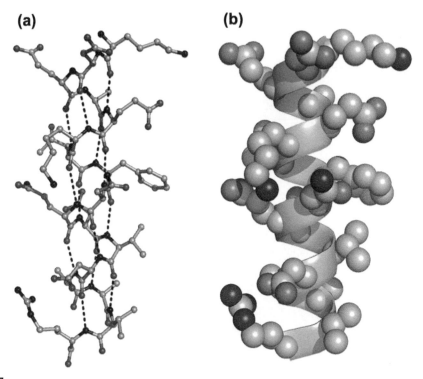

FIGURE 2.7

The α-helix with different R groups decorating the outside. (a) Representation in stick mode showing the H-bond network between the carboxyl and amino groups at every fourth residue along the chain. (b) Representation in ribbon for the C α backbone and space filling for the side chains. (See colour plate.)

Many of these motifs are recognized by specific cofactors which are small organic molecules bound into the active sites of enzymes that are essential to carry out the chemistry required for the enzyme mechanism. These are often vitamins that we require in our diet (see Chapter 1, Table 1.3). One cofactor, nicotinamide adenine dinucleotide (NAD^+), is used by all oxidoreductase enzymes and is recognized by a specific arrangement of α-helices and β-sheets called the Rossmann fold named after Michael Rossmann who solved the first dehydrogenase enzyme structure (*10*). This can be seen in the lactate dehydrogenase structure presented in Figure 2.11. Once a polypeptide chain exceeds 200 amino acids it forms a separate domain. The NAD^+-binding domain is one such example. The rearrangement of such structural domains is a mechanism used for the evolution of new proteins (*11*).

The next level of the structure is the tertiary structure which is the arrangement of the secondary elements in three dimensions. Some proteins are composed of more than one polypeptide chain and the arrangement of the chains with respect to each other is called the quaternary structure.

Proteins are typically composed of 50% water which is essential for the stability and activity of the protein. They have an essential layer of ordered water on their surface and if this is removed they are

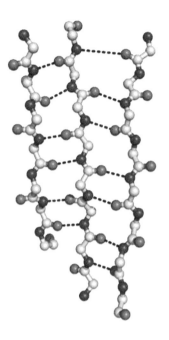

FIGURE 2.8

The β-sheet taken from a sample protein structure. The side chains have been removed. The H-bonding between the carboxyl and amino groups of the different polypeptide chains is shown in black dashed line. The sheet on the right is anti-parallel and the sheets on the left are parallel. (See colour plate.)

unfolded and denatured. Water is involved as an essential component in the mechanism of many hydrolase enzymes such as serine and cysteine proteases (*12*). Hydrophobic amino acids like to be on the inside of the globular proteins and hydrophilic amino acids, such as lysine and arginine, prefer to be on the surface where they make specific interactions with water.

The interactions that hold the tertiary structure of proteins are hydrophobic interactions, ionic interactions (interaction of acidic amino acid side chains with basic amino acid side chains in the 3D structure of the protein, see Figure 2.10), van der Waals interactions, H-bonds (many are present throughout the protein structure) and disulfide bonds in some proteins – the latter being the only covalent interaction. Subunits of multi-subunit proteins are held together by the same weak forces.

Proteins are held together by an accumulation of weak forces and are only marginally stable. They are easily denatured by extremes of pH and most proteins are only stable between pH 5.0 and pH 9.0. They are easily denatured by a rise in temperature and usually work best at the temperature of the environment supporting the organism from which they are isolated. Proteins are not usually stable in organic solvents as this removes the essential water.

The protein fold is based on the arrangement of the amino acids in the primary structure of the protein. Some amino acids like to be in an α-helix region e.g. alanine, while others like to be in the β-sheets; some can be found in both secondary structures. Protein folding is still not fully understood such that it is difficult to predict structure from primary sequence alone and several methods have been

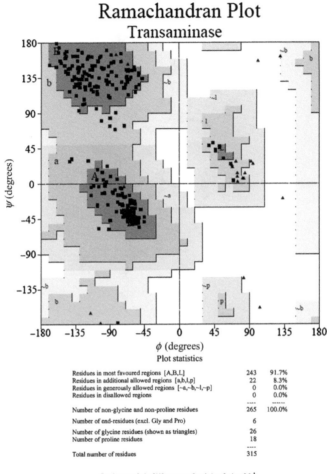

Ramachandran Plot
Transaminase

Plot statistics

Residues in most favoured regions [A,B,L]	243	91.7%
Residues in additional allowed regions [a,b,l,p]	22	8.3%
Residues in generously allowed regions [~a,~b,~l,~p]	0	0.0%
Residues in disallowed regions	0	0.0%
	----	-----
Number of non-glycine and non-proline residues	265	100.0%
Number of end-residues (excl. Gly and Pro)	6	
Number of glycine residues (shown as triangles)	26	
Number of proline residues	18	

Total number of residues	315	

Based on an analysis of 118 structures of resolution of at least 2.0 Å
and R-factor no greater than 20%, a good quality model would be expected
to have over 90% in the most favoured regions.

FIGURE 2.9

An example of a Ramachandran plot produced from PROCHECK (*9*). The red region on the top left is for parallel and anti-parallel β-sheets, the lower left red region for α-helices and the small red region on the right for left-handed helices which are not found in globular proteins; however, some amino acids with important catalytic function can fall into this region. The number of amino acids falling in disallowed regions is indicative of the quality of the crystal structure. (See colour plate.)

developed to tackle this problem (*13*). *In vivo* there are proteins called chaperones that assist in protein folding as the protein is made on the ribosome; chaperones can also refold misfolded proteins.

The Protein Data Bank contains all the information available on protein structures solved to-date and currently contains over 50,000 structures from X-ray crystallography (up to 0.9 Å resolution), from electron microscopy (up to 5 Å resolution) and from nuclear magnetic resonance, which gives

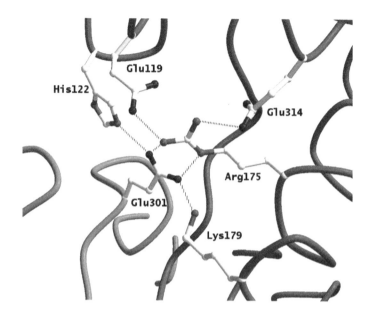

FIGURE 2.10

An ion-pair network on the interface between two protein subunits A (green) and subunit B (blue) of a thermophilic alcohol dehydrogenase enzyme (*24*). (See colour plate.)

an averaged structure as found in solution. This latter method was originally limited to proteins of up to 20,000 MW but with labelling studies and other techniques this can now being extended to larger proteins. The PDB can be freely accessed from the PDB site http://www.rcsb.org/pdb/home/home.do.

2.2.2 Protein folds and structural bioinformatics

While the number of protein structures deposited in the PDB is increasing dramatically each year, it appears that there are only a limited number of protein folds, as first proposed by Chothia and Lesk who suggested an upper limit of about 1000 folds (*14*). This means that some common folds such as the β-barrel structure can contribute to the overall structure of many proteins with a variety of different functions ranging from enzymes to transcription factors. The study of how the amino acid sequence can be understood in the context of protein structure and function makes up the science of bioinformatics. The application of bioinformatics to assign the function of a gene using a variety of software packages can therefore be challenging in many cases with the ever-increasing genomic information available. The continuing increase in the number of available protein structures can provide data to improve the techniques available for gene assignment and protein folding predictions. Generally, it is accepted that if proteins have over 30% sequence identity, they have the same fold. This information is vital for techniques of protein modelling where a known protein structure can be used as a basis to model a protein structure which is currently unknown.

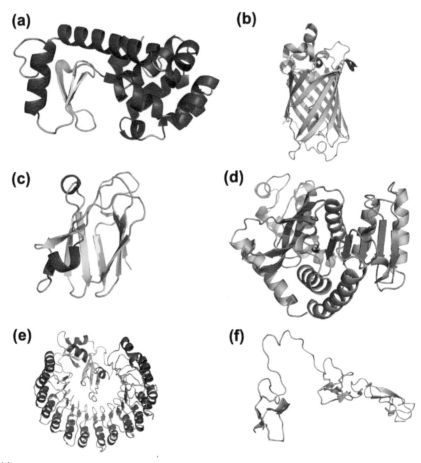

FIGURE 2.11

A figure showing examples of the different types of protein fold as described. (a) Mainly α-helical, lysozyme, (b) and (c) mainly β-sheet, yellow fluorescent protein and plastocyanin (d) αβ-protein, ribonuclease inhibitor (e) α + β protein, lactate dehydrogenase and (f) unstructured proteins with disulfide bridges, EFG domain. (See colour plate.)

Protein structures can be divided into different classes of fold. These were classified as part of the program CATH (*15*). Examples of proteins that fall into these classes are shown in Figure 2.11.

1. All α proteins in which the secondary structure is almost exclusively α helical.
2. All β proteins in which the secondary structure is almost exclusively β sheet and in which the β-sheet strands are parallel.
3. α/β proteins in which the helices and sheets are arranged in β-α-β units.
4. α + β proteins in which the helices and sheets are spatially separated in different parts of the protein.
5. Small disulfide-rich proteins that have few helices or sheets.
6. Relatively unstructured proteins.

Once a model of a selected protein has been made this can be assessed for quality using the Ramachandran Plot described above. Several programs such as Composer (*16*) and Molecular Operating Environment (*17*) can be used to carry out molecular modelling. Some basic modelling can also be carried out on line at http://swissmodel.expasy.org/SWISS-MODEL.html.

The increasing number of structures available for different proteins has resulted in a large amount of data for structural bioinformatics. Many international consortia have been established to try to take a logical structural genomics approach. The limitation to determining all the proteins encoded in a specific genome is the ability to obtain soluble protein that can form well-diffracting protein crystals. The so-called 'low-lying fruit' or easily crystallizable proteins can result in successful structural determination but the more recalcitrant protein targets reduce the overall success rate of the projects. One of the most successful structural genomics programmes is that of the Japanese Thermus thermophilus bacteria project (*18*). Other projects such as the European Structural Genomics Project (*19*) selects human proteins of medical importance. The European Bioinformatics Institute (EBI) is a good source of databases and software analysis in this area and can be accessed at www.ebi.ac.uk. This is also an easy way to access the PDB at http://www.ebi.ac.uk/pdbe/. The Sanger Institute has a site with information of proteins in the human genome and can be accessed at http://sanger.ac.uk. The Pfam software at this site can be searched for different protein families (http://pfam.sanger.ac.uk).

2.3 ENZYMES

Enzymes make up a large proportion of the cell and are the biological catalysts that carry out all the metabolic reactions that take place in the cell.

As described in Chapter 1 (Table 1.4), enzymes can be divided into six main classes based on the reactions they carry out.

- Oxidoreductases – transfer of hydride or H group
- Transferases – group transfer reactions
- Hydrolases – hydrolysis reactions
- Lyases – group elimination to form double bonds
- Isomerases – transfer of group within molecules
- Ligases – formation of C–C, C–S, C–O and C–N bonds.

Every enzyme has an Enzyme Classification number, the first digit of which describes one of the six classes. The remaining three numbers describe the substrate and other features of the reaction. All the known classified enzymes can be located on a useful website called BRENDA http://www.brenda-enzymes.info/. The MACiE database is a useful website to ascertain the most up-to-date mechanism of different known enzymes. This can be accessed from the EBI site http://www.ebi.ac.uk/thornton-srv/databases/MACiE/.

2.3.1 Enzyme inhibitors

Enzymes are the main site of drug interaction in cells. Enzyme inhibitors can work in several ways and can be classified according to whether they directly compete with the substrate for the enzyme, interact with the enzyme substrate complex or bind to a site discrete from the active site and affect the substrate binding.

These different modes of inhibition were described in detail in Chapter 1.12 and can be summarized briefly as follows:

1. *Competitive Inhibition – Reversible.* The drug is similar to the normal substrate of the enzyme and can bind in the active site. Competes with the natural substrate and can be displaced by increasing the substrate concentration.
2. *Competitive Inhibition – Irreversible.* Some competitive inhibitors can be suicide drugs since they partially undergo the reaction and stay bound to the enzyme. One example, cited in Chapter 1, is that of dihydrofolate reductase which is irreversibly inhibited by methotrexate for the treatment of cancer. Another example involves the enzyme cyclooxygenase (COX) which is irreversibly inhibited by aspirin through acetylation of a serine residue in the entrance of the active site of the enzyme (see section 2.3.2.1).
3. *Uncompetitive Inhibitors.* Not affected by increasing the concentration of substrate. Binds to enzyme/substrate complex.
4. *Non-Competitive and Mixed Inhibition.* These inhibitors can bind to both enzyme and enzyme–substrate complex.
5. *Allosteric Inhibition (Non-Michaelis–Menten Kinetics).* Often binds to the enzyme at a site distinct from the active site and affects activity. The effect is usually by a conformational change in the structure of the enzyme – a so-called 'allosteric effect'.

2.3.2 Examples of enzyme inhibition

2.3.2.1 Interaction of aspirin and COX

COX (PDB 1prh) performs the first step in the creation of prostaglandins from a common fatty acid called arachidonic acid. It adds two oxygen molecules to arachidonic acid, beginning a set of reactions that will ultimately create a host of natural, biologically active molecules. Aspirin blocks the binding of arachidonic acid in the COX active site (Figure 2.12). Normal messages are not delivered, so no pain is felt and there is no inflammatory response.

FIGURE 2.12

The aspirin bound to the serine in the active site of the cyclooxygenase. (See colour plate.)

2.3.2.2 Interactions of angiotensin cleaving enzyme with inhibitors

The early inhibitors of Angiotensin Cleaving Enzyme (ACE) were developed based on the known crystal structure of the zinc metalloprotease carboxypeptidase A (20). It was known that ACE cleaved two amino acids from angiotensin whereas carboxypeptidase A cleaved one amino acid from the C-terminus of proteins. The first approved ACE inhibitors such as captopril have been co-crystallized with the ACE enzyme and show the interaction of the thiol (SH) group of the inhibitor with the catalytic zinc ion (21). This ACE complex structure shows that the enzyme has an extended binding pocket that can be used to design improved inhibitors such as lisinopril. A structure of the ACE enzyme with bound lisinopril shows the inhibitor bound into the predicted binding site pocket (Figure 2.13).

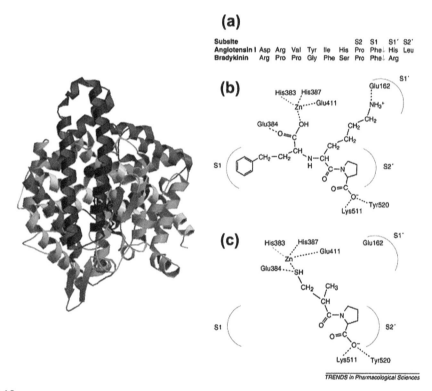

(a)

Subsite						S2	S1	S1'	S2'
Angiotensin I	Asp	Arg	Val	Tyr	Ile	His	Pro	Phe	His	Leu
Bradykinin	Arg	Pro	Pro	Gly	Phe	Ser	Pro	Phe	Arg

FIGURE 2.13

The human ACE enzyme and the active site complexes with inhibitors taken from Ref. (21). (a) The sites of cleavage by ACE in the substrates angiotensin I and bradykinin. (b) The principle interactions between the ACE inhibitor lisinopril and the active site of testicular angiotensin cleaving enzyme (tACE). Non-covalent interactions between enzyme groups and the inhibitor are indicated by broken lines and the sub-sites corresponding to those in (a) are labelled. (c) Projected interactions between captopril and tACE, based on the structure of the ACE–captopril complex with ACE sub-sites indicated. (See colour plate.)

2.4 PROTEIN RECEPTORS

Protein receptors are amongst the most important drug targets. Their roles are to act as a communication between the outside and inside of the cell. As detailed in Chapter 1, small organic molecules such as neurotransmitters and hormones can bind to the receptor on the outside of the cell and produce an effect inside the cell. Other related small drug molecules can (1) act as an agonist and still produce the effect or (2) act as an antagonist and block the effect. Many drugs are agonists or antagonists at different receptors in the cell membrane.

Structurally the membrane proteins can have α-helices or β-sheets as they pass through and interact with the cell membrane. They often have hydrophilic protein domains on either side of the bilipid cell membrane. The protein region inserted into the membrane has hydrophobic amino acid residues on the outside interacting with the membrane lipids (which is opposite to soluble proteins where the hydrophobic residues are located on the inside of the protein). Membrane proteins are difficult to crystallize and have been poorly represented in the PDB. However this situation is improving as methods are continuing to be developed to clone, over-express and crystallize these proteins. In many cases the soluble domains of the receptor are crystallized independently and act to help understand the role of these domains within the intact protein.

2.4.1 Specific examples of protein receptors

As described in Chapter 1, ion channel receptors and G-protein-coupled receptors are two major families of receptors. Some examples of their actual structures are given below.

2.4.1.1 Ion channel receptors

These receptors are also known as 2-TM, 3-TM and 4-TM receptors or ligand-gated ion channels. When the receptor binds the ligand it causes a conformational change that causes an ion channel to open. For example the nicotinic acetylcholine receptor (AChR) is made up of an arrangement of five helical protein subunits (Figure 2.14) which are able to make a pore in the synaptic membrane that can open on the binding of the neurotransmitter acetylcholine; this allows the flow of potassium out of the nerve cell and sodium ions into the nerve cell to develop the action potential or nerve impulse.

2.4.1.2 G-Protein-linked receptors

These receptors (also known as 7-TM receptors) activate a G-protein which initiates a cascade of enzyme activity. Often this results in the production of cyclic adenosine monophosphate as a second messenger. These receptors mediate the action of a variety of hormones and neurotransmitters and at least 800 have been identified in the sequence of the human genome. For example, the muscarinic AChR belongs to the G-protein class of receptor and the structure of an adrenergic receptor–G-protein complex is shown in Figure 2.15. Studies on G-protein receptors that have helped to gain an understanding of their structure and their detailed mode of action have led to the award of the Noble Prize for Chemistry in 2012 to Robert J Lefkowitz and Brian K Kobilka.

2.4.1.3 Kinase-linked receptors (1-TM)

Kinase-linked receptors represent another important family of receptors (Chapter 8). These receptors are activated by peptide hormones, growth factors and cytokines and, in turn, activate enzymes

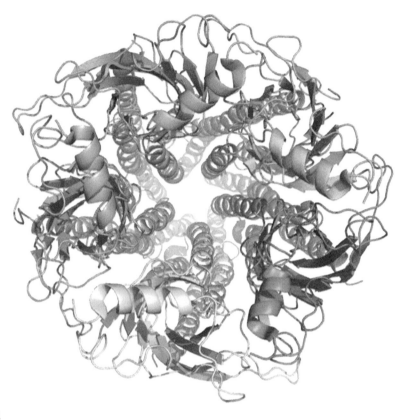

FIGURE 2.14

End on view of the structure of the nicotinic acetylcholine receptor PDB 2BG9 showing the five helical poly-peptides that pass through the membrane in different colours. (See colour plate.)

directly. An example of this kind of receptor is the insulin receptor (see Figure 2.16 and Chapter 6) which activates tyrosine kinase enzymes inside the cell. Structural studies on this receptor have implications for the design of peptide agonists (*22*).

2.5 STRUCTURAL PROTEINS AS DRUG TARGETS

Drugs can bind to structural proteins such as tubulin. An example is the alkaloid Colchicine, used in the treatment of arthritis, which binds tubulin and causes the microtubules to de-polymerize; this stops damaging neutrophils migrating into the joints. However this treatment has many side effects.

Naturally occurring anti-cancer drugs such as vinblastine and taxol can interfere with cell division. Vinblastine prevents polymerization of tubulin to form spindles and taxol stimulates the formation of the microtubules and prevents the depolymerization process.

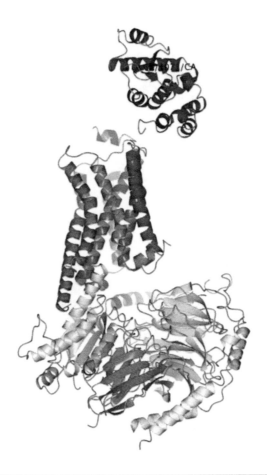

FIGURE 2.15

Structure of the b2 adrenergic receptor–G-protein complex. PDB 3SN6. Figure taken from Ref. (*22*). The structure gives a detailed view of transmembrane signalling. The ternary complex made up of the agonist, the receptor and the G-protein undergoes conformational changes to enhance the affinity of the agonist for the receptor and to favour the hydrolysis of guanosine triphosphate to guanosine diphosphate by the G-protein. At the top shown in purple is an easily crystallizable protein lysozyme which is not a part of this complex but facilitates the crystallization. The agonist binds to the transmembrane helices which contact the G - protein which is a trimer. The protein in red is called a nanobody which is present to aid the crystallization. (See colour plate.)

2.6 PROTEINS AS DRUGS

This topic is discussed in depth in a number of other chapters in this book, so only a fore-taste of the rapidly expanding therapeutic area is given here.

Insulin is one of the most common protein drugs and is used for the treatment of some types of diabetes. Early treatments utilized porcine insulin but more recently human insulin is mostly used; it is produced from the cloned gene and over-expressed for large-scale production.

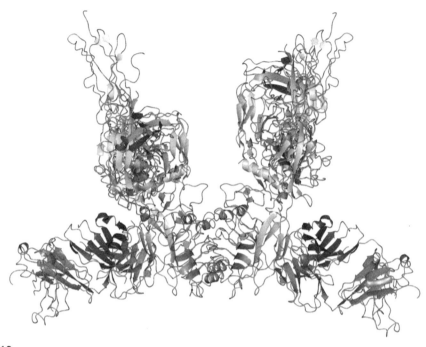

FIGURE 2.16

Part of insulin receptor which binds the insulin protein PDB 3LOH. (See colour plate.)

Human growth hormone was originally isolated from human pituitary glands. This produced a risk of virus transmission such as HIV, so human growth hormone is now produced from over-expression of the cloned human gene.

Therapeutic antibodies are important as drugs. Many therapeutic antibodies are already in the market and many more are in clinical trials. A list of FDA-approved monoclonal antibodies for cancer treatment is set out below (23) (Chapter 9).

- Alemtuzumab (Campath) – Chronic lymphocytic leukaemia
- Bevacizumab (Avastin) – Breast cancer, Colon cancer, Lung cancer
- Cetuximab (Erbitux) – Colon cancer, Head and neck cancers
- Gemtuzumab (Mylotarg) – Acute myelogenous leukaemia
- Ibritumomab (Zevalin) – Non-Hodgkin's lymphoma
- Panitumumab (Vectibix) – Colon cancer
- Rituximab (Rituxan) – Non-Hodgkin's lymphoma
- Tositumomab (Bexxar) – Non-Hodgkin's lymphoma
- Trastuzumab (Herceptin) – Breast cancer

2.7 CONCLUDING REMARKS

Knowledge of the protein target and the molecular details of drug interaction will help the development of new and improved therapeutic agents. An ever-increasing source of data from whole genome sequencing, proteomics and structural bioinformatics will help to achieve these ends. The field of pharmacogenomics will allow development of personalized medicine to deliver a more effective and safer drug administration regime.

References

1. Kaji, Y.; Oshika, T.; Takazawa, Y.; Fukayama, M.; Fujii, N. Pathological Role of D-Amino Acid-Containing Proteins and Advanced Glycation End Products in the Development of Age-Related Macular Degeneration. *Anti-Aging Med.* **2010,** *7,* 107–111.
2. Wolosker, H.; Dumin, E.; Balan, L.; Foltyn, V. N. D-Amino Acids in the Brain: D-Serine in Neurotransmission and Neurodegeneration. *FEBS J.* **2008,** *275,* 3514–3526.
3. Strittmatter, W. J.; Saunders, A. M.; Schmechel, D.; Pericak-Vance, M.; Enghild, J.; Salvesens, G. S.; Roses, A. D. Apolipoprotein E: High-avidity Binding to, Beta-Amyloid and Increased Frequency of Type 4 Allele in Late-Onset Familial Alzheimer Disease. *Proc. Natl. Acad. Sci. USA* **1993,** *90,* 1977–1981.
4. Voet, D. J.; Voet, J. G. *Biochemistry,* 3rd ed.; Wiley: New York, 2004.
5. Schroder, E.; Littlechild, J. A.; Lebedev, A.; Errington, N.; Vagin, A.; Isupov, M. Crystal Structure of Decameric 2-Cys Peroxiredoxin from Human Erythrocytes at 1.7 Å Resolution. *Structure* **2000,** *8,* 605–615.
6. Giles, N.; Watts, A.; Giles, G.; Fry, F.; Littlechild, J. A.; Jacob, C. Metal and Redox Regulation of Cysteine Protein Function. *Chem. Biol.* **2003,** *10,* 677–693.
7. Szabo, K.; Gutowski, N. J.; Holley, J. E.; Littlechild, J. A.; Winyard, P. Redox Control in Human Disease with a Special Emphasis on the Peroxiredoxin-Based Antioxidant System. In *Redox Signaling and Regulation in Biology and Medicine;* Jacob, C., Winyard, P. G., Eds.; Vch Verlagsgesellschaft Mbh, 2009; pp 409–431.
8. Nissen, P.; Hansen, J.; Ban, N.; Moore, P.; Steitz, T. The Structural Basis of Ribosome Activity in Peptide Bond Synthesis. *Science* **2000,** *289,* 920–930.
9. Laskowski, R. A.; MacArthur, M. W.; Moss, D. S.; Thornton, J. M. PROCHECK: A Program to Check the Stereochemical Quality of Protein Structures. *J. Appl. Crystallogr.* **1993,** *26,* 283–291.
10. Rossmann, M. G.; Moras, D.; Olsen, K. W. Chemical and Biological Evolution of Nucleotide Binding Protein. *Nature* **1974,** *250,* 194–199.
11. Bashton, M.; Chothia, C. The Generation of New Protein Functions by Combination of Domains. *Structure* **2007,** *15,* 85–99.
12. Blow, D. M.; Birktoft, J. J.; Hartley, B. S. Role of a Buried Acid Group in the Mechanism of Action of Chymotrypsin. *Nature* **1969,** *221,* 337–340.
13. Lemer, C. M.; Rooman, M. J.; Wodak, S. J. Protein Structure Prediction by Threading Methods: Evaluation of Current Techniques. *Proteins* **1995,** *23,* 337–355.
14. Chothia, C. 1000 Families for the Molecular Biologist. *Nature* **1992,** *357,* 543–544.
15. Orengo, C. A.; Michie, A.; Jones, S.; Jones, D. T.; Thornton, J. M. CATH – A Hierarchic Classification of Protein Domain Structures. *Structure* **1997,** *5,* 1093–1108.
16. Topham, C. M.; Thomas, P.; Overington, J. P.; Johnson, M. S.; Eisenmenger, F.; Blundell, T. L. An Assessment of COMPOSER: A Rule-Based Approach to Modelling Protein Structure. *Biochem. Soc. Symp.* **1990,** *57,* 1–9.
17. Chemical Computing Group Software http://www.chemcomp.com/research-cite.htm.
18. Yokoyama, S.; Hirota, H.; Kigawa, T.; Yabuki, T.; Shirouzu, M.; Terada, T.; Ito, Y.; Matsuo, Y.; Kuroda, Y.; Nishimura, Y.; Kyogoku, Y.; Miki, K.; Masui, R.; Kuramitsu, S. Structural Genomics Projects in Japan, Nature Structural Biology. *Struct. Genomics Suppl.* **2000,** 943–945.

19. Stuart, D.; Jones, Y.; Wilson, K.; Daenke, S. SPINE: Structural Proteomics in Europe – The Best of Both Worlds. *Acta Crystallogr. D* **2006,** *62.* published on line.
20. Natesh, R.; Schwager, S. L. U.; Sturrock, E. D.; Acharya, K. R. *Nature* **2003,** *421,* 551–554.
21. Rasmussen, S. G.; DeVree, B. T.; Zou, Y.; Kruse, A. C.; Chung, K. Y.; Kobilka, T. S.; Thian, F. S.; Chae, P. S.; Pardon, E.; Calinski, D.; Mathiesen, J. M.; Shah, S. T.; Lyons, J. A.; Caffrey, M.; Gellman, S. H.; Steyaert, J.; Skiniotis, G.; Weis, W. I.; Sunahara, R. K.; Kobilka, B. K. *Nature* **2011,** *477,* 549–555.
22. Smith, B.; Huang, K.; Kong, G.; Chan, S.-J.; Nakagawa, S.; Menting, J. G.; Hu, S.-Q.; Whittaker, J.; Steiner, D. F.; Katsoyannis, P.; Ward, C.; Weiss, M.; Lawrence, M. Structural Resolution of a Tandem Hormone-Binding Element in the Insulin Receptor and its Implications for Design of Peptide Agonists. *Proc. Natl. Acad. Sci. USA* **2010,** *107,* 6771–6776.
23. Food and Drug Administration (FDA), Center for Drug Evaluation and Research www.fda.gov/Drugs/default.htm.
24. Littlechild, J. A.; Guy, J. E.; Isupov, M. Hyperthermophilic Dehydrogenase Enzymes. *Biochem. Soc. Trans.* **2004,** *32,* 255–258.

The small molecule drug discovery process – from target selection to candidate selection

Michael Stocks

The School of Pharmacy, The University of Nottingham, Nottingham NG7 2RD, UK

CHAPTER OUTLINE

3.1 Introduction 82
3.2 Where Do Leads Come From? 84
    3.2.1 Definitions 84
    3.2.2 High-throughput screening 85
    3.2.3 Structure-based drug design and virtual screening......................................86
    3.2.4 Natural products 89
    3.2.5 Fragment-based drug discovery..91
    3.2.6 Fast-follower approaches...94
3.3 Lead Generation: Active-to-Hit 95
3.4 Lead Generation: Hit-to-Lead... 96
    3.4.1 Lipophilicity..96
    3.4.2 Solubility..97
    3.4.3 DMPK considerations...98
    3.4.4 Transporter proteins..104
    3.4.5 Plasma protein binding and whole blood potency 105
    3.4.6 Inhibition and induction of CYP enzymes..107
    3.4.7 Time-dependent CYP inhibition 108
    3.4.8 Ligand lipophilicity efficiency – identifying and working at the 'leading edge' 109
    3.4.9 Hit and lead criteria...109
    3.4.10 Case study – IKK-2 inhibitors 111
3.5 LO – Establishing the Screening Cascade and Candidate Biological Target Profile............................ 113
    3.5.1 Initial safety considerations and potential liabilities 114
    3.5.2 Therapeutic ratio/margin of safety 115
    3.5.3 Common toxicities 115
    3.5.4 Cardiovascular side-effects – hERG..115
    3.5.5 Genotoxicity caused by Ames-positive anilines 119
    3.5.6 Phospholipidosis..119
    3.5.7 Pharmacokinetic–pharmacodynamic relationship and dose-to-man prediction.................120
    3.5.8 LO case study – the discovery of aprepitant 121
References and Notes 123

Introduction to Drug Research and Development. http://dx.doi.org/10.1016/B978-0-12-397176-0.00003-0

ABSTRACT

Drug discovery of small molecules from target selection through to clinical evaluation is a very complex, challenging but rewarding area of drug discovery. There are many obstacles along the journey from initial hit-finding activities, through optimization of compounds and eventually to delivery of robust candidate drugs (CDs) for clinical evaluation. This chapter presents key issues and literature solutions with respect to the optimization of hits into CDs. Details of the key hit-finding activities namely high-throughput screening, virtual screening, natural products, fragment-based drug discovery and fast-follower approaches are discussed. Key aspects of compound quality such as lipophilicity, solubility, drug metabolism and pharmacokinetic, plasma protein binding and cytochrome P450 inhibition/induction are discussed as well as potential safety liabilities such as human ether-a-go-go related gene, genotoxicity and phospholipidosis, Finally successful hit-to-lead and lead optimization case studies are presented to illustrate and highlight the key principles.

Keywords/Abbreviations: Drug discovery; Lead generation; Active-to-hit; Hit-to-lead; Natural products; Fast-follower approaches; Absorption, distribution, metabolism & elimination (ADME); Area under the curve (AUC); Candidate drug (CD); Clearance (Cl); Intrinsic clearance (Clint); Maximum concentration (Cmax); Drug metabolism and pharmacokinetic (DMPK); Design-make-test (DMT); Oral bioavailability (F); Fragment-based drug discovery (FBDD); Human ether-a-go-go related gene (hERG); Hit identification (HI); High-throughput screening (HTS); Lead identification (LI); Lead optimization (LO); Minimum effective concentration (MEC); Maximum tolerated concentration (MTC); Pharmacodynamics (PD); Pharmacokinetics (PK); Proof of concept (POC); Proof of mechanism (POM); Proof of principle (POP); Plasma protein binding (PPB); Structure–activity relationship (SAR); Structure-based drug design (SBDD); Time-dependent inhibition (TDI); Target identification (TI); Target validation (TV); Volume of distribution (V_d); Virtual screening (VS).

3.1 INTRODUCTION

The drug discovery process involves the partnership of multiple scientific disciplines united with a common goal of producing medicines for the prevention, treatment or cure of human diseases. Within this partnership (Figure 3.1), medicinal chemistry lies at the core of the drug discovery process and involves the design and synthesis of specific compounds, based on the knowledge and understanding of how they interact with biological targets.

This chapter outlines the various phases of the drug discovery process from target selection through to successful candidate drug (CD) delivery, highlighting key areas within this process.

The modern drug discovery process, generic to most major pharmaceutical companies, is shown (Figure 3.2) and comprises individual areas of overlapping research with the overall aim of delivering a group of high-quality CDs for clinical investigation. The process is both time and resource intensive, with an overall cost for the delivery of a single new drug of approximately £800 million. This coupled with the fact that only 10–20% of projects that deliver CDs for clinical evaluation succeed to launch, necessitates the requirement that the biological target is ideally validated at CD nomination and that the CD has been thoroughly evaluated on the basis of optimal physicochemical properties and selectivity for the chosen biological target.

The drug discovery process begins with target identification (TI) where potential biological targets are selected, generally but not always, in an area of unmet medical need. A biological target can be defined as a protein whose function can be modulated by a therapeutic agent to affect a beneficial

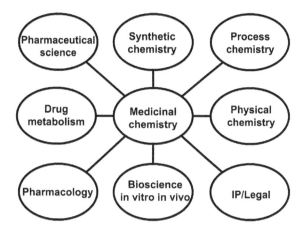

FIGURE 3.1

The modern effective drug discovery team.

patient outcome. The biological target (*1*) may be an enzyme, a receptor, an ion channel or a transporter protein with a strong disease linkage. Once the biological target has been identified, validation of the selected target is required to both improve disease linkage and test the biological hypothesis for efficacy in the disease and decrease attrition due to failure of the hypothesis at a later stage. This is most important as attrition should occur early in the drug discovery process in order to lower the cost burden associated with late-stage failure.

In order for the project to progress and the target to be validated, it is essential that the following key considerations of clinical efficacy, safety and potential genetic variation are investigated. Clinical efficacy requires a clear hypothesis and understanding for disease modulation with the chosen target; a clear demonstration of expression of the target and associated ligands within the disease state that is consistent with the hypothesis and functional validation of the target in a human *in vitro* biological model. Safety considerations require that the target passes a critical safety assessment review by

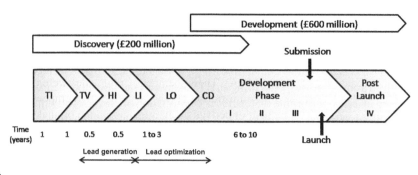

FIGURE 3.2

The modern drug discovery process: TI = target identification; TV = target validation; HI = hit identification; LI = lead identification; LO = lead optimization; CD = candidate drug.

providing the necessary information, e.g. target biology, knockout phenotypes (*2*) and the profile of normal body expression. Genetic considerations require that TI should identify the appropriate polymorphic form for high-throughput screening (HTS) and further *in vitro* biological assay work – this is usually the most prevalent form within the population.

Through establishing primary and secondary confirmatory screens and testing the biological hypothesis with, for example, natural substrates or tool compounds, validation of the target can be achieved. Experimental evidence for the relevance of the biological target in the disease state is sought and a positive outcome from these studies is critical for the project to progress into the next phase – the search for series of small molecules to either inhibit or enhance the effect of the target. Therefore it is essential that scientists involved in TI should be fully integrated throughout the drug discovery process in order to maintain and increase confidence with increased investment in a drug discovery project.

Along with providing target validation (TV), TI/TV bioscientists work with translational research groups to establish biomarker strategies to support the progress of the project into development and beyond. A biomarker is a substance used as an indicator of a biological state. It is a characteristic that is measured and evaluated as an indicator of normal biological processes, pathogenic processes, or pharmacologic responses to a therapeutic intervention. Proof of mechanism (POM) biomarkers with validated methodology are required for each target at the CD nomination stage and proof of concept biomarkers for later stages in the development phase. A POM biomarker is a biomarker that demonstrates a pharmacological effect in man, directly linked to the proposed target mechanism or pathway and is often based on *ex vivo* enzyme or receptor assays in whole blood. A proof of principle biomarker is a biomarker that demonstrates an effect resulting in a biological change that is closely related to the proposed mechanism of action provided there is a clear link between the effect measured and the target disease.

The drug discovery process now enters the next stage, where the search begins to find the small molecule chemical starting points that will eventually evolve through optimization into the CD.

3.2 WHERE DO LEADS COME FROM?

3.2.1 Definitions

The active-to-hit and hit-to-lead processes described within this chapter are based on the definitions of Baxter: (*3*)

- Actives are defined as being the raw output from a high-throughput screen (HTS) and will be in the form of an IC_{50} derived from a dose–response using an HTS assay on a sample solution in dimethyl sulfoxide.
- Hits are defined in terms of a profile that includes confirmation of chemical structure and biological activity as well as physical chemistry data, drug metabolism and pharmacokinetic (DMPK) measurements. DMPK data is used at this early stage of the drug discovery process to avoid failure at a late stage due to poor pharmacokinetics (PK) (metabolic instability and/or poor oral absorption).
- Leads are described in terms of a target profile of chemical and biological data. This profile is chosen so that a lead optimization (LO) programme would have a reasonable chance of producing a CD in a target time of 2 years.

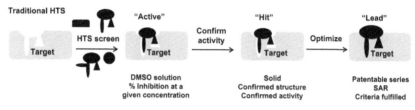

FIGURE 3.3

Schematic showing a traditional HTS.

There are many sources for chemical starting points (hits) (4); however for the purpose of this chapter the main four hit identification methods of HTS, virtual screening (VS), structure-based drug design (SBDD), natural products and fast-follower approaches will be discussed.

3.2.2 High-throughput screening

HTS is 'an industrialized process which brings together validated, tractable targets and chemical diversity to rapidly identify novel lead compounds for early phase drug discovery' and this technique (5) has become the mainstay approach to find hits in modern drug discovery programmes. Progress in robotic and screening technology together with the emergence of combinatorial high-throughput chemistry has led to the widespread implementation in drug discovery of this screening technique. HTS is in effect a numbers game, where on average if you screen enough compounds (either diverse or targeted) then one should obtain a selection of hits for further optimization.

The aim of an HTS is to deliver multiple distinct patentable compound series (different structures) from which the most suitable for rapid development to drug candidates will be selected by medicinal chemists. Importantly, HTS hits should be compounds that specifically inhibit/activate the activity of the biological target; have a defined structure; are chemically tractable and belong to a chemical series with structure–activity relationship (SAR) with 'drug-like' properties. The workflow of an HTS campaign is shown (Figure 3.3). Once the biological target has been identified and a robust biological screen has been established (either binding or functional), it can take as little as a few weeks to test a company's entire compound collection (typically c. 1 million compounds for a large organization).

The actives from the HTS are further profiled in order to confirm the initial activity generating a series of compounds (actives) with a confirmed dose–response correlation (EC_{50}) for the medicinal chemist to evaluate and optimize.

HTS is still seen by many in the pharmaceutical industry as the optimal way to discover new chemical hits for both new and previously explored pharmaceutical targets (6). However, corporate collections are usually composed of a mixture of historical project compounds, natural products and in-house synthesized or acquired compound libraries and, as a consequence, much effort must be expended to ensure that the quality and integrity of the compounds within the screening collection is maintained. A fundamental concern in this lead generation approach is the high initial hit rates that are frequently observed, resulting from many compounds acting through an undesired mechanism. These include compounds that are non-selective (7), reactive (8) or interfere with the

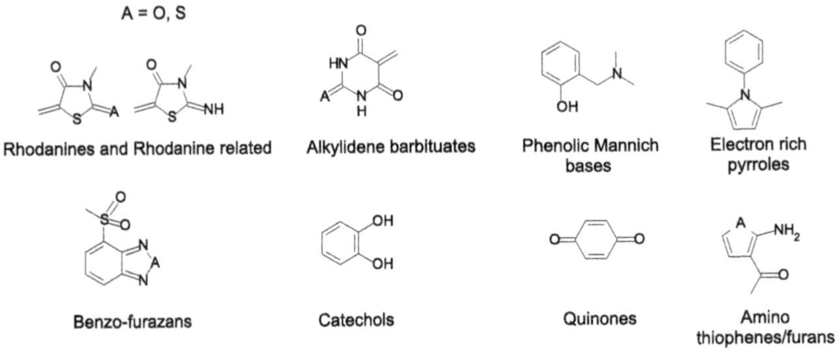

FIGURE 3.4

A selection of common pan assay interference compounds (PAINS).

assay technology through either poor solubility (*9*) or aggregation (*10*). These compounds are called frequent and/or false hitters and there have been a number of studies trying to elucidate the structural features that would allow identification of these frequent and false hitters before significant effort has been expended optimizing the compounds. Recently Baell (*11*) suggested a list of filters that could be used to remove 'pan assay interference compounds (PAINS)' from screening collections. Typically these compound classes contain functionality that either interferes with the screening technology (either as the parent compound or an oxidized intermediate) or interacts non-selectively with the biological target. Some of the most common classes are shown in Figure 3.4 and every effort must be used to identify and remove these compounds from consideration as early as possible.

3.2.3 Structure-based drug design and virtual screening

VS has become an integral part of the drug discovery process (*12*) and is a computational technique to identify those structures which are most likely to bind to a biological target, typically a protein, a receptor or an enzyme. VS can be defined as the automatic evaluation and filtering of very large libraries of compounds (either commercially available or virtually generated from enumerated library ideas) to a manageable number of compounds that can be purchased or readily synthesized and subsequently tested in a biological assay to confirm activity. The aim of VS is to identify molecules of new or under-represented chemical structure that have potential to bind to the biological target of interest. The added benefit of VS is the potential for finding interesting new scaffolds (the core of the molecule from which reagents are attached); therefore low hit rates of interesting scaffolds are preferable to high hit rates of known compound scaffolds. VS can be broadly categorized into two screening techniques: structure-based and ligand-based.

Structure-based VS involves the 'docking' (*13*) of energy-minimized structures into a biological target protein resulting in a scoring function to give an estimate of the likelihood that the ligand will bind to the protein. This technique requires information about the active site of the target and is therefore only applicable where structural information of the target is known (*14*). A virtual screen of a sub-selection of the AstraZeneca compound collection was performed for checkpoint kinase-1

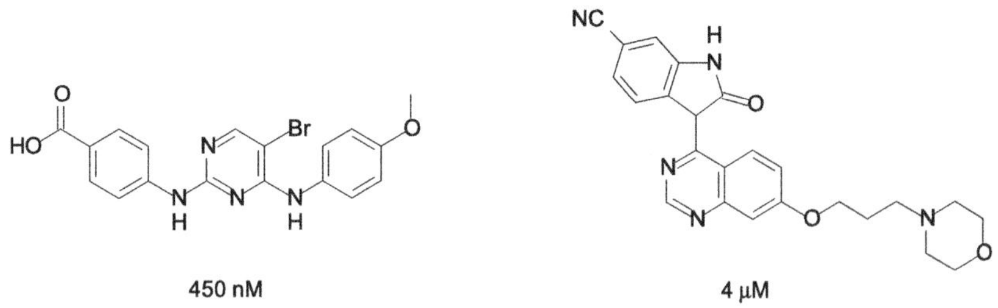

FIGURE 3.5

A selection of hits obtained from the virtual screen on Chk-1 kinase.

(Chk-1 kinase) (15). After filtering of the compound collection by application of generic physical property criteria in order to remove compounds with undesirable chemical functionality, a 3-D pharmacophore screen for compounds with kinase-binding motifs was applied. From this process, a database of approximately 200 K compounds remained for docking into the active site of Chk-1 kinase. After a visual inspection of the top scoring compounds, 103 compounds were ordered for testing in the project assay and from these 36 were found to inhibit the enzyme in a dose–response fashion with IC_{50} values ranging from 110 nM to 68 μM (Figure 3.5).

Ligand-based VS utilizes a series of structurally diverse compounds that are known to bind to a given target protein. Utilizing modelling programmes, the minimized 3-D representation of the molecules are overlaid, guided by atom and/or pharmacophore (16) matching, for example, hydrogen-bonding groups or ring centroids (12). An illustration of this concept is shown (Figure 3.6) between Morphine and Methadone, which are structurally different but present the same pharmacophoric elements for the micro-opioid receptor.

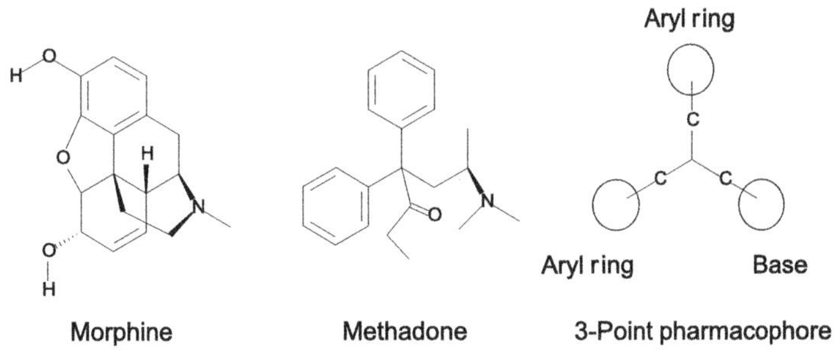

FIGURE 3.6

Structurally diverse morphine and methadone displaying a common pharmacophore model.

From these energy-minimized structures, a pharmacophore model can be built that overlays common structural feature (such as H-bond donor/acceptors, anion, cation and lipophilic groups) within the diverse set of compounds. From this 3-D pharmacophore model, compounds can be virtually profiled to determine those that are compatible with the emerging pharmacophore model and therefore have the potential to bind in the target protein. If there is a proposed bioactive conformation either from a protein–ligand crystal structure or a rigid analogue, overlays may suggest profitable conformations or core group replacements, that could be explored, or ways to stabilize the bioactive conformations. Once generated, the pharmacophore model can be used to search compound databases to look for new structures that match the pharmacophore. The new compound set is screened in a relevant biological assay and the resulting confirmed active compounds are further explored and optimized by the medicinal chemist working in conjunction with the computational chemist (Figure 3.7).

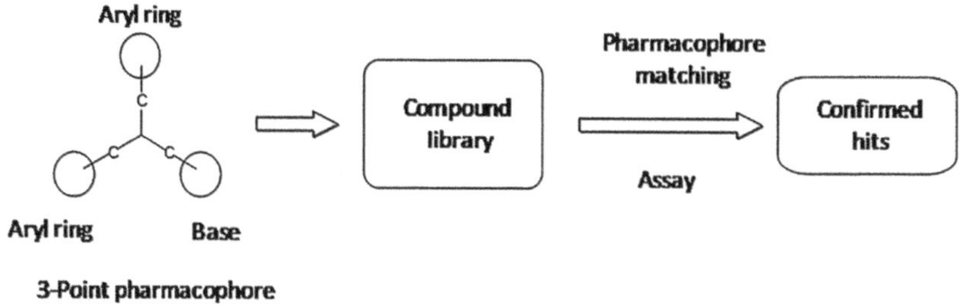

FIGURE 3.7

Scheme showing pharmacophore searching for new hits.

Both the above methods have had major roles in the development of 'targeted library design' (*17*) where virtual compound libraries are designed, based on their potential to interact with a given biological target. In this successful lead generation technique, new or under-represented chemical scaffolds are designed to contain synthetic handles to enable chemical manipulation for the incorporation of diverse sets of reagents containing functional groups rich in the required pharmacophores. An example of this approach was shown in the search for antagonists of the $CXCR_4$ receptor (*18*). In this study, AMD3100, one of the earliest and most potent $CXCR_4$ antagonists to be developed, was used as a reference ligand from which a combinatorial library was derived based on consideration of the pharmacophore requirements for activity. From the virtual screen, predicted $CXCR_4$ potency was used to select compounds for synthesis resulting in several potent hits (Figure 3.8).

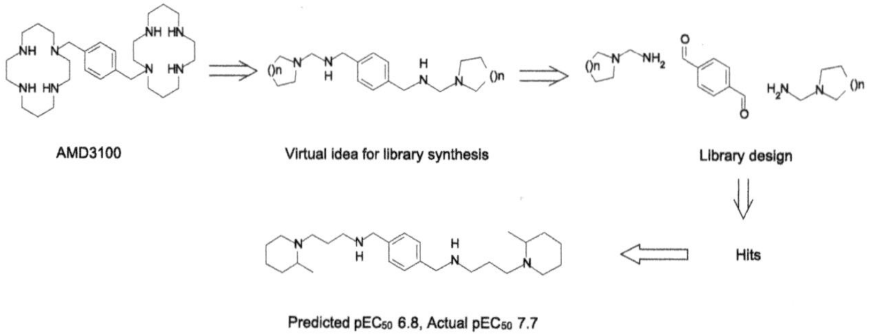

Predicted pEC_{50} 6.8, Actual pEC_{50} 7.7

FIGURE 3.8

Schematic showing the design of a new series of $CXCR_4$ antagonists from AMD3100.

3.2.4 **Natural products**

A natural product is a chemical compound produced by a living organism either directly or by metabolism of another chemical. Natural products have emerged from human (endogenous ligands such as neurotransmitters, hormones and enzyme substrates), animal (venoms and toxins), flora (plants and trees), micro-organism (metabolites) and, more recently, marine (fish and sponge) origins (Figure 3.9).

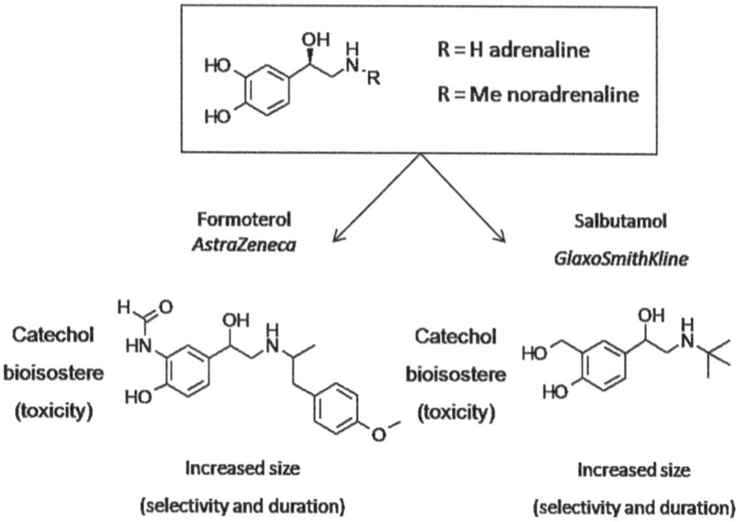

FIGURE 3.9

Drugs derived from natural transmitters adrenaline and noradrenaline.

Natural products are found in nature and are often of great benefit in drug discovery (*19*) as they can offer highly potent tool compounds to probe and validate biological mechanisms as well as offering diverse 3-D chemical structures to kick-start drug discovery programmes with highly potent and often highly selective compounds. Interestingly, a review of the recent FDA list of approvals (*20*) illustrates that these small molecules have provided inspiration for a large proportion of FDA-approved agents and continue to be one of the major sources of stimulation for successful drug discovery.

With the advent of modern drug discovery, where HTS of large targeted compound libraries is becoming the norm, natural products are seldom considered as a drug itself owing to potential unwanted side effects (selectivity), toxicity, poor physical and pharmacodynamic (PD) properties (e.g. solubility or/and high molecular weight (MW) leading to poor absorption) or simply due to supply restrictions and their structural complexity, making rapid progress difficult. However, many successful drug discovery programmes have emerged from natural products and have provided some of the top-selling drugs on the market (Figure 3.10).

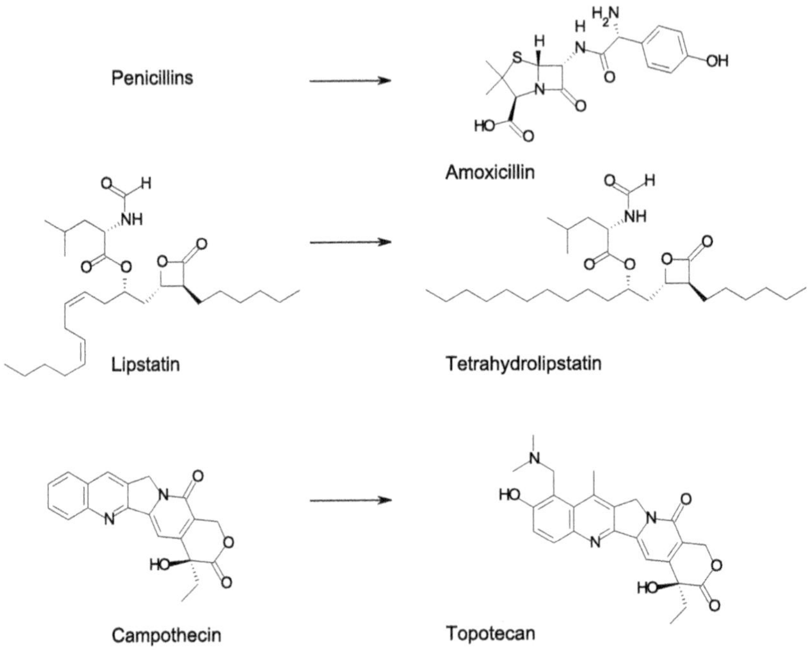

FIGURE 3.10

Amoxicillin is a widely prescribed antibiotic. Tetrahydrolipstatin is an inhibitor of pancreatic lipase used for the treatment of obesity. Topotecan is a topoisomerase inhibitor developed for the treatment of lung, ovarian and colorectal cancers.

In particular, these compounds are important in the treatment of otherwise intractable life-threatening conditions such as organ transplantation (e.g. FK506) and cancer therapy (e.g. taxol) where there are no known small molecule leads.

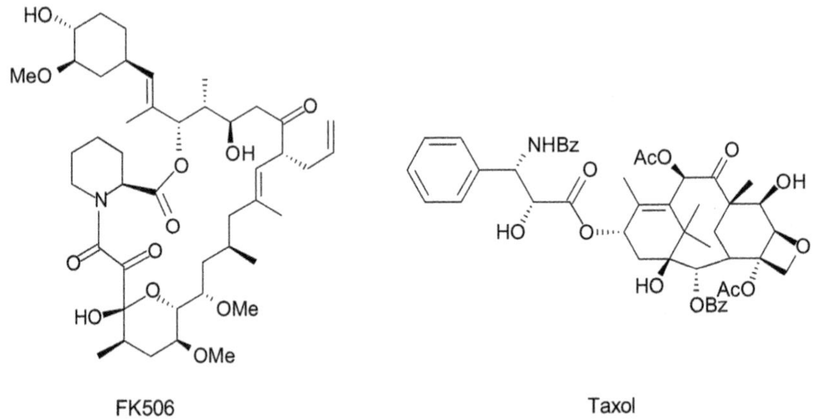

Fragment-based drug discovery – linking fragments

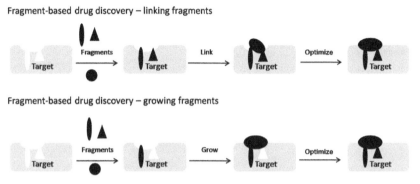

Fragment-based drug discovery – growing fragments

FIGURE 3.11

Schematic showing the concept of linking or growing fragment hits.

3.2.5 Fragment-based drug discovery

Fragment-based drug discovery (FBDD) (*21*) is based on identifying small chemical structure (fragments), which may bind only weakly (at millimolar concentration) to the biological target. The fragments are generally screened at high concentration in a relevant biological assay to confirm specific binding and the fragments are then either 'linked' or 'grown' to produce a lead with a higher affinity (Figure 3.11).

Growing fragments is the stepwise addition of substituents to the fragment core in order to target possible interaction points in the surrounding binding site of the target protein. Growing fragments is the most widely used approach as the other linking approach requires that two fragments bind to adjacent or close by sites on the target protein.

An example of fragment growing was used in the search for phosphoinositide-dependent kinase-1 (PDK1) inhibitors (useful in anti-cancer indications) where a weak aminoindazole fragment was successfully grown with the aid of X-ray crystallography to afford a new series of potent PDK1 inhibitors (Scheme 3.1) (*22*). In this work the composition of the initial fragment library used was biased towards molecules with donor and/or acceptor motifs embedded in an aromatic ring, which could fill the flat lipophilic adenine pocket of ATP (*23*). The library consisted of 1065 fragments which were filtered using a biochemical PDK1 assay at high concentration to afford a selection of 193 compounds for IC_{50} determination. From this, 89 compounds with an $IC_{50} < 400\ \mu M$ were chosen for further study. X-ray crystal soaking experiments were performed with a selection of fragment hits and from these studies aminoindazole proved the most promising. A search based on the aminoindazole structure, with a focus on identifying low MW compounds (<300 Da) as elaborated versions of the actual fragment, identified further analogues. This led to the discovery of the aminopyrimidine derivative, which is a lead-like derivative of the aminoindazole fragment hit (Scheme 3.1).

An example of linking fragments was shown in the identification of inhibitors of *Mycobacterium tuberculosis* pantothenate synthetase (PS) (*24*). In this approach, two weak inhibitors were shown to bind simultaneously at the active site in PS when soaked as a cocktail into crystals of PS, and the compounds were linked successfully through a spacer group to afford a set of inhibitors with enhanced activity (Scheme 3.2).

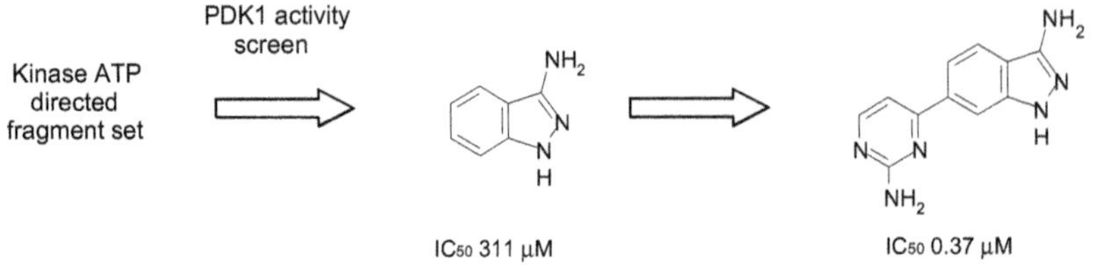

SCHEME 3.1

Growing the aminoindazole fragment hit.

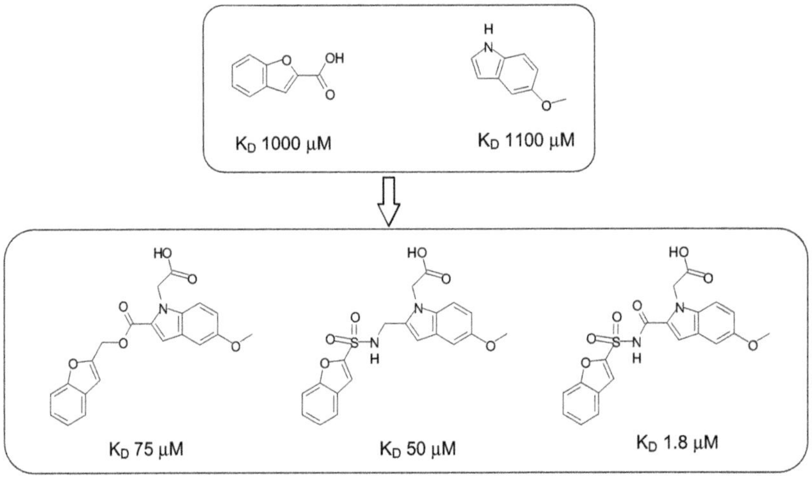

SCHEME 3.2

Fragment linking of two weak hits to afford a series of potent compounds.

In another example of fragment linking (25), two fragments were shown to bind in adjacent pockets of thrombin and they were linked through a –CH$_2$– group and further optimized to afford a series of potent inhibitors (Scheme 3.3).

In the early phase of FBDD, libraries with a few hundred (or thousand) compounds with MWs of around 200–300 Da are screened, and millimolar affinities can be considered useful (26). Since the fragments have relatively low affinity for their targets, they must have high water solubility so that they can be screened at higher concentrations and, as a consequence, it has been proposed that ideal fragments should follow the 'rule of three' (MW < 300, clog P < 3, and the number of hydrogen bond donors and acceptors each should be <3) (27). The small size of the fragment starting points means that they form limited contacts with their biological target and therefore demonstrate relatively low affinity for the target. As a consequence of the low binding affinity, assays need to be able to detect this low affinity and frequently rely on biophysical techniques, such as X-ray crystallography (28), nuclear magnetic resonance (29) or isothermal titration calorimetry (30). The counterpart of the low binding

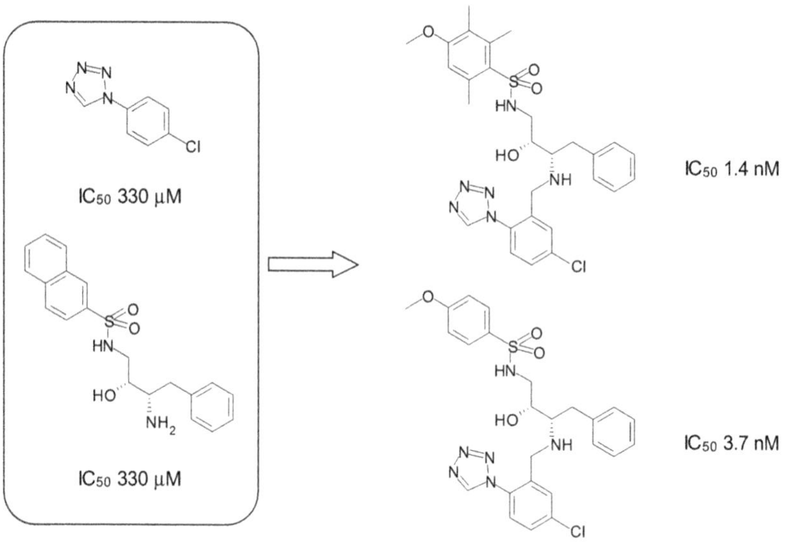

SCHEME 3.3

Linking of two fragments through the incorporation of a $-CH_2-$ group.

affinities observed with a fragment screen is that they by default have a greater chance to form favourable enthalpically driven binding interactions with the receptor-binding site in comparison to larger molecules. As a consequence, fragment screens tend to yield relatively high hit rates. Therefore, as with all hit-finding activities, it is important to confirm binding to the target by more than one method.

As highlighted, there are several challenges in FBDD (*31*), however the benefits of constructing leads (and beyond) from FBDD approaches are becoming more important (*32*) and several CDs have now progressed into clinical evaluation from FBDD, with Indeglitazar (a peroxisome proliferator-activated receptor (PPAR) pan agonist from Plexxikon) being the first drug to receive FDA approval from an FBDD approach (*33*). In the work, a selective pan agonist of all three PPARs, PPARα, PPARγ, and PPARδ, was postulated to be required for use in type II diabetes mellitus. A screening process coupling low-affinity biochemical screening with high-throughput co-crystallography led to a series of compounds that weakly modulated the activities of all three PPARs. From the set, 5-methoxyindole-3-propionic acid possessed weak to barely detectable agonist activity against the three PPARs. However, it was shown from X-ray crystallography that the 5-methoxyindole core bound entirely in one pocket leaving a close second pocket unoccupied. The 5-methoxyindole core was also unique amongst the known PPAR ligands that had been observed in published X-ray crystal structures. Validation of 5-methoxyindole-3-propionic acid as a scaffold for pan-active (i.e. agonists against all the PPARs) compounds was carried out via the synthesis of the *N*-phenyl sulfonamide derivative, which demonstrated significantly improved 100-fold increases in activity and further analogue synthesis afforded Indeglitazar (Scheme 3.4).

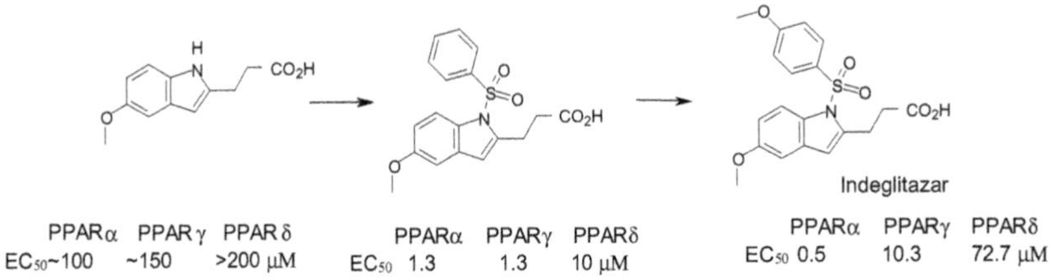

	PPARα	PPARγ	PPARδ
EC$_{50}$	~100	~150	>200 μM

	PPARα	PPARγ	PPARδ
EC$_{50}$	1.3	1.3	10 μM

Indeglitazar

	PPARα	PPARγ	PPARδ
EC$_{50}$	0.5	10.3	72.7 μM

SCHEME 3.4

The discovery of Indeglitazar from an FBDD hit through fragment growing.

3.2.6 Fast-follower approaches

In all commercial areas successful products attract envy from competitors. Fast-follower (pre-launch product) and me-too (drug on the market) approaches are widespread within the industry. Success depends on acquiring maximum information on competitor activity through patent and literature disclosures. Once the validity of a biological target has been confirmed in the clinic and the structure of either the potential CD or compound series has been disclosed, it becomes possible to use this competitor intelligence to start a research programme in the same research area to that of a competitor and it is essential to secure novel intellectual property (IP) by circumventing the scope of the competitor IP. Although this approach has the major advantage that the relevance of the biological target will often have been validated and that the development phase might potentially be more rapid, it is not without risk: the new compound is likely to reach the market place later than the original product, and it will face competition from other 'copies'. Hence, it is important that a fast-follower or me-too drug possesses significant advantages over the original drug (e.g. better potency, improved safety or DMPK profile). A recent example of a fast-follower approach was with the launch of Vardenafil (*34*) from Bayer in 2003, following the launch of Sildenafil by Pfizer in 1998 for erectile dysfunction (*35*) (Scheme 3.5). Interestingly, Pfizer did not claim the core structure present in Vardenafil in their patent application, an observation that was later exploited by Bayer.

Once a series of actives has been obtained, generally from a directed HTS, the next stage of the lead generation process is initiated (Figure 3.12).

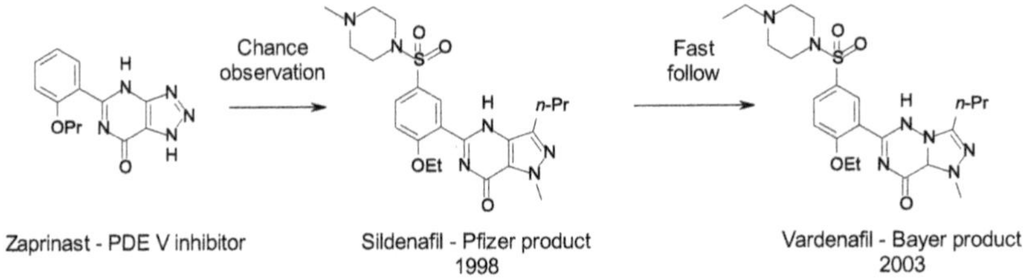

Zaprinast - PDE V inhibitor

Sildenafil - Pfizer product
1998

Vardenafil - Bayer product
2003

SCHEME 3.5

The evolution of Vardenafil from Sildenafil.

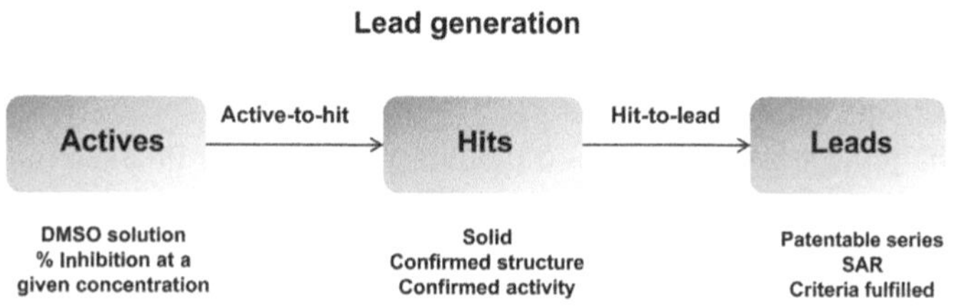

Lead generation

Actives	Active-to-hit →	Hits	Hit-to-lead →	Leads
DMSO solution % Inhibition at a given concentration		Solid Confirmed structure Confirmed activity		Patentable series SAR Criteria fulfilled

FIGURE 3.12

The lead generation process.

3.3 LEAD GENERATION: ACTIVE-TO-HIT

The first phase is called either 'active-to-hit' or 'hit identification' and is concerned with process of extracting maximum value from the HTS providing the upcoming project with the best possible chemical starting points (hits) with known strengths and weaknesses to further exploitation in the hit-to-lead phase.

The active-to-hit workflow is shown (Figure 3.13) and starts with a computational cluster analysis of the actives into groups (or clusters) of compounds with similar structural motifs (*36*). At this stage non-drug-like chemical starting points are removed (PAINS and frequent hitters) and compounds are selected for quality control (for confirmation of compound structure and purity) and follow-on dose response (EC_{50}) determination. At this point, all the testing has been performed on HTS solubilized

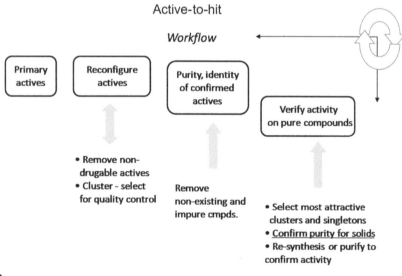

FIGURE 3.13

Active-to-hit workflow.

compound stock and the next stage of the process is to confirm both activity and structure with pure compound derived from a fresh 'solid' sample. This is an essential part of the process as compounds may have degraded in the HTS master plates.

This phase of the project lasts from 3 to 6 months and is critical in signalling the emergence of quality chemical starting points for the next stage of the project and ideally, at least three hit series of different chemical structure should be available for the transition to the hit-to-lead phase.

3.4 LEAD GENERATION: HIT-TO-LEAD

In this stage of the drug discovery process, a consideration of the quality of the compounds is made with respect to both their physicochemical (such as lipophilicity, MW and solubility); drug metabolism (such as metabolic stability, cytochrome P450 (CYP) interaction, transporter efflux potential and the potential to bind to plasma proteins) and potency (whole-cell activity and lipophilic ligand efficiency) requirements.

3.4.1 Lipophilicity

Lipophilicity (fat-liking) is the single most important physical property affecting potency, distribution and elimination of a drug in the body. No other physical property is found to influence so many biological properties as significantly as lipophilicity, and it is consequently the most frequently employed parameter used in drug discovery SAR studies. The property can be readily measured or calculated and lipophilicity has been correlated to many other properties such as solubility, permeability, increases in target potency and toxicity (37). The impact of lipophilicity on drug-like properties can be summarized (Table 3.1) and wherever possible the lipophilicity of potential synthetic targets should be calculated to ensure they fit within the guidelines for oral absorption.

Table 3.1 The Impact of Lipophilicity on Drug-Like Properties

Lipophilicity (Log $D_{7.4}$)	Common Impact on Drug-Like Properties	Common Impact *In vivo*
<1	High solubility Low permeability Low metabolism	Low volume of distribution Low absorption and bioavailability Possible renal clearance
1–3	Moderate solubility Permeability moderate Low metabolism	Balanced volume of distribution Potential for good absorption and bioavailability
3–5	Low solubility High permeability Moderate to high metabolism	Variable oral absorption
>5	Poor solubility High permeability High metabolism	Very high volume of distribution Poor oral absorption

3.4.2 Solubility

Solubility is clearly a very important physicochemical property, as a compound cannot exhibit any useful biological effect if it is totally insoluble in water. It is very important, therefore, to continuously measure the aqueous solubility of the lead series of compounds during the drug discovery process. The solubility of a drug generally increases with a decrease in lipophilicity, but other factors such as hydrogen bonding and crystal packing are also important. The solubility of ionizable compounds (acids or bases) will also increase with increasing the extent of ionization, and solubility is therefore also dependent on pH. As a guideline, aqueous solubility for a potential drug substance should be a minimum of 5 μg/ml but ideally >50 μg/ml. There are several methods by which the solubility of a series of compounds may be attenuated. Lowering the melting point is a potential indicator for, in order to dissolve into a solvent, the compound must first be removed from its crystal lattice. Melting points can be lowered by reducing symmetry, removing planarity and introducing steric effects into the molecules (Figure 3.14).

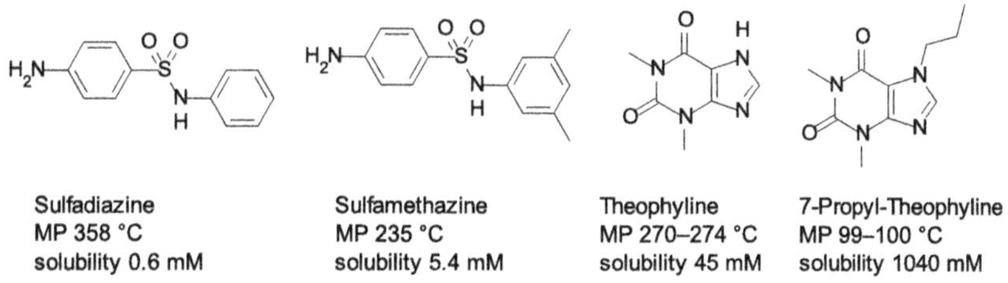

Sulfadiazine
MP 358 °C
solubility 0.6 mM

Sulfamethazine
MP 235 °C
solubility 5.4 mM

Theophyline
MP 270–274 °C
solubility 45 mM

7-Propyl-Theophyline
MP 99–100 °C
solubility 1040 mM

FIGURE 3.14

Examples of increasing solubility.

Further methods to address solubility in drug discovery occur through the incorporation of a solubilizing group into the core structure within the series that either has a beneficial interaction with the binding protein or occupies an aqueous environment either within or outside of the active site. Iressa (*38*) is a potent EGFR(epidermal growth factor receptor)-tyrosine kinase developed from the poorly soluble but highly potent screening hit shown (Figure 3.15).

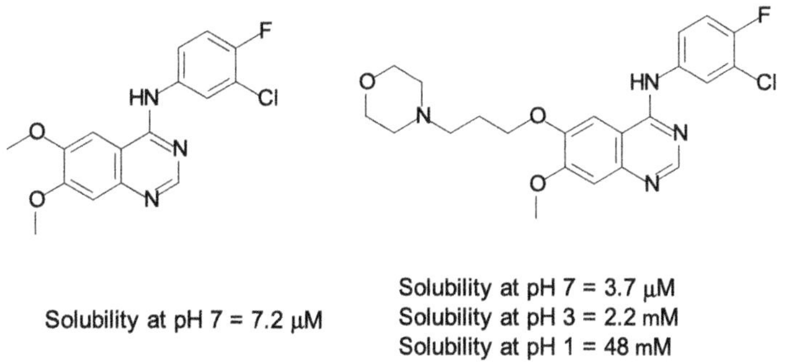

Solubility at pH 7 = 7.2 μM

Solubility at pH 7 = 3.7 μM
Solubility at pH 3 = 2.2 mM
Solubility at pH 1 = 48 mM

FIGURE 3.15

Increase in acidic solubility through the incorporation of a basic solubilizing group.

In the case of an acidic or basic compound, the solubility of the compound can generally be altered by the generation of a salt(s), which improves properties without changing any of the beneficial *in vivo* properties of the parent compound. The ideal properties of a suitable salt are that it has good chemical and solid-state stability, is non-hygroscopic, and has an appropriate dissolution profile, good bioavailability and low counter-ion toxicity. When making a salt, the pKa of the compound is determined to establish if it is possible to form a conjugate acid/base from the parent molecule. The pKa of the acidic and the basic components should differ sufficiently so that it is possible to make a stable salt and as a general rule, a difference of at least 2 pKa units is considered desirable. When considering salt selection it is very important that the counter-ion should not be toxic at the intended dose.

3.4.3 DMPK considerations

Definitions:
- PD is the study of the pharmacological response to the compound, i.e. what the compound does to the body.
- PK is the study of the movement of the compound within the body (absorption, distribution, metabolism & elimination), i.e. what the body does with the compound.
- Absorption is the process by which a compound moves from its site of administration into the blood.
- Distribution is the reversible transfer of a drug, to and away from the blood into tissue.
- Metabolism is the chemical alteration of the compound by the body.
- Elimination is the irreversible excretion of a compound from the blood.

Drug discovery scientists need to appreciate that the ideal drug candidate should possess both good pharmacological activity as well as good PK properties. Without the latter, the efficacy of a highly potent molecule may be short-lived, resulting in the clinical benefits being dramatically reduced. Between 2001 and 2010 the importance of early *in vitro/in vivo* PK screening has become well recognized within the pharmaceutical industry and greater numbers of *in vitro/in vivo* studies are performed at an earlier stage of the drug discovery process in an effort to understand and optimize the PK properties of compounds from the hit stage onwards. It is well recognized that the core of the molecule seldom changes from the initial hit through to the CD, and any problems within the core need to be identified and altered as early as possible in the drug discovery process. As a result of this early upfront loading of DMPK evaluation, attrition due to poor drug PK has been greatly reduced (*39*).

Whilst there are many routes for the administration of drugs, the majority of small molecule compounds are administered orally in a once- or twice-daily tablet form. After oral administration, for the compound to be absorbed from the gastrointestinal tract into the systemic circulation, the compound must first be in solution; only then will it be able to cross the intestinal epithelia and enter the portal vein. Absorption usually occurs by passive diffusion, however the process can be perturbed by slow dissolution, or if absorption and/or dissolution is slow, by the transit time of the gut contents within the small intestine. The compound is absorbed from the gut into the systemic circulation via the hepatic portal vein and then enters the liver where the majority of primary metabolic (first pass) processes occur. The remaining compound then undergoes distribution into the tissues and is located to the site of action (Figure 3.16).

The enzymatic chemical alteration of the compound by the body serves as a primary defence mechanism by which the body attempts to avoid exposure to foreign substances.

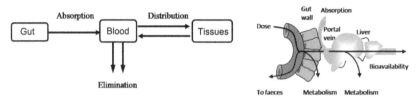

FIGURE 3.16

The drug absorption process and factors affecting a compound's absorption.

There are two types of metabolism as follows:

- Phase I metabolism, which involves the introduction of a new chemical group within the compound by a biochemical reaction such as oxidative processes, e.g. aliphatic or aromatic oxidation, nitrogen or sulfur oxidation and nitrogen, oxygen or sulfur dealkylation; reductive processes, e.g. nitro reduction to hydroxylamine/amine or carbonyl reduction to the corresponding alcohol; hydrolysis, e.g. esters or amides to the corresponding acids and alcohols and amines; or hydrazides to the corresponding acids and substitutes hydrazines. Many Phase I oxidative transformations are carried out by a group of enzymes known as CYPs. CYPs are membrane-bound proteins which are found on the endoplasmic reticulum of cells. The endoplasmic reticulum is a cellular transport system composed of a honeycomb of membrane which pervades the entire cytoplasm. CYPs are haeme-containing proteins – i.e. their active site contains a porphyrin ring co-ordinated to iron. The active species comprises an oxoiron (IV) porphyrin cation radical species (Figure 3.17).

Examples of Phase I processes are shown (Figure 3.18).

- Phase II involves the addition of an endogenous molecule to the compound through glucuronidation (carboxylic acids, alcohols, phenols and amines); sulfation (alcohols, phenols and amines); acetylation (amine); amino acid conjugation (carboxylic acids) or glutathione conjugation (Figure 3.19).

Both Phase I and Phase II routes of metabolism convert compounds to more hydrophilic metabolites, which are subsequently excreted by the body. In a drug discovery project, primary *in vitro* assays

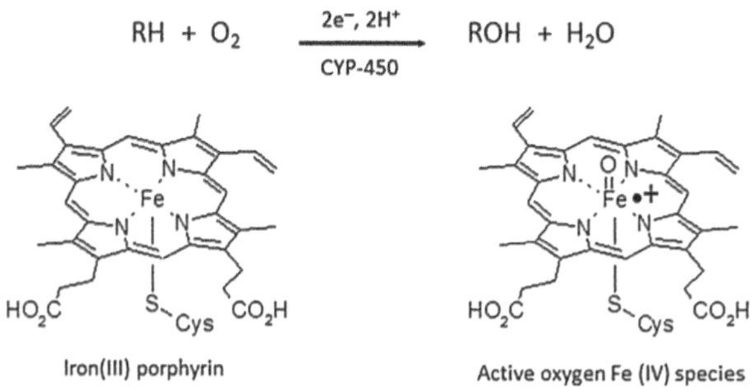

$$RH + O_2 \xrightarrow[\text{CYP-450}]{2e^-, 2H^+} ROH + H_2O$$

Iron(III) porphyrin   Active oxygen Fe (IV) species

FIGURE 3.17

Structure of haeme-containing P450 enzymes.

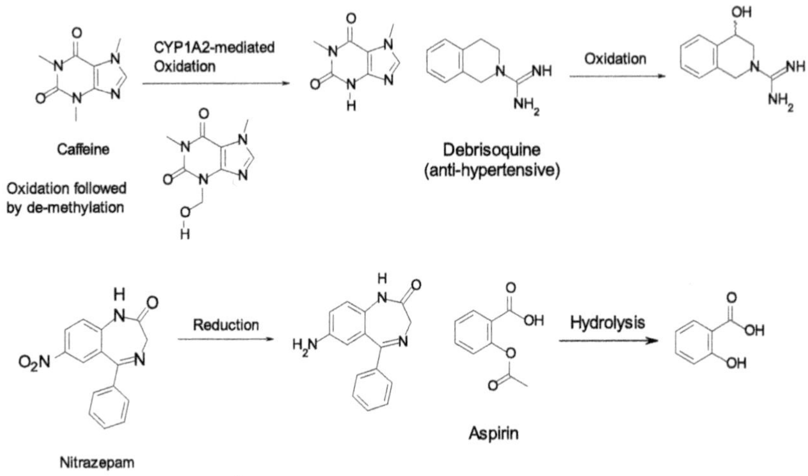

FIGURE 3.18

Examples of Phase I metabolism showing the oxidation of caffeine and debrisoquine, the reduction of nitrazepam and the hydrolysis of aspirin.

used to predict the metabolic fate of a compound are run routinely and act as a predictor of the possible metabolism of the compound after dosing. There are two types of *in vitro* assays that are commonly employed using cells extracted from the liver. The first utilizes microsomes, which are a sub-cellular fraction obtained from centrifugation of liver cells and perform Phase I metabolic processes. The second involves the use of hepatocytes, which are isolated whole liver cells and are capable of performing both Phase I and Phase II metabolism. The resulting rates of metabolism are reported as an intrinsic clearance and can be extrapolated to an *in vivo* prediction of the possible rate of elimination from the body (*40*). Table 3.2 can act as a guide to the extrapolated clearance (Cl) of a compound from *in vitro* assays.

Once confident that a compound has acceptable *in vitro* stability, detailed *in vivo* studies are performed in a rodent species. Quantitative measurement of PK properties enables the comparison of potential drug candidates and facilitates the improvement of their *in vivo* properties.

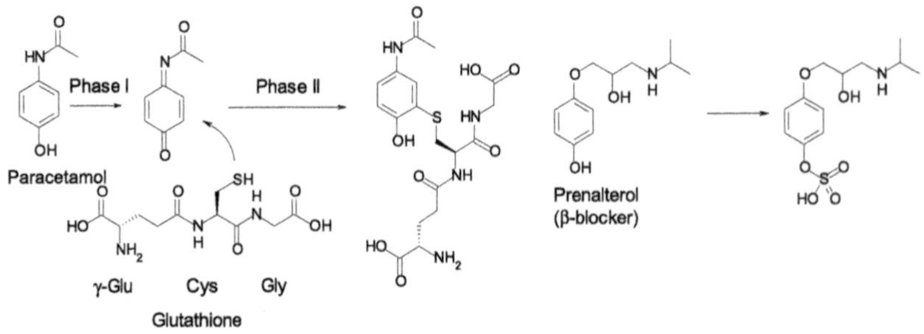

FIGURE 3.19

An example of glutathione Phase II metabolism of paracetamol and the sulfation of prenalterol.

Table 3.2 Clint = Intrinsic Clearance. Hepatocytes (1×10^6 cells/ml): Scaling *In vitro: In vivo*

Species	Rat	Dog	Human	Predicted Percentage Liver Blood Flow
Clint*	Predicted Cl†	Predicted Cl†	Predicted Cl†	Human
69	58	37	18	90
35	50	35	16	80
23	44	33	15	75
17	39	31	14	70
14	35	29	13	65
9	28	26	11	55
7	23	23	9	45
5	17	19	7	35
3	14	16	6	30
2	10	12	5	25

*Units $\mu l/min/10^6$ cells.
†ml/min/kg. The rates of blood flow for the species quoted are rat (70 ml/min/kg), dog (40 ml/min/kg) and human (20 ml/min/kg). Cl = clearance.

Before a discussion of the determination and likely significance of quantitative PK measurements, a few definitions are required:

- Oral bioavailability (F) is the fraction of the dose of drug given orally that is absorbed and survives into the general systemic circulation (i.e. survives first pass metabolism).
- Half-life ($T_{1/2}$) is the time taken for the concentration of drug in the blood or plasma to decline to half its original value.
- Volume of distribution (V_d) is an estimate of the volume that all the drug in the body would occupy if it were present at the same concentration as that found in the blood or plasma.
- Cl is the volume of blood or plasma in which all drug is removed per unit time.
- Area under the curve (AUC) is the total amount of drug that survives into the system circulation.
- Cmax is the maximum concentration of drug achieved in plasma after intravenous or oral dosing.

In a DMPK experiment, a compound is dosed intravenously into an animal (usually a rat) and a graph of plasma concentration as a function of time is plotted (Figure 3.20).

From this graph, a variety of quantitative measurements can be made, the most important of which is the Cl of the compound. $Cl = \dfrac{\text{Dose}}{\text{AUC}}$, where AUC is the total area under the plasma concentration vs. time curve from time zero to infinity. The half-life ($T_{1/2}$) can also be calculated from the plot of concentration vs. time after *i.v.* administration, by plotting the logarithm of concentration vs. time. The resulting straight-line plot may be used to determine the half-life of the drug from any two points on the plot describing a 50% loss in drug concentration. The apparent volume of distribution (V_d) can also be estimated from the following: $V_d = \dfrac{\text{Amount of drug in the body (D)}}{\text{Drug concentration in plasma (C)}}$ and is determined by many

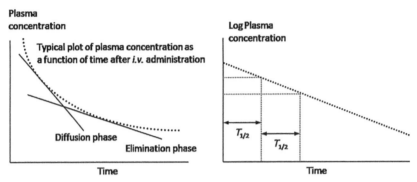

FIGURE 3.20

Plot of plasma concentration as a function of time and estimation of $T_{1/2}$.

factors such as the ability of the compound to distribute into tissues. The rate of elimination (k_e) is a first-order process and is reflected in the terminal half-life of the compound and, as the rate of elimination is a function of the Cl divided by the volume, then the terminal half-life is given by the formula: $k_e = \dfrac{0.693}{t_{1/2}}$, $k_e = \dfrac{\text{Cl}}{V_d}$, and $t_{1/2} = \dfrac{0.693V_d}{\text{Cl}}$. The terminal half-life is directly proportional to the volume of distribution, and inversely proportional to Cl. The terminal half-life of a compound is dependent on both Cl and V_d, therefore, it can be doubled by halving the Cl or by doubling the volume of distribution. Subsequently, another animal is dosed orally and again a graph of plasma concentration as a function of time is plotted (Figure 3.21). From a combination of these two plots several more

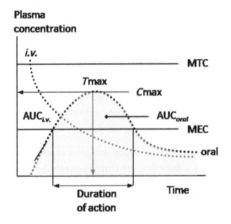

FIGURE 3.21

Plot of plasma concentration as a function of time. The peak time Tmax is the time taken for the drug to reach its peak concentration (Cmax) after the oral dose. The MEC (minimum effective concentration) is the minimum drug concentration needed to achieve a therapeutic effect and the duration of action of the compound is the time spanning the MEC. The maximum tolerated concentration (MTC) is defined as the concentration that would lead to an undesired (toxic) effect.

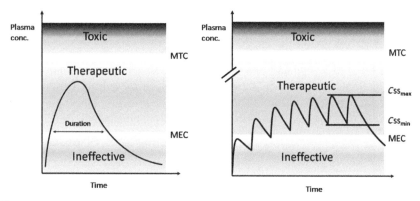

FIGURE 3.22

MTC – maximum tolerated concentration, MEC – minimum effective concentration, Css – steady-state concentration. (For colour version of this figure, the reader is referred to the online version of this book.)

parameters can be calculated, the most important being the compounds' bioavailability (F) calculated by $F = \dfrac{\text{AUC after an oral dose}}{\text{AUC after an equivalent } i.v. \text{ dose}}$.

Most drugs are dosed continuously, either once- or twice-daily, and as a consequence the drug discovery scientist needs to consider the effect of compound levels after multiple dosing (Figure 3.22). The left-hand diagram shows a typical plasma concentration profile over time following an oral dose of a compound. It is important to achieve the concentration of the drug in the blood that is sufficient to give activity at the desired target, to give coverage for a sufficient period of time and to ensure that the drug concentration does not reach levels which might cause side effects or toxicity. The right-hand diagram demonstrates that not all the compounds will have been eliminated from the body by the time of a subsequent dose. Therefore, the concentration of drug will build up over time until it reaches a 'steady state', where the amount eliminated is equal to the amount dosed.

Depending on the physical properties of the compound, two other types of metabolic excretion can occur:

- Renal secretion. As blood flow passes through the kidneys, certain compounds and metabolites are excreted into the urine by glomerular filtration and active secretion by transporters. These result in compound and metabolite elimination, however, renal Cl is generally only significant for hydrophilic, low-MW compounds. Cl specific to the kidneys can be determined by detailed PK studies.
- Biliary excretion, which is sometimes found for higher MW compounds, results in excretion of the compound into the bile. Biliary excretion can be seen when the compound does not scale from *in vitro* to *in vivo* and is illustrated in the case of a series of compounds from the Bristol-Myers Squibb (BMS) endothelin A (ET_A) antagonist project (*41*) (Figure 3.23).

From Figure 3.23, it can be seen that there is a general trend for scaling within the series when rat Cl vs. the rate of microsomal Cl is plotted. Both BMS-187308 and BMS-193884 fit on the correlation whereas a structurally similar compound, BMS-X has a dramatic increase in rat Cl resulting from biliary Cl.

Correlation of in vivo clearance and microsomal rate

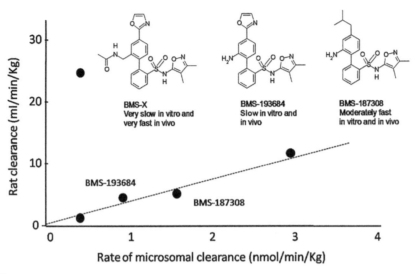

FIGURE 3.23

Rat clearance vs. rate of microsomal metabolism for a set of structurally similar ET_A antagonists.

3.4.4 Transporter proteins

As well as transcellular and paracellular absorption that occurs either into or out of cells, active transport mechanisms can occur which can have a dramatic effect on absorption. Transporters are membrane proteins that carry molecules (such as nutrients, ions and drugs) into and out of cells. Unlike passive diffusion, this process requires energy and can result in compounds being transported either into cells (influx) or out of cells (efflux). A major consequence is that following oral dosing, a compound can be transported back to the intestinal lumen by transporter proteins, such as P-glycoprotein (P-gp) resulting in low bioavailability. There are certain characteristics to molecules that act as P-gp substrates e.g. high MW and lipophilicity as well as amphiphilic compounds with a weak cationic group present (42). An example of structural variation to overcome this issue was demonstrated by Pfizer chemists in their selective neurokinin-2 (NK2) antagonist programme (43), where indications such as irritable bowel syndrome and central nervous system (CNS) disorders could potentially be treated. It was observed that when UK-224,671 was dosed in rats, it showed poor bioavailability and a resulting Caco-2 study (44) showed that the compound was actively transported out of the Caco-2 cells with an efflux ratio of 1/18 (demonstrating the compound possessed high permeability across the cell membrane but was transported out of the cell by a transporter mechanism). Structural variation afforded UK-290,795, which was shown to be a very potent NK2 antagonist. In the Caco-2 assay, UK-290,795 demonstrated very good permeability and was not a substrate for transporter mechanisms (efflux ratio 1:1). UK-224,671 was then tested in a P-gp knockout mouse (a mouse genetically devoid of the P-gp transporter) and in this study the bioavailability of UK-224,671 increased to >20% (Scheme 3.6) (45).

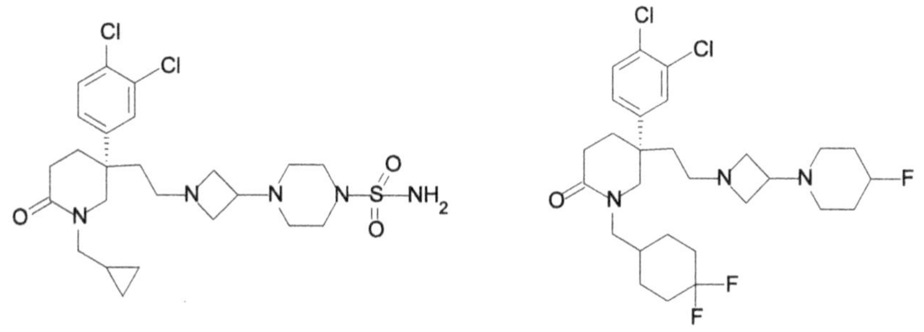

UK-224,671
NK2 pIC$_{50}$ = 8.4
Caco-2 A-B/B-A = 1/18
Rat F = <20%

UK-290,795
NK2 pIC$_{50}$ = 9.4
Caco-2 A-B/B-A = >35/>35
Rat F = >80%

SCHEME 3.6

Addressing P-gp efflux and the evolution of UK-290,795.

3.4.5 Plasma protein binding and whole blood potency

The property of a compound to bind to the blood plasma protein albumin can have a dramatic effect on the *in vivo* efficacy of compounds as it is the free circulating levels of compound that determine efficacy (Figure 3.24).

It is important to note that compounds with high plasma binding are retained in plasma and it is the percent free or fraction unbound that matters such that the difference between 99.9% and 99% (10-fold) is greater than the difference between 90% and 50% (fivefold). As a general rule compounds which are 0–50% bound have negligible effect on *in vivo* (whole blood) potency, 50–90% bound have moderate effect on *in vivo* (whole blood) potency, 90–99% bound have a high effect on *in vivo* (whole blood) potency and >99% bound have a very high effect on *in vivo* (whole blood) potency. The effect of plasma protein binding (PPB) is also affected by lipophilicity and ionization (Figure 3.25) and acidic compounds demonstrate the greatest degree of PPB.

Novartis scientists identified a series of 2-cycloalkylphenoxyacetic acids such as **1** as interesting CRTh2 receptor antagonists (*46*); however the compounds were subsequently shown to have poor functional activity. From a HTS, a new series of 7-azaindole-3-acetic acids **2** were identified that had both good physical and metabolic properties as well as very good potency for the CRTh2 receptor in a titrated prostaglandin D$_2$ (^{3}H-PGD$_2$) radio-ligand-binding assay.

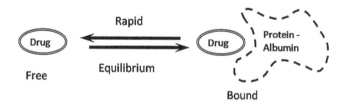

FIGURE 3.24

Drug binding to protein albumin.

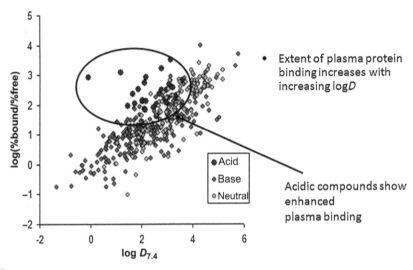

FIGURE 3.25

The effect of plasma protein binding as a function of ionization and lipophilicity.

However, given the potential for carboxylic acids to impact on potency through increased PPB (47), the biological assay was spiked with 0.1% human serum albumin (HSA) as a measure of PPB. Eventually, the assay was performed in more physiologically relevant human whole blood with excellent correlation. From a series of analogues, **3** was identified as a compound with greatly improved functional potency in whole blood, despite similar binding and reduced functional cyclic AMP(cAMP) activity (Table 3.3). From the results **1** (PPB 99.4% bound) had a 24-fold potency shift in the CRTh2 cAMP + HSA assay compared to **3** (PPB 93.4% bound), that had a 3.4-fold shift in activity. The results demonstrate that the reduction of PPB was a key factor in determining whole blood potency.

Table 3.3 Comparison of 7-Azaindole-3-Acetic Acid Compound **3** with Previous 2-Cycloalkylphenoxyacetic Acid Compound **1**. SC = Shape Change. cAMP is a Second Messenger that Amplifies Signal and is Important in Many Biological Processes. cAMP is Derived from Adenosine Triphosphate (ATP) and Used for Intracellular Signal Transduction in Many Different Organisms

Assay	Compound 1	Compound 3
CRTh2 binding K_i (µM)	0.059	0.052
CRTh2 cAMP IC_{50} (µM)	0.148	0.354
CRTh2 cAMP + HSA potency shift	24	3.4
Eosinophil SC IC_{50} (µM)	0.248	0.112
HWB SC IC_{50} (µM)	>10	0.629
Human PPB (bound)	99.4%	93.4%

3.4.6 Inhibition and induction of CYP enzymes

CYP proteins are drug metabolizing enzymes and are membrane-bound enzymes found embedded either in the endoplasmic reticulum of a liver cell or in the gut wall. They contain a haeme group and were originally discovered in rat liver microsomes. The P450 proteins are categorized into families and amino acid subfamilies by their sequence similarities. There are now more than 2500 CYP sequences known, however, there are five main CYPs that the medicinal chemist needs to be aware of CYPs 1A2, 2C9, 2C19, 2D6 and 3A4 (Table 3.4).

It has become increasingly important to screen for either CYP inhibition or CYP induction early on in a drug development programme as, if two co-administered drugs use the same metabolic pathway, they can affect one another (drug–drug interactions). In general, a P450 substrate is any drug that uses the CYP enzyme pathway for its metabolism, and P450 inhibitors generally increase concentrations of other drugs whereas P450 inducers generally decrease concentrations of other drugs. Inhibition of P450 enzymes can lead to the reduction in the rate of metabolism of co-administered drugs and can result in an unpredictable increase in drug levels, which can have undesired side effects. As a general rule, the compound should not inhibit the CYPs at a concentration of <10 µM. Terfenadine was

Table 3.4 The Major Isoforms of CYPs P450

Isoform	Percentage of Total CYP	Substrate Examples	Substrate SAR
1A2	<0.1 (µM)	Caffeine	Planar, moderately basic
2C9	<1 (µM)	Ibuprofen	Weakly acidic, inhibition has caused drug–drug interactions
2C19	<15	Imipramine	Neutral or weakly basic, 2/3 H-bonding groups
2D6	<30	Imipramine	H-bond acceptor (basic) 5 Å from site of metabolism
3A4	>10	Terfenadine	Most promiscuous and will oxidize most substrates. Present also in intestine

a widely used antihistamine and is metabolized by CYP3A4; however it was shown that the metabolism of terfenadine to the active metabolite fexofenadine was inhibited by either grapefruit juice or the antifungal compound ketoconazole (a potent CYP3A4 inhibitor) (48). This led to an increase in the amount of terfenadine resulting ultimately in the potential of cardiac effects, through inhibition of the human ether-a-go-go related gene (hERG) (Section 3.5.4) potassium ion channel (Scheme 3.7).

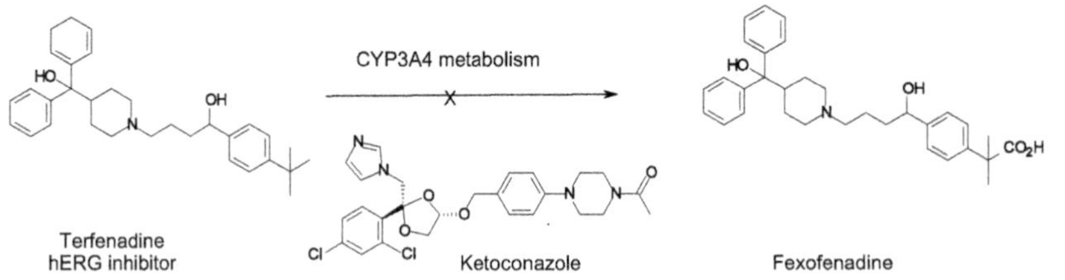

CYP3A4 metabolism

Terfenadine
hERG inhibitor

Ketoconazole

Fexofenadine

SCHEME 3.7

Inhibition of CYP3A4 metabolism by ketoconazole.

3.4.7 Time-dependent CYP inhibition

Time-dependent inhibition (TDI) of CYPs refers to a change in potency either during an *in vitro* incubation or dosing period *in vivo*. Mechanism-based inactivation, an unusual occurrence with most enzymes is observed more often in reactions catalysed by CYPs. The *in vivo* onset of inhibition by time-dependent CYP inhibitors occurs more slowly than with reversible inhibitors and the final effect is generally more profound with inhibition being reversed only by synthesis of new protein (49). The general mechanism of TDI is where the unreactive compound, transformed by normal enzyme catalytic machinery to a species that, prior to release from the active site, inactivates the enzyme (Figure 3.26).

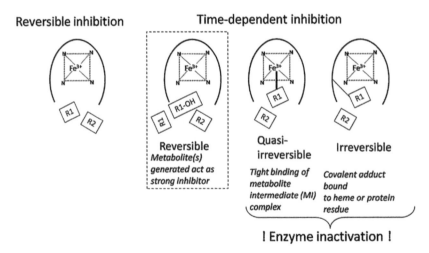

FIGURE 3.26

The mechanism of time-dependent inhibition (TDI).

Table 3.5 Time-Dependent Inhibition Changes of **4** and **5** Using CYP3A4 and BFC (IC_{50}, μM)

Compound	5 min	15 min	30 min	45 min	
4	96	62	33	22	
5	18	21	19	19	4 R = H, 5 R = F

There are several common functional groups that can lead to TDI and these are terminal acetylenes, terminal alkenes, furans and thiophenes, epoxides, secondary and tertiary amines, methylene dioxy compounds, isothiocyanates, thioamides and dithiocarbamates but TDI cannot be predicted from structure alone (*50*). Wu reported that the formation of a reactive intermediate was found to be responsible for CYP3A4 metabolism-dependent inhibition (MDI) in a series of Kv7.2 (a voltage-gated potassium ion channel most probably important in the regulation of neuronal excitability) channel openers. The data suggested that compound **4** is metabolized by CYP3A4 to form a reactive intermediate. This intermediate appears to be restricted to the CYP3A4 active site and covalently binds to the CYP3A4 enzyme, leading to irreversible enzyme inactivation. Extensive structure–3A4 MDI relationship studies resulted in the discovery that the difluoro analogue **5** was both orally bioavailable and free of CYP3A4 MDI (Table 3.5) (*51*).

3.4.8 Ligand lipophilicity efficiency – identifying and working at the 'leading edge'

Ligand lipophilicity efficiency (LLE) is a parameter used in drug discovery to evaluate the quality of research compounds through linking potency and lipophilicity in an attempt to estimate drug-likeness and is defined as the pIC_{50} (or pEC_{50}) of interest minus the log P of the compound (*52*). It is becoming an ever more important measure as it captures both values in a single parameter, and empirical evidence suggests that drug candidates have a high LLE (>6); i.e. a compound with a pIC_{50} of 8 and a clog P of 2. Plotting lipophilicity (clog P) against pIC_{50} for a range of compounds allows a series of compounds to be profiled. The compounds on the leading edge of the plot demonstrate those most interesting (which are not necessarily the most potent) based on potency considerations alone (Figure 3.27).

3.4.9 Hit and lead criteria

At the start of a lead identification (LI) programme the minimum target profile of chemical and biological parameters is defined. An example of such a profile, for an oral project, is illustrated in Table 3.6. The potency target is derived from a combination of factors that reflect the requirement for low dose of compound and selectivity against competing mechanisms. The *in vitro* and *in vivo* DMPK targets have been correlated with Cls of approximately half liver blood flow, detectable oral bioavailability and a measurable free fraction in plasma. In this way, lead compounds should give enough oral exposure to allow an *in vivo* hypothesis test of the target mechanism of action, and provide a solid foundation for an LO project. The physicochemical parameters are a reflection of the fact that lipophilicity and MW generally increase in LO programmes and therefore, to enable further

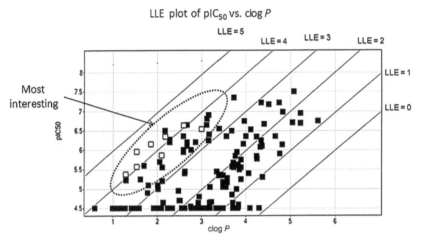

FIGURE 3.27

LLE plot showing the most interesting compounds to explore on the leading edge of activity. For a lead the LLE should be ideally be >4, for a CD LLE is generally >6.

optimization, they are lower than the MW (<500) and lipophilicity (log P < 5) limits recommended by Lipinski for oral drugs (53). Shortly after the introduction of Lipinski's rule of 5, that states that poor oral absorption is likely if a drug candidate has any of the following properties: the MW is over 500; the log P is over 5; there are more than five H-bond donors (sum of OH's and NH's) and there are more

Table 3.6 Example of a Lead Target Profile

Discipline	Parameter	Figure (Unit)
Biology	Binding IC_{50}	<0.1 (μM)
	Functional IC_{50}	<1 (μM)
	Selectivity against other targets	>50 fold
DMPK	Rat hepatocytes Cl	<15 (μl/min/10^6 cells)
	Human microsomes Cl	<30 (μl/min/mg)
	Rat iv Cl	<35 (ml/min/kg)
	Rat iv Vss	>0.5 (l/kg)
	Rat iv $t_{1/2}$	>0.5 (h)
	F	>10 (%)
	PPB	<99.0% (bound)
	P450 inhibition IC_{50}	>10 (μM)
	Caco2 Papp × 10^{-6}	>1 (cm/s)
Physicochemistry	Molecular weight	<450 (g/mol)
	Solubility	>10 (μg/ml)
	Calculated log P	<3.0

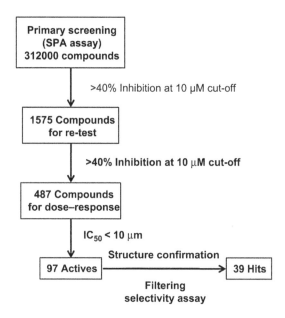

FIGURE 3.28

Summary of the HTS output for IKK-2.

than 10 H-bond acceptors (sum of N's and O's), a new concept of lead-likeness emerged illustrating lead-like properties for compounds at this early screening stage (*54*).

By the end of the LI phase, deficiencies within the series should be addressed so that the resultant lead compounds, or series, fulfil the majority of the criteria for progression.

3.4.10 Case study – IKK-2 inhibitors

A good example of a lead generation study is the work cited by Baxter (*55*) in the discovery of IKK-2 inhibitors. An inhibitor of IKK-2 will block the nuclear factor-kappa B-induced gene expression that contributes to the pathogenesis of inflammatory diseases such as arthritis. The summary of the HTS process and output is outlined in Figure 3.28.

Two related thiophene carboxamide hits, **6** and **7** were identified from the HTS campaign (Figure 3.29) and signalled the emergence of an interesting cluster of compounds for further evaluation.

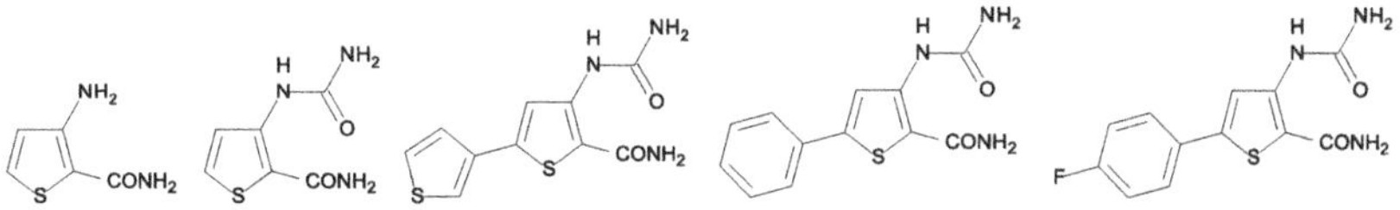

FIGURE 3.29

IKK-2 high-throughput screening hits identified and optimized by Baxter.

Table 3.7 Data for Initial Hits (**6–7**) and Resulting Leads (**8–10**)

Compound	Generic Lead Criteria	6	7	8	9	10
Binding IC$_{50}$ (µM)	<0.1 (µM)	1.6	2	0.063	0.025	0.063
Functional IC$_{50}$ (µM)	<1 (µM)	6.3	>10	0.5	0.25	0.4
Rat hepatocytes Cl (µl/min/10^6 cells)	<15	18	14	21	3	2
Human microsomes Cl (µl/min/mg)	<30	30	49	43	27	8
Molecular weight	<450 (g/mol)	224	185	267	261	279
Calculated log P	<3.0	1.6	0.1	1.7	2.0	2.1
Solubility (µg/ml)	>10	nd	>50	34	6	1.2

The profile of thiophene carboxamides **6** and **7** are shown in Table 3.7. The compounds had been selected on the balance of their potencies, low MW and lipophilicity, acceptable *in vitro* stability and amenability to further modification. Features of **6** and **7** were combined to give analogue **8**, which presented excellent potency in both binding and functional assays. The *in vitro* metabolic stability was improved by replacing the potentially metabolically unstable mono-substituted thiophene ring with a phenyl ring to afford **9** and further optimization afforded **10**.

Further work showed that both the primary amide and the mono-substituted urea were essential for binding and this was subsequently supported by molecular modelling experiments through docking **9** into the X-ray crystal structure (Figure 3.30). In the docking experiment the phenyl ring occupies the sugar-binding pocket and the urea and primary amide hydrogen bond to key residues in the pocket.

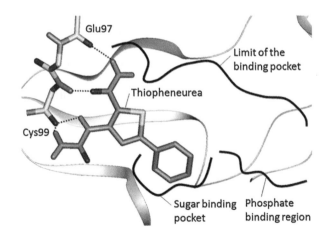

FIGURE 3.30

Human IKK-2 model complexed with compound **9**, showing the key hydrogen bond network.

Table 3.8 Rat Pharmacokinetic Profile of Compounds **8–10**

Compound	Rat Hepatocytes	Cl	V_d	$T_{1/2}$	F	PPB (% Bound)
Generic lead criteria	<15	<35	>0.5	>0.5	>10%	<99%
8	21	51	1.6	0.4	7	93.1
9	3	18	3.2	2.5	27	94.9
10	2	6	0.7	1.4	78	92.5

The analogues **8–10** were assessed in *in vivo* rat PK screens and the results are summarized in Table 3.8 and compound **10** was taken forward for further optimization in the next stage.

3.5 LO – ESTABLISHING THE SCREENING CASCADE AND CANDIDATE BIOLOGICAL TARGET PROFILE

LO is one of the major bottlenecks in the drug discovery process. Creating a new drug is lengthy and complex and, as a consequence, the optimization phase is very much a multidisciplinary team effort involving a close collaboration between medicinal chemists, biologists, pharmacokineticists, pharmaceutical scientists and patent attorneys with the common aim of producing a compound for evaluation in the clinic. The first step in the LO process is the thorough exploration and optimization of the key features contained within the structure of the lead molecules that are essential for pharmacological activity. Essential for the progress of the project is the requirement to establish a robust screening cascade to enable the thorough evaluation of compounds via the design-make-test (DMT) cycle (Figure 3.31).

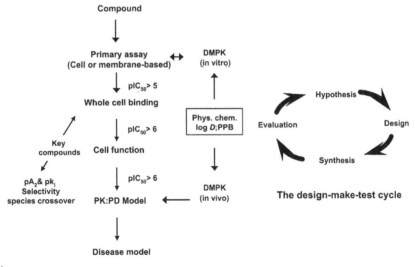

FIGURE 3.31

A typical lead optimisation screening cascade and design-make-test cycle.

Table 3.9 A Typical Example of a Candidate Biological Target Profile

Discipline	Parameter	Figure (Unit)
Biology	Binding (isolated) IC_{50}	<10 nM
	Functional (cell) IC_{50}	<10 nM
	Whole blood potency	<100 nM
	Selectivity against other targets	>100-fold
	hERG	>10 μM
DMPK	Rat hepatocytes Cl	<3 μl/min/10^6 cells
	Human microsomes Cl	<1 μl/min/mg
	Rat *iv* Cl	<20 ml/min/kg
	Rat *iv* Vss	>0.5 l/kg
	Rat *iv* $t_{1/2}$	>2.0 h
	F (in two species)	>30%
	P450 inhibition IC_{50}	>10 μM
	Dose to man prediction	<5 mg/kg
Physicochemistry	Molecular weight	<500 g/mol
	Solubility	>50 μg/ml
	PPB (Bound)	<98%
	Calculated log *P*	<5

Within the DMT stage of the project, key questions are asked and new compounds are designed and tested. By a combination of combinatorial, parallel and classical synthesis, key features contained within the structure of the lead molecules that are essential for pharmacological activity are identified and optimized. The next step is to improve on both the physicochemical properties and the metabolic profile of the compounds. Often, this will be through structural modifications to regions of the molecules that have little or no impact on biological activity. During this process it is essential to evolve the candidate biological target profile, which is a vision of the criteria that the CD must fulfil in order to progress into clinical development (Table 3.9).

3.5.1 Initial safety considerations and potential liabilities

An assessment of the potential safety of a compound needs to be performed before candidate selection. The ideal scenario is that the CD has no side effects associated with it, however this is seldom the case and more often, potential side effects are either observed or predicted in one or more species. The potential toxicity can result from:

• The 'exaggerated' mechanism of action of the compound (such as hypoglycaemia when taking glucose lowering agents or positional hypotension when taking blood-pressure lowering agents) or an undesirable consequence of the pharmacology (such as cytotoxicity in oncology therapy). The observed toxicity can be either acute (i.e. resulting for a single dose) or chronic (i.e. resulting from long-term daily dosing). Rofecoxib (Vioxx) was withdrawn from the market after a clinical study showed that patients receiving a therapeutic dose over a long time (19 months) had a 3.9-fold increase in thromboembolic adverse events (such as heart attack). Whilst the toxicity mechanism is unclear, it is suggested that it was related to the therapeutic target cyclooxygenase-2 (*56*).

Vioxx

- Secondary pharmacology, for example the lack of selectivity against another target.
- Compound-related, which is toxicity due potential liabilities of the compound structure (such as Ames-positive anilines as mutagens).

As a general rule, it is difficult for the drug discovery scientist to resolve issues related to mechanism of action and the likely effects have to be assessed, minimized and an assessment of potential side-effects made against potential patient benefit and this often depends on the seriousness of the disease. However, secondary pharmacology and compound-related effects can be addressed within the drug discovery process and will be discussed further in this chapter.

3.5.2 **Therapeutic ratio/margin of safety**

Within the drug discovery process, the likely side effects associated with a CD have to be identified and minimized. It is important for medicinal chemists to address and take approaches to reduce or remove potential toxicities during the discovery phase as it is during the active phase of analogue synthesis in LO where compound structures can be re-designed to either remove unwanted chemical features or increase selectivity against other biological targets. It is important to reflect that the side effect is based on overall exposure of the compound in a relevant-disease model and not the dose and this leads to the concept of the therapeutic ratio or margin of safety (Figure 3.32).

3.5.3 **Common toxicities**

There are several common toxicities that can now be either predicted based on chemical structure or screened for in the drug discovery process before candidate selection (Table 3.10).

3.5.4 **Cardiovascular side-effects – hERG**

Between 2001 and 2010, the single most common cause of withdrawal or restriction to the use of some FDA (the U.S. Department of Health and Human Services Food and Drug Administration)-approved drugs available on the market has been the discovery that a number of structurally unrelated drugs were implicated in deaths caused by heart malfunction. Since the initial investigations, over 30 marketed drugs have demonstrated such a potential risk (*57*). These drugs, which had generally been marketed

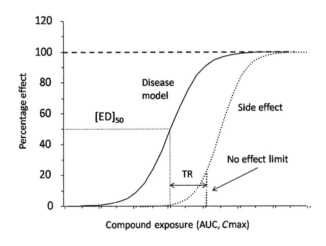

FIGURE 3.32

Plot demonstrating the therapeutic ratio (TR) of the side effect in relation to the ED_{50} in a disease-relevant model. The side effect no effect limit has been set by the project in the CD target profile.

for a range of mainly non-cardiovascular indications, were later shown to demonstrate a common arrhythmia side effect, which was caused by the blocking of hERG channels in heart cells. The hERG channel is an important potassium ion channel that was identified in the late 1980s, and the effect observed is a prolongation of electrical pulses responsible for controlling the heart muscle cells.

Table 3.10 Common Toxicities that Can Now be Either Predicted Based on Chemical Structure or Screened for in the Drug Discovery Process

Type	Effect	Alert
Cardiovascular	Blockade of the hERG potassium channel leading to prolongation of QT interval, arrhythmias and potential death	Binding assays and ion channel electrophysiology
Hepatotoxicity	Formation of glutathione adducts and irreversible CYP P450 inhibition	*In vitro* studies in hepatocytes
Reactive metabolites	Idiosyncratic toxicity derived from mechanistic pathway/metabolites	Metabolite screens
Genetic	Genotoxicity	Mini-Ames and *in vitro* micronucleus tests
Phospholipidosis	Lung and liver toxicity	*In vitro* cellular assays High volume of distribution can be a warning
Drug–drug interactions	P450 inhibition/induction	*In vitro* assays
CNS side effects	Blood–brain penetration, off-target pharmacology	Broad CNS receptor and enzyme screening

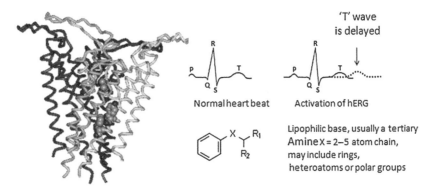

FIGURE 3.33

X-ray crystal structure of the hERG potassium channel. An ECG output demonstrating a healthy heart and after blockade of the hERG channel. The crude hERG pharmacophore, useful for predicting potential hERG activity.

Within the time span of an action potential (about 300 ms) it only requires blocked or partially blocked hERG channels to extend this by 5 or 10 ms to create a potentially dangerous effect. Without control of the rate of heartbeat, dangerous arrhythmias develop, which in some cases can lead to death. The first and last stages of ventricular action potentials detected by ECG are commonly labelled the Q and T waves, thus the 'QT interval' is the critical measurement. Blockade of hERG channels can prolong this QT interval and lead to potentially fatal arrhythmia, including Torsade de Pointes. Due to the diverse array of compounds that have been shown to bind to hERG, a crude pharmacophore has been suggested and we can test for hERG activity in a cloned cell line (Figure 3.33).

There have been several successful strategies employed to reduce or remove hERG activity in compounds which include the following:

- Formation of Zwitterions, although this strategy has the potential issue of lowering permeability, so one needs to target a ΔpKa (difference between acid and base strength) of <4 to reduce this problem.
- Reduction/removal of basic centre, although it has been shown that a basic amine is not a necessity for hERG block.
- Reduction of overall lipophilicity and the incorporation of polar substituents that could disrupt interaction with the lipophilic cavity of hERG.

Terfenadine was shown to be a potent binder to hERG potassium channel and was eventually withdrawn from sale as a widely used antihistamine. During studies, it was noticed that the major metabolite of Terfenadine, whilst retaining its antihistamine activity, did not inhibit hERG. The Zwitterionic nature of this compound is brought about by a combination of the carboxylic acid and amine groups. The resulting metabolite is marketed under the name Fexofenadine. Following on from this, several other companies utilized this approach resulting in Cetirizine that was shown to be inactive at the hERG channel (*58*) (Scheme 3.8).

An example of the reduction in hERG block by the reduction of basicity was demonstrated in the area of 5-HT$_4$ agonists. In this work, Cisapride (a drug used to increase motility in the upper gastrointestinal tract and is therefore used to accelerate gastric emptying) was shown to be a potent

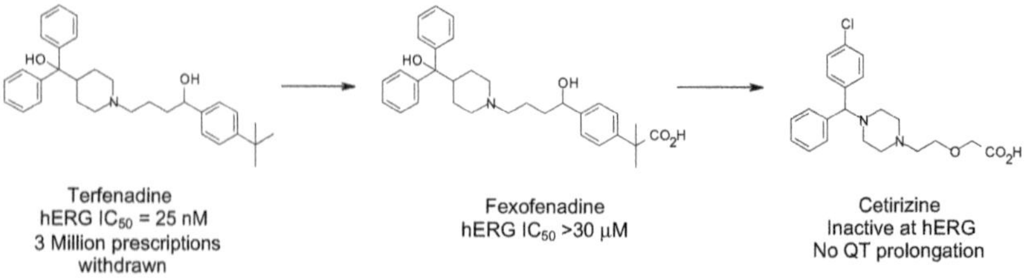

Terfenadine
hERG IC$_{50}$ = 25 nM
3 Million prescriptions
withdrawn

Fexofenadine
hERG IC$_{50}$ >30 μM

Cetirizine
Inactive at hERG
No QT prolongation

SCHEME 3.8

Reduction of hERG channel activity through lowering lipophilicity and incorporation of an acidic group.

hERG binder and through reduction of the basicity through the replacement of the piperidine ring (pKa 7.8) with a morpholine ring (pKa 6.2), a reduction in hERG activity was seen (59) (Scheme 3.9).

Merck chemists working in the area of vascular endothelial growth factor (60) inhibitors demonstrated that a combination of a reduction of basicity and lipophilicity had a beneficial effect (61) (Scheme 3.10).

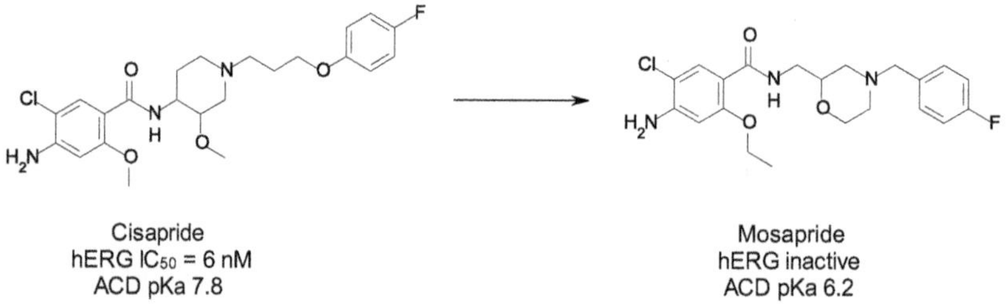

Cisapride
hERG IC$_{50}$ = 6 nM
ACD pKa 7.8

Mosapride
hERG inactive
ACD pKa 6.2

SCHEME 3.9

Reduction of hERG activity through lowering basicity. ACD pKa is an algorithm used to predict the pKa of a compound.

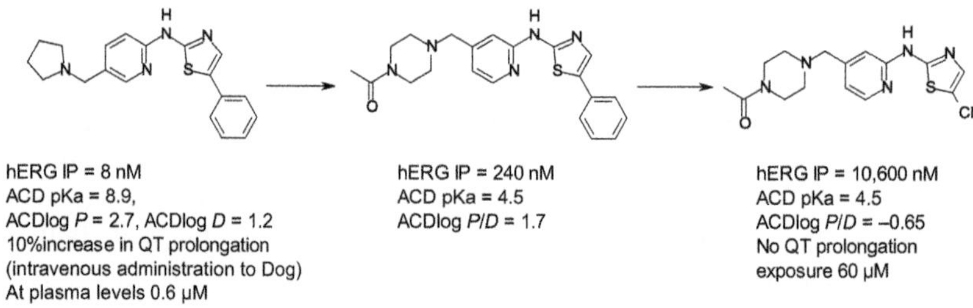

hERG IP = 8 nM
ACD pKa = 8.9,
ACDlog P = 2.7, ACDlog D = 1.2
10%increase in QT prolongation
(intravenous administration to Dog)
At plasma levels 0.6 μM

hERG IP = 240 nM
ACD pKa = 4.5
ACDlog P/D = 1.7

hERG IP = 10,600 nM
ACD pKa = 4.5
ACDlog P/D = −0.65
No QT prolongation
exposure 60 μM

SCHEME 3.10

Reduction of hERG activity through lowering basicity and lipophilicity. IP = inflection point in binding assay that utilises the potent hERG binder (MK499).

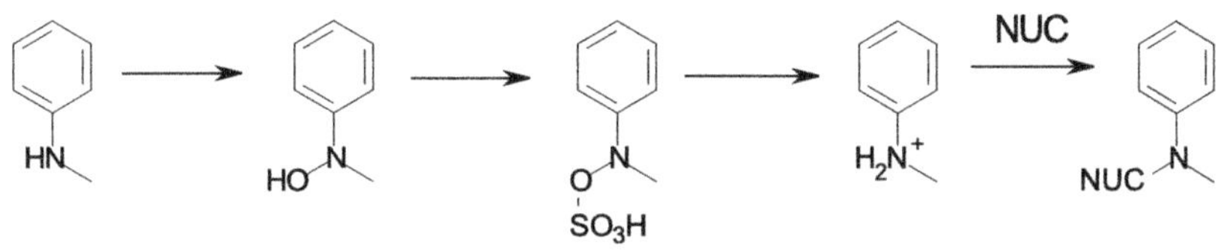

FIGURE 3.34

Formation of *N*-hydroxy aniline leading to the highly reactive nitrenium ion and reaction with a nucleophile (NUC).

3.5.5 Genotoxicity caused by Ames-positive anilines

Primary and secondary amines (especially aromatic amines) form an *N*-hydroxy amine that produces a highly reactive nitrenium intermediate that can react with a nucleophile (such as DNA) to cause genotoxicity (Figure 3.34) (*62*).

It is important to identify and test the embedded amine fragments/potential metabolites and explore alternatives, if found Ames-positive (n.b. Ames is the name given to a test for determining if a chemical is a mutagen and is named after its developer, Bruce Ames). Practolol was found to produce ocular toxicity and Atenolol afforded greater safety through removal of the Ames-positive aniline (Figure 3.35).

3.5.6 Phospholipidosis

Phospholipidosis is a lysosomal storage disorder characterized by the excess accumulation of phospholipids in tissues (*63*). The mechanism of drug-induced phospholipidosis (DIPL) involves the trapping of DIPL compounds within the lysosomes and acidic vesicles of affected cells. It has been reported that many lipophilic cationic amphiphilic compounds, targeting diverse therapies such as antidepressants, anti-malarial, anti-anginal and cholesterol-lowering agents, cause DIPL in animals and humans and therefore it appears to be a compound-specific effect and it is not possible to predict which tissues will be affected by DIPL in animals and humans. As phospholipidosis appears to be compound-specific then there are several approaches that can be employed to reduce the effect through the reduction in lipophilicity and the reduction/removal of basicity in the compound. Scientists at Roche

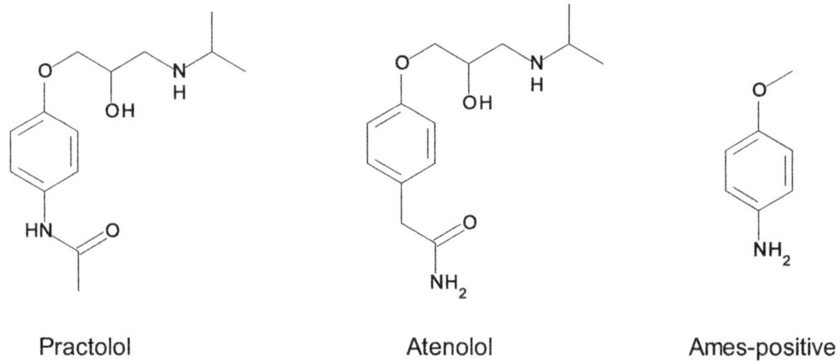

Practolol   Atenolol   Ames-positive

FIGURE 3.35

Discovery of atenolol, devoid of the Ames-positive 4-alkoxy aniline substructure.

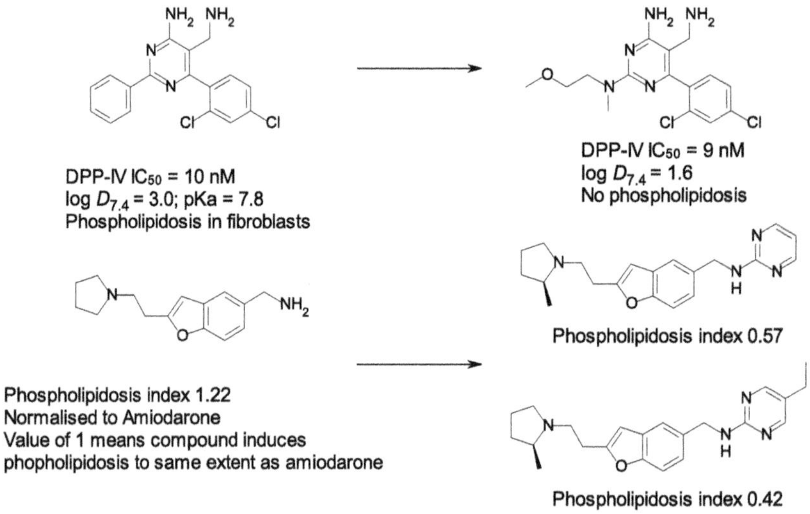

FIGURE 3.36

Examples showing the reduction of phospholipidosis through reduction of lipophilicity or removal of a basic site.

(*64*) showed that in a series of dipeptidyl peptidase IV inhibitors, phospholipidosis could be removed through reduction in lipophilicity, and scientists at Abbott, in a series of histamine H_3-receptor antagonists for use in either sleep disorders or the treatment of neuropathic pain, demonstrated that removal of one basic centre in their lead compound also had a positive effect (Figure 3.36) (*65*).

3.5.7 Pharmacokinetic–pharmacodynamic relationship and dose-to-man prediction

Once a compound has progressed and contains the right balance of potency and metabolic stability, the compound can advance to a relevant animal model. Pharmacokinetic–pharmacodynamic (PK–PD) is the study of the effect/concentration/time relationship in a biologically relevant model. The aim of a PK–PD study is to build confidence that the drug will have both an effect when dosed in man and an indication of the likely dose of drug required to have the required biological effect. Knowledge of PK–PD relationships contributes to selection of the dose and dosing regimen of new compounds and is crucial for the interpretation of PK data in relation to effect and safety. The analysis of PK–PD correlations is very complex with many factors contributing to the outcome of a potential dose-to-man prediction (Figure 3.37). An estimation of likely PK parameters for man (Cl, volume and half-life) from the (PK) of *in vivo* DMPK measurements in pre-clinical species (rat and dog) combined with *in vitro* PK studies in human hepatocytes is made. The correlation 'builds confidence' that the *in vitro* potency observed will be indicative of human *in vivo* dosing and affords an estimate of the concentration of drug in man (how many times more than the *in vitro* IC_{50}) that might be required to afford the biological effect coupled with a safety margin sufficient to progress to initial human studies. The correlation also demonstrates and predicts the pharmacological activity in the presence of PPB i.e. the free fraction activity.

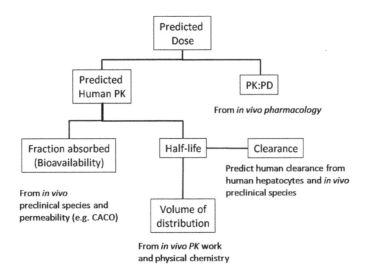

FIGURE 3.37

Measurements contributing to the dose-to-man prediction.

3.5.8 LO case study – the discovery of aprepitant

The discovery of Aprepitant is an excellent illustration of the various considerations and challenges facing medicinal chemists in the LO phase (66). Aprepitant was the first commercially available drug from a new class of agents, the neurokinin-1 (NK-1) receptor antagonists and was indicated for the prevention of acute and delayed chemotherapy-induced nausea and vomiting induced by chemotherapy. Many pharmaceutical companies were searching for selective antagonists at the NK-1 receptor to which the mammalian tachykinin, substance P binds, and it was postulated that a selective antagonist of the CNS NK-1 receptor would be of great interest to treat a multitude of diseases, such as depression, anxiety and nausea (67). Chemists at Pfizer disclosed one of the first non-peptide antagonists of NK-1 (CP-99,994) (68), which was followed by several structurally similar compounds (such as CP-96,345 (69) and L-733,060) (70) (Figure 3.38).

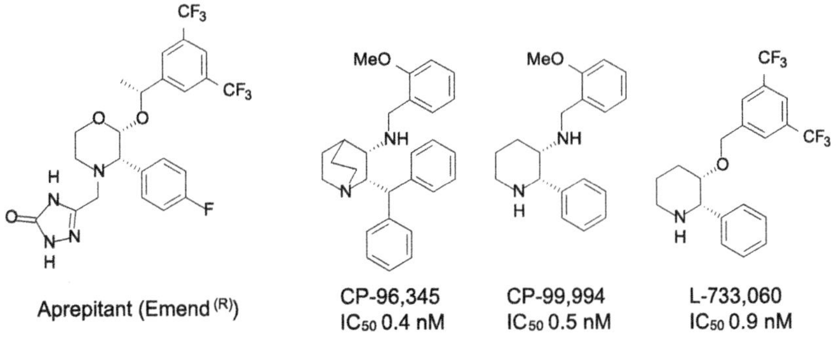

FIGURE 3.38

Aprepitant and initial NK-1 receptor antagonists.

Unfortunately, even though these compounds showed good selectivity against other receptors, they were antagonists of the L-type calcium channel, which was subsequently reported to be responsible for the observed adverse cardiovascular side effects. The affinity of this class of compound for the L-type calcium channel was attributed to the basicity of the quinuclidine (pKa ~9) or piperidine (pKa ~8.7) ring system and indeed structural changes to reduce the basicity of the nitrogen proved beneficial (L-733,060, pKa 8.3). Substituting on the piperidine nitrogen with electron-withdrawing groups, such as $-CH_2CO_2Me$ or $-CH_2CONH_2$ led to compounds with reduced basicity, and as a consequence, markedly decreased L-type channel affinity whilst maintaining high nanomolar levels of potency for the NK-1 receptor.

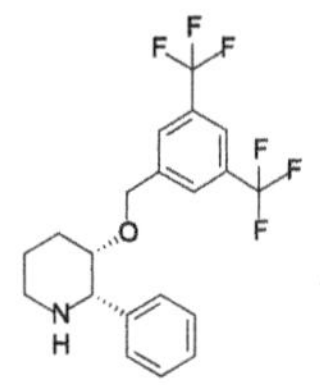

L-733,060
IC$_{50}$ 0.9 nM
Ca^{2+} binding IC$_{50}$ 760 nM
pKa 8.3

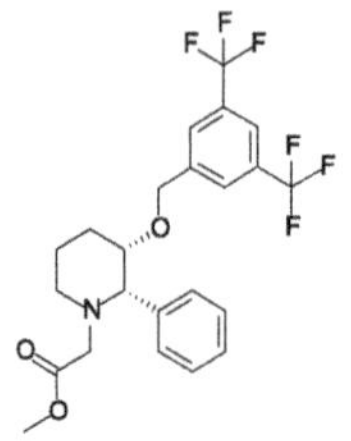

IC$_{50}$ 2.8 nM
Ca^{2+} binding IC$_{50}$ 26 μM

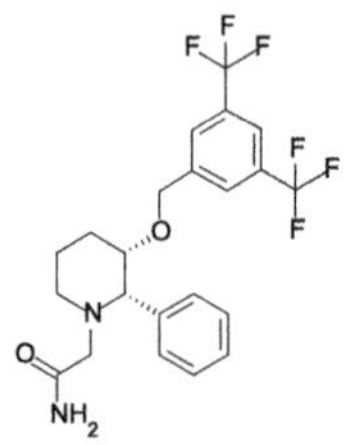

L-736,281
IC$_{50}$ 1.3 nM
Ca^{2+} binding 16% at 5 μM
pKa 5.4

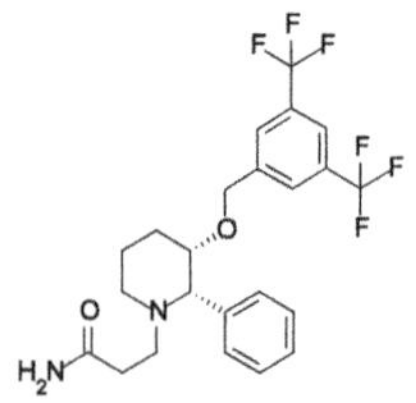

IC$_{50}$ 1.8 nM
Ca^{2+} binding 98% at 5 μM
pKa 6.5

Chemists at Merck explored the approach of reducing basicity further and prepared the morpholine analogues to further reduce affinity for the L-type calcium channel. A brief summary of the SAR obtained is shown in Table 3.11. It was shown that the parent morpholine had good affinity at NK-1, and similar N-substitution with $-CH_2CO_2Me$ and the corresponding primary amide had increased

Table 3.11 Substituent Effects on NK-1 Activity

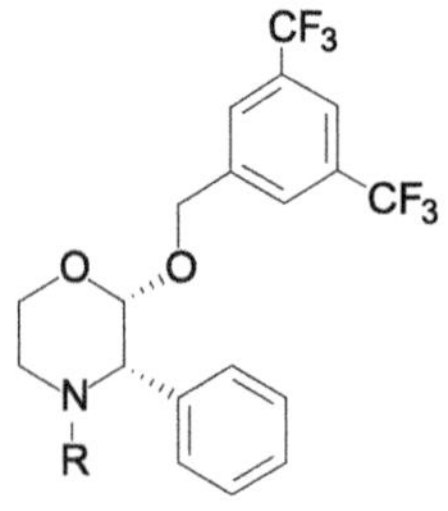

R	IC$_{50}$ (nM)
H	2.4
CH$_2$CO$_2$Me	1.6
CH$_2$CO$_2$H	66
CH$_2$CONH$_2$	1.1
CH$_2$-1,2,4-triazole	0.13
CH$_2$-3-oxo-1,2,4-triazolyl	0.09

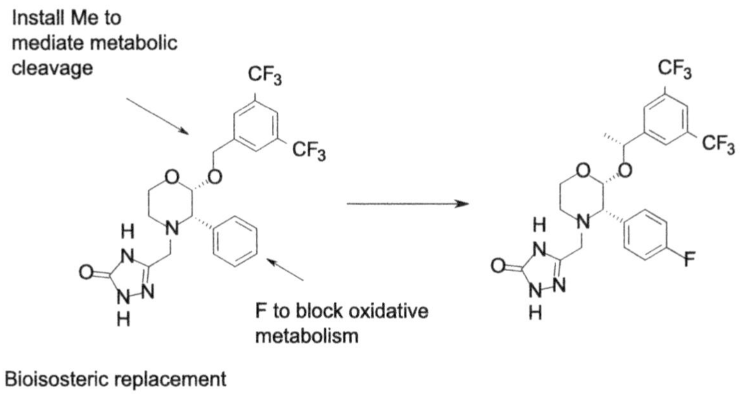

FIGURE 3.39

Lead optimization approaches to mediate metabolic cleavage and oxidation.

affinity whilst the corresponding carboxylic acid had reduced activity. Acid/amide bioisosteres (*71*) were then employed, resulting in the very potent 1,2,4-triazole analogue and the even more potent 3-oxo-1,2,4-triazolyl analogue.

It was shown that only one of the four possible diastereomers, the (2S, 3S)-enantiomer, exhibited very good potency at the NK-1 receptor. *O*-debenzylation was observed *in vitro* as a major metabolite and metabolic stability was then addressed. It was reasoned that substituting one of the benzylic protons with a methyl group would slow either hydrolytic or oxidative debenzylation mechanisms. Finally, whilst it appeared unlikely that the electron-poor 3,5-*bis*(CF$_3$) substituted phenyl ring would be susceptible to oxidative metabolism, it was speculated that such a process could occur with the phenyl ring at C-3 of the morpholine. Substituting the *para*-position of this phenyl ring (the most likely position for metabolism) with a fluorine atom would serve to block both the metabolism at this point as well as slowing other possible oxidative metabolic processes on this ring due to the electron-withdrawing nature of the fluorine atom. These substitutions led to the discovery of Aprepitant (Emend®) that is a highly selective NK-1 antagonist (activity at NK-2 is 7 μM and NK-3 is 150 nM) that has very little affinity for the L-type calcium channel (IC$_{50}$ 11 μM) due mainly to the much-reduced basicity of the morpholine nitrogen (pKa < 3) (Figure 3.39).

References and Notes

1. Gashaw, I.; Ellinghaus, P.; Sommer, A.; Asadullah, K. *Drug Discovery Today* **2012**, *175*, 524–530.
2. A knockout mouse is a genetically engineered mouse where an existing gene has been replaced with an artificial piece of DNA. The loss of gene activity often changes the mouse's phenotype and as a consequence serves as an important animal model to study the role of genes in disease.
3. Baxter, A.; Cooper, A.; Kinchin, E.; Moakes, K.; Unitt, J.; Wallace, A. *Bioorg. Med. Chem. Lett.* **2006**, *16*, 960–963.
4. Abou-Gharbia, M. *J. Med. Chem.* **2009**, *52*, 2–9.
5. Macarron, R.; Banks, M. N.; Bojanic, D.; Burns, D. J.; Cirovic, D. A.; Garyantes, T.; Green, D. V. S.; Hertzberg, R. P.; Janzen, W. P.; Paslay, J. W.; Schopfe, U.; Sittampalam, G. S. *Nat. Rev. Drug Discovery* **2011**, *10*, 188–195.

6. Gribbon, P.; Sewing, A. *Drug Discovery Today* **2005**, *10*, 17–22; Oldenburg, K. R. *Annu. Rep. Med. Chem.* **1998**, *33*, 301–311.

7. McGovern, S. L.; Caselli, E.; Grigorieff, N.; Shoichet, B. K. A. *J. Med. Chem.* **2002**, *45*, 1712–1722.

8. Rishton, G. M. *Drug Discovery Today* **1997**, *2*, 382–384.

9. Lipinski, C. A. *J. Pharmacol. Toxicol. Methods* **2000**, *44* (1), 235–249.

10. Feng, B. Y.; Simeonov, A.; Jadhav, A.; Babaoglu, K.; Inglese, J.; Shoichet, B. K.; Austin, C. P. *J. Med. Chem.* **2007**, *50*, 2385–2390.

11. Baell, J. B.; Holloway, G. A. *J. Med. Chem.* **2010**, *53*, 2719–2740.

12. Leach, A. R.; Gillet, V. J.; Lewis, R. A.; Taylor, R. *J. Med. Chem.* **2010**, *53*, 539–558.

13. Lengauer, T.; Rarey, M. *Curr. Opin. Struct. Biol.* **1996**, *6*, 402–406.

14. Katritch, V.; Jaakola, V.-P.; Lane, J. R.; Lin, J.; IJzerman, A. P.; Yeager, M.; Kufareva, I.; Stevens, R. C.; Abagyan, R. *J. Med. Chem.* **2010**, *53*, 1799–1809.

15. (a) Checkpoint 1 kinase is a serine/threonine kinase that regulates the S and G_2 checkpoints in the cell cycle and is therefore of use in oncology as the majority of tumours are deficient in the G_1-DNA damage checkpoint pathway resulting in reliance on the S and G_2 checkpoints for DNA repair and cell survival. (b) Lyne, P. D.; Kenny, P. W.; Cosgrove, D. A.; Deng, C.; Zabludoff, S.; Wendoloski, J. J.; Ashwell, S. *J. Med. Chem.* **2004**, *47*, 1962–1968.

16. The IUPAC defines a pharmacophore to be 'an ensemble of steric and electronic features that is necessary to ensure the optimal supramolecular interactions with a specific biological target and to trigger (or block) its biological response'.

17. Andrews, K. M.; Cramer, R. D. *J. Med. Chem.* **2000**, *43*, 1723–1740.

18. (a) $CXCR_4$ is a chemokine receptor specific for stromal-derived-factor-1, antagonists of which are useful for e.g. inhibiting HIV entry. (b) Pérez-Nueno, V. I.; Sofia Pettersson, S.; Ritchie, D. W.; Borrell, J. I.; Teixido, L. *J. Chem. Inf. Model.* **2009**, *49*, 810–823.

19. Harvey, A. L. *Drug Discovery Today* **2008**, *13*, 894–901.

20. Ojima, I. *J. Med. Chem.* **2008**, *51*, 2587–2588.

21. Erlanson, D. A. *Top. Curr. Chem.* **2012**, *317*, 1–32.

22. Medina, J. R.; Blackledge, C. W.; Heerding, D. A.; Campobasso, N.; Ward, P.; Briand, J.; Wright, Lois; Axten, J. M. *ACS Med. Chem. Lett.* **2010**, *1*, 439–442.

23. Akritopoulou-Zanze, I.; Hajduk, P. J. *Drug Discovery Today* **2009**, *14*, 291–297.

24. Hung, A. W.; Silvestre, H. L.; Wen, S.; Ciulli, A.; Blundell, T. L.; Abell, C. *Angew. Chem. Int. Ed.* **2009**, *48*, 8452–8456.

25. Howard, N.; Abell, C.; Blakemore, W.; Chessari, G.; Congreve, M.; Howard, S.; Jhoti, H.; Murray, C.; Seavers, L. C. A.; van Montfort, R. L. M. *J. Med. Chem.* **2006**, *49*, 1346–1355.

26. Barelier, S.; Pons, J.; Gehring, K.; Lancelin, J.-M.; Krimm, I. *J. Med. Chem.* **2010**, *53*, 5256–5266.

27. Congreve, M.; Carr, R.; Murray, C.; Jhoti, H. *Drug Discovery Today* **2003**, *8*, 876–877.

28. Blundell, T. L.; Jhoti, H.; Abell, C. *Nat. Rev. Drug Discovery* **2002**, *11*, 45–54.

29. Baurin, N.; Aboul-Ela, F.; Barril, X.; Davis, B.; Drysdale, M.; Dymock, B.; Finch, H.; Fromont, C.; Richardson, C.; Simmonite, H.; Hubbard, R. E. *J. Chem. Inf. Comput. Sci.* **2004**, *44*, 2157–2166.

30. Edink, E.; Jansen, C.; Leurs, R.; de Esch, I. J. P. *Drug Discovery Today: Technologies* **2010**, *7*, e189–e201.

31. Hubbard, R. E. *J. Synchrotron Radiat.* **2008**, *15*, 227–230.

32. Foloppe, N. *Future Med. Chem.* **2011**, *3*, 1111–1115.

33. Artis, D. R.; Lin, J. L.; Zhang, C.; Wang, W.; Mehra, U.; Perreault, M.; Erbe, D.; IKrupka, H. I.; England, B. P.; Arnold, J.; Plotnikov, A. N.; Marimuthu, A.; Nguyen, H.; Will, S.; Signaevsky, M.; Kral, J.; Cantwell, J.; Settachatgull, C.; Yan, D. S.; Fong, D.; Oha, A.; Shi, S.; Womack, P.; Powell, B.; Habets, G.; West, B. L.; Zhang, K. Y. J.; Milburn, M. V.; Vlasuk, G. P.; Hirth, K. P.; Nolop, K.; Bollag, G.; Ibrahim, P. N.; Tobin, J. F. *Proc. Natl. Acad. Sci. U.S.A.* **2009**, *106*, 263–267.

34. Neumeyer, K.; Kirkpatrick, P. *Nat. Rev. Drug Discovery* **2004**, *3*, 295–296.
35. Campbell, S. F. *Clin. Sci.* **2000**, *99*, 255–260.
36. Schnecke, V.; Bostroem, J. *Drug Discovery Today* **2006**, *11*, 43–50.
37. Hann, M. M. *Med. Chem. Commun.* **2011**, *2*, 349–355.
38. Barker, A. J.; Gibson, K. H.; Grundy, W.; Godfrey, A. A.; Barlow, J. J.; Healy, M. P.; Woodburn, J. R.; Ashton, S. E.; Curry, B. J.; Scarlett, L.; Henthorn, L.; Richards, L. *Bioorg. Med. Chem. Lett.* **2001**, *11*, 1911–1914.
39. Kola, I.; Landis, J. *Nat. Rev. Drug Discovery* **2007**, *6*, 881–916.
40. Yamamoto, T.; Itoga, H.; Kohno, Y.; Nagata, K.; Yamazoe, Y. *Xenobiotica* **2005**, *35*, 627–646.
41. (a) An endothelin A antagonist has shown benefit for pulmonary arterial hypertension or anti-cancer indications. (b) Humphreys, W. G.; Obermeier, M. T.; Barrish, J. C.; Chong, S.; Marino, A. M.; Murugesan, N.; Wang-Iverson, D.; Morrison, R. A. *Xenobiotica* **2003**, *33*, 1109–1123.
42. Raub, T. J. *Mol. Pharmacol.* **2006**, *3*, 3–25.
43. MacKenzie, A. R.; Marchington, A. P.; Middleton, D. S.; Newman, S. D.; Jones, B. C. *J. Med. Chem.* **2002**, *45*, 5365–5377.
44. (a) The Caco-2 cell line is a continuous line of human epithelial colorectal adenocarcinoma cells that when cultured under specific conditions differentiate so that they function as enterocytes that line the small intestine. The Caco-2 monolayer is widely used as an in vitro model of the human small intestine mucosa to predict the absorption of orally administered drugs. (b) Artursson, P.; Palm, K.; Luthman, K. *Adv. Drug Delivery Rev.* **2001**, *46*, 27–43.
45. Middleton, D. S.; MacKenzie, A. R.; Newman, S. D.; Corless, M.; Warren, A.; Marchington, A. P.; Jones, B. *Bioorg. Med. Chem. Lett.* **2005**, *15*, 3957–3961.
46. (a) A CRTH$_2$ antagonist inhibits prostaglandin D$_2$-induced eosinophil migration and has potential in allergic diseases such as asthma. (b) Sandham, D. A.; Aldcroft, C.; Baettig, U.; Barker, L.; Beer, D.; Bhalay, G.; Brown, Z.; Dubois, G.; Budd, D.; Bidlake, L.; Campbell, E.; Cox, B.; Everatt, B.; Harrison, D.; Leblanc, C. J.; Manini, J.; Profit, R.; Stringer, R.; Thompson, K. S.; Turner, K. L.; Tweed, M. F.; Walker, C.; Watson, S. J.; Whitebread, S.; Willis, J.; Williams, G.; Wilson, C. *Bioorg. Med. Chem. Lett.* **2007**, *17*, 4347.
47. Young, R. N. *Prog. Med. Chem.* **2001**, *38*, 249.
48. Masimirembwa, C. M.; Otter, C.; Berg, M.; Jonsson, M.; Leidvik, B.; Jonsson, E.; Johansson, T.; Backman, A.; Edlund, A.; Andersson, T. B. *Drug Metab. Dispos.* **1999**, *27*, 1117–1122.
49. Hollenberg, P. F. *Drug Metab. Rev.* **2002**, *34*, 17–35.
50. Fontana, E.; Dansette, P. M.; Poli, S. M. *Curr. Drug Metab.* **2005**, *6*, 413.
51. Wu, Y. J.; Davis, C. D.; Dworetzky, S.; Fitzpatrick, W. C.; Harden, D.; He, H.; Knox, R. J.; Newton, A. E.; Philip, T.; Polson, C.; Sivarao, D. V.; Sun, L.-Q.; Tertyshnikova, S.; Weaver, D.; Yeola, S.; Zoeckler, M.; Sinz, M. W. *J. Med. Chem.* **2003**, *46*, 3778–3781.
52. Ryckmans, T.; Edwards, M. P.; Horne, V. A.; Correia, A. M.; Owen, D. R.; Thompson, L. R.; Tran, I.; Tutt, M. F.; Young, T. *Bioorg. Med. Chem. Lett.* **2009**, *19*, 4406–4409.
53. Lipinski, C. A.; Lombardo, F.; Dominy, B. W.; Feeney, P. J. *Adv. Drug Delivery Rev.* **1997**, *23*, 3–25.
54. Teague, S. J.; Davis, A. M.; Leeson, P. D.; Oprea, T. *Angew. Chem. Int. Ed.* **1999**, *38*, 3743–3748.
55. Baxter, A.; Brough, S.; Cooper, A.; Floettmann, E.; Foster, S.; Harding, C.; Kettle, J.; McInally, T.; Martin, C.; Mobbs, M.; Needham, M.; Newham, P.; Paine, S.; St-Gallay, S.; Salter, S.; Unitt, J.; Xue, Y. *Bioorg. Med. Chem. Lett.* **2004**, *14*, 2817–2822.
56. Dogné, J.-M.; Supuran, C. T.; Pratico, D. *J. Med. Chem.* **2005**, *48*, 2251–2257.
57. Jamieson, C.; Moir, E. M.; Rankovic, Z.; Wishart, G. *J. Med. Chem.* **2006**, *49*, 5029–5046.
58. Rampe, D.; Wible, B.; Brown, A. M.; Dage, R. C. *Mol. Pharmacol.* **1993**, *44*, 1240–1245.
59. Potet, F.; Bouyssou, T.; Escande, E.; Baró, I. *J. Pharmacol. Exp. Ther.* **2001**, 1007–1012.

60. VEGF's normal function is to create new blood vessels however, if VEGF is overexpressed, it can contribute to disease. For example, as solid cancers cannot grow beyond a limited size without an adequate blood supply; cancers that can express VEGF are able to grow and therefore metastasize. It has also been reported that overexpression of VEGF can cause vascular disease in the retina and is being trialled in the clinic for diabetic macular oedema.

61. Bilodeau, M. T.; Balitza, A. E.; Koester, T. L.; Manley, P. J.; Rodman, L. D.; Buser-Doepner, C.; Coll, K. E.; Fernandes, C.; Gibbs, J. B.; Heimbrook, D. C.; Huckle, W. R.; Kohl, N.; Lynch, J. L.; Mao, X.; McFall, R. C.; McLoughlin, D.; Miller-Stein, C. M.; Rickert, K. W.; Sepp-Lorenzino, L.; Shipman, J. M.; Subramanian, R.; Thomas, K. A.; Wong, B. K.; Yu, S.; Hartman, G. D. *J. Med. Chem.* **2004,** *47,* 6363–6372.

62. Nelson, S. D. *Curr. Ther. Res.* **2001,** *62,* 885–899.

63. Reasor, M. J.; Kacew, S. *Exp. Biol. Med.* **2001,** *226,* 825–830.

64. Peters, J.-U.; Hunziker, D.; Fischer, H.; Kansy, M.; Weber, S.; Kritter, S.; Müller, A.; Wallier, A.; Ricklin, F.; Boehringer, M.; Poli, S. M.; Csato, M.; Loeffler, B.-M. *Bioorg. Med. Chem. Lett.* **2004,** *14,* 3575–3578.

65. Sun, M.; Zhao, C.; Gfesser, G. A.; Thiffault, C.; Miller, T. R.; Marsh, K.; Wetter, J.; Curtis, M.; Faghih, R.; Esbenshade, T. A.; Hancock, A. A.; Cowart, M. *J. Med. Chem.* **2005,** *48,* 6482–6490.

66. Hale, J. J.; Mills, S. G.; MacCoss, M.; Finke, P. E.; Cascieri, M. A.; Sadowski, S.; Ber, E.; Chicchi, G. G.; Kurtz, M.; Metzger, J.; Eiermann, G.; Tsou, N. N.; Tattersall, F. D.; Rupniak, N. M. J.; Williams, A. R.; Rycroft, W.; Hargreaves, R.; MacIntyre, D. E. *J. Med. Chem.* **1998,** *41,* 4607–4614.

67. Desai, M. C.; Lefkowitz, S. L. *Bioorg. Med. Chem. Lett.* **1993,** *3,* 2083–2086.

68. Coeker, J. D.; Davies, H. G. *Bioorg. Med. Chem. Lett.* **1996,** *6,* 13–16.

69. Lowe, J. A., III; Drozda, S. E.; Snider, R. M.; Longo, K. P.; Bordner, J. *Bioorg. Med. Chem. Lett.* **1991,** *1,* 129–132.

70. Harrison, T.; Owens, A. P.; Williams, B. J.; Swain, C. J.; Baker, R.; Hutson, P. H.; Sadowski, S.; Cascieri, M. A. *Bioorg. Med. Chem. Lett.* **1995,** *5,* 209–212.

71. Meanwell, N. A. *J. Med. Chem.* **2011,** *54,* 2529–2591.

Further Reading

General

King, F. D. *Medicinal Chemistry, Principles and Practice,* 2nd ed.; Royal Society of Chemistry, 2002. ISBN-13: 978-0854046317.

Vermuth, C. G. *The Practice of Medicinal Chemistry,* 2nd ed.; Academic Press, 2003. ISBN-13: 978-0127444819.

Kerns, E. H.; Di, L. *Drug-Like Properties: Concepts, Structure Design and Methods: From ADME to Toxicity Optimisation;* Academic Press, 2008. ISBN-13: 978-0123695208.

Stocks, M.; Alcaraz, L.; Griffen, E. *On Medicinal Chemistry;* Sci-Ink Ltd, 2007. ISBN-13: 978-0955007231.

Rang, H. P. *Drug Discovery and Development: Technology in Transition;* Churchill Livingstone, 2006. ISBN-13: 978-0443064203.

Kenakin, T. P. *A Pharmacology Primer,* 3rd ed.; Academic Press, 2009. ISBN 978-0-12-374585-9.

Protein therapeutics (introduction to biopharmaceuticals)

4

Jill M. Carton, William R. Strohl

Biologics Research, Janssen R&D, Spring House PA 19477, USA

CHAPTER OUTLINE

4.1 Introduction 128
    4.1.1 Brief history of protein therapeutics 129
    4.1.2 Overview: Protein therapeutics compared to small molecule drugs 131
4.2 Types of Protein Therapeutics I: Regulatory and Enzymatic Activity 133
    4.2.1 Insulin..133
    4.2.2 Interferons...134
    4.2.3 Epoetins..135
    4.2.4 Granulocyte colony-stimulating factor (G-CSF)/granulocyte-macrophage
          colony-stimulating factor (GM-CSF).. 136
    4.2.5 Growth factors..137
    4.2.6 Coagulation and fibrinolytic regulation...138
    4.2.7 Therapeutic enzymes 138
    4.2.8 Peptide therapeutics...140
4.3 Types of Protein Therapeutics II: Targeting Activity 141
    4.3.1 Monoclonal antibodies..142
    4.3.2 Monoclonal antibody therapeutic discovery and engineering 146
    4.3.3 Fc engineering 147
    4.3.4 Fc fusion proteins (FcFPs)...147
    4.3.5 Engineered antibody fragment therapeutics 149
    4.3.6 Non-antibody binding proteins (NABP) 151
4.4 Challenges of Protein Therapeutics 153
    4.4.1 Immunogenicity 153
    4.4.2 Developability and manufacturing...153
    4.4.3 Route of administration..155
4.5 Future Directions for Protein Therapeutics 156
4.6 Biosimilar Protein Therapeutics 157
4.7 Summary and Conclusions 158
References 158

Introduction to Drug Research and Development. http://dx.doi.org/10.1016/B978-0-12-397176-0.00004-2

ABSTRACT

Throughout human history, the morbidity and mortality associated with human disease has driven medical science into an ever-expanding quest for treatment and cure. Over the past century, a therapeutic approach complementing chemical drugs has been developing which uses proteins and peptides in the treatment of disease. Many innovative protein therapeutic platforms are currently being employed and continue to be developed to attain cures in areas of unmet medical need; these include direct copies of natural protein structure and function as well as proteins with completely novel functionality. Today, protein therapeutics represents the fastest growing sector in the pharmaceutical industry and comprises 16% of prescription drug sales in 2011.

Keywords/Abbreviations: Biopharmaceuticals; Food and drug administration (FDA); Protein therapeutics; Monoclonal antibodies (mAb); Compound annual growth rate (CAGR); Interferon (IFN); Multiple sclerosis (MS); Hepatitis C virus (HCV); (recombinant) Polyethylene glycol (rPEG); Non-antibody binding proteins (NABP); Tumour necrosis factor (TNF); Heavy chain (HC); Light chain (LC); Constant region (C); Variable region (V); Light chain variable region (V_L or VL); Heavy chain variable region (V_H or VH); Antibody–antigen binding fragment (FAb); crystallizable antibody fragment (Fc); (neonatal) Fc receptor (FcR(n)); Variable region fragment or variable domain (Fv); Single chain Fv (scFv); Fc fusion protein (FcFP); Insulin; erythropoetin (EPO); Chronic myeloid leukaemia (CML); Chronic granulomatous disease (CGD); Erythroid stimulating agent (ESA); Pure red cell aplasia (PRCA); Granulocyte–(macrophage) colony-stimulating factor (G(M)-CSF); (recombinant) Human growth hormone ((r)hGH); Insulin-like growth factor-1 (IGF-1); Recombinant tissue plasminogen activator (rTPA); Enzyme replacement therapy (ERT); Chinese hamster ovary (CHO); Alpha-glucosidase (GAA); Gonadotropin-releasing hormone (GnRH); Glucagon-like peptide-1 (GLP-1); Immunoglobulin (Ig); Monoclonal antibody (mAb or Mab); Antibody-dependent cellular cytotoxicity (ADCC); Antibody-dependent cellular phagocytosis (ADCP); Human anti-mouse antibody (HAMA); Domain antibody (dAb); Natural killer (NK); Mechanism of action (MOA); Amyloid beta precursor (APPI); Designed ankyrin repeat protein (DARPin); Cryopyrin-associated periodic syndrome (CAPS); Age-related macular degeneration (AMD); Chronic immune thrombocytopenia (ITP).

4.1 INTRODUCTION

Biopharmaceuticals include a collection of protein- or peptide-based therapeutics with diverse compositions and mechanisms of action (MOA). These therapeutic molecules were originally defined by the United States Food and Drug Administration (FDA) as 'a virus, therapeutic serum, toxin, antitoxin, vaccine, blood, blood component or derivative, allergenic product, or analogous product, or arsphenamine or derivative of arsphenamine (or any other trivalent organic arsenic compound), applicable to the prevention, treatment, or cure of a disease or condition of human beings' (42 USC 262-Sec. 262. Regulation of Biological Products). In more modern terms, biopharmaceuticals include a wide range of products such as vaccines, blood and blood components, allergenics, somatic cells, gene therapy, tissues and recombinant therapeutic proteins. This chapter will specifically address protein therapeutics, albeit excluding protein vaccines. Protein-based therapeutics are highly successful in the treatment of a variety of diseases and show significant potential for continued success. Today, biologics are the fastest growing sector of the pharmaceutical industry, with total global market sales in 2011 of about $115 billion (Figure 4.1), which represents about 16% of the gross 2011 prescription drug sales estimated at about $710 billion.

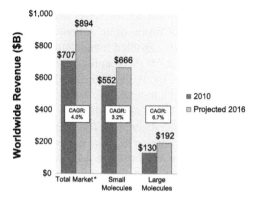

FIGURE 4.1 Biopharmaceuticals continue to be a significant growth driver for the pharmaceutical industry.

Biopharmaceuticals are forecasted to grow at approximately 7% through 2016.

*Source: Evaluate Pharma (published June 2011) reflects approximately 85% of the market, *total market values include vaccines.*

CAG (compound annual growth)

4.1.1 Brief history of protein therapeutics

Proteins are an extraordinary class of macromolecules that play essential and diverse roles in carrying out virtually every cellular process in the body. Proteins were first described as a distinct class of molecules in the eighteenth century based on observed coagulation in acid and heat treatments. The first antibody-based therapeutics date as far back as the 1890s when Behring and Wernick developed serum therapy for diphtheria after key laboratory observations, including that blood isolated from rats resistant to anthrax was toxic to anthrax bacterium. Behring's diphtheria serum therapy resulted in a significant drop in the mortality rate in Paris at the turn of the century and earned him the first Nobel Prize in Medicine and Physiology in 1901.

Further significant progress in the area of protein therapeutics was made in 1922 when the first purified protein therapeutic, insulin, purified from bovine and porcine pancreas, was administered to a young boy in a Toronto hospital for the treatment of severe type 1 diabetes mellitus. Despite dramatic therapeutic efficacy for some patients, numerous problems arose from animal-derived insulin (see Section 4.2.1). A few key biotechnology developments during the late twentieth century were transformational in advancing protein therapeutics from these very early examples to the current state. The development of recombinant DNA technology in the 1970s enabled the engineering and manufacturing of recombinant protein therapeutics. In 1982, the first recombinant protein therapeutic was approved by the US FDA, which was human insulin, sold under the brand name Humulin®, and manufactured by Genentech, Inc. in partnership with Eli Lilly and Company. Because this new insulin was fully human and produced in and purified from *Escherichia coli* fermentation cultures, issues with immunogenicity and limited supply associated with the first-generation animal-derived protein were eliminated. Starting with Humulin® and extending immediately to the generation and manufacturing of recombinant human growth hormone (rhGH), the advent of recombinant DNA technology and manufacturing in microbial fermentation cultures truly revolutionized the development of protein therapeutics.

During the same era in which recombinant DNA technology emerged, a second revolutionizing discovery occurred that significantly impacted the history of protein therapeutics. In 1975, Köhler and Milstein described a method to establish a continuous immortal culture of cells secreting an antibody of predefined specificity through fusion of an antibody-producing B cell with a myeloma cell (*1*). The development of this technology enabled monoclonal antibody (mAb) discovery and led directly to the development of the first mAb therapeutic, muromonab-CD3 (Orthoclone OKT3®; Ortho Biotech (currently Janssen Biotech, Inc.)), in 1986. OKT3 is a murine mAb targeting CD3 on T cells and is approved for use during organ transplant to reduce acute rejection. As was the case for the first-generation insulin therapeutic, OKT3 had its share of problems related to lack of efficacy resulting from rapid clearance due to immunogenicity. Today, mAbs (Section 4.3.1), and related Fc fusion proteins (FcFPs) (Section 4.3.4), combine to make the largest class of protein therapeutics in the market, comprising nearly 50% of the total (Figure 4.2). Advances built on these two seminal discoveries, recombinant DNA technology and hybridoma technology, have driven solutions to the initial challenges of protein therapeutics such that the safety, efficacy and profitability have been demonstrated and are continuing to evolve. In 2001, nine out of ten of the top-selling pharmaceuticals were small molecule drugs, with Procrit/Eprex, which are epoetins used in the treatment of anaemia (see section below), as the only biologic achieving the top 10 status. By 2010, five of the ten top-selling prescription drugs were biologics. Biologics are growing at a current compound annual growth rate (CAGR) of 6.7% as compared with the CAGRs of small molecules and vaccines of 3.2% and 6.3%, respectively. Thus, it is expected that biologics will continue to grow over the next several years.

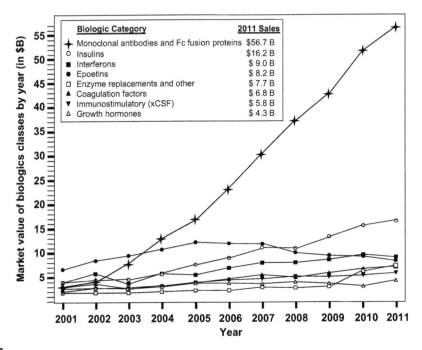

FIGURE 4.2

2011 Biologics total sales by category.

4.1.2 **Overview: Protein therapeutics compared to small molecule drugs**

There are significant differences between a protein therapeutic and a small molecule drug. In Figure 4.3, the molecular structure of aspirin (acetylsalicylic acid), as an example of a synthetic chemical drug, is depicted next to a model of a mAb protein. It is obvious that size and complexity are orders of magnitude different between the molecules. There are many additional complexities associated with protein therapeutics relative to a small molecule drug, some of which arise since biologics are produced in a living cell, through complex cellular processes including transcription, translation, posttranslational modification and protein folding. In contrast, small molecule drugs are chemically synthesized and chemically defined. In many cases, biologics are glycosylated, adding another level of complexity and potential heterogeneity. The inherent molecular differences between small molecule drugs and protein therapeutics drive the unique characteristics that distinguish and define these two very different classes of therapeutics, each with its own advantages and challenges.

Table 4.1 shows a comparison of a protein therapeutic and a small molecule outlining key characteristics. By definition, the compositions of the molecules are different; small molecules are derived from chemically synthesized organic compounds, whereas protein therapeutics, as their name indicates, are produced by living cells. Small molecule drugs are defined by a chemical formula and structure and are produced and purified as single entities. The product definition of a protein therapeutic is much more complex and includes criteria for biological activity, biochemical analysis for heterogeneities such as posttranslational modifications, and the process by which the molecule is produced (bioprocess). Furthermore, small molecules can be used against both extracellular (e.g. G-protein-coupled receptors) and intracellular (e.g. kinase phosphorylation) targets, whereas technology has not yet been developed for intracellular targeting using biologics. Instead, targets of biologics are restricted to extracellular and cell surface molecules due to accessibility limitations. Additionally, immunogenicity, which is the ability of a substance to provoke an immune response, is

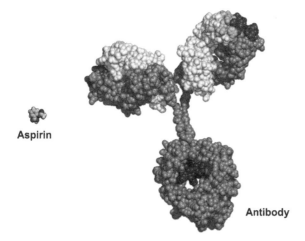

Aspirin

Antibody

FIGURE 4.3 Complexity of a monoclonal antibody (large molecule) compared to aspirin (small molecule).

The figure shows the molecular model of an antibody as an example of a protein therapeutic and the molecular structure of aspirin (Acetylsalicylic acid) to demonstrate the differences in size and molecular complexity associated with the drug classes.

Table 4.1 Comparison Table of Key Characteristics of Protein Therapeutics and Small Molecules

Property	Protein Therapeutic	Small Molecule
Composition	Amino acids; animal-derived or recombinant natural or engineered proteins	Synthetic organic compound
Molecular weight	Most are >700 Da	Most are <700 Da
Product definition	Amino acid sequence and bioactivity and biochemical analysis and the bioprocess used to generate the protein	Chemical identity
Production	A product of living cells	Chemical synthesis
Mode of action	Regulatory or enzyme activity to replace or augment cellular activity; provide novel function through targeting molecular interactions; limited to extracellular and cell surface interactions	Antagonistic and agonistic activity on intracellular and extracellular targets
Delivery	Parenteral, including IV and SC	Usually oral
Half-life	Variable from hours to weeks	Variable from hours to 1 day
Off-target activities	Generally not observed	Drug or metabolites of the drug can be toxic
Immunogenicity	Requires monitoring; can influence efficacy and safety	Generally not observed
Follow-on Drugs	Biosimilars: as similar as possible to the innovator drug, molecular identity is not possible	Generic; molecular identity of the active ingredient, strength, purity and quality

IV, intravenous; SC, subcutaneous.

a unique challenge associated with protein therapeutics. With the obvious exception of vaccines, immunogenicity is undesirable and must be minimized for safety and efficacy of biologics (see Section 4.4.1 below).

Delivery of biologics is currently restricted to various modes of parenteral administration. Proteins are large and complex in composition, they can be challenging to achieve high concentration, and they are unstable unless maintained under specific conditions, typically including continual cold storage. This results in drug delivery and product storage/shipping restrictions which are unique challenges for biologics, adding cost and additional analytical testing, as compared to the small molecule production, formulation, and storage process. The final key difference between biologics and small molecules is that biologics, especially mAbs, may have significantly longer serum half-life values than typical small molecules. Biologics larger than ca. 70 kDa are typically too large to be excreted through the kidney and thus are largely eliminated by the liver, often giving them extended serum residence time. Moreover, immunoglobulin (IgG)-based therapeutics are recycled by the reticuloendothelial system (Section 4.3.1), giving them extended half-life values generally measured in weeks rather than hours. This results in significant differentiation in dosing frequency, with small molecules typically being dosed daily as compared with weekly or monthly dosing for biologics. However, managing a drug with a long half-life can also have clinical consequences, if one needs to remove the drug quickly. This chapter will discuss in detail these and other specific benefits and challenges of protein therapeutics.

Significant structural, functional and pharmacological diversity exists within the protein therapeutic subclass of biopharmaceuticals. Many protein therapeutics mimic directly the natural protein designs that have evolved through millions of years. In addition, technology and know-how, most notably protein engineering, have been developed during the past several decades, enabling rational design of engineered therapeutic approaches taking recombinant therapeutic design beyond what nature has designed. Currently, both natural human serum-derived proteins and novel recombinant engineered protein therapeutics are marketed, and are being developed, albeit with a strong emphasis on the latter. Leader et al. (2) have recently classified protein therapeutics according to pharmacological activity. Using this functional classification, there are four different groups of protein therapeutics: (1) enzymatic or regulatory activity, which includes replacement therapy for defective proteins; (2) targeting activity, including binding and neutralizing or activating a target protein, as well as, targeted delivery of other proteins or drugs, including cytotoxic agents; (3) vaccines, such as prophylactic vaccines that use viral antigens to improve immunity against the pathogen; and (4) diagnostic tools that aid in clinical evaluation. The first two classes as defined by Leader et al. will be addressed below. Protein therapeutics that fit the Leader's category 1, i.e. those with regulatory and enzymatic activity, will be discussed first, followed by descriptions of the second category protein therapeutics that function through targeting activity.

4.2 TYPES OF PROTEIN THERAPEUTICS I: REGULATORY AND ENZYMATIC ACTIVITY

Category 1 therapeutic proteins have enzymatic or regulatory and pharmacological activities involving one of the following: (1) replacing a protein that is deficient or abnormal, (2) augmenting an existing pathway, or (3) providing a novel function or activity.

4.2.1 Insulin

Starting in 1922, insulin was purified from bovine pancreas extracts and used to treat patients with severe type I diabetes. Animal pancreas-derived insulin was used clinically for more than 50 years with some success in controlling disease. However, this approach also resulted in an unacceptable incidence of loss of efficacy due to antidrug immune responses targeting the foreign insulin protein. In 1978, only 4 years after the initial development of recombinant DNA technology, Genentech, Inc. produced recombinant human insulin in *E. coli*. Almost 4 years later, recombinant human insulin, marketed by Eli Lilly and Company as Humulin®, became the first recombinant protein approved for use in patients. The recombinant version of human insulin eliminated the challenges of limited supply and reduced the immunogenicity issues associated with insulin therapy. Since the development of Humulin®, several insulin analogues, including long-lasting and fast-acting insulins, have been developed as second-generation diabetic therapeutics. Through protein engineering and slight variations in the insulin protein amino acid sequence, pharmacokinetic and pharmacodynamic characteristics have been modified to enhance insulin function as a therapeutic protein. These analogues include fast-acting and rapidly absorbed insulins, mimicking endogenous beta cell-derived insulin, such as Eli Lilly and Company insulin lispro (Humalog®), Novo Nordisk, Inc. insulin aspart (Novolog®) and Sanofi-Aventis, L.L.C insulin glulisine (Apidra®). Other versions include recombinant insulins that are absorbed steadily and more slowly after injection, resulting in a longer term maintenance of glucose-lowering activity (3). These longer acting insulin analogues include Novo Nordisk, Inc. version, insulin detemir (Levemir®) and Sanofi-Aventis, L.L.C. insulin glargine (Lantus®) (4).

Today, insulin analogue therapeutics are among the top-selling biologics in total sales, with nine marketed recombinant insulin proteins resulting in 2011 sales of $16 billion. Insulin glargine (Lantus®; Sanofi-Aventis, L.L.C.) is currently the leading recombinant human insulin, with just over $5 billion in sales in 2011. Insulin glargine is a long-acting basal insulin with three amino acid residue changes as compared with wild-type human insulin, resulting in a slow and steady release of insulin throughout the day. Following a once-daily injection, insulin glargine precipitates subcutaneously in a reversible isoelectric transition to form a depot for continuous release of drug. While the basal therapeutic analogues are providing clinical benefit, particularly for type 2 diabetes mellitus, therapy still remains suboptimal for many patient populations. Next-generation insulin analogues currently being investigated in the clinic include approaches to further stabilize insulin in a multimeric structure through the incorporation of zinc as well as the incorporation of glucose-responsive polymers that regulate the release of the insulin protein based on glucose levels.

4.2.2 Interferons

The interferons (IFNs) comprise a major class of protein therapeutics that act by augmenting existing inflammatory pathways for the treatment of a variety of diseases, including multiple sclerosis (MS), viral infections and cancer (5). Native IFNs are widely recognized as integral parts of the immune response and have been developed as therapeutics based on these inherent properties. The IFNs comprise a family of proteins encoded by 14 different genes, with natural immunomodulatory, anti-viral and antitumour activities, that can be grouped into three classes (IFN-α, IFN-β, and IFN-γ), based on their different receptors.

Beta-IFN immunomodulators are the first-line treatment for relapsing forms of MS. The mechanism of action of IFN-β therapy for MS is not well understood but it is hypothesized that IFN-β has a neuroprotective effect by reducing T-cell migration to the central nervous system. IFN-β-1a (Avonex®; Biogen Idec, Inc.), approved by the US FDA in 1996, is the top-selling IFN-β drug with $2.7 billion in sales in 2011. There are currently four IFN-β immunomodulators approved for marketing, differing from each other with respect to dosing frequency and route of administration, either subcutaneous or intramuscular. Total sales of all IFN-β biologics were $6.5 billion in 2011 (Figure 4.2).

The most common indication for IFN-α therapy is the treatment of chronic hepatitis C virus (HCV) infection. Typically, IFN-α therapeutics are given in combination with ribavirin, which is a small molecule purine nucleotide analogue that enhances IFN-α signalling and improves viral clearance. Two IFN-α therapeutics, Pegasys® (peginterferon α-2a; Genentech, Inc.) and Pegintron® (peginterferon α-2b; Merck and Company), have been approved for marketing. As their generic name implies, each of these IFN-αs has been engineered as PEGylated proteins to extend their half-life. The conjugation of a polyethylene glycol (PEG) moiety with the IFN protein increases the drug efficacy by improving exposure to the drug, since the larger hydrodynamic radius of the PEGylated protein allows the IFN to escape renal clearance (see half-life extension in the peptide section below). The improved pharmacokinetics results in once weekly dosing and improved patient adherence. In an alternative model, IFN-α was genetically fused to albumin protein (albinterferon α-2b; Human Genome Sciences); however, safety concerns resulted in withdrawal of the drug application after a number of patients suffered from pulmonary complications. Other indications in which IFN-α therapy has been proved to be effective include combination therapy with tyrosine kinase inhibitors for the treatment of chronic myeloid leukaemia and other malignancies including primarily haematological cancers.

Actimmune® (IFN γ-1b; InterMune, Inc.) is one of the few therapies that have proven effective in the treatment of two rare disorders, chronic granulomatous disease (CGD) and osteopetrosis. CGD is a hereditary chronic immunodeficiency disease in which granulocytes are unable to generate anti-microbial superoxide-based responses, which leads to recurrent fungal and bacterial infections. Actimmune is given as a lifelong prophylaxis for the prevention of severe infections. Osteopetrosis, which literally means 'stone bone', is a rare hereditary disorder of bone resorption by osteoclasts. IFN-γ, and the protein therapeutic Actimmune®, has been shown to normalize osteoclast function and stimulate osteoclasts to generate superoxide, aiding in the resorption of bone.

It is expected that the indications for IFN therapy will continue to expand based on the diverse roles that the different classes of IFN proteins play in our immune functions. The main drawback with current IFN therapy is the side effects, including flu-like symptoms, rashes, and depression, leading to noncompliance in treatment and poor quality of life for long-term treatment. New research and advances in IFN therapy are looking towards satisfying the unmet medical need for better-tolerated IFN therapies. In particular, IFN-λ (Type 3 IFN), initially described in 2003, signals through a distinct receptor that is highly expressed on hepatocytes, but not on many of the cell types where other IFN receptors are expressed. IFN-λ is currently in clinical trials for chronic hepatitis C infection.

4.2.3 Epoetins

Erythropoietin (EPO) is an essential growth factor for the development of red blood cells from bone marrow-derived precursors. Endogenous EPO is underproduced in cancer-induced anaemia, as well as in some late stages of renal disease, leading to low red blood cell counts. Recombinant forms of human EPO, as well as other erythroid stimulating agents (ESA), have demonstrated clinical success in the mentioned indications and notably as supportive care for chemotherapy-induced anaemia (6). Currently, there are nine marketed ESA protein therapeutics. In 2011, the ESA protein therapeutic class had a total sales of $8.2 billion (Figure 4.2), with the top-selling Aranesp® (darbepoetin alfa; Amgen, Inc.) generating $2.3 billion. Two of the marketed proteins, Aranesp® and Mircera® (methoxy polyethylene glycol-epoetin beta; F. Hoffman-La Roche, Ltd), have been engineered to be have longer half-life in serum. The potential risk of augmenting a deficient pathway with a recombinant version of the native protein became evident when patients in Europe developed antibody-mediated pure red cell aplasia (PRCA) while being treated with erythropoietic proteins. The patients developed anti-EPO antibodies that not only neutralized the drug but also crossed to the patient's endogenous EPO leading to severe transfusion-dependent anaemia.

A number of new approaches for stimulating erythropoiesis are being pursued which may address some of the risk associated with the current care. Peginesatide, formerly known as 'hematide', is a peptide-based drug that is linked to PEG for extended half-life. The peptide is unrelated to the sequence of native EPO and has been identified as an agonist that is capable of acting as an ESA. Peginesatide, developed by Affymax, Inc. and Takeda Pharmaceutical Company, Ltd., was approved by the US FDA for treatment of anaemia in March 2012 and is being marketed under the trade name of OMONTYS®. Approaches to erythroid stimulation through regulation of the EPO gene expression by targeting the $3'$ enhancer (Hypoxia inducible factor [HIF] stabilization) and the $5'$ promoter (GATA-2 inhibitor) have also been explored during the past decade. These approaches, although yet to reach the market, are particularly interesting because they lack the need for exogenous ESA therapy.

4.2.4 Granulocyte colony-stimulating factor (G-CSF)/granulocyte-macrophage colony-stimulating factor (GM-CSF)

Granulocyte colony-stimulating factor (G-CSF) and granulocyte-macrophage colony-stimulating factor (GM-CSF) are cytokines that regulate the differentiation of neutrophils, basophils, eosinophils and macrophages. GM-CSF functions just upstream of G-CSF and regulates the differentiation of myeloid progenitor cells to granulocyte/macrophage progenitor cells. G-CSF further differentiates the granulocyte/macrophage progenitors to neutrophils and is essential in maintaining adequate neutrophil levels which is critical for preventing bacterial and fungal infections. Some disease states and certain chemotherapy regimens can reduce neutrophil counts to dangerously low levels. The cloning and large-scale production of recombinant versions of these haematopoietic factors has led directly to their development as protein therapeutics in the treatment of neutropaenia associated with a number of disease states, particularly chemotherapy-related neutropaenia in oncology (7).

Today, multiple G-CSF protein therapeutics are in the market, with total sales of $5.7 billion in 2011 (Figure 4.2). The top-selling G-CSF is Neulasta® (pegfilgrastim; Amgen, Inc.), a second-generation recombinant human protein. Sales of Neulasta® contributed to more than half of the total sales ($3.9 billion) for this family of proteins. Neulasta® is the pegylated version of Neupogen® (filgrastim; Amgen, Inc.), a recombinant human protein produced in *E. coli* bacterial cultures. PEG moieties are attached to the protein to increase the hydrodynamic radius in order to reduce renal secretion of the protein, resulting in a longer circulating half-life. The *in vivo* half-life for Neulasta® in human serum is significantly improved over that of Neupogen®, resulting in a change from once-daily dosing to once-per chemotherapy cycle dosing, significantly simplifying treatment management and compliance for patients.

Like many natural proteins, natural human G-CSF is an *O*-linked glycosylated protein as such its structure includes the attachment of glycoside groups to specific amino acid residues through post-translational modification. Neulasta® and Neupogen® are produced in *E. coli* cells and they do not have the native *O*-linked glycosylation, since *E. coli* cells do not have the cellular machinery for glycosylation processing as found in mammalian cells. This difference in protein expression in mammalian cells and bacterial cells is an important consideration in the manufacturing of protein therapeutics since host cell-specific posttranslational modifications, especially glycosylation, can significantly influence the overall structure and function of the protein. Neutrogin® (lenograstim; Chugai Pharmaceutical Co. and F. Hoffman-La Roche, Ltd) is a G-CSF recombinant therapeutic protein that is produced in a mammalian cell line and thus contains *O*-linked glycosylation. The full impact of the glycosylation moieties on the G-CSF protein is not completely understood, but there is some evidence that suggests that the glycosylation protects the G-CSF from elastase degradation and may help to stabilize Neutrogin® in circulation.

4.2.5 **Growth factors**

Epoetins and G-CSF protein therapeutics, described above, can be considered members of a more generalized group of protein therapeutics called 'growth factors'. In this chapter, they were considered individually because of their significance among the classes of biologics. In addition to these two major families of growth factor therapeutics, other protein therapeutics are developed for treating disease through the regulation of cell differentiation or survival. Table 4.2 describes the subclasses and mechanism of action of marketed growth factor biologics as of 2011. Of significant value in this class, human growth hormone (hGH) is a peptide hormone produced naturally by the pituitary gland that stimulates cell reproduction, regeneration, and growth in humans. rhGH was pioneered for use as a protein therapy by Genentech, Inc. in 1981 and continues to be the standard care for growth hormone (GH) deficiency indications. Currently, there are seven versions of rhGH marketed, including a biosimilar manufactured by Sandoz, Inc. (Omnitrope®). As a class, rhGH therapeutics generated $3.0 billion in total sales in 2011 (Figure 4.2). The leading molecules in this class are Norditropin® (somatropin; Novo Nordisk, Inc.) and Genotropin® (somatropin; Pfizer, Inc.), each being sold in convenient devices for subcutaneous injection. Just downstream in the signalling pathway from hGH, insulin-like growth factor-1 (IGF-1) is synthesized in the liver in response to GH and is part of the cascade leading to normal growth. Increlex® (mecasermin; Ipsen Bio-pharmaceuticals, Inc.) is recombinant human IGF-1 produced in and purified from *E. coli* fermentations and approved by the US FDA in 2005 for treatment of abnormal growth insensitive to hGH treatment. IGF-1 treatment is used as an alternative to hGH, although this approach has been proved to be less efficacious.

Table 4.2 Examples of Growth Factor Therapeutics and Mechanism of Action

Product	Mechanism of Action	Indication
Growth hormone (GH), somatotropin	Human growth hormone receptor agonist	Growth hormone deficiency
Insulin-like growth factor, mecasermin	Stimulates cell growth and proliferation through IGFR-1 agonism	Growth failure
Keratinocyte growth factor, palifermin	FGFR-agonists, stimulates proliferation of keratinocytes	Chemotherapy-induced oral mucositis
Platelet-derived growth factor, becaplermin	PDGFR agonist, promotes chemotactic recruitment, proliferation and differentiation of fibroblasts	Wound healing, diabetic foot ulcers
Bone morphogenic protein 7 (BMP-7), osteogenic protein 1 (OP1)	Agonists, transformation of mesenchymal cells in bone and cartilage	Bone graft and repair

FGFR, fibroblast growth factor receptor; IGFR, insulin-like growth factor receptor; PDGFR, platelet-derived growth factor receptor.

4.2.6 Coagulation and fibrinolytic regulation

The coagulation and fibrinolytic processes oppose each other in a balance that maintains the haemostatic state. A healthy physiological balance between the two processes is essential, so effective protein therapeutics have been developed for rebalancing haemostasis when the proper equilibrium between these processes has been lost. As of 2011, six therapeutic protein coagulation factors, with total market sales of $6.8 billion (Figure 4.2), have been approved for marketing as replacement and supplemental therapy for haemorrhage and haemophilia. The top-selling drug in this class, Advate® (octocog alfa; Baxter) is a recombinant human full-length coagulation factor VIII. Advate® is indicated for the prevention and control of bleeding episodes, particularly in children and adults with haemophilia A. Multiple generations of factor VIII proteins have been developed as therapeutics beginning with plasma-derived human factor VIII, moving to a recombinant version of the full-length factor VIII protein once the gene sequence was determined, and finally moving onto safer production and formulation processes that are completely free of animal or human proteins (8). Each advancement has lent incremental improvements in the critical therapy for bleeding disorders. Aside from coagulation factor VIII, protein replacement therapy has also been approved for coagulation factor VIIa (NovoSeven®; eptacog alfa/Novo Nordisk, Inc.) and coagulation factor IX (BeneFIX®; nonacog alfa; Pfizer, Inc.).

On the fibrinolytic side of the equation, the biologic heparin (highly sulfated glycosaminoglycan polymer, generally manufactured in the 12–15 kDa range) which targets antithrombin III eventually leading to regulation of factor Xa has been a mainstay since the early 1900s. More recently, low-molecular-weight heparins (<2 kDa), which target factor Xa directly, have been developed. The small molecule drug, warfarin (sold under the trade names of Coumadin®, Jantoven®, Marevan®, Warfant®, and Waran®) is also used as standard care since the 1950s, with recent small molecule developments, such as Xarelto® (rivaroxaban; Bayer AG and Janssen Pharmaceuticals, Inc.) and Pradaxa® (dabigatran etexilate; Boehringer Ingleheim GmbH), extending the therapeutic capabilities in this field. However, all the anticoagulant molecules mentioned above prevent the formation of clots, but do not dissolve clots once they are formed. The key biologics marketed in the field of clot dissolution (i.e. thrombolytic activity) are different forms of recombinant tissue plasminogen activator (rTPA), a serine proteinase that cleaves plasminogen to plasmin. Several forms of rTPA have been approved for marketing since the 1990s, including Activase® (alteplase, full-length rTPA; Genentech, Inc.), Retavase® (reteplase; nonglycosylated, truncated form of rTPA; Boehringer Ingleheim GmbH), and TNKase® (tenecteplase, a genetically modified rTPA; Genentech, Inc.) which are indicated for various forms of myocardial infarction. The combined 2011 sales for Activase® and TNKase® were approximately $450 million. While these rTPA biologics are enzymes, they are included here because they are not used strictly as replacement therapeutics and because their mechanism of action relates to maintenance of haemostasis.

4.2.7 Therapeutic enzymes

Therapeutic enzymes are mostly replacement therapies for patients who lack certain enzymatic activities due to genetic defects. Table 4.3 lists the marketed enzyme protein therapeutics as of 2011 (see Chapter 11 for more data). The therapeutic enzyme class includes mostly orphan therapies for rare and chronic diseases with very small patient populations. In total, this market reached just over $3 billion globally in 2011. A few examples of enzyme replacement therapy (ERT) are described here to demonstrate the therapeutic potential and exciting future directions for this protein therapeutic class.

Table 4.3 Examples of Enzyme Replacement Therapeutics and Mechanism of Action

Product	Mechanism of Action	Indication
Cerezyme (imiglucerase)	Beta-glucocerebrosidase replacement, hydrolyses glucocerebroside to glucose and ceramide	Gaucher disease
Lumizyme/myozyme (alglucosidase alfa)	Alglucosidase-alfa replacement, catalyses glycogen	Pompe disease
Elaprase/idursulfase (Iduronate-2-sulfatase)	Iduronate-2-sulfatase enzyme replacement, catalyses glycosaminoglycans (GAG)	Hunter syndrome
Fabrazyme (α-galactosidase A)	Alpha-galactosidase enzyme replacement, hydrolyses glycosphingolipids	Fabry disease and Gaucher disease
Naglazyme/galsulfase (arylsulfatase b)	Arylsulfatase replacement, prevents GAG accumulation	Maroteaux-Lamy syndrome, Mucopolysaccharide disease type 6 (MPS VI)
Aldurazyme (laronidase)	Alpha-L-iduronidase replacement, recombinant human protein, prevents GAG accumulation	Hunter syndrome, Mucopolysaccharide disease type 1 (MPS I)
VPRIV (velaglucerase alfa)	Glucocerebrosidase replacement, recombinant human protein	Gaucher disease

The leader in this class, Cerezyme® (imiglucerase; Sanofi-Aventis, L.L.C./Genzyme Corporation), is used for the treatment of type 1 Gaucher disease. Gaucher disease, which is triggered by the absence of glucocerebrosidase, is the most common lysosomal storage disease resulting in the accumulation of glucocerebroside in macrophages, leading to organomegaly of the liver and spleen as well as dysfunction of the skeletal system, lungs and nervous system. Since 1991, ERT has been proved to be effective in the treatment of this disease. Initially, Gaucher disease was treated with Alglucerase®, which was purified, placental-derived glucocerebrosidase. In 1994, however, it was replaced by Cerezyme®, a recombinant variant of glucocerebrosidase, which is the current standard of care treatment. For many years, inconsistency issues have plagued the manufacturing of Cerezyme®, resulting in short supply of the therapeutic. In 2010, however, an innovative alternative to conventional recombinant protein production, VPRIV® (velaglucerase alfa, Shire plc) was introduced into the manufacturing process (9). VPRIV is a recombinant human protein produced in Chinese hamster ovary (CHO) cells using gene-activation technology which involves introducing a DNA promoter in a precise location upstream of an endogenous gene in a human cell line to drive high expression. This alternative to conventional recombinant DNA technology is also being used in the manufacturing of Replagal® (agalsidase alfa; Shire plc), an ERT for Fabry disease.

Advances in ERT for the treatment of Pompe disease have also proven therapeutic efficacy. Pompe disease is a pathology driven by a deficiency in acid α-glucosidase (GAA), which normally catalyses the breakdown of glycogen into glucose. In the disease state, however, glycogen accumulates to abnormal levels in multiple tissues affecting cardiac, skeletal and smooth muscle functions. The first successful demonstration of ERT in Pompe disease occurred in 1973, when a highly purified placental-derived acid GAA enzyme was prepared and delivered to a patient by intravenous (IV) infusion. In 2006, after a careful comparative clinical study of a recombinant version, Myozyme® (recombinant alglucosidase alfa; Sanofi-Aventis, L.L.C./Genzyme Corporation) was approved by the US FDA. Building on the clinical success observed with this ERT, newer therapeutic strategies are currently being evaluated in the clinic for chaperone therapy to supplement the ERT by protecting nascent enzyme from degradation in addition to the recombinant supplement. The innovative new ERTs described here will certainly be extended to other applications within this unique and diverse class of protein therapeutics.

4.2.8 Peptide therapeutics

Although peptide therapeutics were first introduced into the market over 20 years ago, it has taken many years for this approach to take hold. Activity over the past 5 years has been greater than all of the cumulative previous efforts. Since 2001, more than 10 new peptide therapeutics have entered the market, the top four of which reached global sales over US $1 billion by 2008. These include Copaxone® (glatiramer injection; Teva Pharmaceutical Industries LTD; $3.18 billion), for treatment of relapsing-remitting MS; Lupron® (leuprolide injection; Abbott Laboratories; $2.12 billion), a gonadotropin-releasing hormone (GnRH) analogue receptor agonist for treatment of prostate cancer; Zoladex® (goserelin acetate; AstraZeneca; $1.14 billion), another GnRH analogue receptor agonist for treatment of prostate and breast cancers; and Sandostatin® (octreotide injection; Novartis Corporation; $1.12 billion), a somatostatin analogue used for treatment of patients with acromegaly. Between 2001 and 2008, an average of 16.8 new peptides entered clinical studies per year (*10*).

Peptides are a diverse class of compounds sometimes defined as structures comprising ∼40 amino acid residues or less with limited three-dimensional sequence driven structure. For production, they can be either chemically synthesized through chemical peptide synthesis or expressed recombinantly in cells. In February 2012, the US FDA issued draft guidance for classification of peptides as protein biologics versus peptide chemical entities. Peptides having a length of less than 40 amino acid residues are to be regulated as peptide chemical entities (i.e. like small molecules). Peptides with 100 amino acid residues or greater would be regulated as protein biologics. Peptides with residue lengths between 40 and 99 would be regulated based on the mode of manufacturing; i.e. those generated synthetically would be regulated as peptide chemical entities, whereas those produced in biological systems would be regulated as protein biologics. The potential advantage of the 'biologics' designation would be the 12-year data exclusivity granted in the recent Public Health Service (PHS) Act (Patient Protection and Affordable Care Act [PPACA]; P.L. 111–148). It is expected that this guidance should be further clarified within the next few years.

Peptide therapeutics sources include engineered natural peptides and their derivatives or through *de novo* discovery through selection of synthetic peptide libraries using display technologies. Peptides are an unusual class of protein therapeutics to classify since they cross and encompass many of the traditional means of cataloguing therapeutics; for example, peptide therapeutics can be included in each of the two Leader et al. categories on which this chapter has been organized, some with regulatory and enzymatic activities as well as functioning through targeting activity. Many of the advantages of peptide therapeutics are based on their small size and flexible structure, allowing therapeutic peptides to bind to targets that are historically intractable for larger protein therapeutic molecules. Other beneficial properties of therapeutic peptides include high efficacy, low off-target related toxicities due to high specific binding, and increased tissue penetration for improved target access.

Despite the advantages for application of peptides as therapeutics, there are also significant challenges associated with peptide therapeutics. The *in vivo* stability of peptides is, in general, a challenge since peptides have a short half-life, owing to rapid renal clearance and protease degradation. A number of strategies have been implemented to mitigate the instability of peptides. One approach that is used in several marketed therapeutics is chemical conjugation of the peptides or protein to PEG in order to increase the half-life of the molecule, as was discussed with the IFN therapeutics, Pegasys® and Pegintron®, as well as the G-CSF protein therapeutic, Neulasta®. This approach adds a significantly increased hydrodynamic radius to the small peptide, which can result in a longer circulating half-life by escaping rapid renal clearance. In general, the cutoff for kidney filtration and protein elimination is considered to be the 70 kDa range. Proteins above 70 kDa escape this route of clearance, whereas proteins smaller than 70 kDa are eliminated via kidney filtration based on a sliding scale of size, charge, and shape, with the smallest, most neutral peptides being eliminated fastest, seriously impacting the feasibility of using them as effective therapeutics. Therapeutic proteins and peptides that fall below 70 kDa generally will be removed from circulation rapidly, which may impact the feasibility of using them as effective therapy. Other half-life extension strategies have been employed for peptides, including genetic fusion to naturally long-half-life proteins, such as albumin or human Fc domains (discussed below), or genetic fusion and recombinant production of the peptide candidate and inert polypeptides such as 'recombinant PEG' or homo-amino acid polymer (HAP; HAPylation). In peptide therapeutic design, a balance needs to be struck between the pharmacokinetics required and the value of the small size of the peptide in bio-distribution and access to targets.

Proteolytic instability of natural peptides is a second key limitation associated with using peptides as therapeutics. The flexibility of the peptide structure leaves the amino acid backbone open and accessible to proteolytic cleavage by gastrointestinal, plasma, and tissue peptidases. In many cases, proteolytic cleavage of natural peptides, e.g. cleavage and inactivation of glucagon-like peptide-1 (GLP-1) by dipeptidyl peptidase 4 (DPP4; also known as CD26), is an integral part of the regulation of the biological processes in which those peptides are involved. Thus, for a lead peptide molecule, it is often necessary to engineer the amino acid sequence for protease resistance to provide an improved pharmacokinetic profile. This can include a number of chemical optimizations to the peptide, notably cyclization of the peptide, or the incorporation of nonnatural amino acids or other chemical modifications to protect the peptide from protease degradation.

Recent developments in peptide-based therapeutics are moving into exciting potential areas for next-generation drugs. Cell-penetrating peptides are being explored for intracellular targeting and tissue-specific peptides for local delivery of drug are showing promise at the early stages of development. Although currently still limited to parenteral administration, new formulations and routes of administration, including oral and intranasal delivery, continue to advance. Finally, the advances made in the chemical synthesis including faster production and simpler scaleup are opening the door to better and cheaper manufacturing methods leading to promise of broadening applications for peptide therapeutics in the foreseeable future.

4.3 TYPES OF PROTEIN THERAPEUTICS II: TARGETING ACTIVITY

Therapeutic proteins with special targeting activities comprise a large share of the currently marketed protein therapeutics. These pharmacological activities can involve (1) interfering with the pharmacological activity of molecules or targeting a pathogenic organism or (2) targeted delivery of pharmacologically active compounds or proteins. In the next section, several of the major types of protein therapeutics with targeting activities are described.

4.3.1 Monoclonal antibodies

Among all the protein therapeutics, monoclonal antibodies (mAbs or Mabs) have clearly demonstrated the greatest clinical success and hold the greatest potential yet to be realized. The market for therapeutic mAbs in 2011 was $56.7 billion and is estimated to grow at a rate of approximately 9%–15% through 2015. The total mAb and Fc protein market is expected to reach the $75–$90 billion range by 2015. By the end of 2011, 39 mAb and FcFP therapeutics had been approved and marketed in the United States and European Union. One of the most important new mAbs to be approved for marketing since then is Perjeta® (pertuzumab; Genentech, Inc.), a new anti-HER2 mAb, which was approved for treatment of breast cancer by the FDA in June 2012. The number of new mAbs entering the market is expected to continue to rise based on the fact that 45–50 new first-in-human studies began per year from 2007 to 2009. Moreover, currently there are approximately 55 mAbs and FcFPs that are either post-phase IIb or in phase III clinical trials, indicating a rich late-stage pipeline for the industry. The current leading mAb therapeutics fall into narrow target classes as follows: anti-tumour necrosis factor (TNF)-α mAbs and FcFPs leading with a $24 billion market, anticancer antibodies at $21 billion, and other anti-inflammatory mAbs at $5.8 billion.

Therapeutic antibodies are modelled on our immune system's natural ability to fight disease. Native antibodies are an integral part of our adaptive immune system. The extraordinary binding specificity of antibodies drives their function to circulate, find a target that is recognized as foreign, and bind tightly to the target to neutralize the foreign invader. This binding characteristic is exploited in recombinant therapeutic mAbs to direct interactions with specific targets causing disease-modifying results. Table 4.4 lists examples of many of the marketed antibody therapeutics.

Antibodies are multifunctional, complex heterodimeric glycoproteins (Figure 4.4). The basic structure consists of two identical heavy chains (HCs), approximately 50 kDa each, complexed with two identical light chains (LCs) of approximately 25 kDa, resulting in a 150 kDa functional protein. The N-terminal domains of the HC and LC, which have a high degree of amino acid sequence variation, are called variable regions (V regions). The significant variation in sequence distinguishes the V region, both structurally and functionally, from the rest of the molecule. The variable part of the heavy chain is abbreviated V_H (or VH), the variable part of the light chain is abbreviated V_L (or VL). The remaining portion of the HC and LC is called the constant region, and has a distinct and mostly invariant amino acid sequence that identifies an antibody molecule to an Ig class, IgA, IgD, IgE, IgG and IgM. To date, all recombinant therapeutic antibodies approved are of the IgG isotype. There are two classes of LC, called kappa and lambda, as defined by the light chain constant region amino acid sequence. The LC consists a single constant domain, whereas the HC, which is about twice as long as the LC, consists of 3–4 constant domains, depending on the Ig isotype.

In humans, the IgG class is subdivided into four isotypes, IgG_1, IgG_2, IgG_3 and IgG_4. The alternative form of these abbreviations IgG1, IgG2 etc. is also in common use. The amino acid sequences dictate the structure of the 'tail' region, that is, the Fc domain and the flexibility of the hinge region, which in turn define binding to Fc receptors (FcRs) and complement system components. The Fc-gamma receptors are present on a variety of immune cells including monocytes, macrophages, neutrophils, dendritic cells and natural killer cells. Engagement of the antigen-opsonized IgG with an

Table 4.4 Marketed Antibody, Fc Proteins and Antibody Fragment Therapeutics

Therapeutic	Indication	Target	Scaffold	Company
IgG				
Orthoclone OKT3® (muromonab-CD3)	Transplant rejection	CD3 on T cells	Murine IgG$_{2a}$	Ortho Biotech (Janssen Biotech, Inc.)
Rituxan® (rituximab)	NHL and rheumatoid arthritis	CD20 on B cells	IgG$_1$κ, mouse/human chimeric	Biogen Idec, Inc./Genentech, Inc.
Zenapax® (daclizumab)	Transplant rejection	IL-2Rα (CD25; Tac)	IgG$_1$, humanized	Abbott Laboratories (PDL/ F. Hoffman-La Roche, Ltd)
Synagis® (palivizumab)	RSV infection (infant)	RSV F-protein	IgG$_1$κ, mouse/human chimeric	MedImmune, LLC
Remicade® (infliximab)	Rheumatoid arthritis and Crohn's disease	TNF-α	IgG$_1$κ, mouse/human chimeric	Centocor (Janssen Biotech, Inc.)
Herceptin® (trastuzumab)	Breast cancer	HER2/Neu	IgG$_1$κ, humanized	Genentech, Inc.
Simulect® (basiliximab)	Transplant rejection	IL-2Rα (CD25; tac)	IgG$_1$κ, chimeric	Novartis Corporation
Campath -1H® (alemtuzumab)	Leukaemia	CD52 on B- and T cells	IgG$_1$κ, humanized	ILEX/Millenium Pharmaceuticals
Humira® (adalimumab)	Rheumatoid arthritis and Crohn's disease	TNF-α	IgG$_1$κ, human	CAT Cambridge Antibody Technology Group Plc, Abbott Laboratories
Xolair® (omalizumab)	Asthma	IgE	IgG$_1$κ, humanized	Genentech, Inc.
Raptiva® (efalizumab)	Psoriasis	CD11a, α-subunit of LFA-1	IgG$_1$κ, humanized	Genentech, Inc.
Erbitux® (cetuximab)	Colorectal cancer	EGF-R (HER1, c-ErbB-1)	IgG$_1$κ, chimeric	ImClone Systems/BMS Bristol-Myers Squibb Company
Avastin® (bevacizumab)	Colorectal cancer	VEGF	IgG$_1$, humanized	Genentech, Inc.
Tysabri® (natalizumab)	Multiple sclerosis	α4 subunit of α4β1 or α4β7	IgG$_4$k, humanized	Biogen Idec, Inc./Elan Corporation
Vectibix® (panitumumab)	Colorectal cancer	EGF-R (HER1,c-ErbB-1)	IgG$_2$k, Human	Amgen, Inc.
Soliris® (eculizumab)	Paroxysmal nocturnal haemoglobinuria (PNH)	Complement C5	IgG$_{2/4}$ modified Fc, humanized	Alexion Pharmaceuticals

(continued on next page)

Table 4.4 Marketed Antibody, Fc Proteins and Antibody Fragment Therapeutics *(continued)*

Therapeutic	Indication	Target	Scaffold	Company
Simponi™ (golimumab; CNTO-148)	Rheumatoid arthritis	TNF-α	IgG$_1$, Human	Centocor (Janssen Biotech, Inc.)
Stelara™ (ustekinumab; CNTO-1275)	Psoriasis	P40 subunit of IL-12 & IL-23	IgG$_1$ Human	Centocor (Janssen Biotech, Inc.)
Actemra® (RoActemra in European Union) (tocilizumab)	Castlemans disease and rheumatoid arthritis	IL-6R	IgG$_1$, humanized	F. Hoffman-La Roche, Ltd/ Chugai Pharmaceutical Co.
Ilaris® (canakinumab)	CAPS	IL-1β	IgG$_1$, human	Novartis Corporation
Arzerra™ (ofatumumab)	B-cell chronic leukaemia; non-Hodgkin's lymphoma and rheumatoid arthritis	CD20 on B cells	IgG$_1$, humanized	Genmab/ GlaxoSmithKline
Prolia™ and XGEVA™ (denosumab)	Osteoporosis	RANK ligand	IgG$_2$, Human	Amgen, Inc./ GlaxoSmithKline
Benlysta™ (belimumab)	Lupus (SLE)	BLyS	IgG$_1\lambda$, Human	GlaxoSmithKline/ Human Genome Sciences
Yervoy™ (ipilimumab)	Malignant melanoma	CTLA-4: Cytotoxic T-lymphocyte Antigen 4	IgG$_1$k, Human	Medarex, Inc./ Bristol-Myers Squibb Company
Antibody conjugates				
Mylotarg® (gemtuzumab ozogamicin)	Leukaemia	CD33	Humanized IgG$_4\kappa$- conjugate with ozogamicin	Wyeth Pharmaceuticals, Inc. (Pfizer, Inc.)
Zevalin® (ibritumomab tiuxetan)	NHL	CD20 on B cells	Murine IgG$_1\kappa$ conjugate with ^{90}Y or ^{111}In	Biogen Idec, Inc.
Bexxar® (tositumomab-I131)	Non-Hodgkin's lymphoma	CD20 on B cells	Murine IgG$_{2a}$/λ conjugated with ^{131}I	Corixa Corporation
Adcetris™ (brentuximab vedotin)	Lymphoma	CD30	Chimeric IgG$_1$- conjugated with auristatin derivative	Seattle Genetics, Inc./Takeda Pharmaceutical Company, Ltd.
Bispecific antibodies				
Removab® (Catumaxomab)	Malignant ascites, cancer	EpCAM and CD3	Rat IgG$_{2b}$-mouse IgG$_{2a}$ hybrid	Fresenius/Trion Pharma GmbH

Table 4.4 Marketed Antibody, Fc Proteins and Antibody Fragment Therapeutics *(continued)*

Therapeutic	Indication	Target	Scaffold	Company
Fabs				
ReoPro® (abciximab)	CVD	gPIIb/IIIa on platelets	Chimeric Fab	Centocor (now Janssen Biotech, Inc.)/Eli Lilly and Company
Lucentis® (ranibizumab)	Wet age-related macular degeneration	VEGF-A	Humanized IgG_1k Fab	Genentech, Inc./ Novartis Corporation
Cimzia® (certolizumab pegol)	Rheumatoid arthritis	TNF-α	PEGylated Fab, humanized	UCB, Inc./ Schwartz Pharma AG
Fc fusion proteins				
Enbrel® (etanercept)	RA	TNF-α	IgG_1 Fc fused to TNFR p75 exodomain	Immunex (now Amgen, Inc.)
Amevive® (alefacept)	Psoriasis	CD2	IgG_1 Fc fused to CD2-binding domain of LFA-3	Biogen Idec, Inc.
Orencia® (abatacept)	Rheumatoid arthritis	CD80/CD86	CTLA-4-modified IgG_1 Fc	Bristol-Myers Squibb Company
Arcalyst® (rilonacept)	CAPS, Muckle–Wells syndrome	IL-1β, IL-1α, IL-1RA	IgG_1; with IL-1R & IL-1AP fused inline	Regeneron Pharmaceuticals, Inc.
Nplate® (romiplostim, AMG-531)	Thrombocytopenia	TPO-R	aglucosyl IgG_1 Fc peptide fusion	Amgen Inc.
Nulojix™ (belatacept)	Renal transplantation	CD80/86	CTLA-4 modified IgG_1 Fc fusion	Bristol-Myers Squibb Company

CAPS, cryopyrin-associated periodic syndrome; CVD, cardiovascular disease; NHL, non-Hodgkin's lymphoma; RSV, respiratory syncytial virus; SLE, systemic lupus erythematosus; RA, rheumatoid arthritis

activating FcR can result in antibody-dependent cellular cytotoxicity (ADCC) or antibody-dependent cellular phagocytosis (ADCP) and the release of inflammatory cytokines, as well as reactive nitrogen and oxygen species that can induce apoptosis and further stimulate immune responses. The neonatal Fc receptor (FcRn) is a unique member of the FcR family. Through interactions with this receptor, the circulating IgG molecule escapes default protein degradation pathways and is recycled back into circulation resulting in long serum half-life, typically 14–21 days.

The human IgG Fc contains a single conserved N-linked glycosylation site (N297) important for structure and function. Since N297 glycosidation is required for full engagement of FcRs, most mAb therapeutics are made in host cells capable of protein glycosylation. As shown in Figure 4.4, the glycosylation at N297 resides between the C_H2 domains of the HC homodimer. Heterogeneities in the

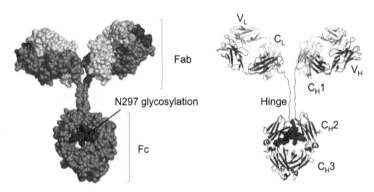

FIGURE 4.4 Structural elements of a monoclonal antibody.

Two molecular models of a monoclonal antibody to represent the key structural features. V_L, Light chain variable region; V_H, Heavy chain variable region; C_L, Light chain constant region; $C_H l$, Heavy chain constant region domain1; $C_H 2$, Heavy chain constant region domain 2; $C_H 3$, Heavy chain constant region domain 3. (For colour version of this figure, the reader is referred to the online version of this book.)

carbohydrate structure influence the overall activities of the antibody Fc, since the glycan is integral to the Fc structure and function. In particular, Fc effector functions through the FcR family, with the exception of FcRn, are strongly impacted by the presence and structure of the glycans. The ratio of the major glycoforms on the therapeutic mAb needs to be monitored carefully during the manufacturing process to ensure consistent product efficacy and safety.

4.3.2 Monoclonal antibody therapeutic discovery and engineering

There are two main functional parts to an IgG antibody: the antibody–antigen binding fragment (FAb) arms, which contain the V regions and provide target binding and specificity, and the Fc region (including the hinge), which provides interaction with the immune system. Each can be engineered separately to provide optimized therapeutic antibody drugs with 'fit-for-purpose' biological functions.

For the V regions, there are four primary sources used today for *de novo* mAb discovery, including (1) nonhuman immunized sources (e.g. mouse or rat hybridomas), followed by humanization and optimization of the sequences; (2) immunization of transgenic animals such as mice or rats in which the natural rodent repertoire has been replaced with the human antibody repertoire (e.g. Medarex or Abgenix transgenic mice producing human antibodies); (3) phage display libraries composed of natural, synthetic, or semisynthetic human antibody genes; or (4) human antibody genes recovered from antibody-producing B or plasma cells obtained from either antigen-naïve or immunized humans.

Methods for mAb discovery via immunization of mice followed by generation of immortalized hybridomas, as first described by Köhler and Milstein in 1975 and concurrent developments in recombinant DNA technologies, including gene cloning into plasmid DNA through restriction enzyme digestion, polymerase chain reaction, and characterization of the mouse and human germ line antibody genes, formed the basis of foundational technologies on which the therapeutic mAb industry was built. Hybridoma methods were used for the discovery of the first US FDA-approved mAb, Muromonab-CD3 (Orthoclone OKT3®). The approval of this antibody was a significant milestone for the biotechnology industry; however, serious issues with human anti-mouse antibody responses limited the success for this fully mouse antibody in the treatment of disease. In 1984, Sheri Morrison described methods for chimerizing an mAb as a means to reduce immunogenicity by replacing the majority of the mouse antibody sequence with a human sequence equivalent (*11*). Today, the humanization technology has advanced such that some rodent-derived antibodies can be engineered to be virtually indistinguishable from human antibodies. The sourcing of fully human antibody V region sequences derived from transgenic rodents producing human antibodies, phage display libraries, or directly from human B cells has increased exponentially in the past decade and will become the method of choice for future antibodies.

4.3.3 Fc engineering

In addition to the identification and engineering of the V region sequences to drive the specificity and affinity for the target, the Fc region of the mAb can be purposefully designed and engineered. The structure and sequence of the Fc region impacts the antibody-dependent engagement of the immune system and, as such, impacts the efficacy and safety of the therapeutic mAb. Historically, only the human IgG_1 isotype was used for incorporation into therapeutic mAb candidates. More recently, efforts have been made to optimize Fc functionality with respect to the desired biological function of the therapeutic antibody, either through the use of alternative human natural isotypes (i.e. IgG_2, IgG_4) or via engineering of the Fc to increase or decrease Fc interactions with the immune system. Jiang et al. (*12*) summarized therapeutic MOAs for antibodies in clinical development into various classes, as follows: class 1, cell-bound antigen binding with depletion; class 2, cell-bound antigen blocking (no depletion); and class 3, soluble antigen blocking. It is the combination of the antigen-binding activity and the role of the Fc that makes mAbs capable of such diverse therapeutic modes of action. For example, in cases in which the target protein is a tumour antigen with highly selective expression on tumour cells, the Fc engineering strategy would be to increase Fc–immune system interactions to drive immune-cell-mediated killing and depletion of that targeted cell. Examples of this type of engineering include engineering of the Fc glycan by removing fucose from the glycan structure, resulting in increased binding to $Fc\gamma RIIIa$ and concomitant increased ADCC. Another example of increasing Fc affinity to $Fc\gamma$ receptors is the engineering of the IgG_1 Fc by mutagenesis resulting in increased binding to all $Fc\gamma$ receptors and/or complement components. On the other hand, if the target protein for the therapeutic antibody is a ubiquitous cell surface receptor that is responsible for upregulating signalling of an immunological pathology, then Fc interaction with the immune system would, be at a minimum, undesirable, and maximally, could pose a significant safety risk. In this case, the appropriate IgG Fc would lack the ability to bind $Fc\gamma$ receptors on immune cells or complement system proteins referred to as a 'silenced Fc'. This would result in target engagement on the surface of the targeted cell but without depletion of that cell (*13*). Finally, protein engineering has led to identifications of Fc variants that influence the interaction with the FcRn, resulting in modulation of the circulating half-life of the mAb.

4.3.4 Fc fusion proteins (FcFPs)

FcFPs are engineered proteins that are composed of a protein or peptide fused to an IgG Fc domain. The first engineered FcFP was described in 1989. This prototype FcFP was called an immunoadhesion since it was a fusion of the ligand-binding domain of CD4 to the N-terminus of an IgG_1 Fc (C_H2 and

C_H3 domains). Although this original Fc fusion molecule was not developed into a therapeutic, this model was successful and provided a template on which most subsequent FcFPs have been based. Etanercept (Enbrel®; Amgen, Inc.) is a blockbuster therapeutic protein composed of the ectodomain of the TNF receptor, TNFR2, genetically fused to the N-terminus of human IgG_1 Fc (Figure 4.5). Enbrel® (etanercept; Amgen, Inc., Pfizer, Inc.) is a leader in the anti-TNF-α biologics drug class in total sales, with $7.9 billion in 2011. Since 1998, when Enbrel® was first approved, six other US FDA-approved FcFPs have entered the market for a variety of indications.

There are several reasons for using an FcFP as a therapeutic. One of the main reasons is for half-life extension for smaller proteins, protein domains, and peptide. IgG Fc domains serve this function particularly well for two reasons, size and FcRn binding activity. As mentioned earlier in this chapter, 70 kDa is the approximate molecular weight cutoff for kidney filtration and protein elimination. Attaching a peptide or protein to each arm of an antibody Fc ($\sim$50 kDa) is generally sufficient to significantly impact circulating half-life. The FcFP also interacts with FcRn, further increasing the serum half-life of the therapeutic. The FcRn receptor binds to Fc-containing proteins in a pH-dependent manner and salvages them from being degraded in endosomes. This mechanism recycles the Fc-containing protein back into the circulation, resulting in a half-life-extending benefit. Marketed FcFPs have a half-life in the 4- to 17-day range.

In addition to extending half-life, the Fc portion of an FcFP can contribute to the pharmacodynamics of a protein therapeutic. Fc-mediated function, such as ADCC, ADCP and CDC (see section above) can be maintained in the FcFP. As in the full IgG molecule, these effector functions are important to the therapeutic design of the FcFP, e.g. maintaining or enhancing ADCC and ADCP in anticancer molecules designed to target and eliminate tumour cells or eliminating or reducing Fc activity in FcFPs that interact with immune cells.

Several other characteristics are often considered advantages of FcFPs. The Fc proteins exists as a homodimer because of the strong interactions of the C_H3 domains, much the same as in a full IgG molecule. Typically, the molecules are designed with an intact hinge, which will further stabilize the

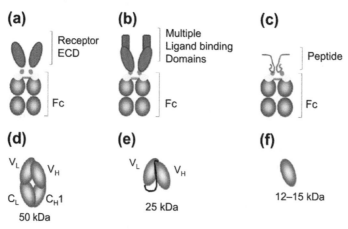

FIGURE 4.5 Examples of antibody fragments.

(a) Receptor Fc fusion; (b) Cytokine trap; (c) Peptide agonist; (d) Fab; (e) Single chain Ab; (f) Single domain Ab.

dimers through the interchain disulfide bond. This format provides the advantage of two binding sites (bivalent) per molecule and improves binding to the target protein which can result in improved efficacy of the therapeutic protein. Furthermore, the Fc fusion provides a convenient and efficient method to purify the protein, since the intact Fc will maintain affinity for Protein A, enabling a simple and efficient chromatography purification method in manufacturing process.

A wide variety of FcFPs can be designed with different modes of activity (*14*). The therapeutic approaches can be classified according to functionality. Like etanercept, soluble receptor FcFPs serve to inhibit a ligand and receptor interaction by operating as a decoy or sink for the ligand and preventing it from activating and signalling through the native receptor. This decoy/regulator concept is also used in nature for protein interaction regulation. Naturally occurring soluble receptors, such as first described for the interleukin (IL)-2R have now been identified for many cytokine receptors. These soluble versions of the receptors are generated by proteolytic cleavage of a surface protein resulting in shedding, or alternatively, by expression of an alternatively mRNA-spliced variant resulting in a receptor-encoding transcript that lacks the membrane-spanning domain. The circulating soluble receptors regulate the ligand/cell surface receptor interaction by acting as a decoy receptor. Using this same mechanism, soluble therapeutic receptor/FcFPs can neutralize ligand receptor interactions that are implicated in disease.

Regeneron Pharmaceuticals, Inc. developed an IL-1-binding FcFP protein which functions via a slightly different mode of action. They call this construct a 'cytokine trap'. High-affinity binding of IL-1 to its receptor requires two proteins, IL-1R1 and the IL-1R accessory protein (IL-1RacP). In the IL-1 cytokine trap FcFP, rilonacept (Arcalyst®; Regeneron Pharmaceuticals, Inc.), the ligand binding domains of each of these two proteins are sequentially genetically fused and fused to the IgG Fc (Figure 4.5). Cytokine trap Fc proteins can provide very high affinity and high potency. Rilonacept is approved by the US FDA for cryopyrin-associated periodic syndrome (CAPS) and is the first cytokine trap Fc protein in the market. Eylea® (aflibercept; Regeneron Pharmaceuticals, Inc.), which was approved in November 2011, for treatment of wet age-related macular degeneration (wet AMD), is the second Fc cytokine trap to be approved for marketing. Aflibercept is an Fc fusion of the second domain of vascular endothelial growth factor receptor-1 (VEGFR1) fused to the third domain of VEGFR2.

A third class of Fc proteins that has demonstrated some clinical success so far is the receptor agonist Fc protein. Many highly active natural ligand peptides are small in size, 50 amino acid residues in length or shorter. Making recombinant therapeutics of these molecules is a challenge because of very rapid disappearance from the serum due to renal clearance and proteolytic instability. This is a rich class of molecules for which FcFP approaches can enable new therapeutic approaches. Advances have been made for these potent metabolic peptides through Fc fusion engineering to stabilize the labile peptides and extend the half-life to improve efficacy. These include mimetibody constructs and peptibodies. Romiplostim (Nplate®; Amgen, Inc.), a thrombopoietin receptor agonist Fc fusion construct for treatment of chronic immune thrombocytopenia, is the first US FDA-approved receptor agonist peptibody. Another exciting peptide FcFP in late-stage clinical trials is LY2189265 (Eli Lilly and Company), a DPP4-resistant version of GLP-1 fused to a partially silenced IgG$_4$ Fc, which is now in phase IIb clinical trials for weekly treatment of type II diabetes.

4.3.5 Engineered antibody fragment therapeutics

Antigen binding fragments, including FAb, single chain Fv (scFv), and single domain antibodies (dAbs), have been engineered and have each demonstrated stable target protein interactions

(Figure 4.5). These molecules complement the therapeutic mAb platform as they maintain the exquisite target binding affinity and specificity of an mAb variable region domain in the context of a smaller and monovalent scaffold which can extend the target repertoire for antibody therapeutics to some which are intractable to full IgG scaffolds (3). There are therapeutic situations in which a shorter circulating half-life and/or completely eliminating Fc effector function could dictate that an antibody scaffold lacking an Fc domain is a more appropriate therapeutic than a full mAb. As an example, Reopro® (abciximab, Centocor Ortho Biotech Products, L.P. (Currently Janssen Biotech, Inc.) and Eli Lilly and Company), which was the first FDA-approved antibody therapeutic, is a FAb protein generated by proteolytic cleavage of the full-length chimeric IgG$_1$ mAb, followed by purification of the released FAb fragments. This FAb binds to the glycoprotein (gp) IIb/IIIa on platelet membranes and is approved for several cardiovascular disease indications as a platelet aggregation inhibitor. Safety concerns dictate FAb design to shorten the time that the Reopro® is found in circulation and to eliminate the chance of platelet depletion through Fc effector function. In addition to Reopro, two other FAb therapeutic proteins, Lucentis® (ranibizumab; Genentech, Inc.) targeting vascular endothelial growth factor (VEGF)-A for treatment of wet AMD and Cimzia® (certolizumab pegol, UCB, Inc./Schwartz Pharma AG) targeting TNF-α for treatment of rheumatoid arthritis and other inflammatory diseases, are currently in the market. Of these three marketed FAb fragments, only Cimzia® is PEGylated to provide a longer serum half-life.

The smaller Ab domain proteins, including scFv and dAbs, have not yet been approved as marketed therapeutics, although their presence in clinical trials is steadily increasing. At least 14 antibody domain proteins of diverse design are in clinical trials as of July 2011. At least nine of these are designs based on scFv and dAbs. scFv molecules are composed of a light chain variable region (V$_L$) domain and a heavy chain variable region (V$_H$) domain genetically fused to each other through a flexible linker region, usually 10–25 amino acid residues in length and frequently (GGGGS)$_3$. Re-engineering natural antibody molecules into the scFv format will typically result in lower affinity; however, some mAbs have been re-engineered to scFv with less than 10-fold reduction in affinity. The scFv format is a desirable format for display technologies such as phage display and *in vitro* display since the single peptide format is very convenient for these technologies. These technologies can enable the discovery of scFv molecules from large antibody libraries. A significant challenge with scFvs is the lack of stability as purified proteins. These molecules tend to aggregate because of exposed hydrophobic regions that are typically buried in the full IgG structure. Additionally, production of scFvs is often hampered by lower expression titres and low solubility as compared with IgGs, resulting in manufacturing challenges related to these poor biophysical characteristics. Despite these challenges, scFv-based molecules have advanced into late stage clinical trials. The most advanced of these is blinatumomab (AMG-103; formerly MT103), which is currently in phase IIb clinical trials for treatment of relapsed and refractory acute lymphoblastic leukaemia (ALL). This therapeutic has a novel design as an anti-CD19/anti-CD3 bispecific T-cell Engager (BiTE), using CD19 target binding to direct the therapeutic to B cells and the CD3 target binding to engage and activate T cells (*15*). This approach is so exciting that it drove the $1.13 billion acquisition of Micromet, the company that developed the BiTE design, by Amgen, Inc. in early 2012.

Domain antibodies (dAbs) are the smallest antibody format that retains binding to a target protein. The advantages and disadvantages of the dAbs over mAb therapeutics are generally the same as for other antibody fragment approaches. These molecules are 12–15 kDa and have a short circulating half-life, unless a half-life extension strategy is employed. However, because of their smaller size, dAbs are expected to have a broader tissue distribution and tumour penetration than a full IgG. This single domain format can be found in nature as part of the immune system in at least two species, camelids and nurse shark. The camelid and nurse shark dAbs have highly conserved residues that lend particular stability and solubility to these naturally occurring dAbs. Ablynx and others have begun developing these natural dAbs into therapeutics through immunizations of llamas followed by molecular cloning of the antibody genes from the antibody-expressing B cells (*16*). On the other hand, human-derived dAbs are unpaired binding domains and as such, the biophysical properties of these dAbs can be poor due to exposed hydrophobic residues. Efforts made to engineer away these hydrophobic regions by introducing hydrophilic mutations, often based on the design and structure of natural camelid dAbs, have resulted in molecules with improved solubility and stability. Human dAbs can also be engineered from hybridoma-derived mice or discovered from synthetic phage display libraries.

4.3.6 Non-antibody binding proteins (NABP)

The great success of antibodies as biotherapeutics spawned the exploration of other forms of molecular recognition proteins that were not built on an antibody scaffold and with properties to address some of the limitations of therapeutic antibodies. These fundamental limitations include limited tissue penetration due to the size of the molecule, constraint on target families due to antibody structure, the cost and complexity associated with development and storage of a glycosylated macromolecule, and the dense intellectual property landscape that surrounded much of the enabling technologies for antibody discovery and development. Non-antibody binding proteins (NABPs) are defined as binding proteins or peptides that do not possess an antibody variable region domain to drive the targeting interaction. It is expected that NABPs will complement, but not replace, antibody therapeutics in the market since antibodies are well understood from therapeutic, regulatory, and manufacturing perspectives, whereas the various NABP platforms are still relatively new and for the most part untested. It is unlikely, for example, for an NABP to compete head-to-head with anti-TNF-α antibodies unless it can provide a significant therapeutic and/or marketing advantage.

To minimize the risk of immunogenicity, NABPs are typically based on naturally occurring human proteins that have structures that lend themselves to a versatile and high-affinity binding potential. They are generally compact with flexible surface loops which are tolerant of varying amino acid sequence and length (Figure 4.6). The scaffolds are often highly soluble, stable to pH and temperature extremes, and resistant to proteolysis. For screening and selection purposes, the scaffolds need to be compatible with a form of display technology such as phage display, ribosome display or other *in vitro* display formats. This often requires a balance between a quickly folding, highly stable protein domain structure and the ability to use a robust display system. Proteins that fold too quickly will not display well using cell display systems relying on the cellular Sec-dependent secretion. The discovery of binding molecules with high affinity and high specificity is reliant on the design and size of the combinatorial library from which the molecule will be selected, the compatibility of the protein scaffold with the display technology, as well as the selection strategy employed. Other than the toxin peptide-derived Prialt® (ziconotide; Azur; see the peptide section above), Kalbitor® (ecallantide, Dyax Corporation), approved for the treatment of hereditary angioedema, is currently the only marketed NABP. Kalbitor® is a 7-kDa protein based on a Kunitz domain, which is a natural serine protease domain typically found as part of a larger protein. Its use as a scaffold for NABP discovery was initiated in early work done at Genentech, Inc. in the mid-1990s in which libraries were designed

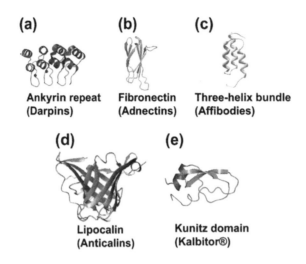

(a) Ankyrin repeat (Darpins)

(b) Fibronectin (Adnectins)

(c) Three-helix bundle (Affibodies)

(d) Lipocalin (Anticalins)

(e) Kunitz domain (Kalbitor®)

FIGURE 4.6

(a–e) Examples of non-antibody binding protein scaffold designs.

that randomized sequences in the binding loops of the Kunitz domain from amyloid beta precursor (APPI). Specific and potent inhibitors of human tissue factor VIIa and human plasma kallikrein were isolated for these early libraries.

There are many different types of alternative scaffold proteins with diverse structures that are being explored as next-generation protein therapeutics, some of which are shown in Figure 4.6. As an example of one of these, the designed ankyrin repeat protein (DARPin) scaffold was discovered at the University of Zürich, and licensed to Molecular Partners for development of the platform for therapeutic applications. The ankyrin repeat proteins are natural receptor proteins that drive protein interactions in a variety of cellular regulatory processes. These proteins consist of four or five repeat motifs arranged as homologous segments of regular repeating segments of β-turn, antiparallel α-helix and a flexible loop reaching to the next β-turn. This forms a very stable single domain of 14–18 kDa with a large target interacting surface. Molecular Partners AG has generated diverse libraries of randomized amino acid sequences in the protein-interacting surfaces. These large libraries can be used for *in vitro* display for selection of DARPins with specific binding properties towards a wide variety of target proteins (*17*).

While there are over a dozen NABP platforms currently being developed as therapeutic protein platforms, the most advanced of these include the adnectins (Bristol-Myers Squibb Company), anticalins (Pieris AG), affibodies (Affibody), and affilins (Scil Proteins Pharma GmbH). Due to their small size, for most applications, these alternative scaffold proteins require a half-life extension strategy as described above for peptide therapeutics. A potential risk associated with all the alternative scaffolds is immunogenicity, which is a general risk that needs to be considered for all protein therapeutics. Several other nontoxin peptide NABP therapeutics are currently in various stages of development, including a few in clinical trials (e.g. CT-322 [BMS-844,203], an anti-VEGFR-2 adnectin from Bristol-Meyers Squibb; PRS-050, an anti-VEGF anticalin from Pieris, and MP0112, an anti-VEGF-A DARPin from Molecular Partners).

4.4 CHALLENGES OF PROTEIN THERAPEUTICS

4.4.1 Immunogenicity

Immunogenicity is a challenge with much greater significance for the biologic therapeutics than for small molecule drugs. Immunogenicity is the potential side effect that is inherent in all protein biologics for the development of an immune response to the therapeutic agent. Immunogenicity can have an impact on therapeutic clearance rate, efficacy and even safety if leading to a severe allergic reaction. The US FDA has set guidelines outlining regulatory expectations for monitoring immunogenicity of therapeutic proteins. It is generally accepted that immunogenicity of replacement proteins is a much more significant issue than it is with binding protein therapeutics such as mAbs, FcFPs and NABPs. The reason for this distinction is that immunogenicity against a replacement protein can also result in immune-mediated clearance of the natural protein which can result in a severe medical condition (e.g. PRCA due to immune responses against a recombinant form of EPO) as opposed to immune responses against binding proteins, which usually only increases clearance and limits exposure to those therapeutics, resulting primarily in reduced efficacy (*18*).

Immunogenicity has been a driving force in some advances made in protein therapeutics, for example, leading mAb state of the art from mouse-derived to chimeric to fully human discovery and engineering. Despite overall correlation of improved immunogenicity with increased 'humanness', it is important to recognize that even human proteins can elicit immunogenic responses, since many factors influence immunogenicity. Product-associated factors that can significantly influence immunogenicity resulting in poor efficacy (and even safety issues) include aggregation, degradation products, posttranslational modification, impurities in the product, and patient-related factors, for example, the disease state, concomitant medications, exposure to the drug (dose and frequency) and the route of administration.

A large effort is ongoing within the industry to develop tools that would allow immunogenicity to be predicted and to reduce the risk of immunogenicity to help guide therapeutic design, because immunogenicity can impact the efficacy and safety of a biologic. Animal models are very poor predictors of human immunogenicity; however, there are some *in vitro* and *in silico* tools that are being used to assess possible immunogenicity risk based on the identification of potential T-cell epitopes. If T-cell epitopes are observed, either through algorithms that predict them or through empirical T-cell stimulation in the laboratory, attempts can be made to avoid them in the final therapeutic through engineering or identifying a different molecule altogether. Perhaps the most effective approach to reducing immunogenicity is together to increase the humanness and the purity level of the therapeutic protein.

4.4.2 Developability and manufacturing

Development and manufacturing methods very clearly distinguish protein therapeutics from small molecules. For most proteins, in order to be physiologically active, they must take on a properly folded three-dimensional structure and undergo appropriate posttranslational modifications. This requires that the proteins are manufactured in cell-based systems. These cell-based expression systems are complex and less well defined than even the most involved chemical synthesis required for a small molecule. On one hand, protein expression and secretion by cells is a complex biological process that is not always precisely controlled in the same manner from culture to culture based on inoculum size, cell regulatory systems governing protein expression, and other cell-based intrinsic factors. On the other hand, extrinsic factors within the reactors such as temperature, pH, ionic strength, aeration, concentration of micronutrients and metals, reactor size and shape, stirring and mixing of the cultures, and other such reactor-related factors can have marked effects on expression levels, protein folding, glycosylation, and protein product stability. Thus, heterogeneity in the protein therapeutic product is inherent, due to the complex cellular processes in which proteins are manufactured.

Most large protein therapeutics, for example mAbs and FcFP, are not a single protein species but rather a heterogeneous population of minor variants due to posttranslational modification variability and other process-induced heterogeneities. Glycosylation is a particularly important posttranslational modification, especially for antibodies, as the glycosylation pattern will influence the interactions of the Fc with the FcR family and regulate Fc effector function. As discussed above, the Fc effector functions can be critically important to the efficacy, safety and half-life of a therapeutic mAb. The presence and pattern of glycosylation on a protein can vary depending on the host cell line and the condition and expression rate that the protein is being generated under. Careful regulation of the glycosylation pattern during therapeutic development is required to ensure consistent therapeutic efficacy from batch to batch. Heterogeneities also can result from other posttranslational modification to amino acid residues, including isomerization, deamidation and oxidation, as well as proteolytic clipping and aggregation.

The manufacturing of a high-quality protein therapeutic involves recombinant DNA technologies, a robust expression system and optimized product purification and formulation methods that result in high yields and high quality in a reproducible and stable manner. Host cells are well-characterized laboratory cell lines that can be transfected with an expression plasmid encoding the genes for a therapeutic resulting in high-quality protein and high-level expression. The transfection leads to the stable insertion of the therapeutic genes in the chromosome of the host cell resulting in a new cell line that will produce the therapeutic protein. There are a variety of mammalian cell types that are considered protein therapeutic host cells. The most widely used today are CHO cells. This cell line was initially established as an immortal laboratory cell line in the 1960s and is well-known for the ability to produce high-quality recombinant protein. It was initially approved as the manufacturing line for tissue plasminogen activator in 1987. Additionally, the CHO host has a strong industrial history and acceptance with regulatory organizations. Other mammalian cell lines that have been used for expression of therapeutic proteins include human embryonic kidney cells 293 (HEK 293 cells), the mouse myeloma cell lines SP2/0 and NS0, the rat myeloma cell line YB2/0, baby hamster kidney cells, and PerC.6, a human embryonic retina cell line transformed with the adenovirus E1 gene.

Simple, nonglycosylated protein therapeutics can be produced in expression systems other than mammalian systems. In addition to the first ever manufactured recombinant protein therapeutic, Humulin®, many recent marketed and clinical candidate biologics, particularly peptides, NABPs, and antibody fragment therapeutics, are currently manufactured in *E. coli* fermentation cultures.

Escherichia coli and other well-characterized expression host bacteria (e.g. *Pseudomonas aeruginosa, Bacillus subtilis*), are generally considered an easier, faster and cheaper manufacturing source than mammalian cells. *Escherichia coli* and other bacteria are not capable of glycosylating proteins, limiting the types of therapeutics that can be generated in *E. coli*, since many therapeutic proteins, including antibodies, are glycoproteins in which properly elaborated glycans play a significant role in their structure and pharmacological function.

Once a manufacturing cell line is identified, a large inventory of the cell line is established and cryopreserved. This process is called cell banking, with both master cell banks and working cell banks generated to ensure maximum consistency in the initiation of each manufacturing process. This serves as a highly consistent source of expressing cell lines for all the clinical and commercial production of the therapeutic protein. For production, the manufacturing cell lines are typically grown in bioreactors or fermentation vessels ranging in size from 500 l to 25,000 l. Safety, availability of consistent supplies, and desired higher consistency in productivity led to the adoption of chemically defined growth media for most protein therapeutic manufacturing cell cultures and bacterial fermentation cultures today. This has increased consistency of the growth conditions on a lot-to-lot basis as well, and media optimization has enhanced productivity, having an impact on cost of goods.

The upstream cell culture or bacterial fermentation process is followed by downstream processes that include recovery or capture of the product from the cells or broths, protein purification and concentration steps, as well as steps to clear any viral load and impurities from the product. These manipulations need to be carefully monitored to minimally impact the integrity of the therapeutic protein. For example, viral clearance is typically accomplished using a low pH hold step which may lead to aggregation and degradation of some proteins. It needs to be established through careful analysis that the viral clearance step and other purification methods will be well tolerated by the therapeutic protein.

Comparability, which is the measure of batch-to-batch consistency in protein structure, function, and purity, is a critical issue in therapeutic protein production processes. Subtle alterations in the manufacturing process, e.g. the extrinsic factors mentioned above, can potentially result in differences in bioactivity, safety, and quality of the protein therapeutic. It is the efficacy and safety of the population of molecules that needs to remain consistent through the product lifetime, since micro-heterogeneity in these molecules is inherent. Thus, comparability assays that assess the key critical attributes of the protein therapeutic molecules with respect to structure, purity, activity, and stability are critical to the manufacturing process (*19*).

4.4.3 **Route of administration**

In order to maintain the biologically active state of a protein therapeutic, delivery to the systemic circulation is currently limited to parenteral routes of administration. Currently, subcutaneous administration is the preferred route of administration, followed by IV infusion. However, some mAb therapeutics are limited to infusion since delivery via the subcutaneous route is generally limited to small volumes (e.g. 1–2 mL/injection). Some protein therapeutics require a higher dose than is possible in a 1.5 ml volume due to limitations on concentration and stability in formulating the protein. The route of administration is a challenge and of significant consideration for protein therapy. Patient acceptance and patient compliance is an important factor, particularly for chronic treatment. The therapy requires a great commitment from patients since they will be receiving an injection or an infusion, rather than swallowing a pill or tablet as is typical for a small molecule drug. Thus, parenteral treatments are generally limited to more serious conditions for which oral alternatives are unavailable or less effective. This can be offset, however, by the considerably lower frequency of delivery often associated with biologics. For example, daily, weekly, and even monthly oral dosing of various bisphosphonates (e.g. Fosamax®, Actonel®, Boniva®, Zometa®) for osteoporosis is being seriously challenged by biannual parenteral dosing of Prolia® (denosumab; Amgen, Inc.).

As a result of the competition with small molecule-based therapeutics, the biologics industry continues to move towards next-generation delivery approaches, including needleless injection devices for subcutaneous delivery, pulmonary delivery, oral and/or buccal adsorption, and transdermal delivery efforts for different types of peptides and proteins. Each approach, however, is fraught with challenges of poor absorption, instability, and formulation requirements that need to be tackled for successful delivery and therapy (20).

4.5 FUTURE DIRECTIONS FOR PROTEIN THERAPEUTICS

The quest for the next-generation protein therapeutics includes focus on enhanced efficacy, expanding target accessibility, improved safety, and drug delivery options. Throughout this chapter, the major strengths of the protein therapeutic classes were described, as well as the significant limitations that remain a challenge. These limitations provide opportunity for the next generation of molecular design. The clinical, scientific and commercial pressures on next-generation protein therapeutics, i.e. improved therapeutic efficacy, efficacy-to-toxicity windows, safety profiles, and/or convenience in dosing regimens, provide the imperative on which these next-generation molecules are being developed. The high specificity and affinity towards a target protein is one of the strongest assets on which to build future protein therapeutics. Extrapolating and building on the successes of the past, many novel applications of protein therapeutics will include targeting and delivery driven through the exquisite binding property. For example, based on today's drugs, improvement in efficacy exists in the area of anticancer antibodies. Combining high-affinity tumour-targeting specificity with enhanced Fc effector function or T-cell engagement are approaches that may lead to improved tumour cell clearing. There are several novel therapeutics already in the clinic that are exploring engineered protein therapeutic designs combining tumour targeting and antibody-driven immunomodulatory activities. Additionally, using protein therapeutic molecules to specifically target tumour cells, while conjugated to highly toxic drugs that can kill the targeted cell population to which it is being delivered, is another approach being explored in the clinic to improve the efficacy of anticancer protein therapies. The first antibody drug conjugate to be approved by the US FDA, gemtuzumab ozogamicin (Mylotarg®; Wyeth Pharmaceuticals), was withdrawn from the market in 2010 for safety reasons and lack of efficacy in postapproval clinical trials. Despite this failure, three toxin- or radionucleotide-conjugated antibody drugs (Zevalin®, Bexxar®, and Adcetris®) are currently approved for marketing and more than 35 others are in various stages of clinical trials. Adcetris® (brentuximab vedotin, Seattle Genetics, Inc./Takeda Pharmaceutical Company, Ltd.), an anti-CD30 mAb conjugated with the potent cytotoxic drug, monomethyl-auristatin E, was approved in August 2011 for treatment of Hodgkin lymphoma. That drug, and trastuzumab-DM1, an anti-HER2 mAb conjugated to $N^{2'}$-deacetyl-$N^{2'}$-(3-mercapto-1-oxopropyl)-maytansine approved for marketing by the FDA in February, 2013, for treatment of breast cancer (see Chapter 9), are leading an exciting new revolution of antibody conjugates that hold enormous promise for improved treatment of cancer.

mAbs, as well as the smaller protein therapeutic scaffolds such as alternative scaffold proteins, are currently being explored for therapeutic application in bispecific designs. The first bispecific antibody, catumaxomab (Removab®, anti-EpCAM/anti-CD3; Fresenius/Trion Pharma GmbH) was approved in Europe in 2009. In this therapeutic design, the anti-EpCAM binding specifically targets the molecule to EpCAM-positive tumour cells, while the anti-CD3 recruits T cells and the IgG Fc recruits Fc effector function, all engineered towards antitumour activities. While catumaxomab itself is severely limited in its therapeutic potential due to its 100% nonhuman design, it provides an important template for design of future IgG-based bispecific antibody therapeutics. As mentioned above, the bispecific scFv-based BiTE design used in blinatumomab (AMG-103) also incorporates the recruitment of T cells through an anti-CD3 arm. In this case, the T cells are drawn into engagement with malignant B cells targeted for elimination for treatment of Acute Lymphoblastic Leukemia (ALL). Another application for bispecific protein therapeutics is to improve disease-modifying efficacy by simultaneously blocking two different targets in a disease pathway. This can be approached either by a single molecule with dual specificity or using multiple therapeutics simultaneously. Multispecificity of mAb- and NABP-based therapeutics is recognized as a significant opportunity for the next generation of protein therapeutics. Until recently, a bottleneck in using this approach has been the challenge of manufacturing these proteins sufficiently and stably. The development of a wide variety of protein therapeutic formats has allowed new engineering approaches to overcome some of the production challenges.

4.6 BIOSIMILAR PROTEIN THERAPEUTICS

Biologics as a class are generally very expensive therapeutics, ranging from $\sim$ $10,000 US dollars per therapeutic cycle (or annually) to over $100,000 per cycle or year. This surge in costly biologics has fuelled a push for lower cost biologic drugs that could provide a tremendous benefit to patients of all economic classes worldwide. Biosimilars are drugs made to be as identical as possible to innovator drugs with the intent of providing a lower cost alternative. Alternatively, 'biobetters' are biologic molecules based on an innovator biologic, but with improvements intended to increase efficacy, potency, marketability, safety or patient compliance. It is important to note that biosimilars cannot be identical to the innovator product due to the complexity and heterogeneities associated with the therapeutic, very unlike generic versions of conventional drugs, whose chemical structure can be exactly duplicated and analysed as such. For small molecule drugs, regulatory approval of generic versions is based on molecular identity of the active ingredient, identity in strength, purity and quality, as well as bioequivalence. A chemical, generic manufacturer can demonstrate bioequivalence and obtain regulatory approval without conducting the full set of trials demonstrating clinical safety and efficacy as required for the innovator. This small molecule generic approach, however, cannot be followed exactly for biologic products. Even if a biosimilar were to possess the exact same primary amino acid sequence, and were produced by the same type of cells, the innovator's exact manufacturing process is not usually in the public domain. Thus, by definition, the biosimilar will be manufactured using different processes from the innovator. These differences in manufacturing process can generate unique heterogeneities within the product which may have different pharmacologic effects on the patient. Furthermore, all therapeutic proteins are potentially immunogenic and the heterogeneities in the product can contribute to the immunogenicity. All these factors require careful consideration for safety and efficacy of a biosimilar drug. In 2006, the first biosimilar, Omnitrope® (somatropin; Sandoz, Inc.), was approved in the European Union, since then many others have been approved globally, with as many as 100 approved in Brazil, India, Russia and China alone. Biosimilars are being watched carefully by the industry and the full impact of biosimilars on the biologics industry, especially with respect to the onslaught of biosimilar mAbs and FcFPs expected over the next decade, is still unknown. There are some expectations that the biosimilar protein therapeutics will drive significant change to the protein therapeutic industry by offering low-cost alternatives to innovator drugs. However, it remains to be seen what the realized cost savings will be with the actual clinical and manufacturing costs. Most importantly, the industry will keep a sharp eye on the safety and efficacy of introducing biosimilars into the protein therapeutic market.

4.7 SUMMARY AND CONCLUSIONS

Protein therapeutics are still relatively new to the pharmaceutical industry, with the vast bulk of protein therapeutics approved since the mid-1990s. Virtually all the major pharmaceutical companies have now embraced biologics as a component of their growth strategy. Interestingly, even with recombinant mAbs only being marketed for the past ca. 15 years, the industry has moved from infancy to maturity in that relatively short period. This is largely due to the enormous advances in technologies fuelled by the successes of protein therapeutics in the market and the need for new growth drivers. The next decade will likely prove to be a sorting out period in which many opposing forces such as intense competition on certain targets, introduction of second- and third-generation molecules and biobetter molecules on validated targets, third-party payor pressures, governmental pricing pressures, and biosimilars all will combine to reshape the biopharmaceutical industry. Even in this intensely competitive environment, there are still enormous opportunities for novel, creative biologic approaches to improve health care for millions of patients.

References

1. Köhler, G.; Milstein, C. Continuous Cultures of Fused Cells Secreting Antibody of Predefined Specificity. *Nature* **1975,** *256,* 495–497.
2. Leader, B.; Baca, Q. J.; Golan, D. E. Protein Therapeutics: A Summary and Pharmacological Classification. *Nat. Rev. Drug Discovery* **2008,** *7,* 21–39.
3. Holliger, P.; Hudson, P. J. Engineered Antibody Fragments and the Rise of Single Domains. *Nat. Biotechnol.* **2005,** *23,* 1126–1136.
4. Berenson, D. F.; Weiss, A. R.; Wan, Z. L.; Weiss, M. A. Insulin Analogs for the Treatment of Diabetes Mellitus: Therapeutic Applications of Protein Engineering. *Ann. N. Y. Acad. Sci.* **2011,** *1243,* E40–E54.
5. George, P. M.; Badiger, R.; Alazawi, W.; Foster, G. R.; Mitchell, J. A. Pharmacology and Therapeutic Potential of Interferons. *Pharmacol. Ther.* **2012,** *135,* 44–53.
6. Oster, H. S.; Neumann, D.; Hoffman, M.; Mittelman, M. Erythropoietin: The Swinging Pendulum. *Leuk. Res.* **2012.**
7. Zsebo, K. M.; Cohen, A. M.; Murdock, D. C.; Boone, T. C.; Inoue, H.; Chazin, V. R.; Hines, D.; Souza, L. M. Recombinant Human Granulocyte Colony Stimulating Factor: Molecular and Biological Characterization. *Immunobiology* **1986,** *172,* 175–184.
8. Suiter, T. M. First and Next Generation Native rFVIII in the Treatment of Hemophilia A. What has been Achieved? Can Patients Be Switched Safely? *Semin. Thromb. Hemostasis* **2002,** *28,* 277–284.
9. Morris, J. L. Velaglucerase Alfa for the Management of Type 1 Gaucher Disease. *Clin. Ther.* **2012,** *34,* 259–271.
10. Latham, P. W. Therapeutic Peptides Revisited. *Nat. Biotechnol.* **1999,** *17,* 755–757.
11. Morrison, S. L.; Johnson, M. J.; Herzenberg, L. A.; Oi, V. T. Chimeric Human Antibody Molecules: Mouse Antigen-Binding Domains with Human Constant Region Domains. *Proc. Natl. Acad. Sci. U.S.A.* **1984,** *81,* 6851–6855.
12. Jiang, X. R.; Song, A.; Bergelson, S.; Arroll, T.; Parekh, B.; May, K.; Chung, S.; Strouse, R.; Mire-Sluis, A.; Schenerman, M. Advances in the Assessment and Control of the Effector Functions of Therapeutic Antibodies. *Nat. Rev. Drug Discovery* **2011,** *10,* 101–111.
13. Reddy, M. P.; Kinney, C. A.; Chaikin, M. A.; Payne, A.; Fishman-Lobell, J.; Tsui, P.; Dal Monte, P. R.; Doyle, M. L.; Brigham-Burke, M. R.; Anderson, D.; Reff, M.; Newman, R.; Hanna, N.; Sweet, R. W.;

Truneh, A. Elimination of Fc Receptor-Dependent Effector Functions of a Modified IgG4 Monoclonal Antibody to Human CD4. *J. Immunol.* **2000,** *164,* 1925–1933.

14. Huang, C. Receptor-Fc Fusion Therapeutics, Traps, and MIMETIBODY Technology. *Curr. Opin. Biotechnol.* **2009,** *20,* 692–699.

15. Loffler, A.; Kufer, P.; Lutterbuse, R.; Zettl, F.; Daniel, P. T.; Schwenkenbecher, J. M.; Riethmuller, G.; Dorken, B.; Bargou, R. C. A Recombinant Bispecific Single-Chain Antibody, CD19 × CD3, Induces Rapid and High Lymphoma-Directed Cytotoxicity By Unstimulated T Lymphocytes. *Blood* **2000,** *95,* 2098–2103.

16. Wolfson, W. Ablynx Makes Nanobodies from Llama Bodies. *Chem. Biol.* **2006,** *13,* 1243–1244.

17. Steiner, D.; Forrer, P.; Pluckthun, A. Efficient Selection of DARPins With Sub-Nanomolar Affinities Using SRP Phage Display. *J. Mol. Biol.* **2008,** *382,* 1211–1227.

18 Buttel, I. C.; Chamberlain, P.; Chowers, Y.; Ehmann, F.; Greinacher, A.; Jefferis, R.; Kramer, D.; Kropshofer, H.; Lloyd, P.; Lubiniecki, A.; Krause, R.; Mire-Sluis, A.; Platts-Mills, T.; Ragheb, J. A.; Reipert, B. M.; Schellekens, H.; Seitz, R.; Stas, P.; Subramanyam, M.; Thorpe, R.; Trouvin, J. H.; Weise, M.; Windisch, J.; Schneider, C. K. Taking Immunogenicity Assessment of Therapeutic Proteins to the Next Level. *Biologicals* **2011,** *39,* 100–109.

19. Lubiniecki, A.; Volkin, D. B.; Federici, M.; Bond, M. D.; Nedved, M. L.; Hendricks, L.; Mehndiratta, P.; Bruner, M.; Burman, S.; Dalmonte, P.; Kline, J.; Ni, A.; Panek, M. E.; Pikounis, B.; Powers, G.; Vafa, O.; Siegel, R. Comparability Assessments of Process and Product Changes Made During Development of Two Different Monoclonal Antibodies. *Biologicals* **2011,** *39,* 9–22.

20. Eppstein, D. A.; Longenecker, J. P. Alternative Delivery Systems for Peptides and Proteins as Drugs. *Crit. Rev. Ther. Drug Carrier Syst.* **1988,** *5,* 99–139.

Further Reading

Books

An, Z. *Therapeutic Monoclonal Antibodies: From Bench to Clinic;* John Wiley and Sons, Inc: Hoboken, NJ, 2009.

Strohl, W. R.; Strohl, L. M. *Therapeutic Antibody Engineering: Current and Future Advances Driving the Strongest Growth Area in the Pharma Industry;* Woodhead Publishing: Cambridge, 2012.

Important Reviews and Original Articles

La Merie Business Intelligence, Top 30 Biologics 2011. R&D Pipeline News [1/2012], **2012,** 1–30.

Aggarwal, S. What's Fueling the Biotech Engine–2010 to 2011. *Nat. Biotechnol.* **2011,** *29* (12), 1083–1089.

Binz, H. K.; Amstutz, P.; Pluckthun, A. Engineering Novel Binding Proteins from Nonimmunoglobulin Domains. *Nat. Biotechnol.* **2005,** *23* (10), 1257–1268.

Birch, J. R.; Racher, A. J. Antibody Production. *Adv. Drug Delivery Rev.* **2006,** *58* (5–6), 671–685.

Carter, P. J. Introduction to Current and Future Protein Therapeutics: A Protein Engineering Perspective. *Exp. Cell. Res.* **2011,** *317* (9), 1261–1269.

Dingermann, T. Recombinant Therapeutic Proteins: Production Platforms and Challenges. *J. Biotechnol.* **2008,** *3* (1), 90–97.

Sato, A. K.; Viswanathan, M.; Kent, R. B.; Wood, C. R. Therapeutic Peptides: Technological Advances Driving Peptides into Development. *Curr. Opin. Biotechnol.* **2006,** *17* (6), 638–642.

Skerra, A. Alternative Non-antibody Scaffolds for Molecular Recognition. *Curr. Opin. Biotechnol.* **2007,** *18* (4), 295–304.

Strohl, W. R.; Knight, D. M. Discovery and Development of Biopharmaceuticals: Current Issues. *Curr. Opin. Biotechnol.* **2009,** *20* (6), 668–672.

Similarities and differences in the discovery and use of biopharmaceuticals and small-molecule chemotherapeutics

James Samanen

James Samanen Consulting LLC, Phoenixville, PA 19460, USA YourEncore, 100 Canal Pointe Blvd, Suite 209 Princeton, NJ 08540, USA Former Discovery Research Portfolio Management GlaxoSmithKline, 709 Swedeland Road, King of Prussia, 19406, USA

CHAPTER OUTLINE

5.1 Introduction 163
5.2 How do SMDs Differ from Biomolecular Drugs? 169
    5.2.1 Structure 169
    5.2.2 Distribution...174
    5.2.3 Metabolism...174
    5.2.4 Serum half-life...174
    5.2.5 Typical dosing regimen 174
    5.2.6 Species reactivity 174
    5.2.7 Antigenicity and hypersensitivity...175
    5.2.8 Clearance mechanisms 175
    5.2.9 Drug–drug interactions..175
    5.2.10 Pharmacology 176
5.3 Historical Changes to the FDA Approach to Handle the Biotech Boom—the Differing Nomenclature for Small Molecules vs Biologics Entering the Clinic 176
5.4 Comparisons of Clinical Metrics—Biologics vs Small Molecules 179
    5.4.1 Overall clinical success rates of biologics vs small molecules 180
    5.4.2 Stage-related success rates for small molecules vs biologics 180
    5.4.3 Stage-related clinical cycle times for small molecules vs biologics...............180
    5.4.4 Comparative cost of R&D for biologics vs small molecules...........................183
    5.4.5 The challenges in comparing small-molecule and biomolecular drug metrics—the influence of biomolecular scaffold................................185
5.5 Are Peptide Drugs Small Molecules or Biologics? 186
5.6 The Manufacture and Supply of SMDs vs Biomolecular Drugs 191

Introduction to Drug Research and Development. http://dx.doi.org/10.1016/B978-0-12-397176-0.00005-4

5.7 The Pricing of SMDs vs Biomolecular Drugs..191
5.8 Comparing Small-Molecule, Peptide and Biomolecular Drugs in the Market........................192
    5.8.1 GPIIb/IIIa antagonists..192
    5.8.2 Her2 inhibitors..194
5.9 Biosimilar Biomolecules vs Generic Small Molecules..195
5.10 Discovery and Preclinical Stages for SMDs vs Biomolecular Drugs—Where the Technologies Differ
    the Most 196
    5.10.1 Target discovery 197
    5.10.2 Lead discovery 197
    5.10.3 Lead optimization...197
    5.10.4 Preclinical evaluation..198
    5.10.5 Clinical phases..198
5.11 Small-Molecule and Biologics Approvals by Therapy Areas.......................................198
5.12 Managing Small-Molecule & Biomolecular Drug R&D in the Same Company 198
5.13 Conclusion..200
References 200

ABSTRACT

Biotechnology has given rise to a broad range of biotherapies or biologics, including biomolecular drugs, vaccines, cell or gene therapies. This chapter focuses on biomolecular drugs, namely monoclonal antibodies (Mabs), cytokines, tissue growth factors and therapeutic proteins. Prior to the US approval of recombinant human insulin in 1982, biomolecular drugs were extracted from natural sources. The tools of molecular biology have dramatically increased the discovery and development of new bio-pharmaceuticals. The most obvious difference between small-molecule drugs (SMDs) and biomolecular drugs is size, like the difference in weight between a bicycle and a business jet. SMDs and biomolecular drugs are compared in this chapter by structure, molecular weight, preparation, physicochemical properties, and route of administration, as well as distribution, metabolism, serum half-life, dosing regimen, species reactivity, antigenicity & hypersensitivity, clearance mechanisms, drug–drug interactions, and pharmacology. This chapter reviews the differences and similarities in the various stages of drug discovery and development, with respect to cost, probability of success and cycle time. The clinical metrics of overall clinical success rate, stage-related success rate, and clinical cycle time are examined for SMDs and biomolecular drugs. The hybrid class of peptide drugs tends to be equated with biologics, due to their amino acid content and because oral activity is rare. But peptides truly bridge the gap between small molecules and biologics, in terms of physical properties, range of therapy areas and means of production. This chapter summarizes the similarities and differences of peptide drugs with SMDs and biomolecular drugs. The manner in which these agents compare as products with respect to manufacturing and pricing are considered. Two case studies are presented—the antagonists where small-molecule, peptide and Mab agents have competed in the market, and Her2 inhibitors where small-molecule and Mab agents may ultimately synergize as a combination product. Biomolecular drugs have levelled the playing field. All the "big Pharma" companies now have the capacity to develop both types of drugs. Conversely the larger biotech companies are developing the capacity for small-molecule synthesis. Now, with many blockbuster biologics nearing patent expiration, biosimilars are on the way. It's no longer a question of "choose which type"—one will need to know how to discover and develop either type of drug.

Keywords/Abbreviations: (Abbreviated) New drug application ((A)NDA); Abciximab; Antigenicity; Biologics license application (BLA); Biologics Price Competition and Innovation Act (BPCIA); Biologics; Biomolecules, biomolecular drugs; Biosimilars/follow-on biologics; Biotechnology; Chemistry, manufacturing and control (CMC); Chimeric Mab; Cytokines; Eptifibatide; Fab; Fc; GPIIB/IIA Antagonists; GlaxoSmithKline (GSK); Hatch-Waxman Act; Her2 Antagonists; Humanized Mab; Hypersensitivity; Insulin; Investigational new drug application (IND); Lapatinib; Monoclonal antibodies (Mabs); New biological entity (NBE); New chemical entity (NCE); New molecular entity (NME); Peptide drugs; Pharmaceutical Research and Manufacturers of America (PhRMA); Probability of success (POS); Recombinant protein (rDNA); Recombinant; Therapeutic proteins; Thrombopoietin (TPO); Tirofiban; Tissue growth factors; Trastuzumab; Uridine $5'$-diphospho (UDP)-glucuronosyl-transferase, US Center for Drugs/Biologics Evaluation and Research (CD/BER); US Food and Drug Administration (FDA).

5.1 INTRODUCTION

Human insulin was the first recombinant biopharmaceutical approved in the US in 1982. Prior to that, protein products approved for use in humans were extracted from natural sources. Thus the discovery of new biopharmaceuticals occurred at a rather leisurely pace. However, the tools of molecular biology that have been developed more recently have dramatically increased the discovery and development of new biopharmaceuticals, such that the number of investigational new drug applications (INDs) for biopharmaceuticals now rivals the number of INDs for SMDs. All the "big Pharma" companies now have the capacity to develop both types of drugs and those that had biologics capacity are increasing this sector. Conversely the larger biotech companies are developing the capacity for small-molecule synthesis and now, with many blockbuster biologics nearing patent expiration, biosimilars are on the way. It's no longer a question of "choose which type"—you will need to know how to discover and develop either type of drug.

Biotechnology has given rise to a broad range of biotherapies or biologics, including vaccines, cell or gene therapies, therapeutic protein hormones, cytokines and tissue growth factors, as well as monoclonal antibodies (Mabs); the alternative abbreviation mAbs is used elsewhere in the book. In this discussion we will primarily focus on the categories of biotherapies that are presently managed by the US Food and Drug Administration (FDA) Center for Drugs Evaluation and Research (CDER) namely Mabs, cytokines, tissue growth factors and therapeutic proteins. These we will refer to as *biomolecular drugs*. Some of the data that we will show includes all biotherapies.

A biotechnology company tends to focus on the discovery and development of biomolecular drugs, but biotech companies can also create products that address a wide range of industries. The term "biopharmaceutical company" is often applied to a company that pursues biologics or biomolecular drugs and not SMDs. Here we will use the term biopharmaceutical (company) to mean the pursuit of both SMDs and biologics or biomolecular drugs. A more precise term may be bio/pharmaceutical, which this author has used previously. Thus, a biopharmaceutical company will have the resources to discover and develop both types of drugs—SMDs (also referred to as new chemical entities (NCEs)) and biomolecular drugs (also called biologics and often referred to as new biological entities (NBEs)).

Since the introduction of recombinant insulin, the market for biomolecular drugs has increased dramatically. By 1997 worldwide sales of biomolecular drugs were over $7 billion. The global sales of biomolecular drugs have continued to rise—Mabs alone in 2006 totalled $4.7 billion (*1*).

A popular misconception is that in the early days most of the new biologics were discovered and developed within biotech companies. Certainly few of the classically NCE-oriented companies entered the NBE arena though the pharmaceutical companies J&J (Ortho Biotech), Roche and Lilly were early players, getting biologics license applications (BLAs) approved in the 1980s (Table 5.1) (*2–4*).

Fifty percent of the BLAs in the 1980s and 45% in the 1990s came from drug companies (data from Table 5.1). Thus while a lot of investment may have gone into biotech startups, it was the previous experience of the drug companies with bringing drugs to market that made them at least equal partners in that aspect of biomolecular R&D. Only 17 drug companies and 16 biotech companies got BLAs in the 1980s and 1990s which is a small subset of the pharmaceutical industry. By 1998 the group Pharmaceutical Research and Manufacturers of America (PhRMA) determined that more than 140 US-based companies were engaged in biomolecular drug development (*2*). Most likely many more pharmaceutical companies were investing in biotech in that period. The investment in biologics was enormous and the payout uncertain. As with the discovery and development of any drug it took years before the new biotechs achieved their first BLA, over 14 years on average (Table 5.2).

While many of the discoveries of new biologics continue to originate in biotech companies, the clinical development of new biologics is increasingly supported by large pharma which, until relatively recently, had been NCE-oriented (*5*).

Indeed, in recent years most of the large pharma have gained an expertise in biologics through entry into the field (sometimes through acquisitions) and are now biopharmaceutical companies (Table 5.3) (*2*). The acquisition of Genzyme by Sanofi-Aventis is a recent example (*6*).

A recent collaborative study by Deloitte and Thomson Reuters showed that the 12 top biopharmaceutical companies all had biologics in their late-stage portfolios, ranging from 21% to 66% of their portfolios (avg. 39%) (*7*).

Prior to the 1980s there were sufficiently few biomolecular drugs that the very term "pharmaceutical" or "drug" was taken to mean small molecule. With the exception of insulin, the few biomolecules approved for human use were administered by a trained health practitioner and were often considered "therapies". Thus one may see the comparison of "SMDs (or pharmaceuticals) vs large-molecule therapies". Here we will consider a large-molecule therapy that is regulated by CDER to be a biomolecular type of drug or pharmaceutical.

The term for first SMD approval, or new molecular entity (NME) could in theory be applied to first biologic approval, but because NME has long been associated with small molecules it is not being associated with first biologic approval—which is simply called a new BLA.

In March 2010 US President Obama signed into law the Biologics Price Competition and Innovation Act (BPCIA) which provides for biosimilar biologic drug approvals as part of the omnibus health care bill. As the FDA develops guidelines for biosimilar approvals and begins to review applications for biosimilars, biologics will begin to enter the large generics market in the US. This topic will be further discussed in a subsequent section.

Prior to recombinant human insulin protein, products approved for use in humans were extracted from natural sources, such as the extraction of gonadotropins from urine and insulin from pigs. It is beyond the scope of this discussion to delve into the details of the processes that give rise to biomolecular drugs or SMDs. The following are good general references that

TABLE 5.1 Early Biotech & Drug Company Biologics Approvals (w/o Diagnostics) (2–4)

US Trade Name	Generic Name	Originating Company	Company Type	US Approval Date	Therapy Area	Protein Type	Decade
Humulin	Insulin	Eli Lilly and Co	Drug	28 October 1982	Endocrine	Insulin	1980
Protropin	Somatropin	Genentech	Biotech	17 October 1985	Endocrine	HGH	1980
Orthoclone OKT3	Muromonab-CD3	Ortho Biotech (J&J)	Drug	1986	Immunological	Mab	1980
Roferon-A	Interferon alpha-2b	Hoffmann-La Roche	Drug	4 June 1986	Antineoplastic	Interferon	1980
Intron A	Interferon alpha-2b	Biogen	Biotech	4 June 1986	Antineoplastic	Interferon	1980
Humatrope	Somatropin	Eli Lilly and Co	Drug	8 March 1987	Endocrine	HGH	1980
Activase	Alteplase	Genentech	Biotech	13 November 1987	Cardiovascular	tPA	1980
Epogen/Procrit	Epoetin alpha	Amgen Inc	Biotech	1 June 1989	Blood cell deficiency	Growth factor	1980
Actimmune	Interferon alpha-1b	InterMune	Biotech	20 December 1990	Anti-infective	Interferon	1990
Neupogen	Filgrastim	Amgen	Biotech	20 February 1991	Blood cell deficiency	Growth factor	1990
Leukine	Sargramostim	Berlex Laboratories	Drug	5 March 1991	Blood cell deficiency	Growth factor	1990
Novolin R	Insulin	Novo Nordisk	Drug	25 June 1991	Endocrine	Insulin	1990
Proleukin	Aldesleukin	Chiron Corp	Biotech	5 May 1992	Antineoplastic	Interleukin	1990
Recombinate	Antihemophilic factor	Baxter Healthcare	Drug	10 December 1992	Haemostasis	Coagulation factor	1990
Kogenate	Antihemophilic factor	Bayer	Drug	25 February 1993	Haemostasis	Coagulation factor	1990

(continued on next page)

TABLE 5.1 Early Biotech & Drug Company Biologics Approvals (w/o Diagnostics) (2–4) *(continued)*

US Trade Name	Generic Name	Originating Company	Company Type	US Approval Date	Therapy Area	Protein Type	Decade
Betaseron	Interferon alpha-1b	Chiron	Biotech	23 July 1993	Immunological	Interferon	1990
Nutropin	Somatropin	Genentech	Biotech	17 November 1993	Endocrine	HGH	1990
Pulmozyme	Dornase alpha	Genentech	Biotech	30 December 1993	Respiratory	Enzyme	1990
ReoPro	Abciximab	Centocor	Biotech	1994	Cardiovascular	Mab	1990
Cerezyme	Imiglucerase	Genzyme	Biotech	23 May 1994	Enzyme replacement	Enzyme	1990
Norditropin	Somatropin	Novo Nordisk	Drug	8 May 1995	Endocrine	HGH	1990
Bio-Tropin	Somatropin	Bio-Technology General	Biotech	25 May 1995	Endocrine	HGH	1990
Genotropin	Somatropin	Pharmacia	Drug	24 August 1995	Endocrine	HGH	1990
Avonex	Interferon alpha-1a	Biogen	Biotech	17 May 1996	Immunological	Interferon	1990
Humalog	Insulin lispro	Eli Lilly and Co	Drug	14 June 1996	Endocrine	Insulin	1990
Serostim	Somatropin	Serono Laboratories	Biotech	23 August 1996	Endocrine	HGH	1990
Retavase	Reteplase	Boehringer Mannheim	Drug	30 October 1996	Cardiovascular	tPA	1990
Rituxan	Rituximab	IDEC	Biotech	1997	Cancer	Mab	1990
Zenapax	Daclizumab	Roche	Drug	1997	Immunological	Mab	1990
BeneFIX	Factor IX therapy	Genetics Institute	Biotech	11 February 1997	Haemostasis	Coagulation factor	1990

Product	Drug name	Company	Type	Date	Therapeutic area	Category	Year
Follistim	rFSH	Organon	Drug	29 September 1997	Endocrine	Fertility hormone	1990
Gonal-F	Follitropin alpha	Serono Laboratories	Biotech	29 September 1997	Endocrine	Fertility hormone	1990
Infergen	Interferon Alfacon-1	InterMune	Biotech	6 October 1997	Anti-infective	Interferon	1990
Neumega	Oprelvekin	Genetics Institute	Biotech	25 November 1997	Blood cell deficiency	Interleukin	1990
Regranex	Becaplermin	Ortho-MacNeil-Jansen	Drug	16 December 1997	Wound healing	Growth factor	1990
Remicade	Infliximab	Centocor	Biotech	1 January 1998	Immunological	Mab	1990
Synagis	Palivizumab	MedImmune, Astra Zeneca	Biotech	1 January 1998	Anti-infective	Mab	1990
Herceptin	Trastuzumab	Genentech	Biotech	1 January 1998	Cancer	Mab	1990
Simulect	Basiliximab	Novartis	Drug	1 January 1998	Immunological	Mab	1990
Refludan	Lepirudin	Hoechst Marion Roussel	Drug	6 March 1998	Haemostasis	Protein inhibitor	1990
GlucaGen	Glucagon	Novo Nordisk	Drug	22 June 1998	Endocrine	Glucagon	1990
Glucagon	Glucagon	Eli Lilly and Co	Drug	11 September 1998	Endocrine	Glucagon	1990
Enbrel	Etanercept	Immunex	Biotech	2 November 1998	Immunological	Protein inhibitor	1990
Ontak	Denileukin diftitox	Seragen	Biotech	5 February 1999	Antineoplastic	Interleukin	1990
NovoSeven	Factor VIIa	Novo Nordisk	Drug	25 March 1999	Haemostasis	Coagulation factor	1990

(continued on next page)

TABLE 5.1 Early Biotech & Drug Company Biologics Approvals (w/o Diagnostics) (2–4) *(continued)*

US Trade Name	Generic Name	Originating Company	Company Type	US Approval Date	Therapy Area	Protein Type	Decade
Mylotarg	Gemtuzumab calicheamicin	Wyeth	Drug	1 January 2000	Cancer	Mab	1990
ReFacto	Factor VIII	Genetics Institute	Biotech	6 March 2000	Haemostasis	Coagulation factor	2000
Lantus	Insulin glargine	Aventis	Drug	20 April 2000	Endocrine	Insulin	2000
TNKase	Tenecteplase	Genentech	Biotech	2 June 2000	Cardiovascular	tPA	2000
NovoLog	Insulin aspart	Novo Nordisk	Drug	7 June 2000	Endocrine	Insulin	2000
Ovidrel	Choriogonadotropin alpha	Serono Laboratories	Biotech	20 September 2000	Endocrine	Fertility hormone	2000
Campath	Alemtuzumab	Genzyme	Biotech	1 January 2001	Cancer	Mab	2000
PEG-Intron	PEG-interferon alpha-2b	Schering	Drug	19 January 2001	Anti-infective	Interferon	2000
Humira	Adalimumab	Cambridge Antibody Technology, BASF Bioresearch	Biotech	1 January 2002	Immunological	Mab	2000
Zevalin	Ibritumomab	Biogen Idec, Schering	Biotech	1 January 2002	Cancer	Mab	2000

TABLE 5.2 Early Biotech Approvals—Years Since Founding

Biotech	Year of First BLA	Founding Year	Years before BLA
Serono Laboratories Inc.	1996	1906*	NA
Genentech Inc.	1985	1976	9
Biogen Inc.	1996	1978	18
Seragen Inc.	1999	1979	20
Amgen Inc.	1989	1980	9
Bio-Technology General Corp	1995	1980	15
Genetics Institute Inc.	1997	1980	17
Chiron Corp	1992	1981	11
Genzyme Corp	1994	1981	13
Immunex Corp	1998	1981	17

*Serono was found in Rome 1906 as Instituto Pharmacologico Serono by Professor Cesare Serono. Early products were human hormones extracted from the urine of postmenopausal women including nuns. Based on data from Ref. (4).

cover the processes involved in the discovery and development of both SMDs and biomolecular drugs (8,9).

Now, more than ever, anyone interested in understanding the biopharmaceutical industry will need to understand both the differences and similarities between small molecules and biologics and their discovery and development as drugs.

5.2 HOW DO SMDs DIFFER FROM BIOMOLECULAR DRUGS?

One has only to consider the size of biomolecular drugs to recognize that the technologies that give rise to biomolecular drugs must be considerably different from the classical SMDs (Figure 5.1). Genentech equates the difference between aspirin (21 atoms) and an antibody ($\sim$25,000 atoms) to the difference in weight between a bicycle ($\sim$20 lbs) and a business jet ($\sim$30,000 lbs) (10).

Let us consider how they differ with respect to distribution, metabolism, serum half-life, typical dosing regimen, toxicity, species reactivity, antigenicity, clearance mechanisms, and drug–drug interactions (especially SMD/biologic drug interactions) (Table 5.4) (2,11).

The differences in structure, molecular weight, preparation, physicochemical properties, and route of administration are likely to be well understood by most who are engaged in the biopharmaceutical industry. The other properties may not be as well understood.

5.2.1 Structure

Small molecules can have any kind of chemical structure. They tend to be what chemists call organic structures, consisting mostly of carbon, nitrogen, and oxygen. There are, of course, inorganic SMDs as well, such as lithium, and gases, for example, nitrous oxide.

TABLE 5.3 Notable Acquisitions and Partnerships Involving Biologics (2,5–6)

Year	Bigger	Smaller	Type	Deal (billions)	Goal
2011	Sanofi-Aventis	Genzyme	Acquisition	$20.10	Biologics products and pipeline
2009	Pfizer	Wyeth	Acquisition	$68	Biologics products and pipeline
2009	Roche	Genentech	Acquisition	$46.80	Biologics products and pipeline
2009	Merck	Schering–Plough	Acquisition	$41.10	Biologics products and pipeline
2009	Cephalon	Arana Therapeutics	Acquisition	$0.20	Inflammatory disease and oncology biologic compounds
2009	Cephalon	Ception	Acquisition	Option	Reslizumab, a humanized Mab for asthma
2008	Eli Lilly	ImClone Systems	Acquisition	$6.50	Biologics products and pipeline
2008	Takeda	Millenium Pharmaceuticals	Acquisition	$8.80	Biologics products and pipeline
2008	Takeda	Alnylam	Partnership	$1	RNAi (RNA interference) therapies for cancer and metabolic disease
2008	Genzyme	Isis	Partnership	$2	Mipomersen, Isis' cholesterol drug targeting RNA
2008	Novartis	Lonza	Partnership	NA	Biologics manufacturing
2007	Schering–Plough	Organon Biosciences	Acquisition	$14	Biologics products and pipeline
2007	Astra Zeneca	Medimmune (Cambridge Antibody Technology)	Acquisition	$16.50	Monoclonal antibody therapies
2007	GlaxoSmithKline (GSK)	OncoMed	Partnership	$1.40	Monoclonal antibody cancer therapies
2006	Genentech	Tanox	Acquisition	$0.90	Xolair
2006	Novartis	NeuTec	Acquisition	$0.57	Mycograb to treat Candida infections
2006	Merck	GlycoFI, Abmaxis	Acquisition	$0.50	Early-stage monoclonal antibody companies
2006	GlaxoSmithKline GSK	Genmab	Partnership	$2.20	Fully human monoclonal antibody Humax CD20
2005	Amgen	Abgenix	Acquisition	$2.20	Full ownership of panitumumab and eliminates a denosumab royalty

NA—Amount not available.

Asprin 21 atoms

Antibody~25,000 atoms

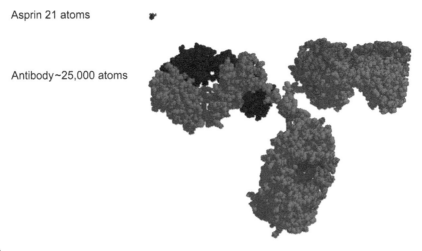

FIGURE 5.1

Difference in size between aspirin and an antibody (*10*). Space-filling models of aspirin and an antibody.

Biomolecules tend to be macromolecular structures composed of polypeptide chains. Strictly speaking they are not biopolymers since they are not created by polymerization. Biomolecules are proteins, primarily composed of peptide chains derived from alpha amino acids (Chapter 2).

In larger peptides and proteins the units of secondary structure (the alpha helices and beta sheets) are important giving rise to relatively large tertiary structures called *domains*. Insulin is presented in a so-called ribbon diagram that traces out the backbone atoms and displays beta sheets or alpha helices in ribbons in Figure 5.2. As described in Chapter 2 the domains can assemble into quaternary structures.

Insulin tends to sit at the interface between SMDs and biomolecular drugs. In Figures 5.2 and 5.3 we see the ribbon diagrams of both insulin and an antibody which highlight the helices and sheets. An antibody contains both alpha-helix and beta-sheet structure. Antibodies figure prominently in this chapter, so it is wise briefly to summarize some of the points from Chapter 4 regarding the domains of an antibody. From the ribbon diagram shown below it is clear that the Y-shaped antibody has three sections or regions and that two regions are more similar to each other than the third (*12*). In Figure 5.3, on the left, we see that the individual domains are differentially coloured. We see on the right that some of the domains are longer than others.

These domains tend to be described by their molecular weight—the long domains are called *heavy chains* and the short domains are called *light chains*. The two similar sections are the antigen-binding regions or Fab regions as shown on the right. These regions are highly variable due to the wide variety of antigens that individual antibodies must recognize. The third region is the constant or Fc region. This region is responsible for the immune response. The Fab regions bind antigen and the Fc region may bind to natural killer cells, macrophages, neutrophils, or mast cells, each of which mediates different aspects of the immune response.

TABLE 5.4 Comparison of Properties of Small-Molecule Drugs and Biomolecular Drugs (2,11)

Property	Small Molecules	Therapeutic Proteins
MW	Low molecular weight (<1000 Da)	High molecular weight ($\gg$1000 Da)
Preparation	Chemical synthesis	Biologically produced, can be engineered
Physicochemical properties	Mostly well defined	Complex—tertiary structure—undergo post transcriptional modifications, e.g. glycosylation
Route of administration	Oral administration usually possible	Usually administered parenterally
	Rapidly enter systemic circulation through blood capillaries	Reach circulation primarily via parenteral route: iv, direct; or sc via lymphatic system
Distribution	To any combination of organs/tissues/cells	Usually limited to plasma and/or extracellular fluids
Metabolism	Metabolized typically by liver and gut CYPs into nonactive and active metabolites	Catabolism by proteolytic degradation to peptides and amino acids
Serum half-life	Short serum half-lives	Relatively long serum half-lives
Typical dosing regimen	Suitable for QD or BID dosing	Dosing usually far less frequent
Toxicity	Can produce specific toxicity due to parent or metabolites (often "off target")	Mostly receptor mediated toxicity, including both super pharmacodynamic responses and biological toxicity (often "exaggerated pharmacology")
Species reactivity	Generally active in multiple animal species	Relevant and irrelevant animals models
Antigenicity & hypersensitivity	Nonantigenic, but can show unpredictable hypersensitivity	Potential for antigenicity (with MW > 10 kDa)
Clearance mechanisms	Small molecules and therapeutic proteins do not share clearance mechanisms—noncompeting, parallel	
- Renal excretion	Yes	Yes
- Hepatic	Biliary excretion	Peptidases
	Metabolism (P450, UGT, Sulfotransferases, etc)	
- Target mediated	Rare	Yes
- Intestinal	CYP, UGT, etc.	No, due to peripheral administration
	Often involves transporters	
Drug–drug interactions	Pharmacokinetic interactions due to competing clearance mechanisms	Less common, less well-defined
	- Decreasing clearance if there is enzyme inhibition	
	- Increasing clearance if there is enzyme induction	
Pharmacology	Tends to mimic or interrupt ligand interaction with its receptor or enzyme	May interact with receptor but the biologic may be the ligand (hormone replacement therapy) or bind to the ligand, it may be a soluble receptor (decoy therapy) or may be an enzyme

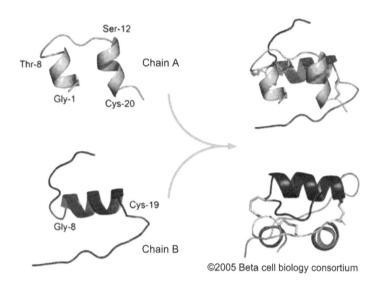

©2005 Beta cell biology consortium

FIGURE 5.2

Secondary and tertiary structure of insulin showing the alpha-helical domains within the A and B chains on the left, which give rise to the overall tertiary structure on the right.

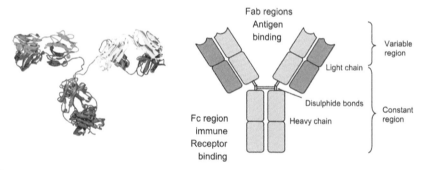

FIGURE 5.3

The variable and constant regions of an antibody.

Enzymes such as papain and pepsin are used to remove part of or all the Fc region. Papain yields individual Fab fragments. Pepsin leaves a bridging region from the Fc region to yield an intact pair of Fab fragments.

Historically, Mabs were created from the antibodies raised in mice. It soon became apparent that the murine structure risked an immune response if administered to humans. The first attempt to resolve this issue was with chimeric Mabs. Chimeric Mabs contain human constant regions with murine variable regions. Later on companies would develop so-called *humanized Mabs*. Humanized Mabs include murine complementarity-determining regions only; the remainder of the variable regions, and the constant regions, are derived from a human source. Most recently, companies have developed genetically manipulated strains of mice that produce fully human antibodies.

It is now timely to consider the other properties of small molecules and biomolecules.

5.2.2 Distribution

In general, size differences tend to influence membrane permeability, thus SMDs are more likely to distribute to any combination of organs, tissues and cells in the body, whereas the distribution of biomolecular drugs tends to be limited to plasma and/or extracellular fluids. There are other factors that give rise to exceptions to this size generalization, but size is clearly an important influence on distribution (see p. 145 in Ref. (9)).

5.2.3 Metabolism

After entering the body drugs may be degraded by various mechanisms. Small molecules tend to be degraded in the liver and the gut particularly by a class of oxidative enzymes called *cytochrome P450s*. In a surprising number of instances what should be a degradative process actually creates a molecule that is more active than the parent drug, a so-called *active metabolite*. An interesting example is clopidogrel, where the active metabolite was discovered years after the drug was on the market (*13*). Biomolecular drugs tend to be degraded by proteases, which act on proteins in a number of ways to reduce them to small peptides and amino acids. Such metabolism can be thwarted by alteration of the protein structure.

5.2.4 Serum half-life

Serum half-life is influenced by many factors. SMDs often display a short serum half-life due to nonspecific binding to plasma proteins, which tends not to happen for biomolecular drugs (p. 149 in Ref. (9)).

5.2.5 Typical dosing regimen

The differences outlined thus far play into the typical dosing regimen. While it is feasible to develop SMDs suitable for once a day or twice a day dosing, biomolecular drugs often need to be administered only biweekly or monthly. With this relatively leisurely dosing regimen the patient is more inclined to visit the physician for a dose of a biomolecular drug rather than be concerned with remembering to take the medication once or twice a day, perhaps with or without a meal, as is the case with an SMD. It is the dosing regimen that provides biomolecular drugs with a competitive advantage over small molecules especially in arenas long dominated by oral small molecules, such as osteoporosis and anti-infectives. With the physician in charge of dose administration, compliance is likely to be less of an issue with a biomolecular drug (*14*).

5.2.6 Species reactivity

SMDs tend to be active in many species, although a comparison of activities in a receptor-binding assay can detect variations in affinity to different species (also called orthologs), which is an important evaluation prior to animal disease model studies (more drug may need to be administered in one species to achieve the same effect in another). The issue can be more dramatic with biologics, especially Mabs. The biomolecular drug has a considerably larger surface area that interacts with its biological target, like comparing a prosthetic hand (biomolecule) in a glove (biological target) vs a thorn (small molecule). Both the small molecule (thorn) and the biomolecule (prosthetic hand) keep the natural ligand (hand) out of the biological target (glove) but species variations (changes in the texture of the glove) in the target is more likely to be sensed by the biomolecule (prosthetic hand) than the small molecule (thorn).

An interesting example of a small molecule-sensing species selectivity is that of Eltrombopag, which binds in the cytoplasmic domain of the thrombopoietin (TPO) receptor. Species variation in that region results in Eltrombopag and similar compounds binding only to the human and chimpanzee TPO receptor (*15*).

5.2.7 Antigenicity and hypersensitivity

Immune surveillance tends to focus on large molecules and cellular fragments. Thus small molecules tend not to elicit an antigenic response. On the other hand drug hypersensitivity and drug allergy is a concern with SMDs. The causes of the latter are dependant on drug type, conversion to reactive metabolites and predisposition to allergic reaction (i.e. many factors) (*16*).

5.2.8 Clearance mechanisms

Small molecules and therapeutic proteins do not share clearance mechanisms. Both suffer renal excretion but by different pathways (*11*). In the liver small molecules can suffer biliary excretion and metabolism by a variety of pathways mediated by P450s, Uridine 5'-diphospho (UDP)-glucuronosyltransferase, sulfotransferases and other enzymes. The metabolism of biomolecular drugs tends to occur via peptidases that naturally degrade proteins into peptides and amino acids. Target-mediated clearance pathways at the site of action can be observed with biomolecules, especially recombinant versions of native proteins. Such clearance mechanisms tend to be rare for small molecules. Orally active SMDs are challenged in the gut by intestinal cytochrome P450s, UDP-glucuronosyltransferase and other enzymes involved in the normal actions of the gut, which can involve the action of transporters. Most biomolecular drugs are not administered orally and therefore do not experience the metabolic activities of the gut.

5.2.9 Drug–drug interactions

Drug–drug interactions are well known with small molecules, mainly due to competitive clearance mechanisms. One drug may decrease the clearance of another if it inhibits a particular enzyme, or increase the clearance of another drug if it induces the activity of a particular enzyme. Increasingly physicians prescribe two or more drugs to act on the same pathology. In areas where drugs already exist for a particular pathology, regulatory authorities will assume that the new drug must show a benefit on top of (in conjunction with) the standard therapy. So it is not surprising that SMDs can influence the activity of biomolecular drugs and vice versa. The evaluation of SMDs and biomolecular drugs is challenged by the lack of relevant activity of many biomolecular drugs in mouse or rat disease models (*17*).

5.2.10 Pharmacology

It is important now to compare the pharmacology of SMDs and biomolecular drugs. Most often the SMD modulates the binding of an endogenous ligand (agonist or enzyme substrate) with an endogenous receptor or enzyme by interacting with the receptor or enzyme (Chapter 1).

In the classical lock and key model, the small molecule binds to the lock to prevent the key from doing its thing. This can also be the case with biomolecules but the pharmacology of biomolecules can be considerably more diverse. The biomolecule may *be* the key, as in enzyme or hormone replacement therapy (Chapters 10 and 11). The biomolecule may bind to the ligand and not the receptor. The biomolecule may be the extracellular portion of a membrane-bound receptor, as in decoy therapy. And the biomolecule can have enzymatic activity. A company's portfolio nomenclature system needs to be able to account for these diverse possibilities.

This brief survey of the differences between SMDs and biomolecular drugs was designed to simply highlight these differences. The reader is encouraged to go into the references to get a more detailed discussion of the differences. A scientist who has worked in one field and is now facing the prospect of leading a project in the other field should become familiar with these differences as they will give rise to issues that he/she may not have faced before.

5.3 HISTORICAL CHANGES TO THE FDA APPROACH TO HANDLE THE BIOTECH BOOM—THE DIFFERING NOMENCLATURE FOR SMALL MOLECULES VS BIOLOGICS ENTERING THE CLINIC

SMDs and biomolecular drugs are described differently over the lifetime of drug discovery and development and marketing, as shown in Figure 5.4.

Biosimilars are also referred to as "follow-on biologics". Phase length is not implied by the size of stage marker. NME relates to the first approvable drug as opposed to second indications or new formulations. The application for a generic small molecule is an "Abbreviated New Drug Application" (ANDA) which does not require clinical trials to prove equivalency. Proposed processes for biosimilars or follow-on biologics are discussed later. A description of European and other regulatory variations in nomenclature are beyond the scope of this chapter.

As discussed earlier, the SMDs that are approved by the FDA are referred to as NCEs while the approved biologics are referred to as NBEs. The new drug application for an NCE is called a New Drug Application (NDA) whereas a new drug application for an NBE is called a BLA. Since NDAs can be granted for secondary indications and formulations of previously approved drugs the term NME describes a drug where the NDA is the first in kind (indication or formulation). This can lead to some confusion, since the FDA can also grant supplemental approvals and BLAs to secondary indications and formulations. But even within the FDA NME continues to be used only for first SMD approvals. First biomolecular drug approvals are simply referred to as "New BLAs". The term NME became important with the Hatch-Waxman Act to provide extra protection in the US to new drug NDAs, which tend to suffer longer approval times than NDAs for new formulations or second indications.

In the early years biotechnology companies focused on producing recombinant versions of proteins that were well characterized and, being on the market, thus had a high probability of gaining regulatory

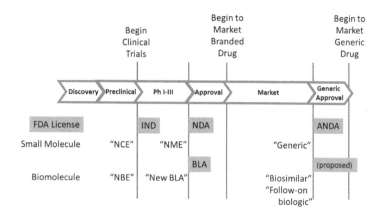

FIGURE 5.4

Small molecules and biomolecules take on different names over the lifetime of a drug.

approval (9). In those early years biologics were reviewed by the Center for Drugs and Biologics (CDB). In the CDB era (1982–1987) only six biologics had been reviewed at CDB.

In 1987 FDA oversight of biologics moved to CDER and the Center for Biologics Evaluation and Research (CBER). Most biologics were reviewed by CBER, but hormonal-type proteins such as insulin and growth hormone were reviewed by CDER. As will be discussed subsequently, this early experience with insulin and growth hormone will complicate the biosimilars issue.

CDER assumed responsibility for all biomolecular drugs in 2003. In the CBER era (1988–2002) 27 biologics had been reviewed. Figure 5.5 displays the relationship of biomolecular drugs to all biologics or biotherapies and shows some examples of important biomolecular drugs. Since 2002 the volume of biologics reviewed at CDER has boomed (4). CBER continues to review vaccines, blood, blood components, toxins for immunization, antitoxins, antivenins, venoms, gene therapies (viral and other vector-based) and products composed of human or animal cells or from physical parts of those cells.

Since the early 1980s the number of INDs per year from NCEs has levelled off while the INDs from NBEs have increased and helped maintain an increasing number of INDs/year, at least up to 1993 (18). A recent Thomson Reuters webinar presentation confirms this trend (19). It is a concern that more drugs appear to go into the clinic than come out of the clinic!

Trusheim et al. and others have studied the number of new SMD approvals (NMEs) in comparison to new biologic drug approvals (new BLAs) in the period between 1988 and 2008 (Figure 5.6; Table 5.5) (20).

In this study, biologics are not restricted to monoclonal antibodies, cytokines, tissue growth factors and therapeutic proteins but include all biotherapies. The last line of Table 5.5 shows therapeutic proteins and MAbs from the Richert study which went to 2003 (1). We extended the tally by Reichert beyond 2003 by adding our own count of Mab and therapeutic protein new BLAs from annual FDA reports through 2008. Mullard (21) and Kneller (22) recently published counts of NMEs and New BLAs which differ somewhat from Trusheim and colleagues (20). We are not in a position to rectify the differences, except to offer a potential explanation - certain small peptide and protein drugs may be considered either biologics or small molecules (Kneller considered such drugs to be biologics).

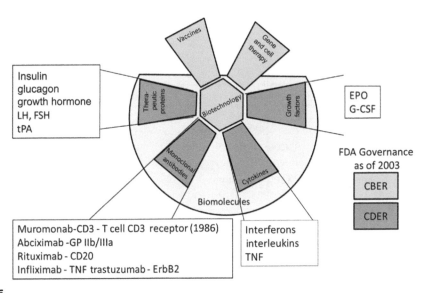

FIGURE 5.5

The relationship of biomolecular drugs to all biotherapies and some examples of biomolecular drugs. (For colour version of this figure, the reader is referred to the online version of this book.)

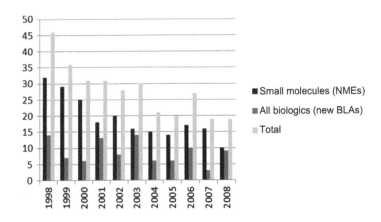

FIGURE 5.6

Numbers of new small-molecule drug approvals per year (NMEs) compared to new biologic drug approvals (new BLAs) 1988–2008 (*20*).

The analysis by Trusheim et al. was not restricted to Mabs, cytokines, tissue growth factors and therapeutic proteins but covers all biotherapies. They found that from 1988 to 2003 the industry averaged 34 NMEs and new BLAs, whereas from 2004 to 2008 the industry averaged only 21 NMEs and new BLAs per year. Within those two periods the percentage of new BLAs was quite similar (31%

TABLE 5.5 Numbers of New Small-Molecule Drug Approvals per Year (NMEs) Compared to New Biologic Drug Approvals (New BLAs) in 1988–2008 (*20*)

	1998	1999	2000	2001	2002	2003	2004	2005	2006	2007	2008
Small molecules (NMEs)	32	29	25	18	20	16	15	14	17	16	10
All biotherapies (new BLAs)	14	7	6	13	8	14	6	6	10	3	9
Mabs and Ther. Proteins	7	2	6	7	7	7	5	2	4	2	4

vs 32%) (*20*). To add some perspective we include the Mabs and therapeutic proteins counted in Reichert (*1*). By the numbers, all biologics are making a substantial contribution to the number of new drugs approved per year.

5.4 COMPARISONS OF CLINICAL METRICS—BIOLOGICS VS SMALL MOLECULES

Only a few biomolecular drugs were approved in the US per year until 1997, when eight were approved in 1 year. From that time onward approvals have been at the rate of over a half dozen per year (*1*). There are now sufficient numbers of biomolecular drugs to begin to allow cross-industry comparisons of metrics between SMDs and biomolecular drugs. The reader may want to read about *Benchmarking through a Consortium* (*23*).

Most of the published comparisons of metrics with respect to SMD vs biomolecular drug discovery or development arise from Dimasi and Reichert at the Tufts Center for the Study of Drug Development (*24*), and Grabowski, at Duke University (*25*). Due to FDA oversight, clinical trial data are available to the public, whereas preclinical data are rarely made available. Thus most comparative metrics data between small molecules and biologics are clinical. A comparison of small molecules and biologics in the discovery and preclinical stages may be found in Section 5.10.

The following discussion will focus on the clinical stages in the following studies of Reichert, DiMasi, Grabowski et al. (*4,18,26–31*). The clinical metrics from these studies will be compared and contrasted, with due regard for the complexity of such comparisons.

These published studies clearly indicate that the metrics between small molecules and biologics can differ. For example, the mean clinical development and approval times were quite different in the 1980s: 5.2 years for biomolecules vs 9 years for small molecules in the 1984–1986 period, but moved to 7.4 years for biomolecules vs 6.9 years for small molecules in the 1999–2001 period (*27*). As noted earlier, by Reichert and Paquette (*4*), the biotech industry initially focused on producing recombinant versions of already approved proteins, likely facilitating both clinical study and approval cycle times (*9*). Since then, the industry has taken on more challenging therapies.

5.4.1 Overall clinical success rates of biologics vs small molecules

A similar trend was observed with success rates, which is the percentage of NCEs that advanced into the next stage (stage-related success rates) or advanced through all stages of development (overall clinical success rates) (Table 5.6).

As seen in Table 5.6, the overall clinical success rates (NBEs achieving BLA) for NBEs initiating clinical trials in the 1980s were substantially higher than for small molecules that initiated trials in the same period, ranging from 48.5% to 60.4% for NBEs (Study A) vs 20.5% to 23.2% for NCEs (Study B, all NDAs). By the 1990s success rates for biologics began to fall into the range of success rates for small molecules. Nonetheless, the success rates for biologics have continued to outpace small molecules in that period—32% for self-originated NBEs vs 13% for self-originated NCEs (29). A most recent study supported by the Biotechnology Industry Organization and BioMed Tracker came to a similar conclusion—a higher overall success rate for NBEs (26% for lead indications) than for NCEs (14% for lead indications) (32).

Dimasi et al. have shown that the success rates for acquired NCEs are higher than the success rates for self-originated NCEs (18,29), most likely due to the fact that assets put up for acquisition/partnering are in some way more promising than those that were not put up for acquisition/partnering. The corresponding analysis of NBEs has yet to be published but would likely follow the same trend.

5.4.2 Stage-related success rates for small molecules vs biologics

In some studies the success rates for individual stages have been determined (27–29). Dimasi et al. determined success rates for individual clinical stages of NCEs and NBEs (Table 5.7). The data in Study A were for agents entering Phase 1 between 1990 and 2003. The data were updated recently, as seen in Study B in Table 5.7, and list success rates for both small molecules and biomolecules entering the clinic between 1993 and 2004. Because the studies sample different eras, the success rates calculated for the cohort of agents in Study A differ from the cohort of agents in Study B.

The success rates for the regulatory stage and Phase 3 are also separate in Study B.

In the clinical stages, the success rates bounced around, with the largest difference being Phase 1. In both studies, biomolecules showing a better success rate than small molecules in Phase 1 (by 13% in Study A and by 21% in Study B), which largely accounts for the better overall clinical success rate for BLAs in both Studies A and B (by 8.7% and 19%).

5.4.3 Stage-related clinical cycle times for small molecules vs biologics

Several studies also determined cycle times (length of time to complete a single stage or several stages) that offer a comparison between small molecules and biologics (Table 5.8) (26,27,31). As mentioned already, due to FDA oversight, clinical trial data are available to the public, whereas preclinical data are only rarely made available. Thus publicly available comparative cycle time data between small molecules and biologics are restricted to clinical stages. As the impact of biologics continues to increase in the biopharmaceutical industry preclinical cycle time data will eventually become available. Small-molecule discovery stage cycle times are discussed in Ref. (33).

As might be expected in the early years, the overall time needed to conduct clinical trials for an NBE was considerably shorter when companies focused on recombinant versions of well-studied proteins that were already on the market. Since that time the clinical cycle times for NBEs have increased and are now comparable to those of NCEs (Table 5.8).

TABLE 5.6 Clinical Success Rates (IND to Launch) for NCEs and NBEs Clustered by Periods of NDA/BLA Approval (4,18,27,29). NCEs are Categorized as to Whether the NDA was Self-originated (In-House) or Obtained by Another Institution (and Subsequently Acquired)

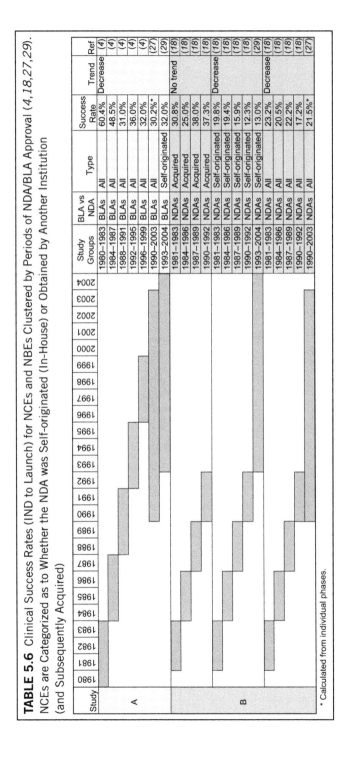

Study	Study Groups	BLA vs NDA	Type	Success Rate	Trend	Ref
A	1980–1983	BLAs	All	60.4%	Decrease	(4)
	1984–1987	BLAs	All	48.5%		(4)
	1988–1991	BLAs	All	31.0%		(4)
	1992–1995	BLAs	All	36.0%		(4)
	1996–1999	BLAs	All	32.0%		(4)
	1990–2003	BLAs	All	30.2%*		(27)
	1993–2004	BLAs	Self-originated	32.0%		(29)
B	1981–1983	NDAs	Acquired	30.8%	No trend	(18)
	1984–1986	NDAs	Acquired	25.0%		(18)
	1987–1989	NDAs	Acquired	38.0%		(18)
	1990–1992	NDAs	Acquired	37.3%		(18)
	1981–1983	NDAs	Self-originated	19.8%	Decrease	(18)
	1984–1986	NDAs	Self-originated	19.4%		(18)
	1987–1989	NDAs	Self-originated	15.9%		(18)
	1990–1992	NDAs	Self-originated	12.3%		(18)
	1993–2004	NDAs	Self-originated	13.0%		(29)
	1981–1983	NDAs	All	23.2%	Decrease	(18)
	1984–1986	NDAs	All	20.5%		(18)
	1987–1989	NDAs	All	22.2%		(18)
	1990–1992	NDAs	All	17.2%		(18)
	1990–2003	NDAs	All	21.5%*		(27)

* Calculated from individual phases.

TABLE 5.7 Comparison of Success Rates and Cycle Times Between Phases of Small Molecule and Biomolecular Drug Discovery and Development

Study Groups - Year when NDA was Filled (bar chart spanning years 90 91 92 93 94 95 96 97 98 99 00 01 02 03 04)

Study	Study Groups	Type	Phase 1	Phase 2	Phase 3	Approval Stage	Total	Self-originated vs Acquired	Trend	Ref
A	1990–2003	BLAs	83.7%	56.3%		64.2%	30.2%	All	Greater	(27)
A	1990–2003	NDAs	71.0%	44.2%		68.5%	21.5%	All		(27)
A	1990–2003	Difference	12.7%	12.1%		-4.3%	8.7%	All		(27)
B	1993–2004	BLAs	84.0%	53.0%	74.0%	96%	32.0%	Self-originated	Greater	(29)
B	1993–2004	NDAs	63.0%	38.0%	64.0%	91%	13.0%	Self-originated		(29)
B	1993–2004	Difference	21.0%	15.0%	10.0%	5.0%	19.0%	Self-originated		(29)
C	1993–2004	BLAs + NDAs	82%	56%	64%	93%	27.3%	Acquired	Greater	(29)
C	1993–2004	BLAs + NDAs	65%	40%	64%	93%	15.5%	Self-originated		(29)
C	1993–2004	BLAs + NDAs	71%	45%	64%	93%	19.0%	All		(29)

TABLE 5.8 Overall and Stage-Related Clinical Cycle Times (months). References are Shown in the Table (32,27,31). Note to study C: the phase results are averages across all compounds that completed the phase, regardless of whether the compound was ultimately approved for marketing (27)

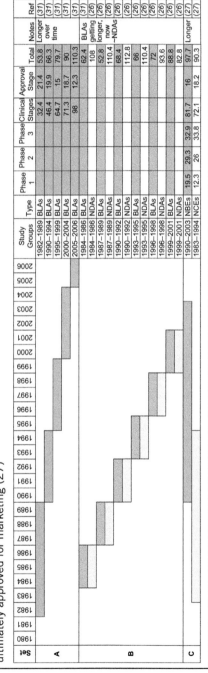

Set	Study Groups	Type	Phase 1	Phase 2	Phase 3	Phase/Clinical Stages	Approval Stage	Total	Notes	Ref
A	1982–1989	BLAs				32.4	21.4	53.8	Longer over time	(31)
A	1990–1994	BLAs	46.4				19.9	66.3		(31)
A	1995–1999	BLAs	64.7				15	79.7		(31)
A	2000–2004	BLAs	71.3				18.7	90		(31)
A	2005–2006	BLAs	98				12.3	110.3		(31)
B	1984–1986	BLAs						62.4	BLAs getting longer, now NDAs	(26)
B	1987–1989	BLAs						108		(26)
B	1990–1992	BLAs						52.8		(26)
B	1990–1992	NDAs						110.4		(26)
B	1993–1995	BLAs						68.4		(26)
B	1993–1995	NDAs						112.8		(26)
B	1996–1998	NDAs						66		(26)
B	1996–1998	BLAs						110.4		(26)
B	1999–2001	BLAs						72		(26)
B	1999–2001	NDAs						93.6		(26)
B		BLAs						88.8		(26)
B		NDAs						82.8		(26)
C	1990–2003	NBEs	19.5	29.3	32.9	81.7	16	97.7	Longer	(27)
C	1983–1994	NCEs	12.3	26	33.8	72.1	18.2	90.3		(27)

TABLE 5.9 Mean (Median) Phase Lengths in Five Disease Areas for Three Types of Drugs: Small-Molecule Drugs (SMD), rDNA Proteins (rDNA) and Monoclonal Antibodies (mAb) Approved 1982–2001 (*30*)

Disease Area	Type of Drug	Clinical Phase (mos)	Approval Phase (mos)	Total (mos)
Anti-infective	SMD (n = 102)	55.8 (47.9)	18.6 (14.0)	74.5 (64.2)
Anti-infective	rDNA (n = 3)	98.3 (85.2)	14.2 (12.9)	112.5 (98.1)
Antineoplastic	SMD (n = 38)	96.8 (81.6)	19.2 (12.8)	116.0 (101.0)
Antineoplastic	rDNA (n = 4)	46.6 (51.3)	27.6 (27.6)	74.2 (71.7)
Antineoplastic	mAb (n = 4)	80.7 (66.0)	9.2 (7.8)	89.9 (71.7)
Cardiovascular	SMD (n = 107)	73.1 (59.6)	30.2 (26.7)	103.3 (91.7)
Cardiovascular	rDNA (n = 4)	55.6 (59.5)	21.1 (17.3)	76.7 (72.5)
Endocrine	SMD (n = 36)	92.2 (75.2)	23.1 (14.9)	115.3 (95.9)
Endocrine	rDNA (n = 10)	51.9 (56.7)	17.1 (13.5)	69.0 (73.7)
Immunological	SMD (n = 11)	90.3 (63.6)	9.8 (8.4)	100.2 (67.7)
Immunological	rDNA (n = 4)	92.1 (96.2)	13.3 (11.9)	105.4 (105.1)
Immunological	mAb (n = 4)	65.4 (68.4)	11.6 (6.9)	77.0 (77.5)

Dimasi and Grabowski determined individual stage-related clinical cycle times for NBEs and NCEs (Table 5.8, Study C) (*26*). It should be noted that the Study C cycle times were calculated for all compounds that completed the phase, regardless of whether the compound was ultimately approved for marketing. For this reason, the overall cycle times shown for that study in Table 5.8 are the sums of these averages and cannot be compared to the overall cycle times in the aforementioned studies (*29,31*), which only covered NBEs and NCEs that achieved launch (i.e. BLAs and NDAs). The Phase 1 cycle times for NBEs are longer than those for NCEs, due to the historical cautions around the potential for immunogenic response of Mabs with sequences that were not completely human or proteins extracted from animal tissues. (With the advent of fully human Mabs and recombinant human proteins the differences in Phase 1 cycle times with NCEs may diminish.)

Reichert compared clinical and approval stage lengths for SMDs, recombinant proteins ("rDNA") and Mabs approved from 1982 to 2001 in the disease areas where biomolecules have had success (Table 5.9) (*30*). Mabs in that period were approved in only two areas, antineoplastic and immunological disorders. Surprising variations in clinical and approval phases were observed. For example, small molecules get through the clinic much faster than biologics in the anti-infectives area, but tended to take longer in other therapy areas.

As the challenge of immunogenicity lessens with fully human Mabs and as more biomolecular drugs are developed in other disease areas, these cycle time studies need to be repeated to see if the observations from these initial studies continue to describe cycle times in the future.

5.4.4 Comparative cost of R&D for biologics vs small molecules

How much do biopharmaceutical companies spend on R&D for biologics vs SMDs? The PhRMA surveyed their members and came up with the totals as shown in Table 5.10 (*34*).

TABLE 5.10 PhRMA Member R&D, 2008

Row	Type	Dollars	Share	Percent 1 and 7
1	**Biotechnology-derived therapeutic proteins**	**$10,542.30**	22.20%	**26.0%**
2	Vaccines	1600.80	3.4	
3	Cell or gene therapy	176.9	0.4	
4	All other biologics	1337.80	2.8	
6	Total biologics	13,657.70	28.8	
7	**Non-biologics**	**30,057.50**	63.4	**74.0%**
8	Uncategorized R&D	3667.90	7.7	
9	Total R&D	$47,383.10	100.00%	
10	Total biotech proteins + non-biologics	$40,599.80		

*Highlighted in **Bold** are Biomolecules ("Biotechnology-Derived Therapeutic Proteins") and Small Molecules ("Nonbiologics") and the Percent of Total Spent on Each (34).*

We are interested in comparing biomolecules to small molecules, i.e. the text highlighted in bold font (rows 1 and 6). As calculated in Table 5.10, the R&D spend on small molecules was three times that of biomolecules. The costs shown in Table 5.10 would be more meaningful if they were related to the number of projects in discovery and development for both small molecules and biologics. Non-project costs such as capitalization costs need to be considered as well.

We learnt in earlier sections that biomolecular drugs and SMDs have different success rates and cycle times. The differences in success rates and cycle times noted above have a knock-on effect on the cost of R&D for biomolecules over small molecules (Table 5.11) (27,35).

Although "preclinical" "out of pocket" expenses were found to be comparable for biomolecular drugs and SMDs, the capitalization costs which are influenced by phase time are much larger for biomolecules, resulting in a larger capitalized cost for these stages. Clinical costs for small-molecule pharmaceuticals suffer in comparison to biomolecules both in terms of "out of pocket" expenses as well as capitalization costs. In total, the overall capitalized costs end up being similar, with

TABLE 5.11 Comparison of Costs between Small-Molecule and Biomolecule Drug Discovery and Development, 2006 (27,35)

	Time of Publication	Overall Average Costs ($million)			"Preclinical" Costs ($million)			Percent	Clinical Costs ($million)			Percent	Ref.
		Out of Pocket	Capitalization Costs	Capitalized	Out of Pocket	Capitalization Costs	Capitalized		Out of Pocket	Capitalization Costs	Capitalized		
Biomolecules	2006	$559	682	$1,241	$198	417	$615	49.6%	$361	265	$626	50.4%	(27)
Small Molecules	2003 adj to 2006	$672	648	$1,318	$150	289	$439	33.3%	$522	357	$879	66.7%	(35)

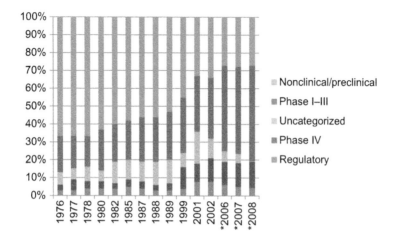

FIGURE 5.7

Pharmaceutical industry percentage allocation of R&D expenditures from 1976 to 2008. Data for 1976–2002 is provided by PhRMA and analyzed by Cohen (*36*). Nonclinical/preclinical is the term by PhRMA for discovery and preclinical research. (See colour plate.)

*Data supplied from PhRMA profiles for 2008–2010 (34,39,40).

biopharmaceuticals perhaps having a small advantage. Because the authors had teased apart the differences in phases, one can see that while overall average costs are similar ($1241 million vs $1318 million) there are important differences within the stages of discovery and development.

The percentage of "preclinical" costs vs clinical costs for both biomolecules and small molecules has been determined. In 2006 roughly 50% of the spend went to "preclinical" for biomolecules but only one-third of the spend went to "preclinical" for small molecules. As will be discussed shortly, this ratio for small molecules is in line with the PhRMA spend per member.

Cohen has shown that the ratio of "preclinical" to clinical costs have changed over the years (Figure 5.7) by aggregating data from past reports of the PhRMA (*40*). We have back-filled the last three available years of data from PhRMA annual profiles. Here discovery and preclinical research are lumped together as "nonclinical/preclinical".

The table shows that in the 1970s the industry spend on discovery and preclinical stages was over 60% of the industry R&D budget. But look where the industry has gone—since 2006 the industry spent <30% of its budget on discovery and preclinical research (presumably both small molecules and biologics). The lion's share of the budget went into the clinical stages of R&D (the same trend was observed by Moses et al. (*37*)). This trend is alarming and is discussed further in Ref. (*38*).

It would be instructive to see if the ratio of preclinical to clinical spends has changed over the years with biomolecules.

5.4.5 The challenges in comparing small-molecule and biomolecular drug metrics—the influence of biomolecular scaffold

The studies on success rates, cycle times and costs discussed above clearly indicate that when it comes to portfolio data analysis, it is exceedingly important to be cautious about intermingling data for

TABLE 5.12 Mean Success Rates for Chimeric and Humanized Mabs; and Mean Clinical and Approval Phases for FDA-Approved Therapeutic Mabs (both Chimeric and Humanized) and Recombinant Proteins (*42*)

Product type	Success Rate	Clinical Phase (months)	Approval Phase (months)	Total (months)
Murine Mab	4%			
Chimeric Mab	22%			
Humanized Mab	21%			
Therapeutic mAb		74.1	9	83.1
Recombinant protein		59.4	16.9	76.3

biomolecules and small molecules. Within the biomolecules, especially as the variety of scaffolds proliferates (from chimeric to fully human Mabs to antibody fragments, diabodies to tetrabodies and peptibodies (*41*)) it is important to contrast the molecular scaffold types of biomolecular drugs. At least some of the processes will differ among scaffold types of biomolecular drugs and these differences could show up in cycle time, success rate, cost and even risk. Reichert found that in the period between January 1994 and February 2002 there was no difference in overall clinical success rates for chimeric and humanized Mabs (22% vs 21%), but understandably murine Mabs had a much lower success rate (4%) due to problems with immunogenicity. Reichert also compared cycle times for "therapeutic Mabs" (murine, chimeric and humanized) and "rDNA" (non-Mab biomolecular drugs) (*42*). She found differences in the length of both the clinical phases and the approval phase when comparing Mabs with rDNA (Table 5.12).

5.5 ARE PEPTIDE DRUGS SMALL MOLECULES OR BIOLOGICS?

This hybrid class of drugs tends to be considered a class of biologics, due to their amino acid content and especially because oral activity is rare among peptide drugs. But peptides truly bridge the gap between small molecules and biologics, in terms of physical properties, range of therapy areas and means of production. Like biologics, peptides structures are partially or completely "natural". We prefer the term *native structure* when referring to those portions of structure that derive from the native amino acids, and *nonnative structure* when referring to those portions of structure that are not derived from native amino acids. We summarize the similarities and differences of peptide drugs with SMDs and biomolecular drugs in Table 5.13, which is a modified version of Table 5.4.

Generally, properties of peptide drugs that are similar to SMDs are highlighted in light grey and relate to distribution, serum half-life, dosing regimen, species reactivity and antigenicity. Peptide drug properties that are similar to biomolecular drugs are highlighted in the darker colour, that is route of administration, target-mediated clearance and drug–drug interactions. Peptide drug properties that are a hybrid of SMDs and biomolecular drugs are highlighted in white and relate to the degree of nonnative structure incorporated into the peptide drug, i.e. the nonnative structural elements tend to increase oral activity and stability to peptidases and the potential for toxicity and hepatic clearance.

TABLE 5.13 Comparing the Properties of Peptide Drugs with Small-Molecule and Biomolecular Drugs

Property	Small Molecules	Biomolecules	Peptides
MW	Low molecular weight (<1000 Da)	High molecular weight (>10,000 Da).	Intermediate (300 Da–10,000 Da)
Preparation	Chemical synthesis	Biologically produced-can be engineered	Smaller peptides via chemical synthesis, larger peptides via recombinant engineering
Physicochemical properties	Mostly well defined	Complex	Fairly defined (small cyclic) to complex
		-Tertiary structure -Undergo post transcriptional Modifications, e.g. glycosylation	-Tertiary structure -Can also undergo post transcriptional modifications, e.g. glycosylation
Route of administration	Oral administration usually possible	Usually administered parenterally	Usually administered parenterally but some oral activity can be seen in small highly modified peptides
	Rapidly enter systemic circulation through blood capillaries	Reach circulation primarily via parenteral route: iv, direct; or sc via lymphatic system	Reach circulation primarily via parenteral route: iv, direct; or sc via lymphatic system
Distribution	To any combination of organs/tissues/cells	Usually limited to plasma and/or extracellular fluids	Small peptides can distribute to most combinations of organs/tissues/cells
Metabolism	Metabolized typically by liver and gut CYP450s into non-active and active metabolites	Catabolism by proteolytic degradation to peptides and amino acids	Peptide with native amino acids, via proteolytic degradation. Nonnative amino acids via liver enzymes
Serum half-life	Short serum half-lives	Relatively long serum half-lives	Short serum half-lives
Typical dosing regimen	Suitable for QD or BID dosing	Dosing usually far less frequent	Suitable for QD or BID dosing
Toxicity	Can produce specific toxicity due to parent or metabolites (often "off target")	Mostly receptor mediated toxicity, including both super pharmacodynamic responses and biological toxicity (often "exaggerated pharmacology")	Peptides with native amino acids tend to be nontoxic. Nonnative amino acids can produce specific toxicity

(continued on next page)

TABLE 5.13 Comparing the Properties of Peptide Drugs with Small-Molecule and Biomolecular Drugs *(continued)*

Property	Small Molecules	Biomolecules	Peptides
Species reactivity	Generally active in multiple animal species	Relevant and irrelevant animals models	Generally active in multiple animal species
Antigenicity	Nonantigenic, but can show unpredictable antigenicity	Potential for antigenicity (with MW >10 kDa)	Tend to be nonantigenic
Clearance mechanisms	Small molecules and therapeutic proteins do not share clearance mechanisms – noncompeting, parallel		Small molecules and therapeutic proteins do not share clearance mechanisms – noncompeting, parallel
-Renal excretion	Yes	Yes	Yes
-Hepatic	Biliary excretion	Peptidases	Peptidases. Hepatic action for nonnative amino acids
	Metabolism (P450, UGT, Sulfotransferases, etc)		
-Target mediated	Rare	Yes	Yes
-Intestinal	CYP, UGT, etc.	No, due to peripheral administration	No, due to peripheral administration
	Often involves transporters		
Drug–drug interactions	Pharmacokinetic interactions due to competing clearance mechanisms	Less common, less well-defined	Less common, less well-defined
	-Decreasing clearance if there is enzyme inhibition		
	-Increasing clearance if there is enzyme induction		

Modified from Ref. (11).

The larger peptides, such as insulin and growth hormone, have been on the market for a long time. Bacitracin, a cyclic peptide first isolated from *Bacillus subtilis*, has been available as an antibiotic since the 1940s (*43*), but it is in recent years that peptide drugs have come to be recognized as distinct from their small and large brethren (*44–46*).

Table 5.14 contains a list of 26 recently approved peptide drugs. Twenty one were approved between 2001 and 2010. Humalog, which tops the list, is a blockbuster drug—it achieved $1.1 billion

TABLE 5.14 Twenty-six Recently Approved Peptide Drugs (44–49). Peptide Type Refers to whether the Peptide Drug is Linear (Single Chain With No Cross-Linking Disulfide Bridges), Cyclic (Single Chain With Disulfide Bridges) or a Polypeptide (Composed of Two or More Linear Peptides)

Peptide Type	Generic Name	Brand Name	Company	Entry Year	Naturally Occurring Peptide	Mechanism of Action and Other Notes
Polypeptide	Insulin lispro	Humalog	Lilly	1996	Insulin	Insulin analogue
Polypeptide	Lepirudin	Refludan	Bayer	1998	Hirudin	Recombinant hirudin
Cyclic	Eptifibatide	Integrilin	Millenium	1998	Analogue of barbourin	GPIIb/IIIa antagonist
Linear	Ganirelix	Antagon	Organon	1999	Gonadotropin-releasing hormone	Gonadotropin-releasing hormone antagonist
Polypeptide	Bivalirudin	Angiomax	Medicines company	2000	Hirudin	Reversible direct thrombin inhibitor
Linear	Cetrorelix	Cetrotide	Merck Serono	2000	Gonadotropin-releasing hormone	Gonadotropin-releasing hormone antagonist
Linear	Triptorelin	Trelstar Depot	Ferring	2000	Gonadotropin-releasing hormone	Gonadotropin-releasing hormone antagonist
Polypeptide	Insulin glargine	Lantus	Sanofi-Aventis	2000	Insulin	Duration of action of 18–26 h
Polypeptide	Insulin aspart recom.	NovoLog (NovoRapid)	Novo Nordisk	2000	Insulin	Increased charge repulsion, prevents hexamers, creating a faster acting insulin
Cyclic	Nesiritide	Natrecor	Scios	2001	B-type natriuretic peptide	Natriuretic peptide receptor-B agonist
Polypeptide	Teriparatide	Forteo	Lilly	2002	Parathyroid analogue	Parathyroid analogue
Polypeptide	Pegvisomant	Somavert	Pfizer	2003	Growth hormone	Growth hormone receptor antagonist
Cyclic	Daptomycin	Cubicin	Cubist pharmaceuticals	2003	Naturally occurring compound found in the soil saprotroph *Streptomyces roseosporus*.	Lipopeptide antibiotic which bind to the membrane and cause rapid depolarization of Gram-positive bacteria only

(continued on next page)

TABLE 5.14 Twenty-six Recently Approved Peptide Drugs (44–49). Peptide Type Refers to whether the Peptide Drug is Linear (Single Chain With No Cross-Linking Disulfide Bridges), Cyclic (Single Chain With Disulfide Bridges) or a Polypeptide (Composed of Two or More Linear Peptides) (continued)

Peptide Type	Entry Year	Generic Name	Brand Name	Company	Naturally Occurring Peptide	Mechanism of Action and Other Notes
Linear	2003	Enfuvirtide	Fuzeon	Roche	gp120	Binds to gp41 preventing the creation of an entry pore
Polypeptide	2004	Insulin glulisine	Apidra	Sanofi-Aventis	Insulin	Appears in the blood earlier and at higher concentrations than human insulin
Cyclic	2005	Octreotide	Sandostatin	Novartis	Somatostatin	Potent inhibitor of growth hormone, glucagon, and insulin
Cyclic	2005	Ziconotide	Prialt	Elan	ω-conotoxin	Selective N-type voltage-gated calcium channel blocker
Linear	2005	Exenatide	Byetta	Amylin	Exendin-4	Amylin is deficient in individuals with diabetes
Polypeptide	2005	Pramlintide	Symlin	Amylin	Amylin	
Polypeptide	2005	Insulin detemir	Levemir	Novo Nordisk	Insulin	Quickly resorbed after which it binds to albumin in the blood through the fatty acid at position B29. It then slowly dissociates from this complex
Polypeptide	2005	Mecasermin rinfabate	Iplex	Insmed	IGF-1 and IGFBP-3	amyotrophic lateral sclerosis, more commonly known as Lou Gehrig's disease
Polypeptide	2005	Mecasermin recombinant	Increlex	Insmed	IGF-1	IGF-1 analogue
Cyclic	2007	Lanreotide	Somatuline Depot	Ipsen	Somatostatin	Somatostatin analogue
Linear	2007	Histrelin	Supprelin LA	Indevus	GnRH	Superagonist resulting in receptor downregulation
Polypeptide	2007	Somatropin	Accretropin	Cangene	Growth hormone	Sustained release formulation
Polypeptide	2009	Ecallantide	Kalbitor	Dyax	Kallikrein	Inhibitor of the protein kallikrein

in worldwide sales in 2004 (*47*). When one considers the field of peptide mimetics which are either partially peptidic (such as the gonadotropin-releasing hormones in Table 5.14), or simply a small molecule behaving like a peptide (*48*), estimates of the peptide drug market can be very large, depending on where one draws the line between a peptidic and nonpeptide SMD. Bear in mind that peptide mimetics can include any type of protease inhibitor, including HIV protease inhibitors (Chapter 13) and ACE inhibitors—which have markets in the billions of dollars.

A challenge to further expansion of peptide drugs is the degree of nonnative character that may need to be built into the drug candidate. Chemical synthesis or recombinant technologies can generate peptides with native sequences at a reasonable cost. However, the cost of synthesis can escalate dramatically depending on the degree of nonnative character that has been incorporated into the drug (*44*). Nonnative structural modifications that were built into the peptide drug to replace native amino acid or even whole peptide fragments can increase the cost of synthesis.

The rate of introduction of peptide drugs is likely to continue to be less than either SMDs or biomolecular drugs, but the value of peptide drugs to patients is undeniable.

5.6 THE MANUFACTURE AND SUPPLY OF SMDs VS BIOMOLECULAR DRUGS

The manufacture and supply of SMDs is commonly referred to as Chemistry, Manufacturing and Control (Regulation) or CMC. Due to the ubiquity of this acronym, CMC tends to be applied to the manufacture and supply of biomolecular drugs as well. Of course, nature performs the chemistry for biomolecular drugs via genetically manipulated highly fermentable cells with the encouragement of molecular biologists who manipulate the cells and elicit fermentation in large bioreactors. There is no similarity between the manufacture of small molecules and biomolecular drugs.

Brewer outlines the differences between the clinical supply chain logistics for SMDs and biomolecular drugs (*50*). The former can be packaged in boxes of blister packs and shipped at room temperature, while the latter are most often packaged in glass vials and shipped at controlled refrigerated conditions. Peptide drugs tend to require packaging in glass vials and shipping at controlled refrigerated conditions. The difference in expense of delivering the drug to the customer will tend to be included in the price.

5.7 THE PRICING OF SMDs VS BIOMOLECULAR DRUGS

The ratio of small vs large applies to the pricing of SMDs vs large biomolecular drugs. Despite the sensational costs of certain biomolecular drug therapies, e.g. for Gaucher's Disease (around $200,000 annually; see Chapter 11 and Ref. (*51*)), it is probably fairest to consider the pricing of drugs to be a continuum as follows:

Small – molecule generic drug $<$ small – molecule branded drug $\leq$ biomolecular branded drug

The Agency for Healthcare Research and Quality estimates the annual costs for rheumatoid arthritis (RA) medications from as low as a few hundred dollars for generic oral small-molecule RA drugs to over $12,000 for biomolecular RA drugs (*52*). Of course the pricing of drugs is determined by many factors, including cost of goods, supply chain costs, and size of market (53). These factors alone tend to elevate the price of biomolecular drugs relative to SMDs.

Trusheim et al. determined that, starting from the year of market entry, the average SMD brings in almost the same amount of sales as the average biomolecular drug (20). Both types of drugs reach the $100 million mark at about the same time (Quarter 21 vs 22) which also happens to be the point where sales growth flattens out. Of course, later on sales of small molecules begin to drop due to generic competition. As will be discussed later in this chapter, once biosimilars are allowed on the market, the sales of biomolecular drugs should drop after a period of exclusivity.

5.8 COMPARING SMALL-MOLECULE, PEPTIDE AND BIOMOLECULAR DRUGS IN THE MARKET

While it is interesting to consider the differences and similarities between SMDs, peptide and biomolecular drugs, what really counts is how they compete against each other in the market place. Of course any drug that reaches the market place has reached there on its own merits. How SMDs, peptide and biomolecular drugs in the same class compete against each other may rest on factors that have nothing to do with whether the drug is, indeed, a small molecule, peptide or biomolecule. This may be illustrated in the following examples of the type of antithrombotic agents known as GPIIb/IIIa antagonists and the type of drugs known as Her2 inhibitors.

5.8.1 GPIIb/IIIa antagonists

GPIIb/IIIa is a dimeric complex of two membrane-bound proteins involved in platelet activation and thrombus formation via binding to plasma fibrinogen. Inhibition of fibrinogen binding to activated GPIIa/IIIa prevents thrombus formation that occurs in heart attacks and stroke (54). The first GPIIb/IIIa antagonist approved for use was Abciximab (ReoPro) developed by Centocor. It is a humanized chimeric Fab fragment (54). Soon after its introduction, COR Therapeutics got FDA approval (in 1998) for their peptide Eptifibatide (Integrelin), which contains the essential Arg/hArg–Gly–Asp tripeptide needed for high-affinity binding to GPIIb/IIIa (55). A few years later, Merck introduced Tirofiban (Aggrastat), a peptide mimetic of the RGD sequence (Figure 5.8) (56).

None of these agents are orally active. While oral GPIIb/IIIa antagonists were subsequently developed for chronic therapy in the prevention of a second heart attack or stroke, such long-term treatment amplifies the bothersome side effect of nuisance bleeding and troublesome risk of hemorrhaging, making them less beneficial than other chronic therapies such as aspirin or clopidogrel (57). Thus oral GPIIb/IIIa antagonists have not been marketed.

Brouse and Roberts prepared a comparison of the three agents for the US VA. Table 5.15 summarizes some of the properties and uses of the agents (58).

Abciximab has the benefit of longer duration of action—days to weeks—due to its tight binding to GPIIb/IIIa. The chemically synthesized Eptifibatide and Tirofiban provide less-expensive therapies, with Tirofiban being the least expensive. Even though there are distinct differences between the three agents, product superiority has not been demonstrated for any of these agents. Berger proposes that the conundrum "will never be solved through a properly executed clinical trial because such a trial would be so expensive that it is very unlikely one will ever be performed" (59).

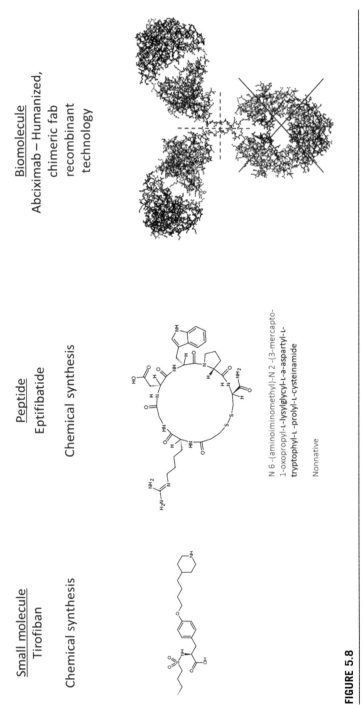

Small molecule
Tirofiban

Chemical synthesis

Peptide
Eptifibatide

Chemical synthesis

N 6 -(aminoiminomethyl)-N 2 -(3-mercapto-
1-oxopropyl-L-lysylglycyl-L-a-aspartyl-L-
tryptophyl-L -prolyl-L-cysteinamide

Nonnative

Biomolecule
Abciximab – Humanized,
chimeric fab
recombinant
technology

FIGURE 5.8

GPIIb/IIIa antagonists for the treatment of thrombosis.

TABLE 5.15 Comparison of Properties of Abciximab, Eptifibatide and Tirofiban (*58*)

Property	Abciximab	Eptifibatide	Tirofiban
Chemical nature	Fab fragment of human/mouse chimeric monoclonal antibody	Cyclic heptapeptide	Nonpeptide
Size	Large (48 kDa)	Small (0.8 kDa)	Small (0.5 kDa)
Antibody formation	Yes (HACA antibodies)	No	No
Binding to platelets	Activated and Nonactivated	Activated only	Activated only
Binding to other platelet receptors	Yes (vitronectin)	No	No
Time to 80% platelet inhibition	<10 min	<15 min	30 min
Half-life (plasma)	30 min	0.83–2.8 h	1.2–2 h
Half-life (platelet-bound)	12–16 h	Seconds	Seconds
Time to normal platelet Function	Within 12 h*	Within 4–6 h	Within 4–8 h
Elimination	Degraded in vascular Space	Renal and nonrenal	Renal
Treatment cost Per day (VA)[†]	$988**	$292[‡]	$249[‡]
Approved indications	Percutaneous coronary intervention	Percutaneous coronary intervention and treatment of acute coronary syndromes	Treatment of acute coronary syndromes

*Abciximab can be found on circulating platelets up to 15 days postinfusion.
[†]Cost estimate based upon a dose for a 70 kg patient.
**Abciximab cost per day calculated for bolus dose and 12-h infusion.
[‡]Eptifibatide & tirofiban cost per day calculated for bolus dose and 24-h infusion; usual treatment duration 72 h for acute coronary syndromes.

5.8.2 Her2 inhibitors

There are several variants of the human epidermal growth factor receptor (EGFR)—EGFR (ErbB-1), HER2/c-neu (ErbB-2), Her 3 (ErbB-3) and Her 4 (ErbB-4). Her2 is a cellular receptor that mediates cell growth, survival, adhesion, migration, and differentiation—functions that are amplified or weakened in cancer cells. Her2 overexpression is observed in 20–30% of breast cancers. Her2 blockade can specifically kill cancer cells that express the Her2 receptor, notably in breast cancer.

The first Her2 inhibitor was Trastuzumab (Herceptin), developed by Genentech and approved for use in 1988. As described in detail in Chapter 9, it is a humanized Mab that binds to the extracellular domain of the Her2 receptor. Being an antibody, Trastuzumab has the added benefit of inducing immune cells to destroy cancer cells (*60*).

Lapatinib (Tykerb and Tyverb) was developed by GSK and was the first small-molecule Her2 inhibitor approved by the FDA (in 2007). It is used in combination with the chemotherapeutic agent, capecitabane.

It targets the ADP binding site of the intracellular kinase domain of both EGFR (ErbB-1) and HER2/c-neu (ErbB-2) (*61*).

Current approved therapy is the use of Trastuzumab as a first treatment for Her2-positive breast cancers. For Trastuzumab patients where remission occurs (usually after 1 year) the use of the Lapatinib/Capecitabane combination is recommended. Recent clinical trials have shown that treatment with both Trastuzumab and Lapatinib may be a better approach than the successive treatment regimen, due to the complementary mechanisms of action of these drugs. It was shown that Lapatinib in combination with Trastuzumab significantly improved progression-free survival and an improved clinical benefit rate vs Lapatinib alone, thus offering a chemotherapy-free option with an acceptable safety profile to patients with ErbB-2 positive metastatic breast cancer (*62*).

At present, there are no peptide-based therapies on the market for Her2-positive breast cancers.

These are but two examples of therapies where small-molecule and biomolecular therapies are effectively competing against each other in the market place. Tirofiban dominates Abciximab in sales due to cost of treatment, while Trastuzumab dominates Lapatinib in sales due being first on the market and its approval as a first-line treatment. But, as featured in the Her2 example, the small-molecule and biomolecular therapies may end up being used together! How wonderful is that!

5.9 BIOSIMILAR BIOMOLECULES VS GENERIC SMALL MOLECULES

The Hatch-Waxman act of 1984 allowed for the sale of generic SMDs, but did not include provisions for biosimilar biologics, mainly because there were so few biologics on the market the demand for a generic type of biologic drug had yet to develop. But we are now at a point where a number of the blockbuster biologics are coming off patent—the demand is there.

So let us consider the differences between generic biomolecular drugs (*aka* biosimilars or follow-on biologics) and generic small molecules. The issues are complex, arising out of the considerable differences between small-molecule and biologic NDAs. Genentech has posted an interesting comparison of SMDs vs biomolecules to explain the challenges to the concept of a generic biomolecular drug (*10*).

As mentioned at the beginning of this Chapter, in 2010 the US introduced health care legislation that included a provision for the approval of biosimilar biologics (*63*). The new legislation allows for approval of biosimilars after makers of the original biologic have had 12 years of patent exclusivity. Although generic versions of biologics have been on sale in Europe for years, it may still be several years before the FDA approves its first biosimilar (*64*). Two commonly held beliefs are (1) that a biosimilar biologic will be more challenging to develop than a generic small molecule and (2) it will be harder to market the biosimilar as equally effective to the original drug due to the fact that (as will be discussed later) a biosimilar is not chemically the same substance as the original drug. Given the huge markets for some of the older biologics, however, there will be many attempts at developing and marketing effective biosimilars.

The generics/biosimilars debate has delineated a regulatory dividing line between small molecules and biomolecules for drugs the size of insulin and growth hormone (*65*). Conveniently, generic versions of these drugs have been around for years (Table 5.2) and have not shown the reproducibility and immunogenicity issues that are of a general concern with respect to biosimilars. Thus, it is likely that drugs of this size will continue to be regarded as small-"ish" drugs which will not need the regulatory management that will be applied to larger biomolecules.

This dividing line provided a convenient resolution to the question of "grandfathering" biomolecules that have been on the market for years, but may not resolve future issues when similar-size biomolecules eventually reach patent expiration.

The difference between generic small molecules and biosimilar biomolecular drugs comes down to the demonstration of equivalency. Most small molecules are sufficiently simple in chemical structure that routine scientific analyses readily establish structural identity between branded pharmaceutical and generic copies. The chemical manufacturing technologies are sufficiently robust that batch-to-batch variations are typically minimal. Biomolecular drugs on the other hand are extremely complex structures that can never be purified to a single chemical structure. The "drug" in the case of a biologic is a collection of similar structures with (hopefully) minor variations in amino acid content and sequence, glycosylation pattern and other structural features. However the purity of a SMD (often >99.5%) can never be achieved in a very large biomolecular drug.

Rader describes the issue as follows. "A product from one manufacturer, made using consistent biological sources (e.g. genes and cell lines), a consistent set of processes, under a consistent set of conditions, using consistent in-process and other controls and assays, and with a consistent set of final specifications constitutes a unique biopharmaceutical product. In this context, regulatory approvals are secret pacts between a manufacturer and the regulatory agency concerning the associated ranges of allowable variations in each of these aspects. Following that paradigm, because manufacturing processes are complex (and never fully publicly disclosed), biopharmaceuticals are considered impossible to exactly replicate by all but their licensed manufacturers (usually their innovators)" (*66*). The BLA can never be exactly replicated. Thus biotech companies argue that for biologics "the process is the product".

To extend this paradigm to manufacturers of biosimilars, each would need to establish its own process and specifications with the FDA without ever knowing what the processes and specifications were employed by the original manufacturer or other manufacturers of biosimilars.

Would-be manufacturers of generic versions of an SMD need only to establish structural identity and equivalency in formulation (achieving equivalent bioavailability) with the FDA. At least some in-human testing will be required with biosimilars to ensure bioequivalency and safety (*67*).

Rader draws the analogy of wine or cheese making, where two manufacturers can follow the same process and create products that the customer can readily distinguish as different. In these markets such diversity is encouraged. Because of the likelihood that any two biosimilars will be different, many prefer the term "follow-on biologic" as a more accurate description of a generic biomolecular drug. The FDA applied the term "follow-on" to the growth hormone products under the previously discussed Hatch-Waxman provisions.

An FDA-approved generic SMD is sold by the generic name. Under the Hatch-Waxman provisions each biomolecular product bears its own brand name and cannot be sold by the generic name. Whether an FDA-approved generic biomolecular drug will ever be sold by the generic name (as provided for by BPCIA (*63*)) will depend on how the debate gets resolved.

5.10 DISCOVERY AND PRECLINICAL STAGES FOR SMDs VS BIOMOLECULAR DRUGS—WHERE THE TECHNOLOGIES DIFFER THE MOST

To summarize and complement the data in Chapter 3 we can outline the differences in each stage of the discovery process for SMDs and biomolecular drugs, following the sequence in Figure 5.9.

FIGURE 5.9

The stages of discovery and development.

Again, one only has to consider the size of biomolecular drugs to recognize that the technologies that give rise to biomolecular drugs must be considerably different from the classical SMDs. The difference in weight between aspirin and an antibody (Figure 5.1) has been equated to the difference between a bicycle ($\sim$20 lbs) and a business jet ($\sim$30,000 lbs) (*10*). It is remarkable that the engine of biopharmaceutical discovery and development can accommodate such differences.

5.10.1 **Target discovery**

Since the focus in this stage is the biological target, not the agent that will provide the therapy, the processes of target discovery are the same for small molecules and biologics. Antibodies are commonly employed tools for target validation. If the project team already intends to develop an Mab agent against the biological target, they might begin the process of Mab generation in this stage.

5.10.2 **Lead discovery**

The focus in this stage is to find agents, referred to as leads or lead series, that interact with the target in the desired manner (agonist, antagonist, etc.). The processes that lead to the discovery of small-molecule or biologic lead series in this stage are starkly different. The discovery of Mab lead series tends to run more quickly than the search for small-molecule lead series with a considerably higher POS. If the team intends to develop a protein agent that is a recombinant variant of a protein that is found in the body, this phase is skipped all together, since the native protein is the lead.

It is in this stage where the team decides to pursue a biologic or small-molecule therapy. It is used to be believed that an oral formulation was more readily marketable than a peripherally administered formulation. The biweekly to monthly dosing schedule of a biologic (*11*) can effectively compete with a more frequently (inconveniently) dosed small molecule. Thus it is no longer a foregone conclusion that chronic therapy necessitates an SMD.

5.10.3 **Lead optimization**

The processes for determining which molecules in the lead series will be the clinical candidate are quite different for small molecules or biologics, yet the goals are similar. The clinical candidate must have sufficient potency to elicit a desired response when administered *in vivo* in an animal model of the disease. The duration of action must be adequate to realize the goals of the therapy. For sufficient duration of action to be achieved the candidate must display a suitable pharmacokinetic profile. The metabolic fate of the candidate must be known. The physical form of the candidate molecule must be amenable to dosage formulation desired in clinical trials and there must not be any perceived impediments towards developing a suitable formulation as a drug.

Despite these similarities, it is fair to assume that it will take longer to discover a small-molecule clinical candidate with a lower POS than to develop a biologic clinical candidate. The comparative data to support these comments are known but held in private consortia (e.g. the KMR Group (http://www.kmrgroup.com/)) and rarely published.

5.10.4 Preclinical evaluation

The advantages gained by biologics over small molecules in the discovery stages may be lost in preclinical evaluation. This is because of the concern for safety, in particular a concern for eliciting an autoimmune response with a biologic, which tends to necessitate lengthy toxicity studies. SMDs are not prone to this issue. With the advent of fully human Mabs these risks are likely to decrease (*68*) and their associated impact on POS and cycle time are likely to diminish between 2011 and 2020.

As noted earlier (Table 5.5) Dimasi (*35*) had found that biologics tend to have larger capitalization costs in the preclinical stage than small molecules.

5.10.5 Clinical phases

As noted earlier, in Phase 1 studies one needs to be cautious with biomolecular drugs in dose escalation studies in the event that unanticipated antigenicity may be revealed for the first time in this stage. Thus, historically the cycle times for biomolecules in Phase 1 have been longer. And also noted earlier, Dimasi found that biomolecules tend to have higher costs in late-stage clinical trials.

5.11 SMALL-MOLECULE AND BIOLOGICS APPROVALS BY THERAPY AREAS

Of the therapy areas that historically have been the domain of SMDs, one may ask in how many biologics are becoming competitive? Biologics have long been associated with the treatment of cancer and immunological disorders. In a recent study conducted by Trusheim et al. 43% of the drugs approved in these therapy areas from 1998 to 2008 were biologics (*20*). With technological advances in the discovery and development of biologics most therapy areas are now amenable to either strategy (*20,68*). Trusheim et al. catalogued biologics approved between 1998 and 2008 into 80% of the World Health Organization therapy area classes (Table 5.16), many more than just cancer and immunological disorders. What may come as a surprise is the number of anti-infectives that were biologics—38% of the drugs approved in these therapy areas between 1998 and 2008 were biologics. Most of the drugs that act on blood and blood-forming organs were biologics (73%).

5.12 MANAGING SMALL-MOLECULE & BIOMOLECULAR DRUG R&D IN THE SAME COMPANY

A company that develops both small molecules and biologics will need to deal with the similarities and differences in the two types of portfolios. The differences have been mentioned in earlier sections and are summarized below (Table 5.17).

TABLE 5.16 Therapeutic Class Composition of Biologics and Small Molecules. New Small-Molecule Drug Approvals (NMEs) and New Biologics Approvals (New BLAs) by the FDA From 1998 to 2008 (*20*). The ATC System Allocates Drugs Into Different Groups According to the Organ or System on Which They Act

ATC Classification*	Biologic	Small Molecule	Total	Percent Biologics
A: Alimentary tract and metabolism	16	26	42	38.1%
B: Blood and blood forming organs	16	6	22	72.7%
C: Cardiovascular system	1	20	21	4.8%
D: Dermatologicals	3	5	8	37.5%
G: Genito-Urinary systems and Sex hormones		13	13	0.0%
H: Systemic hormonal preparations, Excl. Sex hormones and insulins	7		7	100.0%
J: Anti-infectives for Systemic use	23	38	61	37.7%
K: Intravenous Solutions		1	1	0.0%
L: Antineoplastic and Immunomodulating agents	21	28	49	42.9%
M: Musculoskeletal system	4	6	10	40.0%
N: Nervous system	1	36	37	2.7%
P: Antiparasitic products		1	1	0.0%
R: Respiratory system	1	11	12	8.3%
S: Sensory organs	2	10	12	16.7%
V: Various	1	11	12	8.3%
Total	96	212	308	31.2%

*ATC—WHO Anatomical Therapeutic (Chemical) class.

TABLE 5.17 Stage-Related Differences between Small Molecules and Biologics

	Target Discovery	Lead Discovery	Lead Optimization	Preclinical	Phase 1	Phase 2-Launch
Small molecules	Similar	Lower POS	Lower POS	Shorter CT		
Biologics		Shorter CT	Shorter CT	Capitalization costs	Longer CT	Higher costs

As was noted above, with the technological advances in the discovery and development of bio-logics, most therapy areas are now amenable to either strategy (*20,68*). If the company finds that most of its therapy areas are investigating both strategies, the overhead of managing separate small-molecule and biologics portfolios may be unjustifiable. This is particularly important in this era of cost cutting. The fact that both small molecules and biologics can be managed with the same mile-stones and stages argues for treating both strategies in the same portfolio. The biopharmaceutical company that has the resources to discover and develop both types of drugs will inevitably face the challenge of organizing these activities (Chapter 7). Both pursuits engage the same stages of discovery and development and face the same generic issues at the project level. Within the stages, some of the issues will be different (e.g. immunogenicity and oral bioavailability) and so strategies and processes will vary. At the corporate level the similarities argue for a blended organization where the differences are managed in line departments and played out by the specific project teams (Chapter 7). All things being equal, the blended organization can take advantage of similarities between projects and targets.

In a resource-constrained environment, the blended organization may develop biases about adequately resourcing one or the other approach. Within a biopharmaceutical company in earlier eras when the number of biologic approvals was few, the champions of SMDs would challenge the value of allocating resources to biomolecular drug discovery and development. It has only been between 2001 and 2010 that the number of biologics approvals has begun to rival SMD approvals. Senior managers may need to protect resources for the less-favoured approach by forming separate divisions each devoted to the exclusive pursuit of SMDs or biomolecular drugs. Managers may want to avoid splitting platform technologies that serve both divisions, to avoid wasteful resource duplication. The judicious manager that oversees a portfolio with both small molecules and biologics will ensure that the stage-related differences are understood and transparent.

5.13 CONCLUSION

Biologics have levelled the playing field. About as many new biologics (those reviewed at CDER and CBER) are being approved each year as new SMDs (NMEs). All the big pharmaceutical companies now have the capacity to develop both types of drugs. Those that had biologics capacity are increasing it; contrary-wise the larger biotech companies are developing the capacity to develop small molecules. Now, with many blockbuster biologics nearing patent expiration, biosimilars (follow-on biologics) are on the way. For project leaders and portfolio managers in the bio-pharmaceutical industry, it's no longer "choose one or other type"—one will need to know how to discover and develop either type of drug.

References

1. Reichert, J. Biopharmaceutical Approvals in the U.S. Increase. *Reg. Affairs J. Pharma.* **July 2004,** 1–7.
2. Maggon, K. Monoclonal Antibody Gold Rush. *Curr. Med. Chem.* **2007,** *14,* 1978–1987.
3. Thayer, A. M. Great Expectations, Biopharmaceutical Industry Expects Success from New Drug Therapies and a Full Product Development Pipeline. *Chem. Eng. News* **1998,** *76,* 19–31.
4. Reichert, J. M.; Paquette, C. Therapeutic recombinant proteins: Trends in US approvals 1982 to 2002. *Curr. Op. Mol. Ther.* **2003,** *5,* 139–147.
5. PriceWaterhouseCoopers and the National Venture Capital Association. *Biotech – Lifting Big Pharma's Prospects with Biologics, The Money Tree Report,* 2009.
6. Jack, A.; Lemer, J. *Conditional Payment Cure for Pharma Deals, Financial Times.* http://www.ft.com/cms/s/0/a378251c-39af-11e0-8dba-00144feabdc0.html#axzz1vbg84F5a.
7. Deloitte Thompson Reuters. *Is R&D Earning its Investment?.* http://www.deloitte.com/assets/Dcom-UnitedKingdom/.../UK_LS_RD_ROI.pdf.

8. Spilker, B. *Biotechnology, Chapter 12 in Guide to Drug Development, A Comprehensive Review and Assessment;* Wolters Kluwer, Lippincott Williams & Wilkins: New York, 2009.

9. Ng R. Drug Discovery: Large Molecule Drugs. In *Drugs, from Discovery to Approval*, 2nd ed., 2009 (chapter 4).

10. Genentech. Biosimilars. http://www.gene.com/gene/about/views/followon-biologics.html.

11. Klunk, L. J. *Considerations for a Scientifically Rational Approach to Biologics-Small Molecule DDI Studies, New England Drug Metabolism Discussion Group Summer Symposium*, June 9, 2010.

12. Wikipedia. Antibody Structure. http://en.wikipedia.org/wiki/Antibody#Structure.

13. Savi, P.; Pereillo, J. M.; Uzabiaga, M. F.; Combalbert, J.; Picard, C.; Maffrand, J. P.; Pascal, M.; Herbert, J. M. Identification and Biological Activity of the Active Metabolite of Clopidogrel. *Thromb. Haemost.* **2000**, *84*, 891–896.

14. Eisen, S. A.; Miller, D. K.; Woodward, R. S.; Spitznagel, E.; Przybeck, T. R. The Effect of Prescribed Daily Dose Frequency on Patient Medication Compliance. *Arch. Intern. Med.* **1990**, *150*, 1881–1884.

15. Stasi, R. Eltrombopag, The Discovery of a Second Generation Thrombopoietin-Receptor Agonist. *Exp. Opinion Drug Discov.* **2009**, *4*, 85–93.

16. Roujeau, J.-C. Immune Mechanisms in Drug Allergy. *Allergol. Int.* **2006**, *55*, 27–33.

17. Seitz, K.; Zhou, H. Pharmacokinetic Drug–Drug Interaction Potentials for Therapeutic Monoclonal Antibodies: Reality Check. *J. Clin. Pharmacol.* **2007**, *47*, 1104–1118.

18. DiMasi, J. A. New Drug Development in the United States from 1963 to 1999. *Clin. Pharmacol. Ther.* **2001**, *69*, 286–296.

19. Harrison, R. K. *Biopharmaceutical R&D – Where Are We Today?* Thomson Reuters, 2011. PDF March 23, p. 4.

20. Trusheim, M. R.; Aitken, M. L.; Berndt, E. R. Characterizing Markets for Biopharmaceutical Innovations: Do Biologics Differ from Small Molecules? *Forum Health Econ. Policy* **2010**, *13* (1). article 4.

21. Mullard, A. FDA Approvals. *Nat. Rev. Drug Discov.* **2011**, *10*, 82–85.

22. Kneller, R. The importance of New Companies for Drug Discovery: Origins of a Decade of New Drugs. *Nat. Rev. Drug Disc* **2010**, *9*, 867–882.

23. Samanen J. Benchmarking through a Consortium. www.portfoliomanagementsolutions.com.

24. DiMasi at Tufts. http://csdd.tufts.edu/.

25. Grabowski at Duke. http://www.fuqua.duke.edu/faculty_research/faculty_directory/grabowski/.

26. DiMasi, J. A. *The Economics of Pharmaceutical Innovation: Costs, Risks, and Returns;* MIT Sloan School of Management, 2003. Dec. 4.

27. DiMasi, J. A.; Grabowski, H. G. The Cost of Biopharmaceutical R&D: Is Biotech Different? *Manage. Decis. Econ.* **2007**, *28*, 469–479.

28. DiMasi, J. A.; Hansen, R. W.; Grabowski, H. G. The Price of Innovation: New Estimates of Drug Development Costs. *J. Health Econ.* **2003**, *22*, 151–185.

29. DiMasi, J. A.; Feldman, L.; Seckler, A.; Wilson, A. Trends in Risks Associated With New Drug Development: Success Rates for Investigational Drugs. *Clin. Pharmacol. Ther.* **2010**, *87*, 272–277.

30. Reichert, J. Trends in Development and Approval Times for New Therapeutics in the United States. *Nat. Rev. Drug Discov.* **2003**, *2*, 695–702.

31. Grabowski, H. Follow-On Biologics: Data Exclusivity and the Balance between Innovation and Competition. *Nat. Rev. Drug Discov.* **2008**, *7*, 479–488.

32. BIO CEO & Investor Conference, February 15th, 2011BIO.

33. Samanen, J. What Do We Know about Cycle Times in the Industry, http://www.portfoliomanagementsolutions.com

34. Pharmaceutical Research and Manufacturers of America. *Pharmaceutical Industry Profile 2010;* PhRMA: Washington, DC, 2010.

35. DiMasi J. A. Cost & Returns for New Drug Dev. 2006 pdf in www.ftc.gov/be/workshops/pharmaceutical/DiMasi.pdf.

36. Cohen, F. J. Macrotrends in Pharmaceutical Innovation. *Nature Rev. Drug Discov.* **2005,** *4,* 78–84.

37. Moses, H.; Dorsey, E. R.; Matheson, D. H. M.; Their, S. O. Financial Anatomy of Biomedical Research. *JAMA* **2005,** *294,* 1333–1342.

38. Samanen, J. "Why Innovation Has Dried Up, At Least in the U.S." www.portfoliomanagementsolutions.com

39. Pharmaceutical Research and Manufacturers of America. *Pharmaceutical Industry Profile 2008;* PhRMA: Washington, DC, 2008.

40. Pharmaceutical Research and Manufacturers of America. *Pharmaceutical Industry Profile 2009;* PhRMA: Washington, DC, 2009.

41. Hudson, P. J.; Souriau, C. Engineered Antibodies. *Nature Med.* **2003,** *9,* 129–134.

42. Reichert, J. Therapeutic Monoclonal Antibodies: Trends in Development and Approval in the US. *Curr. Opin. Mol. Ther.* **2002, 4,** 110–118.

43. Johnson, B. A.; Anker, H.; Meleney, F. Bacitracin: A New Antibiotic Produced by a Member of the *B. subtilis* Group. *Science* **1945,** *102,* 376–377.

44. Glaser, V. Peptide Drugs Becoming More Prevalent. *GEN Genet. Eng. Biotechnol. News* **2008,** *28* (15). http://www.genengnews.com/gen-articles/peptide-drugs-becoming-more-prevalent/2580/.

45. Peptide Drug Discovery Research Reenergized. *GEN Genet. Eng. Biotechnol. News* **2006, 26** (8). http://www.genengnews.com/gen-articles/peptide-drug-discovery-research-reenergized/1564/.

46. Danho, W.; Swistok, J.; Khan, W.; Chu, X. J.; Cheung, A.; Fry, D.; Sun, H.; Kurylko, G.; Rumennik, L.; Cefalu, J.; Cefalu, G.; Nunn, P. Opportunities and Challenges of Developing Peptide Drugs in the Pharmaceutical Industry. *Adv. Exp. Med. Biol.* **2009,** *611,* 467–469.

47. Christian, N. Watching Peptide Drugs Grow Up. *Chem. Eng. News* **2005,** *83,* 17–24.

48. Mason, J. M. Design and Development of Peptides and Peptide Mimetics as Antagonists for Therapeutic Intervention. *Future Med. Chem* **2010,** *2,* 1813–1822.

49. Frost & Sullivan. *Strategic Analysis of the Therapeutic Peptides Market in Europe, October 2004, # B423–52.*

50. Brewer, T. S. *Clinical Supply Chain Logistics of Small Molecule vs. Biologics – a Provider's Perspective;* ISPE Knowledge Brief. www.ISPE.org.

51. Engelberg, A. B.; Kesselheim, A. S.; Avorn, J. Balancing Innovation, Access, and Profits—Market Exclusivity for Biologics. *N. Engl. J. Med.* **2009,** *361,* 1917–1919.

52. Effective Health Care. *Rheumatoid Arthritis Medicines, A Guide for Adults, Agency for Healthcare Research and Quality.* 08-EHC004-2A, April 2008; AHRQ Publication, www.rheumatology.org/practice/clinical/patients/medications/biologics.pdf.

53. Spilker, B. Biotechnology. In *Guide to Drug Development, A Comprehensive Review and Assessment;* Guide to Drug Development, A Comprehensive Review and Assessment; Wolters Kluwer, Lippincott Williams & Wilkins: New York, 2009; pp 993–1006 (chapter 98).

54. Coller, B. S.; Shattil, S. J. The GPIIb/IIIa (integrin {alpha}IIb{beta}3) Odyssey: A Technology-Driven Saga of a Receptor With Twists, Turns, and Even a Bend. *Blood* **2008,** *112,* 3011–3025.

55. Scarborough, R. Development of Eptifibatide. *Am. Heart J.* **1996,** 1093–1104.

56. Hartman, G. D.; Egbertson, M. S.; Halczenko, W.; Laswell, W. L.; Duggan, M. E.; Smith, R. L.; Naylor, A. M.; Manno, P. D.; Lynch, R. J. Non-peptide Fibrinogen Receptor Antagonists. 1. Discovery and Design of Exosite Inhibitors. *J. Med. Chem.* **1992,** *35,* 4640–4642.

57. Quinn, M. J.; Plow, E. F.; Topol, E. J. Platelet Glycoprotein IIb/IIIa Inhibitors: Recognition of a Two-Edged Sword? *Circulation* **2002,** *106,* 379–385.

58. Brouse, S.; Roberts, K. Drug Class Review: Glycoprotein (GP) IIb/IIIa Receptor Inhibitors For Use in Acute Coronary Syndromes (ACS) & Percutaneous Coronary Intervention (PCI). http://www.pbm.va.gov/Clinical%20Guidance/Drug%20Class%20Reviews/Glycoprotein%20GP%20IIb%20-%20IIIa%20Receptor%20Inhibitors,%20Drug%20Class%20Review.pdf.

59. Berger, P. The Glycoprotein IIb/IIIa Inhibitor Wars, An Update. *J. Am. Coll. Card.* **2010,** *56,* 476–478.

60. Chang, J. C. HER2 Inhibition: From Discovery to Clinical Practice. *Clin. Cancer Res.* **2007,** *13,* 1–3.

61. Moy, B., Kirkpatrick, P., Kar, S., and Goss, P. Lapatinib, Nat. Rev. Drug Discov.., 6, 431-432.

62. Blackwell, K. L.; Burstein, H. J.; Storniolo, A. M.; Rugo, H.; Sledge, G.; Koehler, M.; Ellis, C.; Casey, M.; Vukelja, S.; Bischoff, J.; Baselga, J.; O'Shaughnessy, J. Randomized Study of Lapatinib Alone or in Combination With Trastuzumab in Women With ErbB2-Positive, Trastuzumab-Refractory Metastatic Breast Cancer. *J. Clin. Oncol.* **2010,** *28,* 1124–1130.

63. FDA Lawyers Blog. *The Patient Protection and Affordable Care Act included the Biologics Price Competition and Innovation Act (BPCIA) Which Provides for Biosimilar Biologic Drug Approvals.* http://www.fdalawyersblog.com/2010/03/biosimilars-act-becomes-law-co.html.

64. Medpage Today, 2010. FDA Hears Comments on Biosimilar Approval Pathway, November 2, in http://www.medpagetoday.com/PublicHealthPolicy/FDAGeneral/23122.

65. Torti, F. M. *FDA, 18-September-2008, FDA.* http://www.gene.com/gene/about/views/pdf/FDA_Pallone%20Resp%20dtd%209-18-08.pdf.

66. Rader, R. A. *What Is a Generic Biopharmaceutical? Biogeneric? Follow-On Protein? Biosimilar? Follow-On Biologic?* Bioprocess International, 2007. pp.28–38.

67. Mullard, A. Hearing Shines Spotlight on Biosimilar Controversies. *Nat. Rev. Drug Discov.* **2010,** *9,* 905–906.

68. Nelson, Y. A. L.; Dhimolea, E.; Reichert, J. M. Development Trends for Human Monoclonal Antibody Therapeutics. *Nat. Rev. Drug Discov.* **2010,** *9,* 767–774.

Therapies for type 2 diabetes: modulating the incretin pathway using small molecule peptidase inhibitors or peptide mimetics

Matthew P. Coghlan*, David Fairman[†]

*Cardiovascular & Metabolic Disease Innovative Medicines, MedImmune, Cambridge, UK,
[†]Clinical Pharmacology & DMPK, MedImmune, Cambridge, UK

CHAPTER OUTLINE

6.1 Introduction 206
6.2 Pharmacotherapy of Type 2 Diabetes 210
6.3 The Rationale for Incretin-Based Therapies for Type 2 Diabetes 211
6.4 Discovery and Pharmacokinetics of the Incretin-Based Therapies 214
6.5 Clinical Efficacy of the Incretin-Based Therapies... 216
    6.5.1 Glucose control 216
    6.5.2 Weight loss 218
6.6 Evidence for Disease Modification... 218
6.7 Impact on Cardiovascular Risk 219
6.8 Clinical Safety and Tolerability of Incretin-Based Therapies.................................... 220
    6.8.1 Nausea...220
    6.8.2 Anti-drug antibodies (see Chapter 3) 220
    6.8.3 Injection site reactions..220
    6.8.4 Miscellaneous risks 220
6.9 Conclusions 221
References 222

ABSTRACT

The increasing global prevalence of type 2 diabetes represents a significant burden of disease for afflicted patients and for health care systems. In the developed world poorly controlled diabetes is the leading cause of non-traumatic amputation, blindness and end-stage renal disease requiring dialysis and kidney transplant. Additionally, diabetes represents a significant risk factor for the development of cardiovascular disease with its associated morbidity and premature death. Currently available glucose lowering drugs used to treat type 2 diabetes do not impede progression of the disease. Therefore, as the disease progresses these agents rapidly lose efficacy, first as monotherapy and then in combination, resulting in poorly controlled disease. Clearly, there is a significant need for novel glucose lowering drugs for type 2 diabetes that will deliver sustained efficacy over several years by impeding disease progression. Such agents would reduce the risk of developing the microvascular complications of diabetes that ultimately result in

Introduction to Drug Research and Development. http://dx.doi.org/10.1016/B978-0-12-397176-0.00006-6

amputation, blindness and kidney transplant. Novel glucose lowering drugs should ideally also exhibit a positive impact on the increased cardiovascular risk associated with diabetes. The incretin-based therapies first entered the market in the mid 2000's and were heralded for their potential to impede progression of type 2 diabetes and to reduce cardiovascular risk. Through mimicking the actions of the gut incretin hormone GLP-1, these drugs had been shown to lower blood glucose in clinical trials by potentiating glucose stimulated insulin secretion from pancreatic β-cells. Moreover, data from preclinical rodent disease models and isolated human pancreatic islets suggested that these novel agents could preserve pancreatic β-cell function and thus impede disease progression. Further preclinical and clinical data supported the notion that these drugs could also aid blood glucose control by suppressing glucagon secretion, slowing gastric emptying and by suppressing appetite. The incretin-based drugs have potential to reduce cardiovascular risk through their ability to reduce body weight, blood pressure and atherogenic blood lipids. This chapter will review the incretin-based therapies and consider what impact these new drugs have made to date in the pharmacotherapy of type 2 diabetes. The incretin-based therapies are of particular relevance to this book as this class of drugs is composed of two sub-classes, injectable peptide drugs and oral small molecule drugs. The similarities and differences between these small molecule and peptide drugs are described.

Keywords/Abbreviations: Type 2 diabetes (T2D); Type 1 diabetes (T1D); Incretin; Glucagon-like peptide-1 (GLP-1); GLP-1 receptor agonist (GLP-1RA); Dipeptidyl peptidase-4 inhibitor (DPP4I); Fasting plasma glucose (FPG); G-protein-coupled receptor (GPCR); Hyperglycaemia; Insulin; Insulinotropic; Insulin secretagogue; Insulin sensitizer; Functional β-cell mass; Impaired glucose tolerance (IGT); Oral glucose tolerance test (OGTT); Sulphonylurea drugs (SUs); Homoeostatic model assessment β-cell function index (HOMA-B); Major adverse cardiovascular events (MACE); Thiazolidinedione (TZD).

6.1 INTRODUCTION

Diabetes mellitus is suspected when patients visit their primary care physician with symptoms of constant tiredness, thirst (polydipsia), and passing copious water (polyuria). A diagnosis of diabetes will then be made if the physician confirms that the patient has significantly elevated blood glucose, or hyperglycaemia (Box 6.1) (*1,2*). In the diabetic, hyperglycaemia results as a consequence of the body's inability to absorb dietary sugar (glucose) from the blood into tissues for use as an energy source. It is this fundamental metabolic disturbance of hyperglycaemia that leads to the classical symptoms of tiredness, polydipsia and polyuria. Polyuria results from the inability of the kidneys to fully reabsorb glucose from the blood leading to the production of large volumes of glucose-rich urine.

Diabetes presents a grave burden, both for those individuals afflicted with the disease and for the health care systems required to fund the management of the disease and its related comorbidities and complications. The diagnostic hallmark of diabetes is elevated blood glucose or hyperglycaemia. The primary therapeutic goal in the treatment of diabetes is to maintain blood glucose (glycaemia) as close to the non-diabetic range as possible, not only to provide symptomatic relief but also to stave off the long-term consequences of the disease. Clinical trials have demonstrated that intensive glycaemic control, with anti-hyperglycaemic drugs, will reduce the microvascular complications of the disease (*3*). These complications, which are retinopathy, nephropathy and neuropathy, respectively, represent the major causes of blindness, renal disease requiring kidney transplant and non-traumatic amputations in the developed world. Intensive glycaemic control has also been demonstrated to reduce the

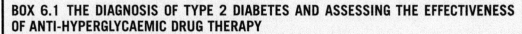

BOX 6.1 THE DIAGNOSIS OF TYPE 2 DIABETES AND ASSESSING THE EFFECTIVENESS OF ANTI-HYPERGLYCAEMIC DRUG THERAPY

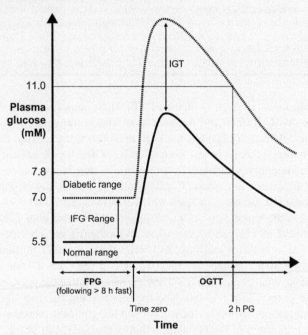

Blood glucose measure	Normal	Prediabetes	Diabetes
FPG - Fasting Plasma Glucose	<5.5 mM	5.6 mM–6.9 mM = Impaired Fasting Glucose (IFG)	>7.0 mM
2 h PG - 2 h plasma glucose in the 75 g glucose oral glucose tolerance test	<5.5 mM	7.8 mM–11.0 mM = Impaired Glucose Tolerance (IGT)	>11.0 mM
HbA1c	<6.1%	5.7–6.4%	>6.5%

Type 2 diabetes is diagnosed by primary care physicians if fasting plasma glucose (FPG), measured after an overnight fast of at least 8 h, exceeds 7 mM (see Table above) (*1*). The prediabetic state of impaired glucose tolerance (IGT) cannot be detected by measuring FPG as this is not elevated above that observed in subjects with normal glucose control. Experimentally, IGT can readily be identified by an oral glucose tolerance test (OGTT) in which the subject is given a 75-g oral solution of glucose to drink (Time zero arrow in Figure) and then the plasma glucose excursion is measured over time. IGT is confirmed if the glucose concentration at 2 h post-glucose challenge exceeds 7.8 mM (2 h plasma glucose arrow in Figure). Note that a 2 h plasma glucose concentration in an OGTT exceeding 11 mM is diagnostic of diabetes. The effectiveness of anti-hyperglycaemic drug therapy to control blood glucose in diabetics is monitored in two ways. Firstly, diabetics are advised to conduct finger prick tests and measure their blood glucose using a glucose meter, on at least a daily basis. Secondly, primary care physicians or diabetologists/endocrinologists, on a half-yearly basis, will sample a patient's blood for a laboratory measurement of glycated haemoglobin A1c (HbA1c). HbA1c is the product of nonenzymatic glycosylation of the erythrocyte protein haemoglobin. HbA1c is considered an excellent surrogate for direct measurements of blood glucose as it has a long half-life and thus presents a composite measure of ambient blood glucose levels over a period of 2 to 3 months. The objective of anti-hyperglycaemic drug therapy in T2D is to achieve an HbA1c level of <7%, where the upper limit of the non-diabetic range is 6.1% (2). Periodic measurement of the patient's HbA1c level allows the physician to adjust the drug therapy in an attempt to maintain glucose control as close to this target as possible.

macrovascular complications of type 1 diabetes (T1D) as manifest by reduced major adverse cardiovascular events (MACE) (fatal and nonfatal heart attack and stroke). However, intensive glycaemic control in type 2 diabetes (T2D) has not been demonstrated to reduce cardiovascular disease. This may reflect that hyperglycaemia in this form of diabetes develops over many years before T2D is diagnosed and pharmacotherapy is initiated, by which time it may be too late for glycaemic control to affect the course of cardiovascular disease. Consequently, it is critically important to treat the non-glycaemic cardiovascular risk factors associated with T2D as this represents 95% of diabetes cases and cardiovascular disease is the major cause of premature death of those with T2D. Cardiovascular risk factors that require control in T2D are hypertension, dyslipidemia and obesity.

The global incidence of diabetes mellitus has been estimated at 371 million people with an estimated 4.8 million dying as a consequence in 2012 (4). In the United States, it is estimated that approximately 10% of the adult population has diabetes, meaning that the prevalence of the disease has doubled in the past 30 years. Clearly, in this time frame, there has been no significant change in the genetics of the US population. However, there has been a significant increase in the US and global population and in the proportion of the population that is overweight or obese. The linkage between T2D and being overweight, or obese, is well established. In the United States, 85% of diabetics are overweight or obese. Moreover, modest weight loss can prevent the onset of T2D in the overweight or obese and lead to a reduced requirement for anti-hyperglycaemic drug therapy in those with established T2D. However, it is important to note that whilst increased body weight is a risk factor for developing T2D, it is not inevitable that everyone who is overweight or obese will develop the disease. Rather, only those overweight and obese individuals with a genetic predisposition will go on to develop T2D.

Two distinct forms of diabetes exist: T1D and T2D. These diseases are characterized by inappropriately elevated blood glucose, or hyperglycaemia, resulting from the absolute (T1D), or relative deficiency (T2D), in endogenous insulin secretion. In healthy subjects, insulin is secreted from the β-cells of the endocrine pancreas in response to rising blood glucose on eating, i.e. post-prandial plasma glucose excursion. Insulin promotes the uptake of glucose into the peripheral tissues, primarily the skeletal muscle, thus returning blood glucose to its usual fasted (post-absorptive) level of approximately 5 mM. Insulin also suppresses hepatic glucose output such that liver-derived glucose does not enter the blood needlessly when dietary glucose is being absorbed by the tissues in an insulin-dependent manner during the post-prandial period. T1D develops when autoimmune destruction of the pancreatic β-cells leads to a catastrophic loss of endogenous insulin secretion. This form of diabetes presents without forewarning, and usually in children, who then require lifetime replacement therapy with exogenous insulin injections to control their blood glucose levels. Hence, T1D has previously been termed juvenile-onset diabetes and also insulin-dependent diabetes. T1D will not be further considered in this chapter as this form of diabetes is not treated with the incretin-based therapies which are the focus of this review.

T2D differs from T1D in that it is a more slowly progressing disease that takes years to develop to the point of diagnosis (Figure 6.1). Consequently, T2D had previously been termed adult-onset diabetes, and, until recently, it had been a disease primarily of the middle-aged and elderly people. However, the obesity-driven epidemic of T2D has been associated with an increased diagnosis of this disease in young people. The first defect in the development of T2D is referred to as 'insulin resistance'. In this condition, insulin has a reduced ability to stimulate uptake of glucose from the blood into the skeletal muscle and to suppress glucose release from the liver. Insulin resistance is associated with the prediabetic state of 'impaired glucose tolerance (IGT)', in which blood glucose levels are maintained within the normal physiological range by a compensatory increase in circulating insulin levels, above that seen in healthy 'insulin-sensitive' subjects (Figure 6.1). The latter is achieved by an increased pancreatic β-cell secretion of insulin in response to a given increase in post-prandial glucose. Experimentally, IGT can readily be identified by an oral glucose tolerance test (OGTT) in which the subject is given an oral solution of glucose to drink and then the blood glucose excursion is measured over time (Box 6.1) (1). However, the OGTT is not deployed in routine clinical practise as it is rather time consuming and labour intensive. Moreover, the OGTT is not used routinely because IGT itself is not a condition that currently merits formal diagnosis and treatment with anti-hyperglycaemic (i.e. anti-diabetic) drugs. In a small proportion of genetically predisposed individuals, IGT will progress to frank T2D. This transition from the prediabetic to the diabetic state results from the

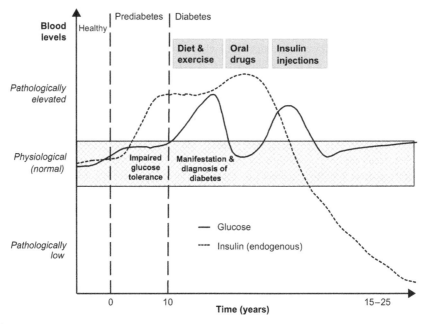

FIGURE 6.1

The natural history of type 2 diabetes.

inability of the pancreas to sustain, over time, a sufficient level of insulin secretion to compensate for an increasing degree of insulin resistance that extends beyond the muscle and ultimately also affects the adipose tissue (fat) and the liver (Figure 6.1).

6.2 PHARMACOTHERAPY OF TYPE 2 DIABETES

Before focussing our attention on the incretin-based drugs for T2D, it is instructive first to present a brief overview of the therapeutic landscape, or 'standard of care', into which these drugs were launched in the past decade as the most recent class of anti-hyperglycaemic drugs. Anti-hyperglycaemic drugs for T2D can be bracketed into two main classes based on their primary mechanism of action. Insulin sensitizer drugs (metformin and the thiazolidinediones (TZDs)) target insulin resistance, thus improving the ability of endogenous insulin to promote glucose disposal from the blood into the tissues. Insulin secretagogue drugs (sulphonylurea drugs or 'SUs') stimulate insulin secretion from the pancreatic β-cells. Therefore, these two classes of drugs target the two pathophysiological defects that cause T2D, that is, insulin resistance and defective insulin secretion from the pancreatic β-cells. However, it is important to note that these drugs were discovered empirically by testing small molecule compounds for their anti-hyperglycaemic efficacy in animal models of diabetes. Only subsequently were their mechanisms of action elucidated to a lesser, or greater, degree. For example, in the 1980s and many years after their launch in the 1950s, research proved that the SUs act by binding to a receptor on the β-cell plasma membrane that closes ATP-sensitive potassium channels leading to membrane depolarization and exocytosis of insulin-containing secretory granules. This mechanism of action, which is independent of glucose, explains why SUs are associated with the risk of inducing hypoglycaemia. The precise molecular mechanism of action of metformin remains somewhat obscure even now, many years after this drug entered routine clinical use for T2D. At the time that the first TZD, Avandia, was launched in 1999, the molecular target of this class of drug had only recently been determined. TZDs work by agonizing the Peroxisome Proliferator Activated Receptor (PPAR)-gamma transcription factor which in turn modulates adipose tissue physiology in a manner that indirectly improves insulin sensitivity in the muscle. It is important to note that, as for any drug, each anti-hyperglycaemic drug has not only certain attributes but also certain disadvantages that determine its utility and define its place in the treatment hierarchy (Table 6.1). For example, SUs are the most effective oral agents for lowering blood glucose but their use is associated with a significantly higher incidence of hypo-glycaemia than with any other oral anti-hyperglycaemic agent. Metformin is considered the first-line drug for diabetes as its use is not associated with weight gain in contrast to the SU and the TZD classes. However, metformin is associated with poor gastrointestinal tolerability that is managed by dose titration to achieve the maximum effective dose.

Newly diagnosed type 2 diabetics, most of whom are overweight or obese, are usually first advised by their doctor to exercise and diet in an attempt to maintain normoglycaemia. It has been proved that modest weight loss improves insulin sensitivity and thus lowers blood glucose levels. However, diet and exercise are usually insufficient alone to control the diabetic's blood glucose within the desired range beyond the short term. Metformin monotherapy is commonly used as the first-line pharmaco-therapy but as the disease progresses, and glucose levels start to drift upwards, an SU is added (2). Whilst SUs deliver good glucose-lowering efficacy, their effects are not usually sustained for more

Table 6.1 Anti-hyperglycaemic Drugs Used to Treat Type 2 Diabetes

Drug	Mechanism of Action	Route	%HbA1c (% Reduction)	Hypoglycaemia Risk	Body Weight
Metformin	Insulin sensitizer	Oral	1–1.5	No	Neutral
SUs	Insulin secretagogue	Oral	1–1.5	Yes	Increases
TZDs	Insulin sensitizer	Oral	0.5–1	No	Increases
DPP4Is	Incretin-based therapy	Oral	0.5–1	No	Neutral
GLP-1RAs	Incretin-based therapy	Injectable	1–2	No	Decreases
Insulin	Insulin replacement therapy	Injectable	Unlimited	Yes	Increases

than 12–18 months. Once combination therapy with metformin and an SU fails to adequately control blood glucose, patients must resort to insulin injections. Therefore, the current standard-of-care pharmacotherapy for T2D does not halt or reverse the disease and over time most patients will progress to late-stage disease in which residual pancreatic β-cell function is so minimal that insulin injection offers the only treatment option (Figure 6.1). In other words, at this point, T2D resembles T1D as there exists an absolute deficiency in endogenous insulin and insulin replacement therapy is required.

The decline in β-cell function that is the 'tipping point' in developing frank T2D represents the critical pathophysiological defect to target in order to discover novel drugs that will modify (i.e. halt or reverse) the course of the disease (5). The decline in β-cell function in T2D is attributed to two defects: firstly, impaired β-cell glucose-dependent insulin secretion and secondly, a reduction in the absolute number of pancreatic β-cells (6). Therefore, a drug that has the potential to rectify either, or preferably both, of these defects would potentially represent a disease-modifying anti-diabetic agent by restoring 'functional β-cell mass' to that which is seen in the prediabetic insulin-resistant but normoglycaemic individual. Preferably, such a drug would be used in conjunction with another intervention (i.e. drug or weight loss) that would improve insulin sensitivity and thus reduce the demand for insulin secretion, which is thought to hasten the decline in β-cell function in T2D (5).

6.3 THE RATIONALE FOR INCRETIN-BASED THERAPIES FOR TYPE 2 DIABETES

Glucose can be infused intravenously in humans to mimic the plasma glucose excursion (glycaemic profile) generated by an oral glucose bolus. In this 'isoglycaemic' experimental protocol, the increase in plasma insulin is markedly greater in subjects who receive oral glucose than in those who receive intravenous (IV) glucose (Figure 6.2). In fact, two- to threefold more insulin is secreted in response to the oral glucose bolus. This phenomenon is called the 'incretin effect'. The incretin effect is mediated by the incretin hormones glucagon-like peptide-1 (GLP-1) and glucose-dependent insulinotropic peptide (GIP). GLP-1 and GIP are secreted from the L and K enteroendocrine cells lining the gut

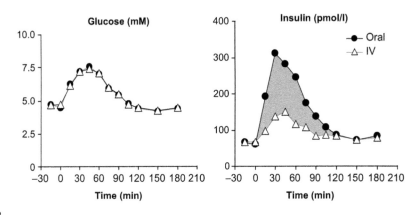

FIGURE 6.2

The incretin effect.

following absorption of glucose. However, the precise mechanism by which glucose stimulates their release is not currently understood. Also, the rapidity with which GLP-1 and GIP levels rise in the plasma during feeding suggests that mechanisms in addition to direct food contact with the gut epithelium may be involved.

GLP-1 and GIP stimulate insulin secretion from isolated pancreatic islets or β-cells in a glucose-dependent manner. The glucose dependence of their insulinotropic activity distinguishes these peptides from SUs which stimulate insulin secretion independent of glucose; hence the use of SUs is associated with an increased risk of hypoglycaemia in humans. GLP-1 and GIP bind to related G-protein-coupled receptors on pancreatic β-cells and activate adenylate cyclase leading to raised cellular cyclic adenosine monophosphate and the activation of a number of downstream signalling events that extend beyond those that stimulate insulin secretion (7). The effects of GLP-1 on pancreatic β-cell biology and function have been extensively studied using rodent/human *in vitro* systems (isolated pancreatic islets and β-cell lines) and in rodent models of IGT and T2D. In these studies, GLP-1 has been demonstrated to improve β-cell function by stimulating the expression of insulin and other genes essential for maintenance of the cellular machinery required to support sustained glucose-dependent insulin secretion. This is in contrast with the SUs which do not affect β-cell gene expression and thus are incapable of sustaining insulin secretion in the long term in rodents or in type 2 diabetics. Moreover, GLP-1 has been shown to preserve and increase functional β-cell mass in rodent models and in rodent/human *in vitro* systems: GLP-1 protects β-cells from high glucose/lipid/cytokine-induced apoptosis and stimulates β-cell proliferation; GLP-1 induces the differentiation of new β-cells from progenitor cells in the pancreatic duct. Collectively, these data support the notion that GLP-1, or drugs that mimic its activity, may have potential as disease-modifying drugs for the treatment of T2D (Table 6.2).

Beyond the effects on β-cell mass, and insulin secretion, GLP-1 possesses a number of biological activities that should further contribute towards improved glucose control in T2D (Table 6.2). GLP-1 suppresses glucagon secretion from pancreatic α-cells, but does so in a glucose-dependent manner such that it does not interfere with glucagon release in response to hypoglycaemia. Additionally, GLP-1 slows gastric emptying which delays the absorption of glucose from meals and enhances

Table 6.2 Pleiotropic Pharmacological Effects of GLP-1 on Metabolic Parameters in Preclinical Models of T2D and in Type 2 Diabetic Subjects

Characteristic Feature of T2D	Pharmacological Effect of GLP-1 Treatment	Impact on Disease State	Level of Evidence
Defective glucose-stimulated insulin secretion	Acutely improves defective pancreatic β-cell glucose-stimulated insulin secretion	Lowers blood glucose towards the non-diabetic range	Demonstrated *in vitro* using isolated rodent/human islets/β-cells, in rodent T2D models and in type 2 diabetic subjects
Reduced functional β-cell mass due to increased rate of β-cell apoptosis in the diabetic milieu of high glucose, high lipid and inflammatory cytokines	Restores functional β-cell mass by inhibition of apoptosis and stimulation of β-cell proliferation and neogenesis	Disease modification that halts and/or reverses disease progression as evidenced by normalization of blood glucose even after wash-out of GLP-1	*In vitro* evidence in isolated rodent/human islets/β-cells and in rodent T2D models. However, acute and subacute infusion of GLP-1 or chronic (3 years) GLP-1RA injections in T2D humans, followed by wash-out, has not demonstrated sustained effects on blood glucose required for evidence of disease modification
Hyperglucagonaemia	Suppresses pancreatic α-cell glucagon secretion and thus reduces hepatic glucose output	Lowers blood glucose towards the non-diabetic range	Demonstrated in rodent models and type 2 diabetic subjects
Disturbed gastric emptying	Slows gastric emptying and hence slows glucose absorption and reduces post-prandial hyperglycaemia	Lowers blood glucose towards the non-diabetic range	Demonstrated in rodent models and in type 2 diabetic subjects
Excessive food intake (hyperphagia) associated with obese type 2 diabetics	Induces satiety and hence suppresses appetite and food intake	Lowers blood glucose by reducing food intake. Reduces body weight which is known to improve glucose control	Demonstrated in rodent models and in type 2 diabetic subjects

satiety which inhibits food intake. These factors combine with the enhanced insulin release to significantly reduce the magnitude of the post-prandial glucose excursion. Crucially, the reduced food intake is sustained over the longer term (months to years) and translates into clinically relevant reductions in body weight. The mechanisms for these effects of GLP-1 remain poorly elucidated. For example, the effect of GLP-1 on satiety and food intake is presumed to be mediated by modulation of hypothalamic neuronal networks. However, it is not known to what extent these neurons are directly modulated by circulating GLP-1 accessing the hypothalamus versus peripheral GLP-1 stimulation of vagal afferents.

In T2D, the incretin effect is reduced or absent (8). The secretion of GIP and GLP-1 is preserved in T2D but the insulinotropic activity of GIP is lost owing to downregulation of the GIP receptor on pancreatic β-cells. In contrast, when administered at doses that deliver supraphysiological levels in the plasma, GLP-1 retains its insulinotropic activity. Consequently, therapeutic attention has focused on developing GLP-1 mimetics to restore the incretin effect and thus lower hyperglycaemia in T2D.

6.4 DISCOVERY AND PHARMACOKINETICS OF THE INCRETIN-BASED THERAPIES

GLP-1 is N-terminally degraded in the blood within minutes ($t_{1/2} = 2$ min), to an inactive species, predominantly by dipeptidyl peptidase-4 (DPP4). The plasma membrane protein DPP4 is widely expressed on endothelial and epithelial cells throughout the body. In addition, DPP4 is expressed on circulating T cells and is also known as T-cell antigen CD26. DPP4 selectively removes the N-terminal dipeptide from a number of bioactive peptides including GLP-1, GIP and a wide range of chemokines. DPP4-knockout mice have increased levels of plasma-intact GLP-1 and this is associated with increased insulin secretion and a lower blood glucose excursion following an oral glucose challenge. That is, the knockout mice exhibit improved glucose tolerance compared to wild-type mice (9). Preclinical studies have also demonstrated that small molecule selective inhibitors of DPP4 prevent the degradation of GLP-1 *in vivo*, leading to increased insulin secretion and improved glucose control.

Given its rapid clearance, native GLP-1 must be administered continuously to achieve a sustained pharmacological effect. Studies have utilized delivery by IV infusion or by pump-mediated subcutaneous (SC) infusion, in order to observe its beneficial effects on blood glucose control in T2D subjects (9). Whilst this approach was used in experimental studies to establish clinical proof of principle for GLP-1 in T2D, it does not represent a feasible therapeutic modality for the T2D population. Therefore, starting in the 1990s, two approaches were adopted in parallel by the biopharmaceutical industry to discover and develop drugs that were predicted to mimic GLP-1. These approaches focussed on developing protease-resistant peptide agonists of the GLP-1 receptor and small molecule chemical inhibitors of DPP4. As GLP-1R agonists (GLP-1RAs) and DPP4 inhibitors (DPP4Is) have been in the market since 2005, it is now possible to review whether these agents have delivered 'on their promise' and to compare the peptide and small molecule agents.

Sitagliptin (Januvia®, Merck & Co. Ltd., USA), the first DPP4I was launched for T2D in 2006. Subsequently, three other selective DPP4Is have been launched for T2D (Table 6.3). These orally available small molecule 'gliptins' are chemically distinct but possess low nanomolar potency (IC50s) and broadly similar pharmacokinetic properties. At their approved doses, each of these agents inhibits plasma DPP4 protease activity by >80% at Cmax and by >70% at the dose interval (i.e. immediately before a successive dose) (10). Consequently, the 'gliptins' effect an enhancement of both the postprandial rise in GLP-1 and the basal GLP in the post-absorptive or fasted state over 24 h (Figure 6.3).

Exenatide (Byetta®, Amylin Pharmaceuticals, USA) was the first incretin-based therapy to be launched for T2D in 2005. Exenatide was discovered in a search for biologically active peptides in the

Table 6.3 GLP-1 Receptor Agonists & DPP4 Inhibitors

Drug	Mechanism	Dose and Frequency*	Dosage Route	Need for Titration	Molecular Weight (Da)
Byetta® (exenatide)	GLP-1RA (SA)	10 µg bid	SC†	2–4 weeks at 5 µg bid	4187
Bydureon® (exenatide)	GLP-1RA (LA)	2 mg qw	SC	Dosing regimen leads to gradual increase in plasma levels over ~10 weeks	4187
Victoza® (liraglutide)	GLP-1RA (LA)	1.8 mg qd	SC	Starting dose 0.6 mg and increased by 0.6 mg weekly	3751
Januvia® (sitagliptin)	DPP4I	100 mg qd	Oral	None	407
Galvus® (vildagliptin)	DPP4I	50 mg bid	Oral	None	303
Tradjenta® (linagliptin)	DPP4I	5 mg qd	Oral	None	473
Onglyza® (saxagliptin)	DPP4I	5 mg qd	Oral	None	315

SA, short-acting; LA, long-acting; qd, once daily; bid, twice daily; qw, once weekly.
*Approved maintenance dose for diabetes.
†Subcutaneous injection.

venom of the Gila monster lizard. This peptide shares 53% sequence homology with human GLP-1 and is equipotent at the GLP-1R. However, exenatide is resistant to DPP4-mediated proteolytic inactivation by virtue of an alanine to glycine amino acid substitution at position 2. Therefore, exenatide has a longer half-life than GLP-1, of approximately 3 h, that allows for it to be administered by twice daily SC injection before breakfast and dinner (Figure 6.3). Exenatide is cleared from the circulation to below biologically active concentrations within 6 or 8 h of its injection. In 2011, Bydureon® (Amylin Pharmaceuticals, USA), a microsphere extended-release formulation of exenatide was launched that provides continuous exenatide exposure with once-weekly SC administration. Bydureon is a biodegradable formulation that slowly releases exenatide into the SC space over approximately 8–10 weeks. Once absorbed into the systemic circulation, the pharmacokinetic properties of Bydureon-derived exenatide are identical to those of Byetta-derived exenatide.

In 2009, liraglutide (Victoza®, Novo Nordisk, Denmark) was the second GLP-1RA to be launched. Liraglutide is a GLP-1 analogue in which lysine 34 is replaced with arginine and a C-16 fatty acid chain is added via a spacer glutamic acid residue at position 26. The fatty acylation promotes non-covalent binding of liraglutide to serum albumin (99–99.5% bound in plasma) that 'protects' the peptide from exposure to DPP4 and prolongs its terminal half-life to approximately 13.5 h. In addition, Victoza® demonstrates slow absorption over 6–12 h following SC administration. The resulting

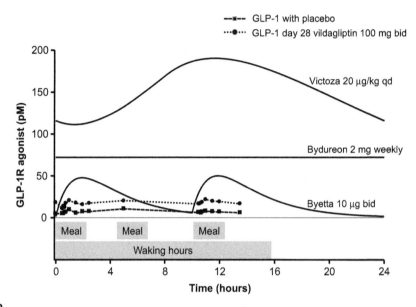

FIGURE 6.3

Endogenous GLP-1 concentrations and exogenous GLP-1 receptor agonist concentrations.

pharmacokinetic profile has a low 'peak-to-trough' ratio and delivers continuous activation of the GLP-1R with a once daily (qd) regimen (Figure 6.3). As a consequence of the albumin binding reducing the amount of Victoza® available to activate the GLP-1R at any time, the estimated free peptide concentration is shown for comparison to the exenatide exposures (Byetta and Bydureon) and endogenous GLP-1 levels (vildagliptin and placebo) in Figure 6.3.

6.5 CLINICAL EFFICACY OF THE INCRETIN-BASED THERAPIES

6.5.1 Glucose control

There is an extensive literature describing the glucose-lowering efficacy of GLP-1RAs and DPP4Is in randomized controlled trials in type 2 diabetic subjects (11–13). These studies include the 12- to 52-week duration phase 3 clinical trials that formed part of the submission for regulatory approval dossiers for each drug. Most of these phase 3 trials evaluated the incretin-based drugs as an 'add-on' to metformin, the recognized first-line pharmacotherapy. However, these drugs have also been trialled, and approved, as monotherapy and in combination with the other standard agents listed in Table 6.1. Head-to-head comparisons of the incretin-based drugs, with each other and with agents from other classes, have also been conducted in phase 3 trials and/or in smaller experimental clinical studies. The clinical efficacy data for the incretin-based drugs confirms the proof-of-principle data that was originally generated by infusing native GLP-1 into type 2 diabetics. Both DPP4Is and GLP-1RAs are known to increase glucose-dependent insulin secretion in diabetics and thus promote more effective

Table 6.4 Clinical Efficacy of Incretin-Based Drugs

Drug Class	HbA1c* (% Reduction)	FPG* (mM Reduction)	Body Weight (kg Reduction)
DPP4I	0.5–1.0	0.5–1.5	0
GLP-1RA (short-acting)	1.0–1.5	1.0–1.5	3–5
GLP-1RA (long-acting)	1.5–2.0	1.5–2.5	3–5

Figures represent percentage reduction from baseline at highest maintenance doses evaluated in published meta-analyses and systematic reviews of 12- to 52-week duration randomized controlled trials (11,12).

insulin-stimulated uptake of glucose into the skeletal muscle. Additionally, both classes suppress hyperglucagonaemia in type 2 diabetics and thus reduce inappropriate release of hepatic glucose into the blood in the presence of post-prandial or fasting hyperglycaemia. However, the magnitude of the glucose-lowering efficacy differs between the DPP4I and GLP-1RA classes (Table 6.4). The GLP-1RAs exhibit significantly greater glucose-lowering efficacy than the DPP4Is. The difference can be rationalized based on the respective 'drive on target' delivered by DPP4Is and GLP-1RAs. The DPP4Is modestly raise endogenous plasma GLP-1 to approximately three- to five-fold the level observed in placebo control subjects (Figure 6.3). In contrast, the GLP-1RAs, which possess very similar *in vitro* potencies to GLP-1 at the human GLP-1R, achieve free plasma concentrations that represent 10- to 50-fold of endogenous plasma GLP-1 levels. Within the DPP4I class, there is little difference in efficacy from one 'gliptin' to the next as these agents all possess similar pharmacokinetic exposure profiles and effect similar increases in endogenous plasma GLP-1. However, the GLP-1RAs can be classified into the 'short-acting' Byetta and the 'long-acting', or 'continuous acting', Bydureon and Victoza (Table 6.4). Thus, Byetta achieves plasma concentrations adequate for GLP-1R activation during the waking hours following its twice-daily injections before breakfast and dinner. Bydureon and Victoza achieve plasma concentrations that activate the GLP-1R continuously. The continuous activation of GLP-1R by Bydureon and Victoza is reflected by the greater glucose lowering efficacy of these agents compared to Byetta. This difference is primarily attributed to the more significant effects of the 'long-acting' GLP-1RAs on fasting plasma glucose (Table 6.4) (*13*).

The observation that both DPP4Is and GLP-1RAs reduce fasting plasma glucose, albeit to varying degrees, may appear somewhat unexpected in view of the fact that GLP-1 is secreted and hence elevated postprandially. However, preclinical and clinical data confirm that basal GLP-1 tone does indeed influence fasting plasma glucose. Thus, GLP-1-receptor-knockout mice exhibit elevated fasting plasma glucose; a peptide antagonist of the GLP-1 receptor increases fasting plasma glucose in humans and continuous infusion of GLP-1 over 24 h in diabetic humans provides more effective glucose control than does a 16-h daytime infusion.

In addition to the direct effects on insulin-stimulated glucose uptake, infusion of GLP-1 to supraphysiological plasma levels is known to suppress gastric emptying during feeding and thus slow the rate at which glucose is absorbed from the gut. Together, these effects act to significantly reduce post-prandial glycaemic excursions at each meal. Interestingly, despite greater GLP-1R activation, the effects of Bydureon and Victoza on post-prandial glucose concentrations are less marked than with Byetta following chronic dosing while the differential effects on fasting plasma glucose are maintained. This apparent disconnect between fasting and post-prandial glucose has been linked to tachyphylaxis of the GLP-1R-induced slowing of gastric emptying. Commonly, agonists exhibit some form of desensitization after chronic receptor activation. Due to its rapid onset (hours), this effect is thought to be occurring at the level of the vagus nerve (*13*).

6.5.2 Weight loss

GLP-1 infusion over several weeks in diabetics leads to a modest weight loss and this itself is known to improve insulin sensitivity and whole body glucose control. The GLP-1RAs resemble native GLP-1 as they slow gastric emptying, induce satiety, reduce food intake and produce modest but significant weight loss that can be sustained for at least 2 years with continuous dosing. Interestingly, both Bydureon and Victoza appear to have a dose-response profile for weight loss that is right shifted when compared to that for glucose lowering, indicating that the level of GLP-1R activation required to induce weight loss is higher than that required to sensitize pancreatic β-cells to glucose. For example, despite near-identical effects on fasting plasma glucose, Bydureon at 0.8 mg weekly dose has no effect on weight loss, while 2 mg leads to a significant loss compared to placebo (*14*). DPP4Is are not associated with weight loss consistent with the lower GLP-1R activation achieved with these agents compared with GLP-1RAs.

All GLP-1RAs are associated with nausea and this can be thought of as an extension of their pharmacological role in regulating gastric emptying and feelings of satiety. The occurrence and management of this will be discussed in Section 6.8.

6.6 EVIDENCE FOR DISEASE MODIFICATION

Preclinical data in rodent/human isolated pancreatic islets and β-cells as well as in rodent disease models raised the prospect that incretin-based therapies might demonstrate disease-modifying effects in T2D subjects by preserving and/or regenerating functional pancreatic β-cell mass (see Section 6.3 above and Table 6.2). In humans, pancreatic β-cell function can be assessed by static and dynamic measurements of insulin secretion. Static or fasting measures include the homoeostatic model assessment β-cell function index (HOMA-B) and the plasma proinsulin to insulin ratio. HOMA-B represents a simple equation that relates plasma insulin to the plasma glucose concentration and thus provides a single time point measure of β-cell insulin secretion. Proinsulin is processed within insulin secretory granules to mature insulin and is co-secreted with insulin. Thus an increase in the plasma proinsulin to insulin ratio is considered a surrogate marker for defective β-cell function. However, the β-cell is primarily active in the post-prandial period following ingestion of glucose at meal times. Therefore, dynamic measures of insulin secretion over time in response to an oral or IV glucose challenge are the preferred method for assessing β-cell function. The effects of GLP-1RAs and DPP4Is on β-cell function in T2D subjects have been studied extensively using static and dynamic measurements of insulin secretion. These agents effect stable improvements in β-cell function in type 2 diabetics that are observed following each administration of the drug and for the duration of the on-treatment phase of such studies. However, there is at present no evidence that either GLP-1RAs or DPP4Is have durable effects on β-cell function that persist after cessation of therapy. For example, Byetta improved β-cell function, and significantly lowered blood glucose, in type 2 diabetics when administered twice daily for 1 year. However, following 4 weeks of drug washout from week 53 to week 56, β-cell function and blood glucose returned to their day zero baseline diabetic levels. Similar data has been reported following 3 years treatment with Byetta and in up to 2 year studies with DPP4Is (*15,16*).

The absence of durable effects of the incretin-drugs on wash-out is interpreted as evidence that in humans these drugs do not regenerate lost functional β-cell mass. In contrast, in rodent models of T2D, GLP-1 itself, the GLP-1RAs and DPP4Is have been shown to increase β-cell mass (*16*). In these studies, a durable improvement in β-cell function has been reported on wash-out of the drug indicating a disease-modifying effect. It is not clear why these drugs are able to modify the course of disease in preclinical models but not in T2D subjects. However, the age of the animals used may not accurately reflect the target human patients. Hence, exenatide has been shown to increase β-cell mass in young mice (6 weeks old) but not in old mice (7–8 months). It is also possible that the drug levels achieved in animal studies that deliver effects on β-cell mass are not achieved in the clinic. Additionally, the precise mechanism by which these drugs increase functional β-cell mass in animal models may not translate to humans. That is to say, the composite effect of increased β-cell proliferation and decreased β-cell apoptosis observed in animals may not be achievable in humans. In contrast, the effects of these drugs on insulin biosynthesis observed preclinically may be recapitulated in humans as these drugs do deliver stable improvements in β-cell function over 2–3 years at least. This sustained effect is important as it differentiates the incretin-based drugs from the other classes of insulin secretagogue drugs, the SUs, that typically lose efficacy by 12–18 months in humans.

6.7 IMPACT ON CARDIOVASCULAR RISK

Individuals with T2D have a significantly elevated risk of cardiovascular disease and its associated morbidity and mortality. Traditional risk factors for cardiovascular disease are elevated blood glucose, an atherogenic blood lipid profile, elevated blood pressure and obesity. The incretin-based drugs have been demonstrated to positively impact all these risk factors in preclinical disease models and in type 2 diabetics (*17*). The exception is that the DPP4Is do not lead to body weight loss. However, DPP4Is are weight neutral in contrast to insulin, SUs and TZD anti-diabetic agents, which promote weight gain. The precise mechanisms by which the incretin-based drugs improve blood pressure are not fully understood. Both direct effects on endothelial/vascular tone and indirect effects mediated by the vagus and sympathetic nervous system have been reported in preclinical models. It is intriguing to note that some of the effects of these agents on the cardiovascular system may be independent of the GLP-1 receptor as they are preserved, at least in part, in GLP-1-receptor-knockout mice and by the inactive proteolytic product of GLP-1. It should be noted that the GLP-1RAs have been associated with mild and transient increases in heart rate in the clinic but the significance of this observation remains to be determined. The rather modest beneficial effects of these drugs on blood lipids in type 2 diabetics may be a consequence of improved metabolic control and/or reduced body weight. Recently published data on the incidence of MACE (fatal and nonfatal heart attack and stroke) in T2D patients receiving incretin-based therapy appears consistent with the positive impact of these agents on cardiovascular risk factors. A meta-analysis of 42,000 patients in DPP4I clinical trials, and an analysis of real-world data for 39,000 patients on GLP-1RA therapy, reported that these agents significantly reduced the incidence of MACE compared to comparator or placebo (*18,19*). Whilst these data are most encouraging, definitive conclusions regarding the impact of incretin-based therapies on MACE must await the results of the ongoing cardiovascular outcome studies for each drug.

6.8 CLINICAL SAFETY AND TOLERABILITY OF INCRETIN-BASED THERAPIES

6.8.1 Nausea

Acute delivery of the GLP-1RAs is associated with a dose-dependent increase in nausea and, at supraclinical doses, vomiting. In the clinical setting, this effect is managed via dose titration of all the agents. For example, to enable delivery of tolerated doses of Victoza high enough to impact diabetes (1.8 mg or greater), the drug is titrated from 0.6 mg qd for the first week and increased by 0.6 mg increments weekly until the desired dose is reached. This gradual titration is essentially 'built in' to the Bydureon regimen as its controlled-release delivery leads to a gradual increase in the level of GLP-1 activation over approximately 10 weeks. This has the benefit of reducing the incidence of nausea compared to Byetta despite higher steady-state concentrations of drug and activity on fasting plasma glucose.

6.8.2 Anti-drug antibodies (see Chapter 3)

Both exenatide (either via Byetta or Bydureon) and Victoza show the presence of anti-drug antibodies (ADA) in the serum following chronic dosing ranging from ~10 to 50% incidence. The presence of these antibodies is not strongly associated with the outcome of treatment indicating that the antibodies are predominantly non-neutralizing.

Importantly, ADA-positive serum samples from human subjects undergoing GLP-1RA treatment revealed no significant cross-reactivity with either glucagon or GLP-1. Consistent with this, the presence of ADA was not associated with impaired glucose control or counter-regulation. These data indicate that treatment-induced neutralization of endogenous peptides is a very low risk with this class.

6.8.3 Injection site reactions

Injection site reactions, such as injection site rash and erythema, are generally minor in nature and of low incidence with the GLP-1RAs. In clinical trials, Bydureon appears to show a higher incidence of reactions (17.1%) than Byetta (12.7%) and Victoza (2%). This may be due in part to the microsphere extended-release formulation delivery technology that gives Bydureon its convenient once-weekly dosing regimen. All the agents are associated with a very low incidence of withdrawal due to these effects (Bydureon and Victoza, 1 and 0.2%, respectively). Site reactions are of a higher incidence in patients that are positive for ADA but are not associated with any increased severity.

6.8.4 Miscellaneous risks

Despite the widespread expression of DPP4 throughout the body, and its extensive list of substrates, inhibitors of DPP4 have been proved to be remarkably benign in terms of their safety profile in humans (*13*). Only upper respiratory tract infection, nasopharyngitis and headache were reported as adverse events more commonly observed with DPP4Is than with placebo. Acute pancreatitis has been occasionally reported as part of postmarketing safety surveillance for both DPP4Is and GLP-1RAs. However, studies using controlled health care claims databases concluded that the incidence of acute pancreatitis in patients receiving incretin-based therapy is no different from that observed in patients using other anti-diabetic drugs. In rodent toxicology studies, high-dose GLP-1RA exposure results in increased secretion of calcitonin from thyroid C cells and C-cell hyperplasia and carcinoma. In non-human primates and humans, GLP-1 receptor expression in C cells is very low. Liraglutide neither induced C-cell proliferation in non-human primates nor did it increase plasma calcitonin in humans. On the basis of these species-specific differences, the US Food and Drug Administration considered the risk of developing medullary thyroid cancer as a result of GLP-1RA therapy to be low in humans. However, liraglutide is contraindicated in patients with a personal or family history of medullary thyroid cancer and in patients with multiple endocrine neoplasia syndrome.

6.9 CONCLUSIONS

Incretin-based therapies represent the first class of anti-hyperglycaemic drug that has been discovered based on targeting a known pathophysiological defect in T2D. The class has several advantages over most of the other anti-hyperglycaemic drugs as it is not associated with an increased risk of hypoglycaemia or weight gain (Table 6.1). In fact, the GLP-1RA subclass delivers a clinically meaningful and sustained reduction in body weight. Importantly, the incretin-based drugs appear to deliver sustained efficacy over at least 2–3 years in contrast to the current front-line regime of metformin and/or sulphonylureas (5,15,16). Therefore, whilst clinical data from 'drug wash-out' studies does not provide evidence that the class effects disease modification, by regenerating lost functional β-cell mass, there is evidence that these agents are able to preserve residual β-cell function. The results of ongoing cardiovascular outcomes studies are awaited before definitive conclusions can be reached regarding the effects of the incretin-based therapies on major adverse cardiovascular events (MACE). However, it is encouraging that recent analysis of real-world data and meta-analyses of clinical trial data indicates that this class significantly reduced the incidence of MACE compared to comparator or placebo (18,19).

Within the class, DPP4Is and GLP-1RAs display a complementary profile of advantages and disadvantages. DPP4Is exhibit less glucose-lowering efficacy than the GLP-1RAs and do not effect any change in body weight but have the advantage of being orally administered. The GLP-1RAs more effectively control hyperglycaemia and significantly reduce body weight but are disadvantaged by a parenteral route of administration and an association with nausea. The former disadvantage will be addressed to a degree when new long-acting GLP-1RAs reach the market that require injections weekly or less frequently (20). Two such medicines are currently in phase 3 clinical development. Albiglutide is a fusion protein composed of two molecules of a GLP-1 analogue covalently bound to human serum albumin resulting in a half-life of 6–8 days. Dulaglutide is a GLP-1 peptide fused to an Fc antibody fragment that extends its half-life to approximately 90 h. Therefore, albiglutide may be compatible with dosing less frequently than dulaglutide which is dosed weekly. In contrast to Bydureon, which requires a large-gauge needle for injection owing to its microsphere formulation, both albiglutide and dulaglutide are injected using a small-gauge needle which should minimize injection-related pain and reactions. Before leaving the subject of 'next-generation' versions of the incretin-based therapies, it should be noted that the first two DPP4Is to be launched, sitagliptin

(Januvia®) and saxagliptin (Onglyza®), are both now available as fixed-dose single-pill combinations with metformin. These drugs are marketed as Janumet® (Merck & Co. Ltd., USA) and Kombiglyze (AstraZeneca/Bristol-Myers Squibb).

The most recently issued algorithm on the medical management of T2D recommends diet and exercise measures to decrease body weight and improve glycaemic control (2). If glucose levels are insufficiently controlled by lifestyle intervention, the treatment algorithm advises the introduction of metformin as the first-line drug therapy. When metformin monotherapy fails to achieve effective glucose control, a sulphonylurea or insulin is recommended in combination with metformin. Lifestyle intervention, metformin, insulin and sulphonylureas are described in the guidelines as 'Tier 1: well-validated core' therapies for T2D. The TZDs and the GLP-1RAs represent alternative options for combination therapy with metformin, but are described as 'Tier 2: less-well-validated' therapies. However, it is important to note that the prescription of TZDs has been significantly limited by concerns regarding their potential adverse effects on cardiovascular risk and that the use of insulin is limited by the inherent risk and fear of hypoglycaemia (2,5). Therefore, in the real world, the treatment hierarchy may currently be more accurately summarized as metformin with or without SU followed by metformin with either a DPP4I or a GLP-1RA. Commercial forecasts for the next 10 years project that the incretin-based therapies will secure their position as third-line pharmacotherapy behind metformin and SUs (21). This will be aided by the fact that the first agents in this class will become generic during this period thus making these agents more affordable and cost-effective for health care payers. It will be interesting to see whether generic versions of the incretin-based therapies ultimately displace SUs as the preferred combination therapy with metformin. From a pathophysiological perspective, it has been argued that the preferred front-line combination should be an insulin sensitizer (metformin and/or TZD) plus an incretin-based drug (5). This combination is predicted to preserve β-cell function and thus provide more durable glycaemic control than use of an SU which do not preserve β-cell function. Another emerging use for the GLP-1RAs is their combination with insulin. For example, using a short-acting GLP-1RA instead of a short-acting insulin to provide meal time glucose control in combination with basal insulin (that provides overnight glucose control) is predicted to have benefits in terms of a lower risk of hypoglycaemia and no, or less, body weight gain (20).

In summary, since their launch in the mid-2000s, the incretin-based therapies have secured a prominent place in the treatment regimen for T2D as safe and effective drugs. The complementary efficacy and tolerability profiles of the DPP4Is and the GLP-1RAs offer physicians a means to select the appropriate drug for the individual patient based on their disease status (HbA1c percentage), body weight and willingness to self-administer an injectable drug. The position of the class is likely to be further strengthened as early entrants become available generically and as improved versions, requiring less-frequent administration, are launched.

References

1. American Diabetes Association Position Statement. Diagnosis and Classification of Diabetes Mellitus. *Diabetes Care* **2012,** *35,* S64–S71.
2. Nathan, D. M.; Holman, R. R.; Buse, J. B.; Sherwin, R.; Davidson, M. B.; Zinman, B.; Ferrannini, E. Medical Management of Hyperglycemia in Type 2 Diabetes: A Consensus Algorithm for the Initiation and Adjustment of Therapy. *Diabetes Care* **2009,** *32,* 193–203.

3. Turner, R.; Cull, C.; Holman, R. United Kingdom Prospective Diabetes Study: A 9-year Update of a Randomised, Controlled Trial on the Effect of Improved Metabolic Control on Complications in Non-insulin Dependent Diabetes Mellitus. *Ann. Intern. Med.* **1996,** *124,* 136–145.

4. International Diabetes Foundation Diabetes Atlas, 5th ed., 2012.

5. DeFronzo, R. A. From the Triumvirate to the Ominous Octet: A New Paradigm for the Treatment of Type 2 Diabetes Mellitus. *Diabetes* **2009,** *58,* 773–795.

6. Butler, A. E.; Janson, J.; Bonner-Weir, S.; Ritzel, R.; Rizza, R. A.; Butler, P. C. β-Cell Deficit and Increased β-Cell Apoptosis in Humans with Type 2 Diabetes. *Diabetes* **2003,** *52,* 102–110.

7. Yabe, D.; Seino, Y. Two Incretin Hormones GLP-1 and GIP: Comparison of their Actions in Insulin Secretion and B Cell Preservation. *Prog. Biophys. Mol. Biol.* **2011,** *107,* 248–256.

8. Nauck, M. A. Incretin-Based Therapies for Type 2 Diabetes Mellitus: Properties, Functions, and Clinical Implications. *Am. J. Med.* **2011,** *124,* S3–S18.

9. Deacon, C. F. Incretin-Based Treatment of Type 2 Diabetes: Glucagon-Like Peptide-1 Receptor Agonists and Dipeptidyl Peptidase-4 Inhibitors. *Diabetes Obes. Metab.* **2007,** *9,* 23–31.

10. Deacon, C. F. Dipeptidyl Peptidase-4 Inhibitors in the Treatment of Type 2 Diabetes: A Comparative Review. *Diabetes Obes. Metab.* **2011,** *13,* 7–18.

11. Aroda, V. R.; Henry, R. R.; Han, J.; Huang, W.; DeYoung, M. B.; Darsow, T.; Hoogwerf, B. J. Efficacy of GLP-1 Receptor Agonists and DPP-4 Inhibitors: Meta-analysis and Systematic Review. *Clin. Ther.* **2012,** *34,* 1247–1258.

12. Deacon, C. F.; Mannucci, E.; Ahren, B. Glycaemic Efficacy of Glucagon-like Peptide-1 Receptor Agonists and Dipeptidyl Peptidase-4 Inhibitors as add-on Therapy to Metformin in Subjects with Type 2 Diabetes – A Review and Meta Analysis. *Diabetes Obes. Metab.* **2012,** *14,* 762–767.

13. Fineman, M. S.; Cirincione, B. B.; Maggs, D.; Diamant, M. GLP-1 Based Therapies: Differential Effects on Fasting and Postprandial Glucose. *Diabetes Obes. Metab.* **2012,** *14,* 675–688.

14. Fineman, M. S.; Flanagan, S.; Taylor, K.; Aisporna, M.; Shen, L. Z.; Mace, K. F.; Walsh, B.; Diamant, M.; Cirincione, B.; Kothare, P.; Li, W.; MacConell, L. Pharmacokinetics and Pharmacodynamics of Exenatide Extended-Release after Single and Multiple Dosing. *Clin. Pharmacokinet.* **2011,** *50,* 65–74.

15. Bunck, M. C.; Taskinen, M.; Corner, A.; Smith, U.; Eliasson, B.; Yki-Jarvinen, H.; Heine, R. J.; Diamant, M.; Shaginian, R. M. Effects of Exenatide on Measures of β-Cell Function after 3 years in Metformin-Treated Patients with Type 2 Diabetes. *Diabetes Care* **2011,** *34,* 2041–2047.

16. van Genugten, R. E.; van Raalte, D. H.; Diamant, M. Dipeptidyl Peptidase-4 Inhibitors and Preservation of Pancreatic Islet-Cell Function: A Critical Appraisal of the Evidence. *Diabetes Obes. Metab.* **2012,** *14,* 101–111.

17. Sivertsen, J.; Rosenmeier, J.; Holst, J. J.; Vilsboll, T. The Effect of Glucagon-Like Peptide 1 on Cardio-vascular Risk. *Nat. Rev. Cardiol.* **2012,** *9,* 209–222.

18. Monami, M.; Ahren, B.; Dicembrini, I.; Mannucci, E. Dipeptidyl Peptidase-4 Inhibitors and Cardiovascular Risk: A Meta-analysis of Randomized Clinical Trials. *Diabetes Obes. Metab.* **2013,** *15,* 112–120.

19. Best, J. H.; Smith, D. B.; Hoogwerf, B. J.; Wenten, M.; Herman, W. H.; Hussein, M. A.; Pelletier, E. M. Risk of Cardiovascular Disease Events in Patients with Type 2 Diabetes Prescribed the Glucagon-Like Peptide 1 (GLP-1) Receptor Agonist Exenatide Twice Daily or Other Glucose-lowering Therapies. *Diabetes Care* **2011,** *34,* 90–95.

20. Meier, J. J. GLP-1 Receptor Agonists for Individualized Treatment of Type 2 Diabetes Mellitus. *Nat. Rev. Endocrinol.* **2012,** *8,* 728–742.

21. Decision Resources Report on Type 2 Diabetes Therapy Market (February 2012).

The structure and business of biopharmaceutical companies including the management of risks and resources

7

James Samanen

James Samanen Consulting LLC, Phoenixville, PA 19460, USA, YourEncore, 100 Canal Pointe Blvd, Suite 209 Princeton, NJ 08540, USA, Former Discovery Research Portfolio Management GlaxoSmithKline, 709 Swedeland Road, King of Prussia, 19406, USA

CHAPTER OUTLINE

7.1 Introduction 227
    7.1.1 The business of biopharmaceutical research and development 227
        7.1.1.1 The successful drugs also pay for the failed drugs 227
        7.1.1.2 Revenue challenges 227
        7.1.1.3 Revenue opportunities 230
    7.1.2 The science of biopharmaceutical R&D 230
7.2 The Organization of Biopharmaceutical R&D 231
    7.2.1 Biopharmaceutical R&D is drug discovery and drug development...................231
    7.2.2 The logical organization of biopharmaceutical R&D 231
    7.2.3 Stage-gate organization – the project pipeline 232
        7.2.3.1 The project pipeline..233
        7.2.3.2 Not all projects start at the first stage...233
    7.2.4 Stage-related goals..234
    7.2.5 Attrition..235
        7.2.5.1 The Attrition-based pipeline..236
    7.2.6 Risk – how it influences POS, cost, value and commitment 237
    7.2.7 Resource – who does what, when and where 238
        7.2.7.1 Association of resource with stage – matrix of work............................238
        7.2.7.2 Association of resource with business units – overarching organization...............238
    7.2.8 In-house vs. in-license 240
    7.2.9 In-house vs. out-license 241
    7.2.10 Big vs. small...242
    7.2.11 The necessity of standard operating procedures 243
7.3 Project Management in Biopharmaceutical R&D... 244
7.4 Portfolio Management in Biopharmaceutical R&D... 244

Introduction to Drug Research and Development. http://dx.doi.org/10.1016/B978-0-12-397176-0.00007-8

7.5 Cost Reduction Experiments in the Business of Biopharmaceutical Discovery and Development 245
    7.5.1 Process improvement..245
    7.5.2 Reducing resource – vertical disintegration 245
        7.5.2.1 The need to downsize... 245
        7.5.2.2 Vertical integration vs. vertical disintegration (Decentralization)............................ 245
    7.5.3 Increasing success through changes in the organizational model 249
        7.5.3.1 Big and small at the same time.. 249
        7.5.3.2 Planning for failure – POC at chorus (Lilly) and flexion therapeutics..................... 250
        7.5.3.3 Open innovation 250
7.6 Conclusion... 251
References 251

ABSTRACT

Successful drugs have a good return on investment by bringing in considerably more revenue than the expenses of discovery, development, and manufacturing. Successful drugs pay for all drug projects, those that fail and those that have yet to fail or succeed. Most research and development (R&D) projects fail. Since R&D is the future of the company, a lot is at stake in the business of R&D. This chapter considers the organization of biopharmaceutical R&D, as well as various organizational experiments, that are already under way, that deal with the enormous risk and cost of biopharmaceutical R&D. There is a fairly uniform sequence of events involved in the discovery and development of biopharmaceuticals. The Stage-Gate Organization of the project pipeline is described along with stage-related goals. The high attrition in the industry is examined as well as reasons for project failure, particularly in the clinic. The fact that most projects fail in the biopharmaceutical industry means that risk, the probability that a project will fail, influences a number of key behaviours in biopharmaceutical R&D. The manner in which risk influences probability of success, cost, value and corporate commitment is considered. Not all discoveries occur within a company – many are in-licenced. Reduced revenues challenge a company's ability to develop all its assets, increasing the demands on project and portfolio management, and for out-licencing or partnering. In large biopharmaceutical companies, resource tends to be organized into business units, therapy areas, line departments, and platform technology groups. In the new era of reduced profits many companies are moving away from vertical integration towards decentralization, performing many to most functions in other companies, and in the extreme, towards virtual drug discovery and development. The risks and benefits with the external allocation of resource via outsourcing and partnering are discussed. Experiments with the organizational model of biopharmaceutical R&D are explored which aim to reduce risk, increase success and efficiency, including attempts to be big and small at the same time, planning for failure, and open innovation. There are also external revenue challenges, including generics competition and third-party payer constraints. On the upside are a number of opportunities to increase revenue, including new biologics and new areas of exploration – epigenetics and gene therapy – and by expanding markets into rapidly developing countries. Managers face complex challenges to the business of biopharmaceutical R&D. But, regardless of the type of company or set of partnered companies, academic institutions and service organizations that perform biopharmaceutical R&D, to a large extent the sequence of events in which a drug is discovered and developed will always be the same. And as long as the industry can continue to find new therapies that positively impact the lives of patients, it will continue to be an exciting and challenging industry.

Keywords/Abbreviations: Return on investment (ROI); New molecular entity (NME); Innovation gap; Epigenetics; RNA interference (RNAi); Stage-Gate (milestone) concept; Life cycle management; Project/pipeline/portfolio; Paralogue synergy; Mergers and acquisitions; Proof of concept (POC); Probability of success (POS); Estimated present net value (EPNV); Pharmaceutical Research and Manufacturers of America (PhRMA); Standard operating procedure (SOP); Good clinical/manufacturing/laboratory practice (GC/M/LP); Vertical (dis)integration; Contract Research Organization (CRO); Virtual drug discovery and development (VD&D); Centres of Excellence in Drug Discovery (CEDDs); Open innovation; Biopharmaceutical; Generic drugs; Brazil, Russia, India, and China (BRIC); Biologics; Gene therapy; Stage-Gate Organization; Paralogue; Target discovery; Lead discovery; Lead optimization; Preclinical evaluation; Full development; Registration & launch; Reimbursement; Attrition; Pipeline; Risk; Cost; Value; Resource; Therapy area; Platform technologies; In-licence; Out-license; Repurposing; Matrix of work; Intellectual property (IP); Project management; Portfolio management; Process improvement; Vertical disintegration; Vertical integration; Decentralization; Outsourcing; Partnering; Virtual pipeline; Chorus; Flexion therapeutics; Open innovation; Precompetitive research.

7.1 INTRODUCTION

In this chapter, unless a distinction is made in the text, the term biopharmaceutical industry is used to cover both the conventional pharmaceutical industry and the manufacturers of biopharmaceuticals.

7.1.1 The business of biopharmaceutical research and development

The biopharmaceutical industry is a big business. As ranked by percentage return on revenues, the pharmaceutical industry ranked second in 2008, behind the mining & gas production industry (*1*). Many of the products are big money winners. Table 7.1 lists the top 10 best-selling drugs in the US in 2010.

Companies that make a lot of money want to continue to make a lot of money, but as we will see, it's a tough business.

7.1.1.1 The successful drugs also pay for the failed drugs

In any company, the cost of research and development (R&D) is paid for by the drugs that bring in revenue for the company. As will be discussed later in this chapter (Section 7.2.5) most R&D projects fail. A successful drug is one that brings in considerably more revenue than the expenses from the discovery, development, manufacture and sales of the drug. Such drugs have a good return on investment (ROI). The successful drugs pay for all drug projects, those that fail and those that have yet to fail or succeed. Since R&D is the future of the company, there is a lot at stake in the business of R&D.

7.1.1.2 Revenue challenges

Ideally the cost of developing the drug should be less than the amount of money that will be made on the drug. This can be displayed in a cash flow diagram (Figure 7.1), where the expenses incurred during the R&D of the drug are displayed as negative cash flow and revenues from the sales of the drug create positive cash flow. But there are many challenges to the outcome that revenues will exceed expenses.

Until product launch, the cash flow is negative, in terms of the expenses incurred from the R&D of the drug. Upon launch the sales of the drug bring in revenues and hopefully the revenues are greater than the expenses and cash flow becomes positive. It is also to be hoped that the volume under the cash

Table 7.1 The Top 10 Best-Selling Drugs in the US (2–10)

	Drug	$ Billion	Type	Company	Date of Patent Expiration	Reference
1	Lipitor	7.2	Cholesterol-lowering statin drug	Pfizer	2011	(3)
2	Nexium	6.3	Antacid drug	AZ	2015	(4)
3	Plavix	6.1	Blood thinner	BMS	2012	(5)
4	Advair Diskus	4.7	Asthma inhaler	GSK	2010	(6)
5	Abilify	4.6	Antipsychotic drug	Otsuka/BMS	2014	(7)
6	Seroquel	4.4	Antipsychotic drug	AZ	2012	(8)
7	Singulair	4.1	Oral asthma drug	Merck	2012	(3)
8	Crestor	3.8	Cholesterol-lowering statin drug	AZ	2016	(9)
9	Actos	3.5	Diabetes drug	Takeda	2012	(8)
10	Epogen	3.3	Injectable anaemia drug	Amgen	2013	(10)

flow positive curve is considerably larger than the volume under the cash flow negative curve to make up for the expense of discovery. Optimally the volume under the cash flow positive curve is larger than the negative cash flow of at least some of the failed projects. Additional Phase III clinical trials required by regulatory agencies can significantly increase the prelaunch expense. Astoundingly, only two out of 10 marketed drugs return revenues that match or exceed R&D costs (12).

The launched drug should have one or more patents that protect the drug from generic competition. The period of patent protection, however, is initiated well back in the drug discovery phase when the drug is initially synthesized and shown to have significant biological activity. Lengthy clinical trials reduce the

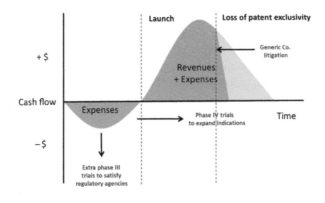

FIGURE 7.1

A hypothetical cash flow vs. time plot (11). Cash flow is the difference between revenue and expense.

period of patent protection. Companies typically run clinical trials postlaunch to expand the set of indications for a drug (Phase IV), which could add to revenues but cut into the period of patent protection.

Upon patent expiration, other companies may enter the market with considerably cheaper generic nonbranded drugs, which rapidly reduce revenues of the branded drug. Generics share of the market was 49% in 2000, rising to 74% in 2009 (*12*). The 10 best-selling drugs in Table 7.1 will all lose patent protection by 2016.

In the US, congress created a 'period of exclusivity' to ensure that at least some of the revenue stream is protected from generic competition. It has been estimated that the branded pharmaceutical companies will lose $100 billion dollars in the US sales between 2010 and 2013 (*13*).

In recent years third-party payers have become the most important force in determining the market for a new drug. They tend to focus more on the 'value' (i.e. cost) of a drug over the innovation or patient need (*14*). (More attention could be paid to the financial aspects of the business, but we must consider such detail to be beyond the scope of this chapter.)

Working against the relentless challenge of patent expirations, biopharmaceutical companies strive to find new drugs. Innovation feeds the engine. But here too, there are problems. A simple measure of innovation is the number of 'new molecular entities' (NMEs) that the FDA approves each year (first-ever approval of a drug). As seen in Figure 7.2 (*15,16,17*), pharmaceutical companies have spent more on R&D each year, while except for a spike in the late 1990s, the number of NMEs has remained essentially the same. The increasing distance between NMEs and Pharma Spend in Figure 7.2 is often referred to as the 'innovation gap', e.g. Ref. (*18*).

One may argue that while NME is perhaps the simplest measure of innovation or productivity, since it removes the subsequent filings and approvals on a drug, NME rarely defines the most valuable version (indication and/or formulation) of an asset. Better formulations and more lucrative indications tend to be further back in the pipeline and reach approval after the NME. Nonetheless it is the measure of innovation or productivity most employed by industry analysts.

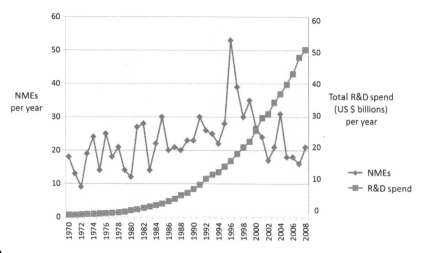

FIGURE 7.2

NMEs per year vs. total R&D spend in billions of dollars (US$ 2008) (prior to 1984 first New Chemical Entity (NCE) approvals).

Risk is enormous in the pharmaceutical industry. There is only a 2% probability of success (POS) that a project will go all the way through discovery and development (*19*). Discovery and development times are always long. It takes on an average about 14 years for a project to go all the way through discovery and development (*20*).

7.1.1.3 *Revenue opportunities*

On the upside are a number of opportunities to increase revenue:

Expanding Markets – Brazil, Russia, India and China Countries. Beyond the industrialized economies that have long been the main market for pharmaceuticals, there are a number of countries whose economies are rapidly expanding, such as BRIC. The market for pharmaceuticals in these countries has been estimated to be over $130 billion, collectively larger than Japan, and growing rapidly (*21*). China alone is predicted to become the second largest pharmaceutical market by 2016. Many drugs have yet to be marketed in these countries.

Biologics. Biologics present new opportunities in challenging therapy areas (TAs). As discussed in Chapter 5, biologics are on the rise as an important contributor to the biopharmaceutical pharmacopeia, such that 43% of the drugs approved in these TAs from 1998 to 2008 were biologics (*22*). With technological advances in the discovery and development of biologics, 80% (most!) of the TAs are now amenable to either strategy (*22,23*). The current treatments for type 2 diabetes discussed in Chapter 6 illustrate this point.

New Approaches – Epigenetics and Gene Therapy. Science marches on! New opportunities present themselves over time that were either unheard of or too speculative to consider in earlier eras. Two areas that are receiving a lot of attention currently are the areas of epigenetics and RNA interference (RNAi).

Epigenetics is the study of factors that modulate genetic transcription without changing the DNA or RNA of a cell. This study has led to the identification of at least two mechanisms which can be modified by drugs – DNA methylation and histone modification. Drugs that involve these mechanisms have shown efficacy in cancer (*24*). Other indications are likely to emerge as well.

RNAi. It has long been known that short strands of RNA can bind to native RNA and interfere with genetic transcription (*25*). It has taken a long time to figure out how to create drugs that mimic RNAi or to develop formulations that allow for short strands of interfering RNA to get through the cellular and nuclear walls and deliver the interfering RNA into the nucleus. Such a formulation recently demonstrated success in human patients with solid tumour (*26*). A truly revolutionary use for RNAi is to block the transcription of abnormal proteins arising from abnormal genes.

It is too early to determine the full extent of indications that will come from drugs with mechanisms involving epigenetics or gene therapy, but the potential shines bright for drug discovery in these areas.

The other major area where biopharmaceutical companies can modulate their revenue challenges is through restructuring the way in which they conduct R&D. Before we discuss these changes we need to understand the organization of biopharmaceutical R&D. Then the various experiments that are under way to deal with the enormous risk and cost of biopharmaceutical R&D will be considered.

7.1.2 **The science of biopharmaceutical R&D**

Biopharmaceutical R&D is all about science – whether it's the discovery of a new protein implicated in tumour growth or it's the discovery of a genetic biomarker implicating a better outcome for patients. Different kinds of science are involved in each of these discoveries. The nature of the work in biopharmaceutical R&D is highly varied. It may require the hands and observational insight of a highly trained PhD or MD, or a robot that performs repetitive tasks in a highly precise manner. Neither work is less valuable.

There is a common misperception that pharmaceutical companies are about small-molecule drugs and biotechnology companies are about biologics or biomolecular drugs. The term bio-pharmaceutical is often applied to companies that pursue biologics or biomolecular drugs and not small-molecule drugs. As adumbrated above, here the term biopharmaceutical will be used to mean both – small-molecule drugs and biologics or biomolecular drugs. A more precise term may be bio/pharmaceutical, which this author has used previously.

This chapter does not intend to discuss the enormous subject of the science of biopharmaceutical R&D. A good start for that subject is the recent book by Rick Ng (*11*) and other chapters in this book. This chapter will discuss the organization and management of biopharmaceutical R&D with a particular emphasis on discovery. A useful guide on managing the science may be found in the recently revised book by Bamfield (*27*).

7.2 THE ORGANIZATION OF BIOPHARMACEUTICAL R&D

There is a sequence of events leading to the discovery and development of biopharmaceuticals. Until recently, all these events were performed by employees of the large biopharmaceutical companies. As will be discussed subsequently, many of these activities are now performed in a variety of institutions external to the biopharmaceutical company. At the end of this chapter we will consider some of the experiments that companies are conducting on both the organization and the process of bio-pharmaceutical R&D. It will take a long time before the results of the experiments become evident. But, by in large, the sequence of events is essentially the same regardless of the organization of the work.

7.2.1 Biopharmaceutical R&D is drug discovery and drug development

Drug Discovery encompasses all scientific exploration giving rise to the discovery of a clinical candidate. The clinical candidate is a chemical or biological substance which can be produced in large quantity and has been shown to impact a specific disease mechanism in cellular and animal disease models, suggesting therapeutic benefit to patients.

Drug Development encompasses all scientific exploration of the clinical candidate to prove its safety and efficacy in humans and its therapeutic benefit to patients. There is considerable feedback between development and discovery, such that observations in the clinic can lead to the identification of new clinical candidates that take advantage of these observations.

Discovery and development encompass different stages in the R&D process, yet they share common goals to discover and develop drugs that all R&D can support, with minimal expense and with maximum value.

7.2.2 The logical organization of biopharmaceutical R&D

Biopharmaceutical R&D is organized by what happens first and what happens next in a logical order of events (Figure 7.3). Regardless of the type of company or set of partnered companies and service organizations that perform biopharmaceutical R&D, to a large extent the manner in which a drug is

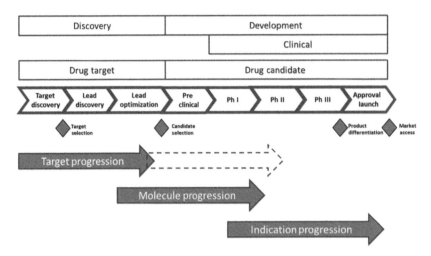

FIGURE 7.3

Logical organization of drug discovery and development. This figure displays eight stages of R&D. Stage names may vary among companies. Different companies may split stages, e.g. in Chapter 3, target discovery is shown as target identification and target validation: Some companies combine lead discovery and lead optimization.

discovered and developed will always be the same. By analogy, it will always make more sense to install the components of a motherboard and then the components that connect to it, before a computer is surrounded with its casing. What may change in biopharmaceutical R&D is the organization that supports the process.

If one considers a disease mechanism to involve a sequence of interactions between proteins and DNA or RNA, the *biological target* is the entity in that disease mechanism which is to be inhibited or augmented through interaction with a drug. At the beginning of drug discovery work is focussed on target progression or target validation, which is the development of a body of evidence that correlates the biological target to a desired potential therapy. For some disease areas, such as antibacterial and antiviral therapy, one only needs to kill the bacterium or virus *in vitro*. For these therapies target validation is achieved early in target discovery. Target validation for other disease areas, such as psychiatric disorders, is only truly achieved when the drug is proven to work in the relevant clinical population. Late discovery and early development is engaged in molecule progression – identifying and developing agents which are proven efficacious and safe in humans. It is only in the clinic where the true medical indication for the biopharmaceutical agent is elaborated.

7.2.3 Stage-gate organization – the project pipeline

The organization and application of work in biopharmaceutical R&D is best managed in a stage-gate organizational structure (*28*). Each stage encompasses a body of work that involves several goals and several disciplines – a work matrix. Each gate (milestone) is an organizational decision to commit resource to the next stage. Figure 7.4 displays eight stages starting with target discovery and ending in registration & launch, as well as typical gates or decision points where progress is assessed (the

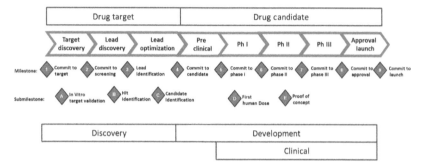

FIGURE 7.4

Stage-Gate (Milestone) Organization of biopharmaceutical R&D – the pipeline.

vertical arrows in Figure 7.4). A ninth stage could be added for life cycle management, which relates to the activities around optimizing the market potential of the drug. Most biopharmaceutical companies find it useful to have one or more committees focus the organization to a decision at each milestone. There are often sub-milestones as well, some of which are shown in Figure 7.4. The goals of each stage are discussed in Section 7.2.4.

In drug discovery the focus is on the drug target, whereas in drug development the focus shifts to the drug candidate identified in lead optimization.

7.2.3.1 The project pipeline

A biopharmaceutical company will work to discover and develop more than one drug at a time. Each drug discovery or development effort is referred to as a *project*. The set of projects is referred to as the *portfolio*. All projects in the company are expected to go through all these stages of work. When viewed by the various stages of work, the portfolio of projects is called the *pipeline*, as outlined in Figure 7.4. A subset of projects in the portfolio may be called a programme, e.g. the cancer programme.

7.2.3.2 Not all projects start at the first stage

There can be many situations where a particular company finds it can start a project at some stage other than the first stage of target discovery.

In-Licencing. If a company or nonprofit laboratory has developed a particular target (and any related small molecules or biologicals) through one or more stages and publishes the information that led to project progression or offers its work for sale, another company may be enticed to initiate in-house work at that point in the pipeline (Section 7.2.8).

Mergers and Acquisitions. Upon a merger or acquisition the assets of two companies are joined. To the new company at least some of the projects will seem to have come into the middle of the pipeline. Of course, these projects began in a predecessor company.

Paralogue Synergy. Targets often bear genetic similarities to other biological targets, such targets are called *paralogues*. Thus work on Target A1 could allow a company to readily initiate work on Target A2, which is its genetic sibling, starting at or near the same stage as Target A1, depending on how readily transferable the work is from Target A1 to Target A2. It is often the case that the project team working on Target A1 will develop selectivity assays to ensure that they develop agents that act on Target A1 and not Target A2. In the process they may discover agents that are more efficacious against Target A2 than A1. Such agents would be shelved by the A1 project team, but should be made available to a subsequent A2 project team that forms around Target A2 should a relevant potential disease indication be envisioned. This synergy within a company is particularly useful.

7.2.4 Stage-related goals

Different work is performed in each stage of discovery and development. There are overarching goals for each stage, which must be met to progress to the next stage. Some work may be deferred to later stages, e.g. see Section 7.5.3.2. Below are the typical goals of each stage (see also the discussion in Chapter 3).

In *target discovery*, the project team strives to improve (1) target validation: define and execute a set of experiments that establishes the target in relation to proposed therapeutic intervention and (2) target tractability: express and purify the target protein; ensure that suitable quantities can be obtained from a stable cell line for future assay work; determine suitability of native protein for assay development; and identify tool agents to assist in assay development.

In *lead discovery (target to lead)*, the processes that identify small-molecule leads are considerably different from those that identify biomolecular leads (see Chapter 5 on Small Molecules vs. Biomolecules).

For small molecules, the project team strives to (1) develop an *in vitro* assay with a tool agent that will identify other agents which interact with the target through the desired mechanism of action; (2) create variants of the target and assay amenable for screening preferentially in high-throughput mode; and (3) identify first agents (hits) through screening. The various screening strategies have been discussed elsewhere in this book.

Once potential hits have been identified medicinal chemists get involved in determining which families of hits are sufficiently druggable to justify chemical exploration of the series to identify lead series for lead optimization. See following chapters 8, 3, 15 and Refs (*11,29*).

Lead Discovery for Biologics. The processes that identify biomolecular leads are considerably different and vary according to the type of macromolecule. In general, this stage is considerably shorter for biomolecules. It is beyond the scope of this discussion to delve into the process of biomolecular discovery. The reader may wish to refer to Refs (*30–32*).

The mode of action of a biomolecular drug can differ considerably from the mode of action of a small-molecule drug. A small molecule invariably is designed to interact with a biomolecular target, as is the case with biomolecular drugs, but the pharmacology of biomolecular drugs is considerably more diverse. In the classical lock-and-key model, the biomolecular drug could be the key, as in hormone replacement therapy, it could bind to the key (ligand binding) or it could be an extracellular portion of the receptor, as in decoy therapy, and it can even have enzymatic activity, e.g. Estrogen Replacement Therapy (ERT) (Chapter 11) and tissue plasminogen activator. A company's portfolio nomenclature system needs to account for these diverse possibilities.

Small molecules often mimic a biomolecule that is the natural ligand for the target of interest, at least in the manner of binding to the target. There are a number of marketed drugs that are natural ligands for targets (*33*).

In *Lead Optimization (Lead to Candidate)*, the project team strives to (1) achieve desired levels of potency, efficacy, selectivity and pharmacokinetics; (2) prepare analogues (variants) of the initial leads and assay for potency and selectivity, developing a structure–activity relationship to catalyse the design of future analogues with enhanced properties; (3) evaluate better analogues in animal disease models and (4) narrow down to a single drug candidate to recommend for clinical trials. Both the small-molecule project team and the biomolecular project team perform similar activities in this stage, with respect to generating and evaluating variants of the lead but the actual goals, strategies and processes vary considerably.

In *Preclinical Evaluation (Candidate to Phase 1)*, the project team strives to (1) perform more animal toxicity studies, importantly by the route of administration intended for clinical trials; (2) prepare sufficient quantity of drug substance for at least Phase 1 clinical trial and (3) formulate the drug for the route of administration intended (a) for clinical trials and (b) for the final product (or begin investigations).

In *proof of concept (POC) (Phases 1 and 2)*, the project team strives to (1) in Phase 1 – determine safety in normal healthy volunteers, starting with minimal dose, escalating to desired dose and (2) in Phase 2 – evaluate drug in patient population and determine efficacy (POC). Increasingly, patients are included in Phase 1 to seek POC at an early stage. Anti-infective agents achieve POC in Phase 1.

In *full development (Phases 3 and 4)*, the drug candidate enters Phase 3 where it is fully evaluated in the patient population to determine final approvable indications and conduct any final trials required for approval by regulatory agencies. Thereafter Phase 4 clinical trials are conducted that might expand the indications of the drug. Phase 4 trials occur after registration and launch.

While these are the goals for each of these stages of development, the work that must be accomplished for each project may vary significantly, depending upon the TA, and corporate strategy. Some work may be deferred to later stages, e.g. see Section 7.5.3.2.

In some companies, the TAs manage a project up to POC and other business units take over the asset from there as product development.

In *Registration and Launch (Reimbursement)*, the project team gears up to (1) register the drug with the regulatory agencies for approval to market; (2) get approval for reimbursement from third-party providers and (3) prepare for launch. Active management of the postlaunch period is critical to maximize revenues and minimize expenses. Many of the line departments that were involved in the development of the drug and even some of the departments involved in discovery may get involved in postlaunch activities.

7.2.5 Attrition

Everyone is most interested in the projects that succeed. The challenge is that most projects in biopharmaceutical R&D do not succeed. Figure 7.5 outlines the high attrition in the industry.

The FDA determined that a 10% improvement in compound attrition would save $100 million per drug (*35*). The reasons for attrition in discovery have yet to be tabulated across the industry. There have been studies on attrition in development however. It was shown in 1991 that 60% of attrition in Development arises from poor pharmacokinetics and safety (human and animal) (Figure 7.6) (*36*). These are issues that arise in Development but can be prevented prior to clinical trials. By 2000 PK issues had dropped below 10% (*36*), but 45% of attrition still arose from issues that are preventable prior to clinical trials.

Each drug that reaches the market bears the discovery and development costs of all the other project failures. The true cost of a drug reaching the market has been estimated to be US$ 0.8–1.0 billion and is likely higher today (*37,38*).

Analyses of attrition in discovery likely occur within individual biopharmaceutical companies and within private consortiums of biopharmaceutical companies. It is unfortunate that industry-wide data

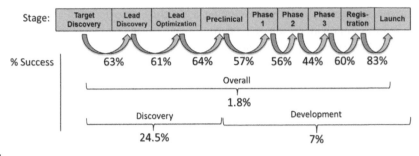

FIGURE 7.5

Attrition across the pipeline (*19,34*).

Data from leading pharma (19). Gilbert et al. show essentially the same level of pipeline attrition (34).

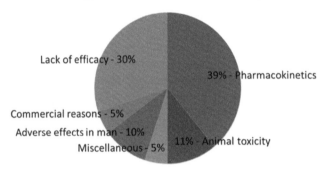

FIGURE 7.6

Reasons for project failure in the clinic.

Data from Ref. (36).

on attrition in discovery is not available to the public. Attrition cannot be reduced unless it is tracked and acted upon.

7.2.5.1 The Attrition-based pipeline

If a company wants its internal R&D to supply new drugs to market, year-on-year, it must account for attrition in its pipeline. To adequately account for attrition, the R&D organization should, in principle, make sure that each stage is at least larger than the subsequent stage by the percentage of attrition that typically occurs at each transition. Figure 7.7 shows a hypothetical pipeline, using industry attrition rates, that would provide one launch in a given year.

In this scenario, the discovery pipeline would need to be considerably larger than the development pipeline to keep the development pipeline filled. In fact, 78% of the projects in the entire pipeline would need to be in discovery.

Unfortunately, the pipeline is not scalable – if four launches are needed due to future patent expiration of marketed blockbusters, a company with this hypothetical pipeline would need to begin with 228 projects in target discovery (by simple multiplication). One would wonder if the company would have the capacity for such an increase in the pipeline. Historically acquisitions made up the difference, but reducing attrition would have dramatic impact.

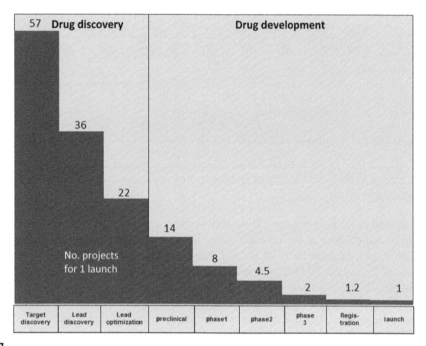

FIGURE 7.7

Hypothetical pipeline based on industry attrition and one launch per year, based on the data in Figure 7.5.

One could call the attrition-based pipeline a sustainable pipeline. As we have discussed in The Sustainable Pipeline Myth, the big biopharmaceutical companies seem to have abandoned the sustainable pipeline some time ago, and fill the gap with in-licencing (*39*).

7.2.6 Risk – how it influences POS, cost, value and commitment

The fact that most projects fail in the biopharmaceutical industry means that risk, the probability that a project will fail, influences a number of key behaviours in biopharmaceutical R&D.

Risk in Discovery and Development. At the beginning stages of discovery, there is a scant evidence that a proposed therapy might work. Those blockbusters that command billions of dollars in sales today were in with the pack of wild ideas at the beginning of drug discovery. Risk that the proposed therapy might not come to fruition is so pervasive in drug discovery, it is often not discussed as such – it's seen as 'work'. At the beginning of discovery the risk of failure in the clinic is part of the aggregate risk that the new project faces, even though work to reduce the clinical risk generally does not occur until a candidate is identified. Every experiment or trial performed along the way addresses a single component of the aggregate project risk. In development (especially late stage), there is a mounting body of evidence that the proposed therapy might succeed. Risk is less obvious, so it needs special formal attention. However, it can be tightly managed. Thus as work is performed in a project to remove risk, the POS understandably increases, see Figure 7.8.

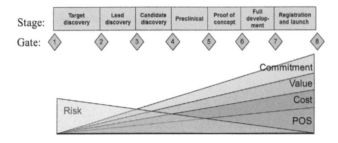

FIGURE 7.8

The relationship between risk, POS, cost, value and commitment as a project progresses through the stages of discovery and development.

Value in Discovery vs. Development. Discovery is the start of the value chain, but given the low POS, quantitative estimations of the potential value of the project are meaningless. In development (especially late stage), because most of the risk has been dispensed and is more readily defined, determinations of Estimated Present Net Value (EPNV) are important. The potential value of the discovery portfolio, however, is not ignored, and can be dealt with in a qualitative manner.

Cost, Value and Commitment in Discovery and Development. As work progresses on a project in discovery and development, the overall cost of the project understandably increases. But as POS increases, the potential value of the project also increases and consequently corporate commitment to the project increases. In other sections we will see how these factors influence key behaviours within the organization of discovery and development.

7.2.7 Resource – who does what, when and where

Resource, the modern euphemism for employees, is a huge part of the organization of bio-pharmaceutical R&D. Who does what and when is the constant part of resource, dictated by the logical organization of biopharmaceutical R&D. Where the work gets done varies widely between companies.

7.2.7.1 Association of resource with stage – matrix of work

At each stage of discovery and development, different disciplines work at different and often multiple stages (Table 7.2). These disciplines are organized into *line departments*, some of which work in specific stages, while other disciplines are needed at all stages of R&D – human resources, finance, corporate intellectual property (IP), licensing (Section 7.2.3.2 (In-Licencing)), alliance management, and marketing (In discovery, marketing influences the selection of biological targets through the desired product profile). More will be discussed about the application of resource in later sections.

7.2.7.2 Association of resource with business units – overarching organization

The line departments that perform stage-specific work on projects are organized into various business units that have their own budgets and operate relatively autonomously of other business units.

Typically R&D is divided into discovery and development, as shown in Figure 7.3. These can be individual divisions each with a head that reports to the head of R&D.

Occasionally a company may acquire a small company that focusses on a particular disease area, or class of biological target, or platform technology and keep the companies as solely owned subsidiaries.

Table 7.2 Application of Disciplines to Multiple Stages of Discovery and Development. The Table Exemplifies the Types Disciplines that may be Needed in a Project at Different Stages and is Neither Complete nor Correct for Any Particular Company. Job Titles may Vary for the Same Discipline Within the Same Company and between Companies

Resource Level: Less · More

Discipline	Task	Discovery				Development		
		Target discovery	Lead discovery	Candidate discovery	preclinical evaluation	Proof of concept	Full development	Registration & launch
Cell biology	Gene cloning	Yes	Yes					
Cell biology	Gene expression	Yes	Yes			Yes	Yes	
Cell biology	Protein synthesis	Yes	Yes					
Cell biology	Protein purification	Yes	Yes					
Cell biology	(Polyclonal/monoclonal) antibody generation and purification	Yes	Yes					
Biology	In vitro assay development		Yes	Yes				
Biology	Cell or tissue preparation		Yes	Yes				
Biology	In vitro screen performance		Yes	Yes				
Biology	In vivo assay development		Yes	Yes				
Biology	Animal preparation and care			Yes	Yes			
Biology	In vivo screen performance			Yes				
Biology	Screen analysis		Yes	Yes				
Mathematics	Screen performance		Yes	Yes	Yes			
Mathematics	Screen analysis		Yes	Yes	Yes			
Mathematics	Assay statistics		Yes	Yes				
Mathematics	Clinical statistics					Yes	Yes	
Chemistry	Tool compound design	Yes	Yes	Yes	Yes			
Chemistry	Tool compound synthesis	Yes	Yes	Yes	Yes			
Chemistry	Analog design		Yes	Yes				
Chemistry	Analog synthesis		Yes	Yes				
Chemistry	Bulk synthesis			Yes	Yes	Yes	Yes	Yes
Chemistry	Process optimization				Yes	Yes	Yes	Yes
Medical & clinical care	Clinical trial execution					Yes	Yes	
Finance and marketing	Product development			Yes	Yes	Yes	Yes	Yes

Some recent examples within the GlaxoSmithKline (GSK) Empire include the following companies which became autonomous business units upon acquisition: Stiefel Laboratories, a business unit focussed on dermatology; Sirtris, a business unit focussed on sirtuins and Domantis, a business unit focussed on a unique type of protein that mimics antibody structure.

7.2.7.2.1 Therapy areas

The bulk of the organization in biopharmaceutical R&D tends to be organized under TAs. TAs are clusters of diseases for which the company is pursuing therapies, e.g. allergy & respiratory; cardiovascular and metabolic diseases; inflammation; infectious diseases; gastrointestinal; neuroscience (and pain); oncology; and ophthalmology. Each TA has budgetary and resource control over the majority of resources within each area.

The resource within each TA is typically organized into line departments based on areas of expertise, e.g. medicinal chemistry, biology, drug metabolism/pharmacokinetics, formulations, toxicology, clinical, etc. Most R&D organizations establish a 3–5 year business plan for each TA, containing the long-term goals of the TA (40). At the end of that period the organization may make organizational adjustments that could include re-clustering of diseases in the TAs, and initiating or eliminating work on certain diseases. At least once a year each TA reviews its progress towards its annual goals and towards the completion of its business plan. This review is often referred to as the portfolio review. Typically, in that review the budget for the next year is established.

7.2.7.2.2 Nonaligned resource – platform technologies

For some areas of expertise it may be more economical or efficient to avoid the duplication of resource across TAs and business units. Such areas of expertise may be organized as service organizations to the TAs.

Other areas of expertise may be focussed on performing all the work in specific stages, e.g. the departments needed to perform lead discovery and preclinical evaluation may be organized into individual organizations that have responsibilities for managing projects in those stages.

These tend to include but are not limited to (1) biology and biochemistry staff engaged in screening activities and protein crystallography, (2) chemistry staff focussed on early lead discovery, analytical chemistry, computational chemistry and molecular modelling, and bulk preparation of clinical trial materials, (3) cell biology staff engaged in genetic/genomic disease–target association and biomarker activities, (4) biology and biochemistry staff engaged in preclinical and clinical drug metabolism studies, pharmacokinetics, and toxicology and (5) finance, legal, alliance management, information management, IT support, and operations management. Inevitably, however, such line departments will need representatives to each TA or dedicate certain staff to each TA. There will always be tension between the TA's and nonaligned resource.

7.2.8 In-house vs. in-license

Big Pharma have had the capability to do it all in-house. Attractive biological targets or clinical candidates may be in-licenced at any stage (Figure 7.9). In-licencing costs will vary and are likely to be higher in later stages.

Deloitte and Thomson Reuters recently published a study of the internal rate of return for the top 12 biopharmaceutical companies (41). Within their data set was a chart showing the origin of late-stage

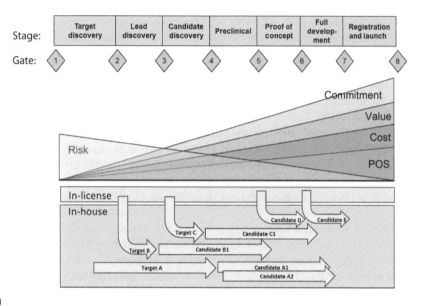

FIGURE 7.9

In-licencing can occur at any stage.

projects. The percentage of 'self-originated' (not acquired) projects in the late-stage pipelines of these companies ranged from 0% to 30% – with an average of only 12%.

Dimasi et al. found that acquired assets have higher success rates in Phases 1 and 2 than self-originated assets (*42*). Thus, when determining corporate performance, it may be better to separate projects that originated via in-licencing from those that began in target discovery.

7.2.9 In-house vs. out-license

Big Pharma have the resources to take a project all the way from target discovery through to product launch. There are several circumstances in which it may be attractive to out-license the project to keep developing value.

Insufficient Funds. No corporate budget is so large that a company can do everything. Every company no matter how large has to make choices, prioritize, and possibly put aside some assets at almost any stage of discovery or development. It is typically assumed that the assets that 'don't make the cut' in a portfolio funding exercise have issues that make them less attractive than those that do make the cut. There are circumstances, however, where valuable assets do not make the cut: (1) the company has to decide between two similar assets, both of which are similar in value and (2) the company has made a strategic determination not to pursue certain kinds of assets (e.g. a few years ago it was anti-infective agents and cardiovascular agents, more recently it has been neurological targets). Out-licencing could potentially save and enhance the value that has been created in such assets.

An interesting out-licencing example is that of Tadalifil, marketed by Lilly as Cialis in the US. As with Viagra, the former GlaxoWellcome (GW), in partnership with ICOS, found some unique properties in its phophodiesterase V (PDE-V) inhibitor during clinical trials as an antihypertensive agent. The GW management chose to not develop erectile dysfunction agents, and terminated the agreement with ICOS, who went off and formed an alliance with Lilly to develop the drug (*43*). As the market for such agents proved to be enormous, years later the subsequent GSK agreed to codevelop Vardenafil, the Bayer PDE-5 inhibitor, and marketed it as Levitra. An interesting exercise would be to calculate how much value was captured by GSK in this class of drugs compared to Lilly and Pfizer.

Repurposing. As a potential drug moves through the pipeline, the set of indications initially envisioned for the drug is ultimately narrowed to one or two indications by the time the drug is proposed to regulators. Often the first proposed indication is the one most likely to gain approval, with follow-on studies planned to expand the number of indications for the drug later on. Even when more than one indication is ultimately approved for a drug, the set of indications tends to fall within the same TA.

While the company may find it attractive to go after disease indications that fall into other TAs, it may be more attractive to out-license the drug to another company to explore such alternate indications. Such activities may be seen as 'squeezing water from a stone' within the parent company, especially if the market for the alternate indications is smaller than had been experienced with the initial indications. But those smaller markets may be perceived as a fresh opportunity for another company.

Repurposing can include totally serendipitous discoveries that had not been envisioned for the drug, e.g. the aforementioned realignment of PDE-V inhibitors from hypertension to erectile dysfunction. The discovery that thalidomide works in leprosy is another classic example (*44*). The Pfizer PDE-V example is a case where the repurposing occurred within the parent company, and the GW PDE-V example is a case where changing partners played a role. Chong and Sullivan argue that the screening of old drugs for new purposes ought to occur via a public collection of all 9900 drugs approved since 1938 (*45*).

Small Start-up Companies. Of course, small start-up companies typically lack the resources to take a project all the way from target discovery through to product launch. If the proposed indication falls into a niche market where larger pharma lack the marketing expertise, it may behove the start-up to build in the resources to take the project to launch. But most often the small start-up will see out-licencing as a huge win, even if it means dissolution of the company to keep the asset moving, as most of the employees will benefit from the sale of the company. Larger companies will have a life after out-licencing and will use the revenues from sale of an asset to keep other assets moving through the pipeline.

7.2.10 Big vs. small

Large and small biopharmaceutical companies are typically characterized as polar opposites. The large companies are characterized as sluggish behemoths and the small companies as agile but unstable. Medium size may be better in many respects, but are presently considered to be (1) unable to survive late-stage failures of important development projects or (2) overly exposed to patent expiration of key products.

The Tendency to Get Big. Biopharmaceutical companies tend to get larger for two reasons. Success in the clinic tends to bring considerable new revenue to the company. The smaller companies tend to invest the new revenue back into the company either through increasing the spend on R&D or through acquisitions. The larger companies use acquisitions to fill gaps in the pipeline or in corporate strategy and bring new revenues into the company, and reduced the impact of patent expirations (*46*). Since 1988 mergers and acquisitions have been responsible for the number of biopharmaceutical companies in the Pharmaceutical Research and Manufacturers of America (PHRMA) to shrink from 42 to 11 (*46*). Many consider mergers to be a zero sum game that enrich the banks and lawyers that participate in the merger activities but from an industry perspective reduce competition and jobs (*47,48*).

7.2.11 The necessity of standard operating procedures

A certain amount of bureaucracy is necessary in a biopharmaceutical company, especially if it intends to market a product in the Western world. Governmental regulatory authorities approve and oversee the manufacture and marketing of biopharmaceuticals through a number of approval processes and standard operating procedures (SOPs). A licenced company is expected to follow the SOPs agreed to at the time of approval. The set of SOPs that regulate biopharmaceutical companies are given the acronym GXPs, because of the variations – there are standard clinical procedures in Good Clinical Practices, standard manufacturing practices in good manufacturing practices (GMPs), and standard analytical procedures in Good Laboratory Practices (GLPs).

While the structure of SOPs is necessarily embedded in a biopharmaceutical company, there is a tendency to develop SOPs for nonregulated procedures to enhance efficiency of corporate approval processes that may also increase the bureaucratic burden of the company. The need for SOPs tends to be proportional to the size of corporate structure and the size of business units within the corporate structure.

SOPs are intended to streamline standardized approval processes with the assumption that if the SOP is followed the regulatory or authorizing body does not need to be informed. The SOP presumably provides a safe, low-risk, reproducible procedure that if followed exactly will ensure successful completion of the task (e.g. certain chemical or biological processes). Under SOP-regulated processes the regulatory or authorizing body needs only to be informed if conditions prevent the SOP from being followed. Under those circumstances the SOP may be revised or approval for a temporary variance from the SOP may be sought.

An important aspect of the approval process is the element of presumed risk inherent in the process being approved. If the risk is low, verbal approval is sufficient. If the risk is high a standardized approval process with subsequent SOPs may help to mitigate risk. Most daily tasks are low risk. Some daily tasks that are high risk (e.g. work with dangerous chemicals or microorganisms) may need SOPs to ensure safe and successful completion.

Needless to say, but importantly, there is also a factor of decision-making speed that is important in this continuum of approval processes. Little to no progress would be made if all decisions required formal approval by a senior manager. The word bureaucracy tends to be applied to systems that are inordinately slowed by the decision-making process.

The nonstandard processes tend to occur in the discovery and preclinical stages. Even then, such work is likely performed with the awareness of a need for creating a GMP or GLP standard process in the clinical stages and is therefore likely to be well documented. Clear documentation in the discovery phase is also very important in case, perhaps many years later, a patent dispute requires clear identification of certain timelines (for example the first preparation of a particular compound).

7.3 PROJECT MANAGEMENT IN BIOPHARMACEUTICAL R&D

Programs and projects are the life blood of the company. They must be managed properly. A *project* is a way of organizing resource. It is a group of individuals who are assembled to perform different tasks on a common set of objectives for a defined period of time (*49*). Projects need a leader who can define the work objectives and criteria for success and recruit staff from all relevant areas of expertise. The need to organize a project is most apparent when more than two departments contribute resource at the same time.

Oversight/management of the matrix of work, on a particular asset at each stage in R&D is performed by a project leader. Rarely does a project leader manage the project through all stages of discovery and development. Keeping on a project leader into the next stage requires familiarity with the nature of work and decision making, which varies dramatically at each stage. Familiarity with relevant line department leaders is useful. There will be a greater tendency to allow project leaders to stay on in smaller companies where such familiarity is possible. Professional project leaders, if given sufficiently empowered staff and authority, could in principle manage a project through all stages.

The level of professionalism in project management tends to be stage-related. In the stages of drug discovery, where, as noted earlier, risk is high and potential value is low, project leaders tend to be part-time volunteers that may lack the requisite skill set to manage a project. Late-stage development projects, where most of the risk has been dispensed and actual value is determinable, tend to be run by certified professional project managers (*50*). Senior managers must ensure each project has an adequately trained and fully empowered project leader.

7.4 PORTFOLIO MANAGEMENT IN BIOPHARMACEUTICAL R&D

The Project Management Institute maintains The Standard for Portfolio Management (e.g. second edition, 2008) which is a trans-industry guide. The book lists all executive functions that impact the corporate project portfolio which include the following functions (*51*).

- Categorize and provide portfolio views of projects
- Evaluate project progress against milestone criteria
- Select and approve projects for inclusion in the active portfolio
- Approve projects for advancement into the next stage
- Select and approve projects for removal from the active portfolio
- Identify and analyse portfolio risks
- Develop and monitor portfolio risk responses
- Prioritize projects
- Balance or align the project portfolio to current prioritization and strategy
- Develop and execute portfolio adjustment action plan
- Review and report portfolio performance with respect to prioritization and current strategy
- Review and report portfolio resource alignment with prioritization and strategy

Portfolio management is what senior management makes of it. If one or more persons in a company manage these functions, then the company practices portfolio management. The persons performing these tasks may not have a portfolio manager job title or work in a department of portfolio management. There may be duplicate positions based on stage of discovery or development. Some of these persons may be located in finance, corporate strategic planning, resource management, or report directly to the head of R&D. Some of these functions may have been delegated to managers of line departments. It is important that the individuals who perform these tasks communicate with each other. If no one in the group of managers can say that his or her group performs most of these functions, there could be alignment problems with portfolio management in the company (*52*).

7.5 COST REDUCTION EXPERIMENTS IN THE BUSINESS OF BIOPHARMACEUTICAL DISCOVERY AND DEVELOPMENT

At the beginning of this chapter, challenges to the business of biopharmaceutical discovery and development were outlined. Critics of the industry have been sounding the alarm for years. Many of the large biopharmaceutical companies are now working on ways to remedy the situation. We call them experiments because there are risks associated with all the approaches being taken to reduce cost, reduce risk and increase success and the outcomes have yet to be demonstrated.

7.5.1 Process improvement

Within the line departments, removing inefficiencies can help reduce cost. The principles of Lean Six Sigma (*53*) can be applied to many of the processes in drug discovery (*54*). While it is important that at least one staff member is trained in the principles, that one individual can foster the creation and mentoring of process improvement project teams throughout the company.

Process improvements are attractive because they can be measured in weeks and months. More challenging experiments are the organizational changes that will take years to come to fruition.

7.5.2 Reducing resource – vertical disintegration

7.5.2.1 The need to downsize

In previous decades biopharmaceutical companies were sufficiently profitable that the line departments in R&D could staff to over-capacity. That way they could handle fluctuations in demand. Full-time employees, however, not only get paid even in periods of low demand but the company also pays for sick time, vacations, medical expenses, and contribute to their pensions. As the years go by most employees get salary increases and bonuses. In terms of corporate finance, employees are an ever-increasing 'fixed cost'. In previous decades, employees, especially those with many years of service, were considered to be the life blood of R&D because the wild increases in revenue that this fixed cost created were greatly appreciated. But between 2001 and 2010 as profits declined and new revenues were harder to come by, the fixed cost became more of a burden.

7.5.2.2 Vertical integration vs. vertical disintegration (Decentralization)

The resource for each discipline in biopharmaceutical R&D may reside within a single company or several-to-many companies. In-house location of resource simplifies coordination, fosters synergy, but increases fixed cost. Historically, Big Pharma tended to execute all stages of R&D in-house. Such an organization engages in vertical integration (VI). VI is the process of performing all R&D, manufacturing and distribution within a particular company.

As just noted, as fixed cost becomes a burden to big pharma outsourcing becomes attractive.

Vertical disintegration is the move towards performing many to most of these processes in other companies (55). Thus vertical disintegration is another way of describing outsourcing, or externalization or decentralization. As will be described later in this section, if most of the functions of R&D are externalized, then the company practices virtual discovery and development (Figure 7.10).

A number of industries have gone through the transformation – e.g. automobiles, aerospace, Information Technology (IT) and the motion picture industry. The Wikipedia notes that one major reason for vertical disintegration is to share risk with other companies. Noted Andrew Witty (GSK CEO), 'Externalising R&D enables GSK to capture scientific diversity and balance expenditure and risk in drug development. In the future, we believe that up to 50% of GSK's drug discovery could be sourced from outside the company' (56).

7.5.2.2.1 Outsourcing and partnering – what and why

Outsourcing is the external allocation of resource through for-profit contract research organizations (CROs) and nonprofit laboratories funded by academia or philanthropic organizations. Outsourcing usually involves the contracting out of a business function – commonly one previously performed in-house – to an external provider. The outsourced work is usually governed by a contract where the parent company defines the work to be done in detail. Some of the potential benefits to outsourcing can include cost savings, access to greater expertise and improved quality, although the realization of these benefits needs to be demonstrated.

A partnership is usually broader in scope than outsourced work. With a partnership the partner owns the assets it creates but receives funding from a company to help the partner conduct the work and to give the company first right of refusal to the acquisition of assets once the work is completed. Some of the potential benefits to partnering can include access to talent, enhanced capacity for innovation (a presumed talent of the partner) and risk reduction (the partner absorbs the risk).

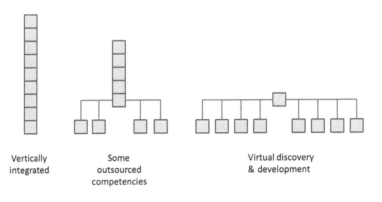

Vertically integrated

Some outsourced competencies

Virtual discovery & development

FIGURE 7.10

From a vertically integrated company to one that practices virtual discovery & development. Each box represents an area of expertise.

Theoretically any expertise may be outsourcing and any strategic area may be partnered. One could argue that certain functions must remain in-house to keep the company intact, such as operations, finance, marketing, legal, project management, and portfolio management. Besides these functions, the company must identify so-called core competencies, which are essential to defining the business strategy.

Particular in-house expertise may be lacking in any company, big or small. Expertise, with respect to knowledge, experience, talent, flexibility, and ability to handle all work, is roughly proportional to staff size – knowledge, experience, talent, flexibility, and ability to handle all work. Outsourcing and partnering can augment in-house expertise. An important consideration in weighing which areas of expertise to outsource is to identify those areas where close collaboration is required.

Since 2000, the pharmaceutical industry has cut almost 300,000 jobs, about as many people as currently work at the three largest drug makers – Pfizer, Merck, and GSK combined (*57*). These layoffs tend to be attributed to actual or anticipated decreases in earnings. Pfizer and Merck also had to contend with their respective takeovers of Wyeth and Schering-Plough.

Coincident with these layoffs is an activity that explains how the remaining staff has dealt with the loss of headcount – increased utilization of CROs, collaborations with academia, small start-ups, and even with peer pharma, through what has come to be known as precompetitive research. Precompetitive research may be defined as noncompetitive, cooperative research which leads the way to competitive development in the future by addressing key requirements of a new technology. An example is the recent agreement between Lilly, Merck, and Pfizer to share genomics data on lung and gastric cancers in Asia (*58*).

This externalization of effort has reached 50% of the R&D budget at GSK (*46,59*). If the GSK model is followed by all big pharma, we can expect a continuation of these dramatic layoffs well into the new decade. We can also expect further externalization of R&D spend by big pharma.

When the extent of externalization reaches 50%, it is safe to assume that all line departments are forced into a position of rationalizing the in-house need for every skill set within R&D. Any competency that can be found outside the company by contract or collaboration will be removed from the company.

Dixon, Lawton and Machin argue that 'insufficient overall investment is available today to support the growth of new vertically integrated companies from a research base, as occurred in past decades' and that 'the process of creating and selling a drug can no longer be optimally accommodated within one vertically integrated business, and that vertical disintegration into "discovery", "development" and "marketing and sales" businesses is needed' (*60*). Clearly, the business factors that favoured VI are no longer operating in the biopharmaceutical industry.

Another challenge with VI comes with the need to invest in new technologies. When a company desires to embrace a new technology it requires an investment in new equipment and staff. The company may need to sell off equipment and let go of staff involved with old technology to afford the new technology. If a particular technology is thriving in company A and company B cannot afford to invest in the technology to the extent of company A, company B would be better off partnering with company A rather than trying to duplicate the services of A within B. In an era of booming technology, a company that partners with new technology firms may be able to harness more technology than a company that feels compelled to bring it in-house.

VI demands a large internal budget, which is composed largely of fixed costs – laboratories and staff. Company loyalty was important with VI, since the staff was viewed as irreplaceable talent. Staff that stays within a company for decades becomes more expensive as salaries increase, and as medical expenses increase. Big pharma have long been accustomed to these expenses. The recent downturn in profitability (only 0.7% profit growth in 2009 over the previous year, compared to 10.4% in 2008, 15.9% in 2007, 8.1% in 2006 and 2005 – among Fortune 500 companies) (*1*) brings a challenge to the assumption that such costs are necessary. Big pharma discovery and development must now argue for its existence as an internal competency.

Outsourcing may lower cost, but can add contract issues (e.g. the need to pay for contracted work even if a project terminates). Big pharma are presently moving to outsource, while small pharma and CROs move to add staff to accommodate more work. CROs tend to specialize in no more than a few stages of R&D.

7.5.2.2.2 Virtual drug discovery and development

A maximally outsourced company is one in which the only competencies remaining are those that keep the company intact, such as operations, finance, marketing, legal, project management, and portfolio management. All the other competencies are outsourced: biology, chemistry, toxicology, clinical sciences, statistics, and analytical sciences. Such a company engages in *virtual drug discovery and development* (VD&D). This type of company may be difficult to distinguish from a CRO, especially those that offer a wide variety of services, such as Quintiles and Covance.

A company that engages in VD&D is one that displays the following characteristics.

- The employees are limited to a small core group responsible for strategic (project & portfolio) management, IP management, electronic data capture, regulatory strategy, and financial control.
- Project teams consist of a project manager that oversees the work of a collection of academic and nonprofit groups and CROs.
- The company maintains control over IP, financing, electronic data capture and data submission to regulatory authorities.

The last characteristic is the one that most readily distinguishes a VD&D company from academic and nonprofit groups and CROs – it holds the reins. It owns the patents. It is the financial conduit through which its collaborators get reimbursed for their services. It makes the final submission to regulatory authorities. Depending on the clout of its financial partners, a VD&D company may be less visible that it's major partners.

A company that engages in virtual discovery in-licences its assets from discovery organizations. A company that engages in virtual development out-licences or partners with the companies that are capable of performing clinical trials.

A small VD&D company may simply be an extension of a venture capital company with office space shared with the Venture Capital (VC). If the company has less than a dozen ongoing projects, it only needs a set of offices to house its set of project managers.

7.5.2.2.3 The virtual pipeline

Today, for the big biopharmaceutical companies, where only 12% of projects are self-originated, the R&D pipeline is a virtual pipeline in which the trajectory of work still starts in target discovery and proceeds through to approval and launch but that thread of work winds its way through a number of companies. This is depicted in Figure 7.11 below in a portfolio of five targets the drugs of which are ultimately marketed by Company 1. In this scenario, Company 1 took only one target through discovery to a clinical candidate and all the way through development to approval and launch. The rest

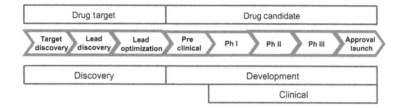

FIGURE 7.11

An example of a portfolio of five targets that are taken through to approval and launch by Company 1. In this example only Target A was worked on in Company 1 through all the stages of R&D. The other four targets were brought through discovery and most or all the stages of development by other companies. The drug targets and their subsequent drug candidates were transitioned to subsequent companies via out-licencings, mergers, acquisitions and partnerships.

of its portfolio came from one or more companies through mergers, acquisitions or in-licencing. With Target E, Company 1 partnered with another company to gain approval and launch in order to co-market the drug from Target E.

In a virtual pipeline, the project team becomes a virtual project team or more accurately a series of project teams that may never interact with each other.

7.5.3 Increasing success through changes in the organizational model

7.5.3.1 Big and small at the same time

Big companies tend to be considered overly bureaucratic and slow to adapt to changes in the market place or to the latest technological advances. At least some of the large pharma are experimenting with reducing the bureaucracy of the large company by being 'big and small at the same time' (*61*). This quote was from Tachi Yamada, who was then head of GSK R&D and managed the reorganization of corporate R&D into Centres of Excellence in Drug Discovery (CEDDs), each staffed by roughly 350 researchers. In 2008 GSK took the concept a step further by splitting the CEDDs into discovery performance units (DPUs) of roughly 60 researchers, to get closer to the agility of a small start-up (*59,62,63*). By 2009 there were 35 DPUs. Each DPU had full control over its own budget which could be spent internally or externally to deliver their best-possible pipeline. The goals in each of these moves were to drive accountability deeper into the organization, reduce central oversight as much as necessary and foster a 'biotech-like' atmosphere in a very large bio-pharmaceutical company. Time will tell whether the DPU experiments works as being big and small at the same time.

7.5.3.2 Planning for failure – POC at chorus (Lilly) and flexion therapeutics

The overall order in which things are done in biopharmaceutical R&D are logically based on what needs to happen before something else can happen. There are some activities, however, which are performed in preclinical investigation and Phase I on the premise that that the project will be successful. As noted earlier, however, the clinical stages of R&D are the most expensive stages in the pipeline. Thus, if a reason for failure can be discovered earlier in the pipeline than in the clinic, then remediating the situation or terminating the project at an earlier stage will save the company a lot of money.

The POC milestone is critical for any project. With that decision the project team now knows it is working with a potential commercial product. Subsequent studies will help to determine the magnitude of its commercial value. For projects where POC is highly questionable, the team may decide to postpone some work that would normally be conducted in Phase I or II to Phase III in order to get to an early decision. Types of deferred work could include defining the best formulation, extended toxicology studies, manufacturing process optimization and pilot studies to identify other indications. Eli Lilly created a separate division, Chorus, to do just that (*64*). It also fostered the creation of Flexion therapeutics which only conducts Phase I and II trials with the goal of getting to POC in a short period of time (29 months vs. 40 months) at lower cost ($4 million vs. $22.8 million). Of course the Phase III trials for projects coming out of Chorus or Flexion may be longer than usual to perform the work that had been deferred from Phase I or II.

Industry estimates of attrition suggest that more projects fail at POC than at any other development milestone. Projects conducted in the manner of Chorus may not get to launch quicker, but failure at POC would have been determined in a shorter time frame with less cost, freeing up resources to work on other projects. And since the successful drugs pay for those that failed before launch the overall cost of bringing a drug to market could be reduced.

7.5.3.3 Open innovation

Prior to the sequencing of the human genome, genetic information that was relevant to biopharmaceutical discovery was considered to be unique and highly patentable. Companies prided themselves on their own genetic research that provided unique insights into disease. Since the sequencing of the human genome once private insights are now of public knowledge. Biopharmaceutical companies have realized that the new insights are more likely to occur outside their walls than inside.

New technologies that significantly augment the multifaceted discovery and development process are becoming available at an ever-increasing rate. Rather than moving a company forward, the purchasing of new equipment may actually prevent the company from accessing even newer technologies in a few years hence. So partnering with a company with a unique technology may be a more cost-effective and agile strategy for keeping up with new technologies.

Open innovation and precompetitive research collaborations (described earlier) are strategies that are now seen as strategies that let biopharmaceutical companies explore new avenues of biomedical research at a fraction of the cost of bringing in the requisite staff and equipment to conduct the research in-house. Hunter and Stephens note that such strategies do not eliminate the possibility of IP protection. They describe IP as the currency of open innovation (*65*).

Of course the price tag for innovation is related to where it impacts the biopharmaceutical pipeline. Advances that give rise to new discoveries in target discovery will not be as valuable as a Phase III drug that is ready for marketing.

7.6 **CONCLUSION**

Managers of biopharmaceutical drug discovery and development face complex challenges in the business of R&D. While many aspects of the way the business of R&D is conducted will change in the coming years it is unlikely that the basic R&D process outlined in Figure 7.3 will change dramatically. As long as the industry can continue to find new therapies that positively impact the lives of patients, it will continue to be an exciting and a challenging industry.

References

1. Fortune. *Global 500 Rankings 2009*. http://money.cnn.com/magazines/fortune/global500/2009/performers/industries/fastgrowers/profit1yr.html.
2. DeNoon, D. J. *Most-Prescribed Drug List Differs from List of Drugs With Biggest Market Share*. http://www.webmd.com/news/20110420/the-10-most-prescribed-drugs.
3. Breazzano, S. *The Patent Cliff: What Big Pharma Investors Need to Know*. http://seekingalpha.com/article/257762-the-patent-cliff-what-big-pharma-investors-need-to-know.
4. Drug Patent Watch Nexium, **2009**. http://drugpatentwatch.com/ultimate/preview/tradename/index.php?query=NEXIUM.
5. UPDATE 1-FDA Extends Exclusivity for Bristol, Sanofi's Plavix, Reuters, January 25, **2011**. http://www.reuters.com/article/2011/01/25/bristolmyers-idUSSGE70O0EE20110125.
6. Wikipedia, Fluticasone/Salmeterol. http://en.wikipedia.org/wiki/Fluticasone/salmeterol.
7. Wikipedia, Aripiprazole. http://en.wikipedia.org/wiki/Aripiprazole.
8. Houlton, S. *Pharma Braces for Patent Cliff Impact, RSC. org*. http://www.rsc.org/chemistryworld/News/2011/December/pharma-braces-patent-cliff-impact.asp.
9. Wikipedia, Crestor. http://en.wikipedia.org/wiki/Crestor.
10. Wiki Analysis, Amgen. http://www.wikinvest.com/stock/Amgen_(AMGN).
11. Ng, R. *Drugs, from Discovery to Approval,* 2nd ed.;.
12. Pharmaceutical Research and Manufacturers of America. *Pharmaceutical Industry Profile 2010;* PHRMA: Washington, DC, 2010.
13. Edwards, J. Off a Cliff: $100 Billion in Revenues Will Disappear from Drug Business by 2013, CBS Money Watch, June 16, **2011**.
14. Hughes, B. Payers Growing Influence on R&D Decision Making. *Nat. Rev. Drug Discov.* **2008,** 7, 2–3.
15. Cohen, F. J. Macro Trends in Pharmaceutical Innovation. *Nat. Rev. Drug Discov.* **2005,** 4, 78–84.
16. Kaitin, K. J. Deconstructing the Drug Development Process. *Clin. Pharmacol. Ther.* **2010,** 87, 356–361.
17. Pharmaceutical Research and Manufacturers of America. *Pharmaceutical Industry Profile 2009;* PHRMA: Washington, DC, 2009.
18. Campbell, N. Mega Mergers – Are they Turning Pharma Companies into Zombies. *Pharma Focus Asia;* 8–14.
19. Brown, D.; Superti-Furga, G. Rediscovering the Sweet Spot in Drug Discovery. *Drug Discov. Today* **2003,** 8, 106–1077.
20. Dimasi, J. A. New Drug Development in the United States from 1963 to 1999. *Clin. Pharmacol. Ther.* **2001,** 69, 286–296.
21. Espicom. *The Market for Pharmaceuticals in Brazil, Russia, India & China*. http://www.espicom.com/prodcat2.nsf/Product_ID_Lookup/00000939?OpenDocument.

22. Trusheim, M. R.; Aitken, M. L.; Berndt, E. R. Characterizing Markets for Biopharmaceutical Innovations: Do Biologics Differ from Small Molecules? *Forum for Health Economics & Policy* **2010**, *13* (1). Article 4.

23. Nelson, Y. A. L.; Dhimolea, E.; Reichert, J. M. Development Trends for Human Monoclonal Antibody Therapeutics. *Nat. Rev. Drug Discov.* **2010**, *9*, 767–774.

24. Yoo, C. B.; Jones, P. A. Epigenetic Therapy of Cancer: Past, Present and Future. *Nat. Rev. Drug Discov.* **2006**, *5*, 37–50.

25. Whitehead, K. A.; Langer, R.; Anderson, D. G. Knocking Down Barriers: Advances in siRNA Delivery. *Nat. Rev. Drug Discov.* **2009**, *8*, 129–138.

26. Crunkhorn, S. RNA Interference: Clinical Gene-Silencing Success. *Nat. Rev. Drug. Discov.* **2010**, *9*, 359.

27. Bamfield, P. *Research and Development in the Chemical and Pharmaceutical Industry*, 3rd ed.; Wiley-VCH Verlag GmbH & Co. KgaA: Weinheim, 2006.

28. Stage Gate International. http://www.stage-gate.com/index.php/.

29. Spilker, B. *Guide to Drug Development, A Comprehensive Review and Assessment;* Wolters Kluwer, Lippincott Williams & Wilkins: New York, 2009.

30. Gill, D. S.; Damle, N. K. Biopharmaceutical Drug Discovery Using Novel Protein Scaffolds. *Curr. Opin. Biotechnol.* **2006**, *17*, 653–658.

31. Spilker, B. *"Biotechnology", Chapter 12 in Guide to Drug Development, A Comprehensive Review and Assessment;* Wolters Kluwer, Lippincott Williams & Wilkins: New York, 2009.

32. Ng, R. *"Drug Discovery: Large Molecule Drugs", Chapter 4 in Drugs, from Discovery to Approval,* 2nd ed.; Wiley-Blackwell, 2009.

33. Ng, R. *"Cytokines" Chapter 4.4, and "Hormones" Chapter 4.5 in Drugs, from Discovery to Approval,* 2nd ed.; Wiley-Blackwell, 2009.

34. Gilbert, J.; Henske, P.; Singh, A. *Rebuilding Big Pharma's Business Model;* Bain & Company. http://www.bain.com.

35. FDA white paper. *Innovation or Stagnation.* http://www.fda.gov/ohrms/dockets/ac/07/briefing/2007-4329b_02_04_Critical%20Path%20Report%202004.pdf?utm_campaign=Google2&utm_source=fdaSearch&utm_medium=website&utm_term=innovation or stagnation&utm_content=3.

36. Khola, I.; Landis, J. Can the Pharmaceutical Industry Reduce Attrition Rates? *Nat. Rev. Drug Discov.* **2004**, *3*, 711.

37. DiMasi, J. A.; Grabowski, H. Economics of New Oncology Drug Development. *J. Clin. Oncol.* **2007**, *10*, 209–216.

38. DiMasi, J. A.; Hansen, R. W.; Grabowski, H. The Price of Innovation: New Estimates of Drug Development Costs. *J. Health Econ.* **2003**, *22*, 151–185.

39. Samanen, J. The Sustainable Pipeline Myth. http://www.discoverymanagementsolutions.com/the-organization-of-biopharmaceutical-rd/the-sustainable-pipeline-myth/.

40. The Project Management Institute; *Identify Components: Inputs, Section 4.1.1.1, The Standard for Portfolio Management.* 2nd ed., Vol. 51; Project Management Institute: USA, 2008.

41. Deloitte and Thomson Reuters. *R&D Value Measurement, Is R&D Earning Its Investment?* www.deloitte.com/assets/Dcom-UnitedKingdom/./UK_LS_RD_ROI.pdf.

42. DiMasi, J. A.; Feldman, L.; Seckler, A.; Wilson, A. Trends in Risks Associated with New Drug Development: Success Rates for Investigational Drugs. *Clin. Pharmacol. Ther.* **2010**, *87*, 272–277. Fig. 3, p. 276.

43. Wikipedia, Tadalafil. http://en.wikipedia.org/wiki/Tadalafil.

44. Teo, S. K. Thalidomide as a Novel Therapeutic Agent: New Uses for an Old Product. *Drug Discov. Today* **2005**, *10*, 107–114.

45. Chong, C. R.; Sullivan, D. J. New Uses for Old Drugs. *Nature* **2007**, *448*, 645–646.

46. Arrowsmith, J. A Decade of Change. *Nat. Rev. Drug Discov.* **2012,** *11,* 17–18.
47. LaMattina, J. The Impact of Mergers on Pharmaceutical R&D. *Nat. Rev. Drug Discov.* **2011,** *10,* 559–560.
48. Jessop, N. Behind the Scenes of Pharma Mergers and Acquisitions. *Pharm. Tech. Eur.* **Sep 1, 2010,** *22* (9). http://pharmtech.findpharma.com/pharmtech/Manufacturing/Behind-The-Scenes-Of-Pharma-Mergers-And-Acquisitio/ArticleStandard/Article/detail/685056.
49. Project Management Institute. *A Guide to the Project Management Body of Knowledge (PMBOK Guide),* 4th ed.; 434.
50. Samanen, J. Project/Program Authority versus Organizational Structure. http://www.projectleadersolutions.com/project-program-management-in-biopharmaceutical-rd/programproject-authority-versus-organizational-structure/.
51. The Project Management Institute. Portfolio Management Processes, Chapter 3.0. p. 35. In *The Standard for Portfolio Management;* The Standard for Portfolio Management, 2nd ed.; Project Management Institute, 2008; pp 33–44
52. Samanen, J. Who Does Portfolio Management in Your Company? http://www.portfoliomanagementsolutions.com/.
53. George, M.; Rowlands, D.; Kastle, B. *What Is Lean Six Sigma?* McGraw-Hill: New York, 2004.
54. Ullman, F.; Boutellier, R. A Case Study of Lean Drug Discovery: From Project Driven Research to Innovation Studio and Process Factories. *Drug Discov. Today* **2008,** *13,* 543–550.
55. Wikipedia, Vertical Disintegration. http://en.wikipedia.org/wiki/Vertical_disintegration.
56. Witty A. quoted in Drugs.com, 2008 GSK Sets Out New Strategic Priorities, Investor Presentation. http://www.drugs.com/news/gsk-sets-out-new-strategic-priorities-8477.html
57. Herper, M. *A Decade in Drug Industry Layoffs, Forbes.com.* http://www.forbes.com/sites/matthewherper/2011/04/13/a-decade-in-drug-industry-layoffs/.
58. Genomeweb.comLilly, Merck, Pfizer Agree to Share Pre-competitive Genomics Data on Lung, Gastric Cancers in Asia, 2010. http://www.genomeweb.com/dxpgx/lilly-merck-pfizer-agree-share-pre-competitive-genomics-data-lung-gastric-cancer
59. Greenstreet, Y. Change Broker. *The Scientist* **2009,** *23,* 28–36.
60. Dixon, J.; Lawton, G.; Machin, P. Vertical Disintegration: A Strategy for Pharmaceutical Businesses in 2009? *Nat. Rev. Drug Discov.* **2009,** *8,* 435.
61. Money.com Tachi Yamada, as Head of GSK R&D. http//money.cnn.com/magazines/business2/business2_archive/2004/04/01/366201/index.html, Apr. 1, 2004
62. Nature News Online. The Future of Pharma: GSK's Research Leaders Answer Nature's Questions About Where Their Company—and their Industry—Is Headed. *Nature News Online* **Oct. 9, 2008.**
63. Nature.com An Audience with Patrick Vallance. *Nat. Rev. Drug Discov.* **2010,** *9,* 834.
64. Longman, R. Lillys Chorus Experiment. *In Vivo* **2007,** *25,* 37–39. www.flexiontherapeutics.com/docs/In%20Vivo%20article%20Lillys%20Chorus%20Experiment%20May%202007.pdf.
65. Hunter, J.; Stephens, S. Is Open Innovation the Way Forward for Big Pharma? *Nat. Rev. Drug Discov.* **2010,** *9,* 87–88.

CHAPTER

8

Discovery and development of the anticancer agent gefitinib, an inhibitor of the epidermal growth factor receptor tyrosine kinase

Andy Barker*, David Andrews†

*Scientific Consultant, †Associate Director of Oncology Chemistry, Astrazeneca, Alderley Park, Macclesfield, Cheshire SK10 4TG

CHAPTER OUTLINE

8.1 Introduction and Biological Background 256
8.2 Biological and Chemical Approach to EGFR Tyrosine Kinase Inhibitors 258
8.2.1 Isoflavones 258
8.2.2 Anthelmintics 259
8.2.3 Anilinoquinazolines 259
8.3 Structure–Activity Studies 262
8.4 *In vivo* Studies 267
8.5 Early Development 271
8.6 Development and Clinical Studies 273
8.7 Next-Generation Approaches 277
8.7.1 Dual-ErbB inhibitors 277
8.7.2 First-generation irreversible inhibitors 277
8.7.3 Second-generation irreversible inhibitors 279
8.7.4 EGFR monoclonal antibodies 279
References 279

ABSTRACT

The epidermal growth factor receptor (EGFR) tyrosine kinase is a target for cancer chemotherapy. This chapter describes the discovery of small-molecule inhibitors of the kinase, their characterization and the medicinal chemistry programme that resulted in the identification of gefitinib. The development and studies undertaken to progress the drug in the clinic are presented together with a brief summary of other inhibitors of the EGFR kinase and their properties.

Keywords/Abbreviations: Cancer; Gefitinib; Kinase; Metabolism; Adenosine triphosphate (ATP); vascular endothelial growth factor receptor 1 (c-flt); transcription factor belonging to the immediate early response gene family (c-fos); Calculated log P (c log P); Deoxyribonucleic acid (DNA); Epidermal growth factor (EGF); Epidermal growth factor receptor (EGFR); Epidermal growth factor receptor tyrosine kinase (EGFRTK); Epidermal growth factor receptor (ErbB1); Epidermal growth factor receptor 2 (ErbB2);

Introduction to Drug Research and Development. http://dx.doi.org/10.1016/B978-0-12-397176-0.00008-X

Prototype mitogen activated protein kinase (ERK2); US Federal Drug Administration (FDA); Fibroblast growth factor (FGF); Adapter protein involved in signal transduction (GRB2); Epidermal growth factor receptor (HER1); Epidermal growth factor receptor 2 (HER2); Human umbilical vein endothelial cells (HUVEC); Phase III randomised; open label; first line study of gefitinib vs carboplatin/paclitaxel (IPASS); Phase III clinical study comparing gefitinib with placebo in pre-treated patients with advanced NSCLC (ISEL); Vascular endothelial growth factor receptor 2 (KDR); GTPase enzyme component of signal transduction pathways (K-ras); Logarithm of the octanol-water partition coefficient that takes account of the extent of ionisation of a molecule (log D); Logarithm of the octanol-water partition coefficient of a molecule (log P); hepatocyte growth factor receptor tyrosine kinase (MET); Dual specificity kinase that phosphorylates ERK1 and 2 (MEK-1); (3-(4,5-Dimethylthiazol-2-yl)-2,5-diphenyltetrazolium bromide (MTT); Non small cell lung cancer (NSCLC); Polymerase chain reaction (PCR); Negative logarithm of the acid dissociation constant of a molecule (pKa); Protein kinase C (PKC); Progression free survival (PFS); Kinase component of the MAP kinase cascade (Raf); Ribonucleic acid (RNA); Guanine nucleotide exchange factor involved in Ras signalling (SOS); Tyrosine kinase inhibitor (TKI); Vascular endothelial growth factor (VEGF).

8.1 INTRODUCTION AND BIOLOGICAL BACKGROUND

Conventional chemotherapy relies on the use of cytotoxic agents to kill the rapidly dividing cells present in tumours. Many of these agents damage or inhibit DNA synthesis and replication and result in cell death from the accumulated damage. Whilst these drugs can be effective against rapidly dividing cell populations, many common solid tumours only have a small proportion of their cell mass that is actively dividing and there are populations of normal cells that also undergo rapid replication. This results in side effects which have a significant impact on the patient, such as gut toxicity, immunosuppression and hair loss; these effects limit the doses of chemotherapeutic that can be given and the overall efficacy that can be achieved in many situations. Thus, whilst cytotoxic chemotherapy is useful in some types of tumour, the results for many of the common solid tumours are relatively disappointing and agents that are effective against these solid tumours and which are better tolerated would represent a major step forward in the fight against cancer.

Epidermal growth factor (EGF) and its associated receptor (EGFR) had been discovered in the 1950s and 1960s by Cohen and co-workers. EGF stimulates cell growth, proliferation and differentiation through binding to the receptor and initiating a series of signal transduction events inside the cell. In the late 1970s and early 1980s advances in biology increased the understanding of the mechanisms by which growth factors such as EGF produced their effects (1). This deeper understanding of how cells communicate with the external environment and respond to signals for growth provided insights into how aberrations or damage to these sensitive systems could result in the uncontrolled cellular growth characteristic of many tumours and highlighted the possibility of a new approach to the treatment of cancer by modification of these signal transduction systems. Some precedent for this approach was available in that signalling proteins had proven to be successful targets in the treatment of hormonally driven tumours such as prostate and breast cancer.

EGF is a 53 amino acid peptide that binds to the external domain of the EGF receptor, a transmembrane protein possessing an internal tyrosine kinase domain that is normally of low activity. When the ligand binds to the external domain it causes dimerization of the receptor, which allows tyrosine residues close to the kinase active site to be autophosphorylated by the proximal kinase. Conformational changes subsequently cause full activation of the kinase enzyme which recruits and phosphorylates a variety of substrates. Many of these substrates are kinase enzymes themselves which

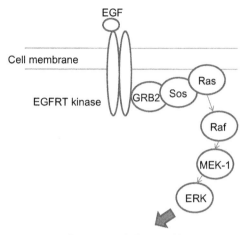

FIGURE 8.1

Epidermal growth factor receptor tyrosine kinase (EGFRTK) signal transduction cascade. (For colour version of this figure, the reader is referred to the online version of this book.)

subsequently also become activated resulting in a large amplification of the initial signal. Multiple such events eventually result in interactions with transcription factors and gene transcription, which is responsible for the pharmacological effects such as cell growth and division associated with EGF (Figure 8.1). The whole system is tightly controlled and regulated.

It became evident that many common tumours of epithelial origin expressed EGFR and that patient survival correlated with the expression level (2,3). Interfering with signalling through the EGF/EGFR system thus became an attractive target for a new way to treat cancer.

Tyrosine kinase enzymes act by transferring the gamma phosphate of ATP to the phenolic hydroxyl group of tyrosine residues in substrate peptides (Figure 8.2) and inhibiting the tyrosine kinase activity of the EGFR was identified as an appropriate target for a small molecule approach to controlling aberrant signal transduction. Key to success would be the ability to achieve selectivity for the kinase since multiple other growth factors controlling important biochemical and physiological functions also

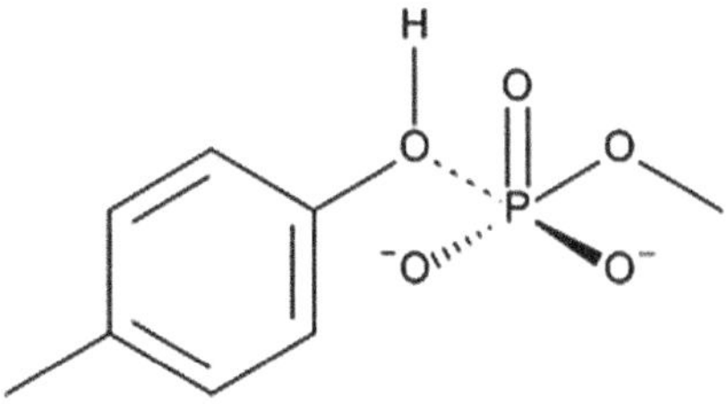

FIGURE 8.2

Transition state between ATP and tyrosine residue on substrate peptide.

utilized tyrosine kinase enzymes as a part of the signal transduction process and inhibition of these other kinases could produce unacceptable side effects.

8.2 BIOLOGICAL AND CHEMICAL APPROACH TO EGFR TYROSINE KINASE INHIBITORS

An assay was established to detect tyrosine kinase activity. It consisted of a membrane preparation from A431 human vulvar carcinoma cells, which have high levels of the EGFR tyrosine kinase, and a synthetic peptide containing multiple tyrosine residues. ATP containing a radiolabelled gamma phosphate group was incorporated and the enzyme transferred the radiolabelled phosphate to the tyrosine residues and this was detected by standard isolation and scintillation techniques. In the presence of an effective inhibitor the transfer of the radiolabel would be prevented (*4*).

A collection of approximately 2000 compounds was initially screened in the assay. The compounds tested were a structurally diverse set of molecules and inhibitors were followed up by screening structurally similar compounds from a larger collection. In conjunction with this a search of the compound collection was made to identify compounds that possessed structural features associated with the gamma phosphate–tyrosine intermediate, which is considered to be a transition state in the enzyme mechanism and these compounds were also screened. Hits were rescreened and the structures of hits were confirmed.

A cellular screen was also established using a human cell line, the KB oral cell carcinoma (*4*). Cells were seeded in plates containing a medium where natural growth factors had been removed by treatment with charcoal and incubated for 72 h in the absence or presence of 10 ng/ml of EGF. After this period the cells were incubated for a further hour with 3-(4,5-dimethylthiazol-2-yl)-2,5-diphenyltetrazolium bromide (MTT) and the absorbance at 540 nm measured as a surrogate for viable cell number. The two arms of this experiment allowed effects on EGF-stimulated cell growth and 'basal' cell growth to be measured. By adding other growth factors instead of EGF the effect of compounds on other signal transduction processes could also be investigated. The ideal compound would prevent the additional growth seen in the presence of EGF but not inhibit basal growth or growth stimulated by other growth factors such as insulin. Compounds active in the enzyme screen were tested in the cell assays.

A relatively high hit rate was seen in the enzyme assay and a diverse set of structures was identified as hits. These included isoflavones, anthelmintic structures and anilinoquinazolines and all were considered potential start points for a medicinal chemistry programme.

8.2.1 Isoflavones

The isoflavones identified from the enzyme screen were typified by the fluorine-substituted compound **1** and the natural product **2** (Figure 8.3, Table 8.1) isolated from a Florida orange bush. There were concerns about such polyhydroxylated compounds and their chelating and redox properties but activity against the kinase was submicromolar and this translated to activity against EGF-stimulated KB cell growth. Some synthetic modifications were made to these molecules and activity improvements kinase activity observed (see compound **3** in Table 8.1). The improvement in enzyme activity was not always reflected in similar gains in the cellular screen and the permeability of these compounds was identified as a potential issue. The difficulty of the chemistry and evidence for lack of selectivity for EGF-driven growth vs. other growth factors in the cellular screen resulted in further work in this series being halted.

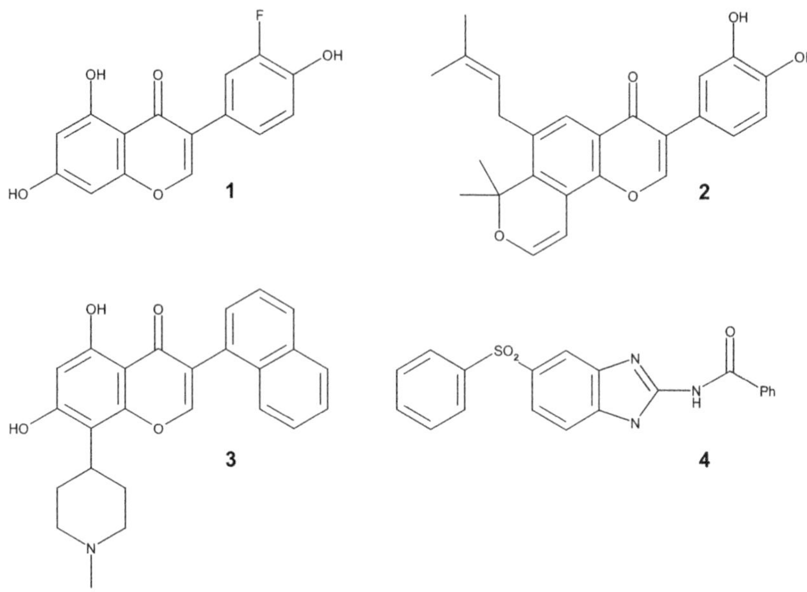

FIGURE 8.3

Hits from enzyme screening: isoflavones and anthelmintics.

8.2.2 Anthelmintics

Several anthelmintic structures were identified as inhibitors of the kinase and had respectable KB cell inhibitory activity, e.g. compound **4** (Figure 8.3, Table 8.1). Such molecules were synthetically accessible and the best potency obtained against the enzyme was approximately 50 nM. However it became clear that these molecules interacted with multiple targets, including non-kinases and further work was stopped.

8.2.3 Anilinoquinazolines

Of most interest from the initial screens were the anilinoquinazoline compounds. Two compounds **5** and **6** were identified (Figure 8.4) which inhibited the EGF kinase with good potency and which also displayed activity against the EGF-stimulated growth of KB cells (Table 8.2). Importantly the

Table 8.1 *In vitro* Data for Compounds **1–4**

Compound	EGF Kinase IC$_{50}$ μM	EGF-Stimulated Cell Growth IC$_{50}$ μM
1	0.7	2.7
2	~5	2.5
3	0.03	8
4	0.16	0.6

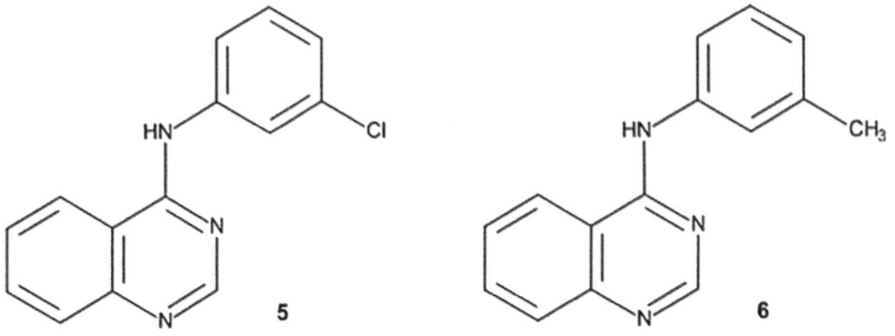

FIGURE 8.4

Hits from enzyme screening: anilinoquinazolines.

compound **5** had little effect on 'basal' growth of KB cells until much higher concentrations than those inhibiting EGF-stimulated growth were reached (Figure 8.5). Compound **5** had little or no effect on the insulin-stimulated growth of KB cells at similar concentrations demonstrating that selectivity for the EGF signal transduction pathway could be achieved. Similar results were obtained in other tumour cell lines and in HUVECs (human umbilical vein endothelial cells). The cellular effects were reversible – when the compound was removed cellular growth resumed after a period. It was also shown that receptor autophosphorylation was inhibited by the compound. This was achieved by treating cells with compound, stimulating with EGF and then lysing the tumour cells, isolating the EGFR by electrophoresis and Western blotting and measuring the level of tyrosine phosphorylation with anti-phosphotyrosine antibodies. Inhibition of receptor autophosphorylation was demonstrated in a variety of tumour cell lines including HT29 (colon), KB (oral squamous carcinoma), Du145 (prostate) and A549 (lung), showing clearly that inhibition of the kinase prevented activation of the receptor signal transduction cascade. Further work on the 4-(3-chlorophenylamino)-quinazoline **5** demonstrated that it was competitive with ATP in binding to the enzyme (K_i 2.1 ± 0.2 nM) and non-competitive with substrate. This demonstration of selectivity was important since there were concerns that any molecule binding at the ATP site of kinases would not be selective and would inhibit many biochemical systems with multiple pharmacological side effects.

The data above led to the selection of the anilinoquinazolines as the lead series. Synthesis of these compounds was achieved by reaction of substituted anilines with a variety of 4-chloroquinazolines in isopropanol. The 4-chloroquinazolines were in turn prepared from 4-quinazolones by reaction with phosphoryl chloride or thionyl chloride (Figure 8.6). This chemistry provided quick access to a wide

Table 8.2 *In vitro* Data for Compounds **5** and **6**		
Compound	EGF Kinase IC$_{50}$ μM	EGF-Stimulated Cell Growth IC$_{50}$ μM
5	0.04	1.2
6	0.18	2.4

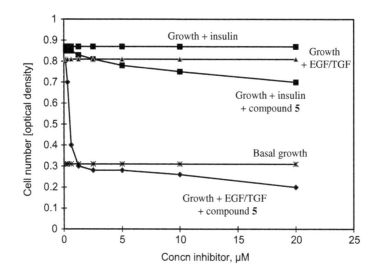

FIGURE 8.5

Effect of compound **5** on stimulated and non-stimulated KB cell growth.

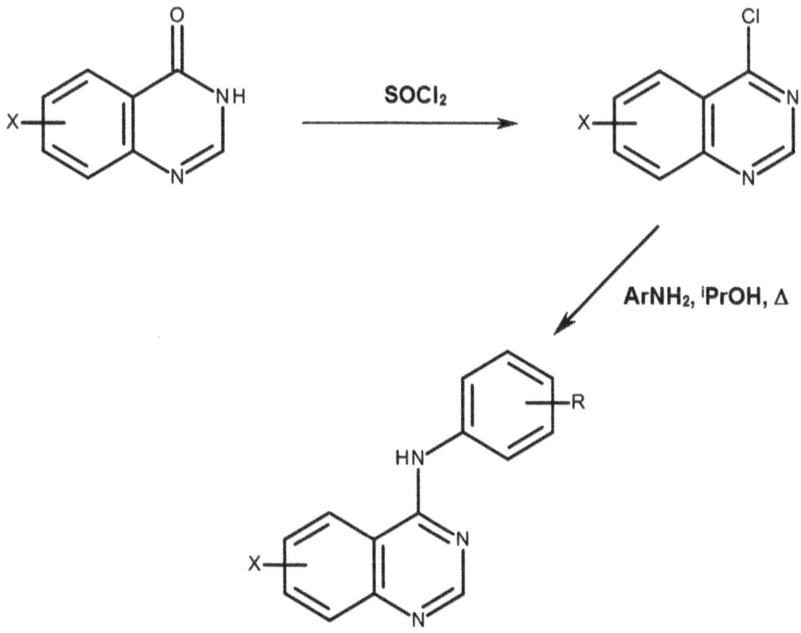

FIGURE 8.6

Synthetic route to 4-anilinoquinazolines.

variety of target structures and allowed a rapid assessment of structure–activity relationships at the enzyme and in cellular environments (5).

8.3 STRUCTURE–ACTIVITY STUDIES

Removal of the N1 or N3 atoms to give the isoquinoline and quinoline structures **7** and **8** (Figure 8.7) resulted in significant loss of activity (Table 8.3) and indicated the likely importance of these atoms in the interaction with the kinase so work was focussed on the quinazoline pharmacophore. It was also quickly established that substitution of the linking NH with a methyl group resulted in poor activity, most likely due to conformational effects on the side chain.

Using the 6,7-dimethoxyquinazoline as a base structure a series of compounds was synthesized to investigate the effect of substitution in the aniline side chain. A comparison of the ortho-, meta- and para-chloro aniline compounds highlighted the likely importance of the conformations available to the aniline side chain (Figure 8.8, Table 8.4).

The ortho-chloro compound **9** lost significant activity in comparison to the meta and para analogues suggesting that the ortho-substituted compound could not adopt the required conformation for effective binding to the kinase target. The meta- **10** and para-chloro **11** compounds showed good activity against the enzyme which translated into effective inhibition of EGF-driven KB cell growth. Since the meta-substituent appeared to offer good potency a variety of other substituents was placed in this position. A wide variety of substituents was tolerated and gave good enzyme potency. These ranged from the powerfully electron withdrawing nitro substituent in compound **12** to the electron

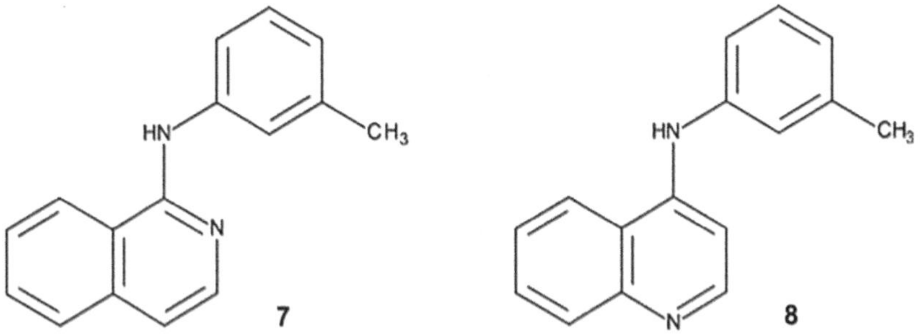

FIGURE 8.7

Structures of isoquinoline and quinoline analogues.

Table 8.3 Enzyme Data for Isoquinoline and Quinoline Analogues

Compound	EGF Kinase IC$_{50}$ μM
7	>5
8	>5

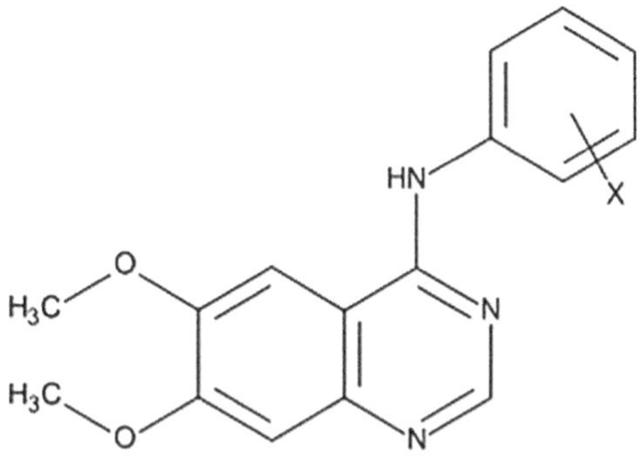

FIGURE 8.8

Structures of compounds **9–17**.

donating amino group in compound **17**. Whilst the phenyl-substituted compound **14** retained activity the phenoxy compound **16** did not, possibly indicating a size restriction in this area of the pharmacophore. The enzyme activity of these compounds translated into cellular activity for the most part but the relationship was not linear suggesting that membrane permeability or enzyme kinetics of individual compounds influenced cellular potency. Relatively small and mildly electron-withdrawing substituents in the meta and para positions of the aniline were identified as the best substituents for good enzyme and cell activity.

Investigation of substituents on the quinazoline quickly showed that groups in the 5- and 8-positions of the heterocycle reduced EGF kinase activity (Figure 8.9, Table 8.5). Comparing the four possible chlorine substitutions (compounds **18–21**) showed the 6- and 7-substitution patterns (compounds **19** and **20**) to provide the best enzyme activity. There was a modest reduction in potency with the 5-chloroquinazoline **18** whilst the 8-substituted compound **21** displayed a more significant drop-off in enzyme potency. Further investigation of this effect with compounds **22–24** reinforced these observations. The 8-methoxy compound **24** was significantly less active against the kinase and whilst electronic effects of this substituent are probably significant the conclusion was drawn that the

Table 8.4 *In vitro* Data for Compounds **9–17**

Compound	X	EGF Kinase IC$_{50}$ μM	EGF-Stimulated Cell Growth IC$_{50}$ μM
9	o-Cl	>1	1.06
10	m-Cl	0.002	0.007
11	p-Cl	0.010	0.19
12	m-NO$_2$	0.001	0.2
13	m-CF$_3$	0.009	0.28
14	m-Ph	0.020	0.46
15	m-CH$_3$	0.005	0.05
16	m-OPh	>1	–
17	m-NH$_2$	0.022	1.1

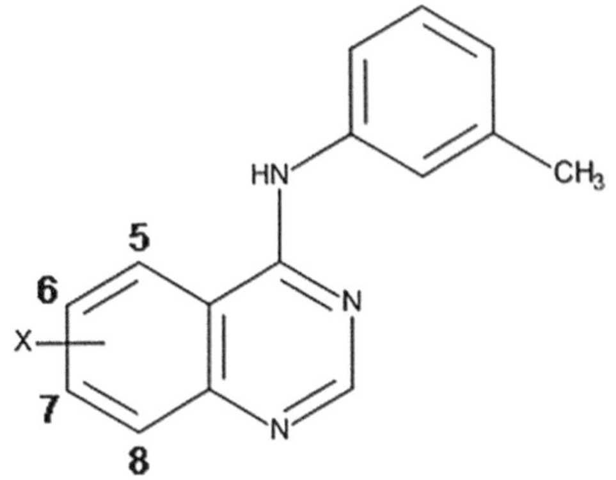

FIGURE 8.9

Structures of compounds **6** and **18–30**.

proximity of the 8-substituent to the N1 centre had a significant steric effect on the ability of the N1 centre to enter into effective hydrogen bonding interactions with the target kinase; substituent electronic effects were also likely to affect the basicity of N1. The 5-nitro compound **23** supported the pattern observed with the 5-chloro compound **18**. There was a significant loss of enzyme activity and again, whilst electronic effects could not be ruled out, the conclusion was drawn that the 5-substituents had an effect on the conformational flexibility of the aniline side chain. This was supported by the observation that the NMR spectrum of compound **23** displayed two clear rotamers at room temperature strongly suggesting that conformational flexibility was limited.

Table 8.5 *In vitro* Data for Compounds **6** and **18–30**

Compound	X	EGF Kinase IC_{50} µM	EGF-Stimulated Cell Growth IC_{50} µM
6	H	0.18	2.4
18	5-Cl	0.5	10.5
19	6-Cl	0.1	1.56
20	7-Cl	0.09	2.8
21	8-Cl	1.3	12.5
22	6,7-(OMe)$_2$	0.005	0.05
23	5-NO$_2$ 6,7-(OMe)$_2$	1.45	7.6
24	6,7,8-(OMe)$_3$	>5	>12.5
25	6-NO$_2$	0.75	5.5
26	6-Ph	0.055	0.71
27	6-NH$_2$	0.52	1.0
28	6-OMe	0.04	0.5
29	6-NHCOCH$_2$Cl	<0.005	0.014
30	6-NHCH$_2$CH$_2$OMe	0.01	0.2

Focus on the 6- and 7-substituents revealed that a wide variety of functional groups was compatible with good enzyme and cellular potency. Strongly electron-withdrawing substituents such as the 6-nitro compound **25** reduced enzyme activity; however, 6-methoxy substitution **28** improved potency suggesting that the mesomeric effect of the oxygen in the latter substituent was important and may have effects on the basicity of the quinazoline N1 centre. However the 6,7-dimethoxy-substituted quinazoline **22** was even more potent and presumably conformational effects in this substitution pattern prevent effective mesomeric contributions for the oxygen atoms to the basicity of the quinazoline nitrogen atom but this may be counterbalanced by increasing interactions of the aromatic ring with residues in the enzyme. Relatively large substituents such as phenyl **26** were tolerated and this provided the potential for a degree of freedom in design of future molecules. Interestingly the 6-chloroacetamide group **29** provided a very potent molecule and it was postulated that this moiety could be taking part in a covalent interaction with a cysteine residue close to the ATP-binding site resulting in an irreversible inhibition of the enzyme. This was not pursued but it is interesting to note that several other research groups have subsequently followed up with irreversible inhibitors of the EGF kinase enzyme with some clinical success.

It was clear that both 6- and 7-substitution was well tolerated and that a variety of mildly electron-donating groups resulted in good activity and that quite large groups could be incorporated into the molecule without losing activity. The latter was believed to be due to these groups being located towards the solvent surface of the enzyme and this provided scope for further investigations.

Throughout these investigations it was important to confirm that the activity seen in cellular assays was, in fact, the result of inhibition of the EGF kinase and not through other targets. Various techniques were used to investigate this aspect. First, plotting enzyme activity against stimulated KB cell potency revealed a predominantly linear relationship (Figure 8.10). In a smaller study the correlation between cellular potency and the effect of the compounds on the autophosphorylation of the receptor kinase also

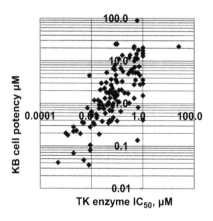

FIGURE 8.10

Relationship between KB cell potency and enzyme activity. (For colour version of this figure, the reader is referred to the online version of this book.)

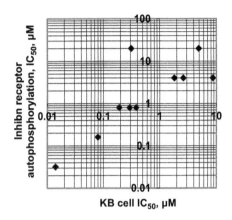

FIGURE 8.11

Relationship between cell potency and receptor autophosphorylation. (For colour version of this figure, the reader is referred to the online version of this book.)

revealed a predominantly linear relationship (Figure 8.11). In conjunction with selectivity assays against a range of other kinases and the observation that the compounds had little or no effect on basal cell growth or cell growth stimulated by other growth factors such as insulin-like growth factor (IGF)-1 and insulin gave confidence that activity in cellular environments was due to EGFR kinase inhibition.

One of the most important observations was that although the inhibitors were competitive with ATP and that the ATP-binding site is highly conserved across most kinases, selectivity for the EGF kinase could be achieved. We postulated that there were several reasons for this. Although it was likely that many small molecules made the same interactions with the peptide backbone through the N1 quinazoline atom, 'locating' the molecule, flexible side chains could interact with non-conserved amino-acid residues close to the ATP-binding site and achieve selectivity through these non-specific interactions. This phenomenon also explained the importance and effect of quinazoline 5-substituents and ortho substitution on the aniline. Both substituent patterns make significant effects

Table 8.6 Selectivity Data for Compounds **5** and **30**

	Compound 5	Compound 30
EGFRTK IC_{50} μM	0.04	0.01
Raf IC_{50} μM	–	100
MEK-1 IC_{50} μM	–	>100
ErbB2 IC_{50} μM	–	1.2–3.7
c-flt IC_{50} μM	21% inhibition at 12.5 μM	>100
KB Cells + EGF IC_{50} μM	1.2	0.2
KB Cells − EGF IC_{50} μM	15	11.6

Raf − kinase component of the MAP kinase cascade; MEK-1 − dual specificity kinase that phosphorylates ERK1 and 2; ErbB2 − epidermal growth factor receptor 2 kinase; c-flt − vascular endothelial growth factor receptor 1 kinase.

on the aniline side chain conformation and can result in the required conformation for effective interaction in the 'selectivity pocket' being unattainable or high energy (*6*).

Examples of the selectivity that could be achieved with these molecules are illustrated by the compounds **5** and **30**. Both compounds exhibited good selectivity against other class 1 receptor tyrosine kinases such as erbB2 and against a variety of other kinases from the serine/threonine families. This translated into good margins between their effects on EGF-stimulated cell growth and basal growth and effects on KB cell growth stimulated by other growth factors (Table 8.6).

8.4 *IN VIVO* STUDIES

Having achieved good potency and selectivity at the enzyme and cellular level we began to explore translating this activity to an *in vivo* model of cancer.

The *in vivo* model used was a xenograft comprising human tumour tissue in athymic ('nude', immunocompromised) mice. The A431 human vulvar squamous cell carcinoma cell line was implanted in mice and allowed to establish; compounds were dosed orally once daily and growth rates compared with untreated mice bearing the same tumour.

An early compound tested in this assay was the dimethoxyquinazoline **22**. This compound was potent *in vitro* and had some *in vivo* activity in the xenograft model. Analysis of blood samples from mice dosed with this compound (Figure 8.12) showed that it was quickly cleared, mainly due to metabolism and two major metabolites were observed. Isolation of these compounds revealed the metabolites to be the result of oxidation at the para position of the aniline and on the meta-methyl group resulting in the compounds **31** and **32** (Figure 8.13). Since continuous inhibition of the kinase was believed to be required for effective *in vivo* activity modifications to compound **22** were investigated to increase *in vivo* exposure. Isosteric replacements of the methyl group with chlorine and the para hydrogen with a fluorine atom were made to block these sites of metabolism resulting in compound **33** (Table 8.7). This compound retained its *in vitro* potency and on oral dosing to mice, blood levels and exposure were significantly higher and resulted in improved activity in the disease model (Figure 8.12).

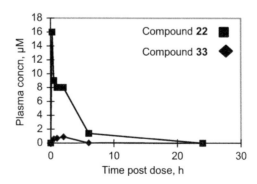

FIGURE 8.12

Pharmacokinetic profiles of compounds **22** and **33** following 50 mg/kg oral dose to mice.

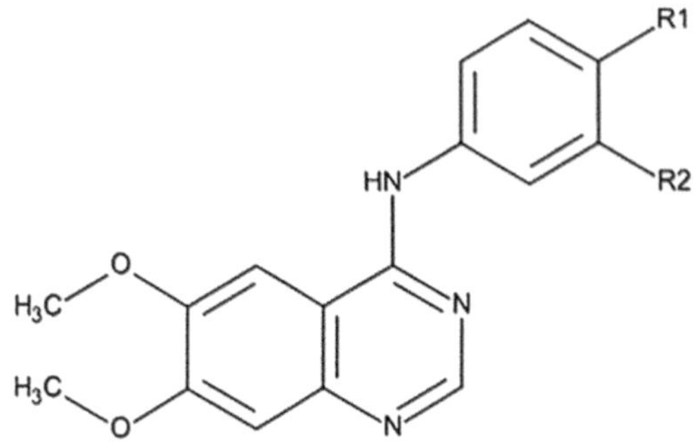

31	R1	OH	R2	CH_3
32	R1	H	R2	CH_2OH
33	R1	F	R2	Cl

FIGURE 8.13

Structures of compounds **31–33**.

In making these substitutions the physicochemical properties of the compound (clog P 4.8) resulted in high binding to protein, effectively reducing exposure of free drug to the target and impacting aqueous solubility. We believed that continuous inhibition of the kinase would be necessary for good activity based on earlier experiments and the nature of the signal transduction process. To achieve this we would need a potent compound showing good exposure over an extended period and concluded that we would need to modify the physicochemical properties of the compounds to provide good aqueous solubility, reduce the potential for metabolic clearance and provide good levels of non-protein-bound drug.

Table 8.7 *In vitro* and Exposure Data for Compounds **33–40**

Compound	EGF Kinase IC_{50} µM	Stimulated KB Cell IC_{50} µM	Blood Level at 2 h µM	Blood Level at 6 h µM	Blood Level at 24 h µM
33	0.009	0.08			
34	0.023	0.08	6.8	38	5.7
35	0.01	0.09	7.1	2.4	
36	0.049	0.15		34	
37	0.071	0.12		8.8	3.7
38	0.079	0.12		5.2	2.3
39	0.095	0.16		23	3
40	0.067	0.13		15	0.45

Blood levels in mice following oral doses of 200 mg/kg.

To support this hypothesis we examined the blood levels of candidate molecules 2 h after oral dosing to mice and their aqueous solubilities alongside the log P of the molecules (Figure 8.14). Solubility, as expected, increased with reductions in lipophilicity whilst 2 h blood levels were low at both high and low log P but appeared to be optimal at log P values between 2 and 3.5 and this became our target.

Supporting this hypothesis was the observation that the 6-amino compound **27**, which was not particularly potent in the *in vitro* assays, displayed excellent activity in the *in vivo* model when dosed at 200 mg/kg/day to mice. Particularly relevant in this observation was that with a log P of 2.9 this molecule had good aqueous solubility and relatively low protein binding resulting in good exposure over an extended period in the animal model. Equally encouraging was that the compound was well tolerated when dosed continuously at this level for 70 days, supporting our hypothesis that EGFR kinase inhibitors would be better tolerated than conventional chemotherapeutic agents.

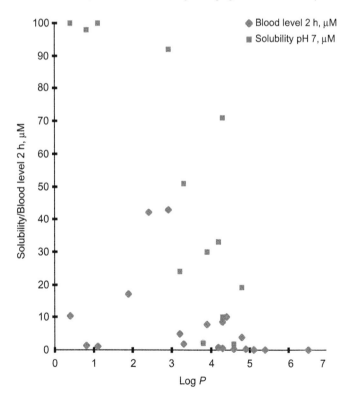

FIGURE 8.14

Relationships between aqueous solubility, blood levels in mice and lipophilicity. (For colour version of this figure, the reader is referred to the online version of this book.)

The previous observation that a wide variety of substituents was tolerated in the 6- and 7-positions of the quinazoline led us to explore modification of the physicochemical properties of the molecule by incorporating a range of amines in the 6-methoxy substituent to improve physicochemical properties and exposure whilst retaining *in vitro* potency (Figure 8.15, compounds; Table 8.7).

Highly basic compounds were cleared relatively quickly but less basic amines provided sufficient aqueous solubility, levels of free drug and resulted in good levels of extended exposure. The 3-morpholinopropoxy derivative **34** was identified as possessing the required physicochemical properties for good oral exposure and potency and selected for full development.

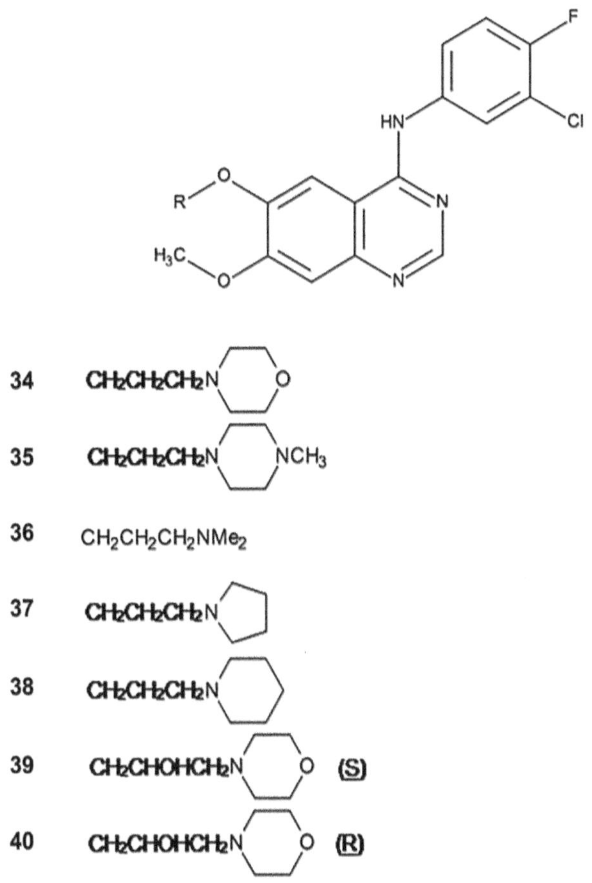

34	CH₂CH₂CH₂N⟨morpholine⟩
35	CH₂CH₂CH₂N⟨NCH₃ piperazine⟩
36	CH₂CH₂CH₂NMe₂
37	CH₂CH₂CH₂N⟨pyrrolidine⟩
38	CH₂CH₂CH₂N⟨piperidine⟩
39	CH₂CHOHCH₂N⟨morpholine⟩ (S)
40	CH₂CHOHCH₂N⟨morpholine⟩ (R)

FIGURE 8.15

Structures of compounds **34–40**.

8.5 EARLY DEVELOPMENT

The compound chosen for full development **34** (Figure 8.16) was a potent EGF kinase inhibitor, competitive with ATP binding and non-competitive with the substrate. It displayed good selectivity against a variety of other kinase enzymes and selectively inhibited EGF-stimulated cell growth at concentrations 45 times lower than those which affected basal growth. Inhibition of cellular growth correlated well with the inhibition of autophosphorylation of the receptor and the cellular effects were also reversible on removal of compound (Table 8.8) after 14 days exposure to 15 μM of compound **34**.

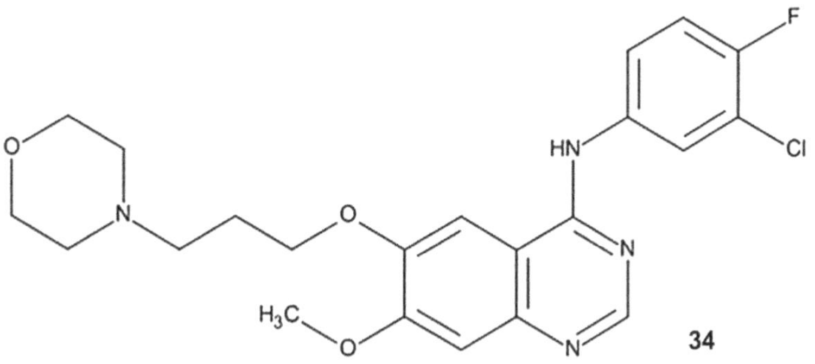

FIGURE 8.16

Structure of gefitinib.

Table 8.8 Properties of Compound **34**

EGFRTK IC_{50} μM	0.023
ErbB2 kinase IC_{50} μM	~2
KDR IC_{50} μM	~20
c-Flt, PKC, raf, MEK-1, ERK2 IC_{50} μM	>100
EGF-stimulated KB cell growth IC_{50} μM	0.054
Basal KB cell growth IC_{50} μM	8.8
EGF-stimulated HUVECs IC_{50} μM	0.03–0.1
FGF-stimulated HUVECs IC_{50} μM	1–3
VEGF-stimulated HUVECs IC_{50} μM	1–3
Inhibition of EGF-stimulated EGFR autophosphorylation in Du145 cells	Complete inhibition at 0.16 μM
pKa	7.17
Aqueous solubility at pH 7.4 (μM)	4
Log D	3.3
Free drug in 100% human plasma	3%

Refer to legend of Table 8.6: PKC – protein kinase C, ERK2 – mitogen-activated protein kinase, HUVEC – human umbilical vein endothelial cell, FGF – fibroblast growth factor, VEGF – vascular endothelial growth factor, log D – logarithm of the octanol–water partition coefficient that takes account of the extent of ionization of a molecule.

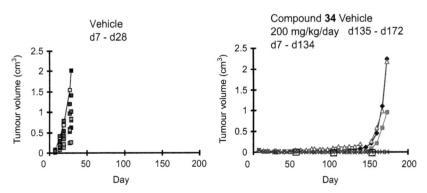

FIGURE 8.17

Activity of compound **34** against A431 xenograft in nude mice. (For colour version of this figure, the reader is referred to the online version of this book.)

The physicochemical properties of the compound provided for good oral exposure with a pKa of 7.17 and an aqueous solubility of 4 μM. With a log D of 3.3 free levels of the drug were 3% in human plasma. These properties resulted in 24 h exposure in mice following doses of 200 mg/kg/day and allowed us to demonstrate good activity in the mouse models of human tumours.

In the A431 xenograft a dose of 200 mg/kg/day completely inhibited tumour growth for 127 days whilst dosing continued (Figure 8.17). When dosing was stopped some tumours returned to growth. Importantly the compound had the effect of shrinking advanced tumours (established for 1 month before treatment) in a similar experiment (Figure 8.18). Activity was seen in a wide range of other

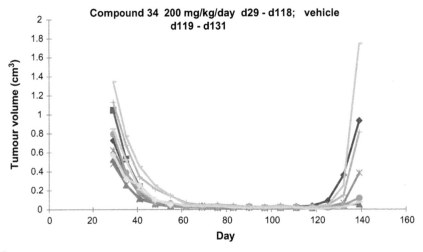

FIGURE 8.18

Activity of compound **34** against advanced A431 tumours.

Table 8.9 Relationship between Xenograft Activity and c-fos Inhibition

Dose of 34, mg/kg p.o.	% Inhibition of A431 Tumour Volume	c-fos m-RNA (% Control)
12.5	53	53
50	77	6
200	99	0.4

Dosing was for 21 days in xenograft model; c-fos measurements were made 6 h after the last of four doses in a similar experiment.

human tumour xenografts driven by EGF at doses ranging from 12.5 to 200 mg/kg/day. These tumours included A549 (non-small-cell lung carcinoma), various colorectal tumours (CR10, HT29, and LoVo), DU145 prostate tumours, HX62 ovarian tumours and MCF-7 breast tumour xenografts. Activity was related to exposure to compound and the compound had an impact on the levels of the immediate early response gene, c-fos, which is normally elevated in response to growth factor signal transduction processes (Table 8.9). Levels of c-fos were measured in A431 tumours treated with compound **34** by excising the tumour 6 h after the last dose, extracting the mRNA and quantifying the levels of c-fos mRNA by reverse-transcription PCR. This resulted in a clear relationship between dose of compound, tumour growth inhibition and levels of the early response gene that supported the hypothesis that EGF kinase inhibition resulted in tumour growth inhibition in the *in vivo* setting and provided the possibility of using c-fos levels as a surrogate marker for activity in the *in vivo* and clinical settings.

The maximum dose used in the *in vivo* studies (200 mg/kg/day p.o.) did not result in any body weight loss or other signs of toxicity in the dosed mice whilst doses of 50 mg/kg/day to rats resulted in minor haematopoietic and lymphoid changes.

8.6 DEVELOPMENT AND CLINICAL STUDIES

The synthetic route established to produce compound **34** was slightly modified and utilized the sequence outlined in Figure 8.19.

The compound taken into development was designated ZD1839 and later known as gefitinib. A number of Phase I studies were carried out, progressively evaluating single dose, then multiple dose tolerability and pharmacokinetics (7). One of the first such studies by Ranson and co-workers was an open-label, escalating dose study (7). ZD1839 was administered orally once daily for 14 consecutive days, followed by a 2 week wash-out period. Doses ranging from 50 mg per patient per day up to 925 mg/day were explored. The most frequently observed events were grade 1 and 2 acne-like rash, nausea and diarrhoea, with grade 3 and 4 events being uncommon. The observed terminal half-life was 48 h and exposure was dose-proportional. Encouragingly, four of the 16 patients with non-small-cell lung cancer (NSCLC) had objective partial responses when dosed in the range 300–700 mg/day. Overall 16 patients remained on study for ≥3 months and five NSCLC patients (three of whom were in the above partial response group) remained on study for ≥6 months.

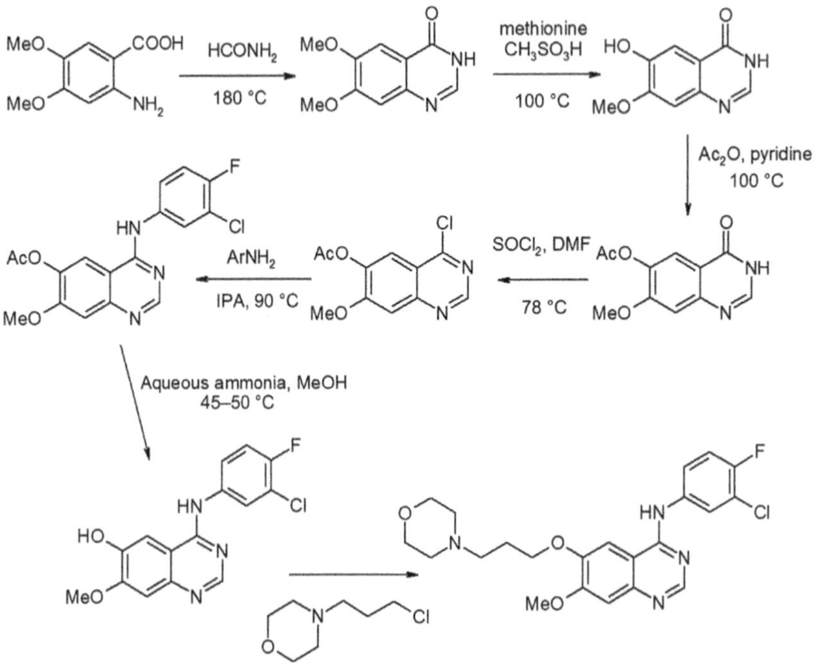

FIGURE 8.19

Synthetic route to gefitinib.

Gefitinib went on to be studied in a large number of different clinical studies, evaluating its use as a single agent, and in combination against a variety of tumour types. In the phase III Iressa survival evaluation in lung cancer (ISEL) study comparing gefitinib with placebo in pre-treated patients with advanced NSCLC, subset analyses identified greater benefit with gefitinib in patients who were never-smokers or of Asian origin (8). These subgroups also have a relatively high incidence of EGFR mutations and it has been shown that NSCLC patients with tumours containing mutations in the EGFR gene were highly responsive to gefitinib (9). To evaluate the hypothesis that first line therapy with a tyrosine kinase inhibitor (TKI) would be at least as effective as chemotherapy, the IPASS study was conducted. One thousand two hundred and seventeen previously untreated patients in East Asia who had advanced pulmonary adenocarcinoma and who were non- or former light smokers were randomly assigned to receive gefitinib (250 mg/day) or carboplatin plus paclitaxel as outlined in Figure 8.20. The primary end point was progression-free survival (PFS – the length of time during and after treatment that a patient lives with the disease but it does not get worse).

The IPASS study exceeded its primary objective of non-inferiority and demonstrated superiority of gefitinib relative to carboplatin/paclitaxel for PFS. Mutation of the EGFR gene was a strong predictor of a better outcome with gefitinib. A subgroup of 261 patients who were EGFR mutation positive showed significantly longer PFS on gefitinib than those who received carboplatin–paclitaxel (Figure 8.21). Hazard ratio for progression or death was 0.48; 95% CI, 0.36–0.64; $P < 0.001$, whereas

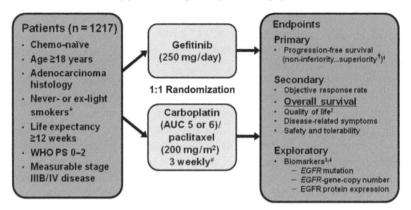

IPASS: study design

Conducted in China, Japan, Thailand, Taiwan, Indonesia, Malaysia, Philippines, Hong Kong and Singapore

'Never-smokers, <100 cigarettes in lifetime; ex-light smokers,
stopped ≥15 years ago and smoked ≤10 pack-years; #limited to a maximum of 6 cycles;
†If the primary objective of non-inferiority was reached, then superiority could be assessed
Carboplatin/paclitaxel was offered to gefitinib patients upon progression

FIGURE 8.20

Iressa pan-Asia study (IPASS) study.

IPASS design (10).

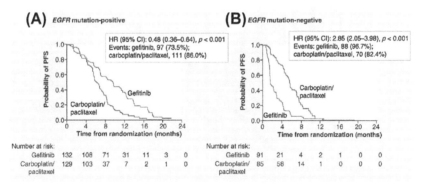

FIGURE 8.21

Kaplan–Meier curves for progression-free survival in patients who were positive for the EGFR mutation (Panel A), and in patients who were negative for the EGFR mutation (Panel B).

Data from the IPASS study and reproduced from Ref. (11).

the converse was true for the 176 patients who were EGFR wild-type. In this latter group, the hazard ratio for progression or death with gefitinib was 2.85; 95% CI, 2.05–3.98; $P < 0.001$. A hazard ratio of <1 implies a lower risk of progression with gefitinib than with carboplatin/paclitaxel. IPASS was a landmark study in that it was the first to confirm EGFR mutation as a clinically relevant predictive marker for the efficacy of an antitumour drug in NSCLC.

A number of other clinical studies in NSCLC further researched gefitinib in the treatment of EGFR-mutant tumours (12) and this agent has now been approved in excess of 80 countries worldwide.

Away from the carefully controlled environment of clinical trials, the methodology for the detection of EGFR mutations in tumours is still being refined and standardization of techniques to avoid false positives or negatives is necessary to identify those patients who would benefit from treatment with gefitinib. The collection of sufficient tissue for diagnosis is also challenging, since tumours are often relatively inaccessible or the biopsies taken are relatively small. With this in mind, the use of surrogate samples including serum and plasma has also been explored (13).

Gefitinib 250 mg/day is generally well tolerated across studies of both pre-treated and treatment naïve patients, with mild to moderate skin rashes and diarrhoea the most commonly reported treatment-related adverse events (12). Patients in IPASS treated with first-line gefitinib experienced higher incidence of skin rash or acne (66%) compared with patients in the carboplatin/paclitaxel arm of the study (22%), and the incidence of diarrhoea was 45% in the gefitinib arm vs. 22% in the carboplatin/paclitaxel arm. The incidence of neutropaenia (1% vs. 65%), leukopaenia (0% vs. 34%), anaemia (2% vs. 9%), thrombocytopaenia (1% vs. 5%), nausea (12% vs. 44%) and vomiting (10% vs. 33%) was higher in patients who received chemotherapy (Figure 8.22). Relatively few patients in the gefitinib arm of the study (6.9%) discontinued treatment because of adverse events compared with

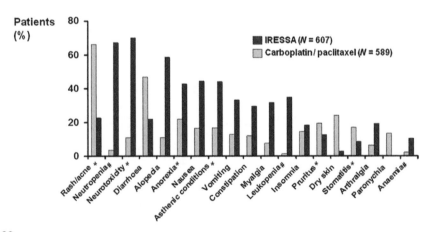

FIGURE 8.22

IPASS: most common adverse events ($\geq$10% on either treatment) with >3% difference between treatments. #Absolute neutrophil count, white blood cell count, or haemoglobin worsened from baseline to CTC grade 3/4; IRESSA $N = 599$, carboplatin/paclitaxel $N = 577$. *Grouped term (sum of several preferred terms).

Most common adverse events (all grades) occurring at a frequency of $\geq$10% on either treatment arm, with >3% difference between treatments in the IPASS study (10).

13.6% in the chemotherapy arm. In addition, a clinically relevant improvement in health-related quality of life was seen in significantly more patients overall receiving gefitinib than carboplatin/ paclitaxel (*10*), with differences in the EGFR mutation-positive subgroup (favouring gefitinib) and EGFR mutation-negative subgroup (favouring carboplatin/paclitaxel) (*14*). Symptom improvement rates were similar in both arms, but were significantly higher with gefitinib in the EGFR mutation-positive subgroup and with carboplatin/paclitaxel in the EGFR mutation-negative subgroup (*10,14*).

Following the recognition that EGFR mutation status was the critical determinant of compound efficacy, a landmark was achieved in 2009, when this agent was approved as Iressa in the EU for the treatment of adults with locally advanced or metastatic NSCLC with activating mutations of EGFRTK across all lines of therapy, as distinct from second-line therapy in pre-treated patients.

8.7 NEXT-GENERATION APPROACHES

Gefitinib **34** and erlotinib **41** have undoubtedly brought treatment benefit to a selected patient population, however the emergence of secondary resistance mutations, coupled with developing hypotheses of how to improve the safety and tolerability of EGFR-targeted agents has led to the development of a number of next-generation therapies. Dual ErbB receptor inhibitors, first- and second-generation irreversible inhibitors and EGFR monoclonal antibodies have all been explored. This area has been extensively reviewed (*15–18*) and so only a selected subset of illustrative agents is described below.

8.7.1 Dual-ErbB inhibitors

Many human cancers express multiple ErbB2 receptor tyrosine kinases, suggesting that in some settings, dual- or pan-ErbB inhibitors could have therapeutic utility. Lapatinib (**42**; Figure 8.23 and Table 8.10) is a dual reversible inhibitor or both EGFR/ErbB1/HER1 and ErbB/ErbB2/HER2 with a slow receptor kinetic off-rate. This compound has been evaluated in multiple phase I and II clinical trials in patients with multiple tumour types. Encouraging data in metastatic breast cancer led to the compound being approved by the US FDA in 2007, initially for use in combination with the 5-fluorouracil prodrug cepacitabine in patients with advanced or metastatic disease and whose tumours overexpress HER2 (*19*).

8.7.2 First-generation irreversible inhibitors

Most patients responding to the EGFR inhibitors gefitinib and erlotinib eventually become non-respondent as their tumours acquire resistance to these agents (*20*). Secondary resistance mutations in the EGFR gene are notable pathways, as is amplification of the gene encoding hepatocyte growth factor receptor (MET) tyrosine kinase. One of the best characterized resistance mutations is observed in roughly half of patients with NSCLC and acquired resistance to EGFR inhibitors and is a T790M point mutation in exon 20. It is thought that this mutation confers resistance by both reducing the affinity of gefitinib for the binding site as well as increasing its affinity for ATP (*21*). A number of researchers exploited irreversible inhibitors either as a means of targeting mutated EGFR or as a strategy for tackling resistance or to secure pan-ErbB inhibition.

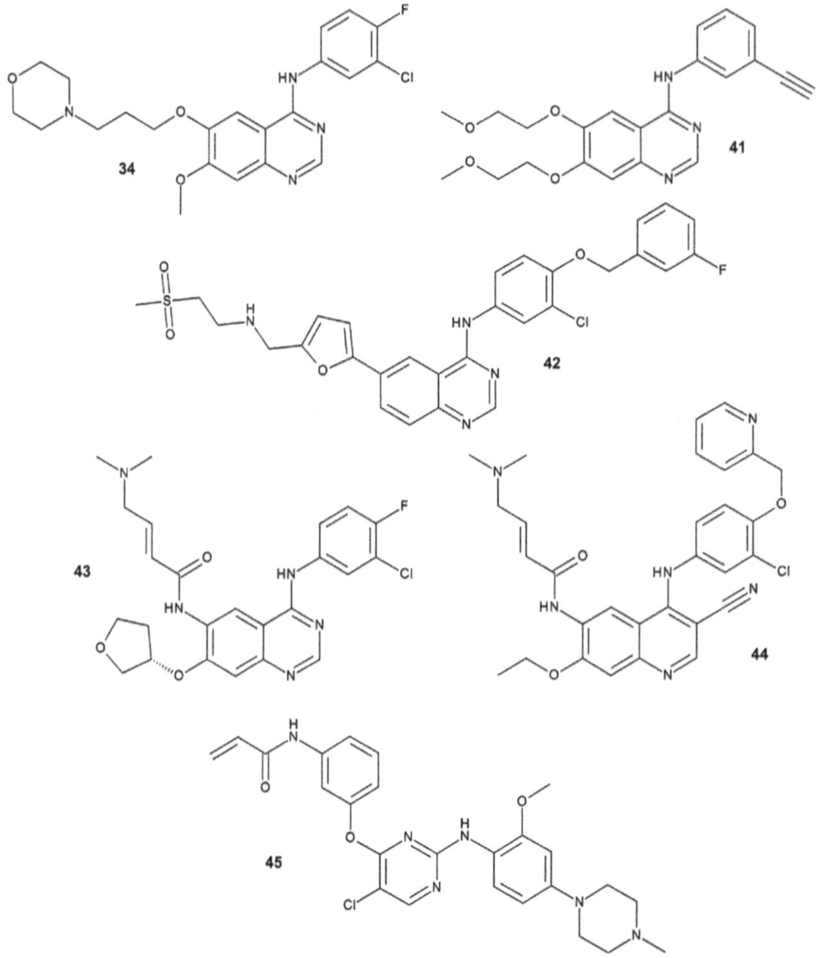

FIGURE 8.23

Structures of EGFR inhibitors.

Table 8.10 Selected ErbB RTK Inhibitors

Name	EGFR Kinase IC$_{50}$ μM	HER2 Kinase IC$_{50}$ μM
Gefitinib **34**	0.001	0.24
Erlotinib **41**	0.001	0.76
Lapatinib **42**	0.01	0.009
BIBW2992 **43**	0.0005	0.014
HKI-272 **44**	0.092	0.059

Values are biochemical assay isolated enzyme inhibition IC$_{50}$s in micrometre.
Source: Reproduced from Ref. (18).

The first-generation inhibitors were based upon the 4-anilinoquinazoline core of gefitinib bearing a pendant electrophilic group that undergoes Michael addition to a conserved cysteine residue in the kinase – Cys797. However, experimental studies showed that the concentration of these inhibitors required to inhibit the T790M-resistant·mutant EGFR also effectively inhibited wild-type EGFR. In patients, this concurrent inhibition led to skin rash and diarrhoea and these safety concerns limited the ability to achieve plasma concentrations sufficient to inhibit the resistance mutated EGFR. As a result, only restricted clinical efficacy of first-generation inhibitors such as BIBW2992 **43** and HKI-272 **44** was seen.

8.7.3 Second-generation irreversible inhibitors

It has been suggested that the quinazoline scaffold may not be best suited to potent inhibition of T790M EGFR because the larger methionine gatekeeper residue of the mutant enzyme induces a steric clash with this relatively inflexible template; this is not observed with the smaller wild-type threonine residue (*22*). Additionally, the lipophilic nature of the methionine is probably also important; with the wild-type threonine, a putative hydrogen bond can be imagined with the quinazoline core, and this is not possible in the methionine case. These factors have led to the exploration of a more flexible pyrimidine series of compounds, exemplified by WZ4002 (**45**, Figure 8.23). Initial reports have focussed on the *in vitro* optimization of wild-type sparing compounds (*23*); the disclosure of further pre-clinical *in vivo* data is keenly awaited.

8.7.4 EGFR monoclonal antibodies

A number of EGFR monoclonal antibodies have been reported (see Chapter 9 for a detailed survey). One of the most extensively studied is cetuximab, first approved in Europe in 2008. This agent is a chimeric (human/mouse) monoclonal antibody to EGFR. Unlike small-molecule EGFR inhibitors, cetuximab is administered intravenously, binding the tyrosine kinase receptor at the cell surface and preventing activation by dimerization. Licensed indications currently include head & neck cancer as well as colorectal cancer, either as single agent or in combination. Personalized medicine is again being explored with this agent as some clinical trials of patients with advanced or metastatic colorectal cancer have found that cetuximab did not work in patients whose tumours had mutations of the K-ras gene (*24*). Interestingly, in April 2012, Bristol-Myers Squibb (BMS) and Lilly received a complete response letter from the FDA regarding the filing of cetuximab in first-line NSCLC; the companies reported no plans to resubmit a filing (*25*).

Modulation of aberrant signal transduction pathways has proven to be an effective way to treat many tumours. Further work in this field is likely to provide additional novel and effective treatments for patients with cancer.

References

1. Cohen, P. The Role of Protein Phosphorylation in Human Health and Disease. *Eur. J. Biochem.* **2001,** *268,* 5001–5010.
2. Ullrich, A.; Schlessinger, J. EGF and TGF Alpha – Implications in Tumours of Epithelial Origin. *Cell.* **1990,** *61,* 203.

3. Arteaga, C. L. Dependence of Human Tumours on Epidermal Growth Factor Receptor. *Oncologist* **2002,** *7* (Suppl. 4), 31–39.

4. Wakeling, A. E.; Barker, A. J.; Davies, D. H.; Brown, D. S.; Green, L. R.; Cartlidge, S. A.; Woodburn, J. R. Specific Inhibition of Epidermal Growth Factor Receptor Tyrosine Kinase by 4-Anilinoquinazolines. *Breast Cancer Res. Treat.* **1996,** *38,* 67–73.

5. Barker, A. J.; Gibson, K. H.; Grundy, W.; Godfrey, A. A.; Barlow, J. J.; Healy, M. P.; Woodburn, J. R.; Ashton, S. E.; Curry, B. J.; Scarlett, L.; Henthorn, L.; Richards, L. Studies Leading to the Identification of Iressa. *Bioorg. Med. Chem. Lett.* **2001,** *11,* 1911–1914.

6. Liao, J. J.-L. Molecular Recognition of Protein Kinase Binding Pockets for Design of Potent and Selective Kinase Inhibitors. *J. Med. Chem.* **2007,** *50,* 409–424.

7. Ranson, M.; Hammond, L. A.; Ferry, D.; Kris, M.; Tullo, A.; Murray, P. I.; Miller, V.; Averbuch, S.; Ochs, J.; Morris, C.; Feyereislova, A.; Swaisland, H.; Rowinsky, E. K. ZD1839, a Selective Oral Epidermal Growth Factor Receptor-Tyrosine Kinase Inhibitor Is Well Tolerated and Active in Patients with Solid Malignant Tumours: Results of a Phase I Trial. *J. Clin. Oncol.* **2002,** *20,* 2240–2250.

8. Thatcher, N.; Chang, A.; Parikh, P.; Rodrigues Pereira, J.; Ciuleanu, T.; von Pawel, J.; Thongprasert, S.; Huat Tan, E.; Pemberton, K.; Archer, V.; Carroll, K. Gefitinib Plus Best Supportive Care in Previously Treated Patients with Refractory Advanced Non-Small-Cell Lung Cancer: Results from a Randomised, Placebo-Controlled, Multicentre Study (Iressa Survival Evaluation in Lung Cancer). *Lancet* **2005,** *366,* 1527–1537.

9. Paez, J. G.; Jänne, P. A.; Lee, J. C.; Tracy, S.; Greulich, H.; Gabriel, S.; Herman, P.; Kaye, F. J.; Lindeman, N.; Boggon, T. J.; Naoki, K.; Sasaki, H.; Fujii, Y.; Eck, M. J.; Sellers, W. R.; Johnson, B. E.; Meyerson, M. EGFR Mutations in Lung Cancer: Correlation with Clinical Response to Gefitinib Therapy. *Science* **2004,** *304,* 1497–1500.

10. Mok, T. S.; Wu, Y.-L.; Thongprasert, S.; Yang, C.-H.; Chu, D.-T.; Saijo, N.; Sunpaweravong, P.; Han, B.; Margono, B.; Ichinose, Y.; Nishiwaki, Y.; Ohe, Y.; Yang, J.-J.; Chewaskulyong, B.; Jiang, H.; Duffield, E. L.; Watkins, C. L.; Armour, A. A.; Fukuoka, M. Gefitinib or Carboplatin–Paclitaxel in Pulmonary Adenocarcinoma. *N. Engl. J. Med.* **2009,** *361,* 947–957.

11. Reck, M. A Major Step Towards Individualised Therapy of Lung Cancer with Gefitinib: The IPASS Trial and Beyond. *Expert Rev. Anticancer Ther.* **2010,** *10,* 955–965.

12. Sanford, M.; Scott, L. J. Gefitinib: A Review of Its Use in the Treatment of Locally Advanced/Metastatic Non-Small Cell Lung Cancer. *Drugs* **2009,** *69,* 2303–2328.

13. Goto, K.; Ichinose, Y.; Ohe, Y.; Yamamoto, N.; Negoro, S.; Nishio, K.; Itoh, Y.; Jiang, H.; Duffield, E.; McCormack, R.; Saijo, N.; Mok, T. S. K.; Fukuoka, M. Epidermal Growth Factor Receptor Mutation Status in Circulating Free DNA in Serum: From IPASS, a Phase III Study of Gefitinib or Carboplatin/ Paclitaxel in Non-Small Cell Lung Cancer. *J. Thorac. Oncol.* **2012,** *7,* 115–121.

14. Thongprasert, S.; Duffield, E.; Saijo, N.; Wu, Y.-L.; Yang, J. C.-H.; Chu, D.-T.; Liao, M.; Chen, Y.-M.; Kuo, H.-P.; Negoro, S.; Lam, K. C.; Armour, A. A.; Magill, P.; Fukuoka, M. Health-Related Quality-of-Life in a Randomized Phase III First-Line Study of Gefitinib versus Carboplatin/Paclitaxel in Clinically Selected Patients from Asia with Advanced NSCLC (IPASS). *J. Thorac. Oncol.* **2011,** *6,* 1872–1880.

15. Albanell, J.; Gascon, P. Small Molecules with EGFR-TK Inhibitor Activity. *Curr. Drug Targets* **2005,** *6,* 259–274.

16. Hartmann, J. T.; Haap, M.; Kopp, H.-G.; Lipp, H.-P. Tyrosine Kinase Inhibitors – A Review on Pharmacology, Metabolism and Side Effects. *Curr. Drug Metab.* **2009,** *10,* 470–481.

17. Mukherji, D.; Spicer, J. Second-Generation Epidermal Growth Factor Tyrosine Kinase Inhibitors in Non-Small Cell Lung Cancer. *Expert Opin. Invest. Drugs* **2009,** *18,* 293–301.

18. Sharma, P. S.; Sharma, R.; Tyagi, T. Receptor Tyrosine Kinase Inhibitors as Potent Weapons in War against Cancers. *Curr. Pharm. Des.* **2009,** *15,* 758–776.

19. Higa, G. M.; Abraham, J. Lapatinib in the Treatment of Breast Cancer. *Expert Rev. Anticancer Ther.* **2007,** *7,* 1183–1192.

20. Kobayashi, S.; Boggon, T. J.; Dayaram, T.; Jänne, P. A.; Kocher, O.; Meyerson, M.; Johnson, B. E.; Eck, M. J.; Tenen, D. G.; Halmos, B. EGFR Mutation and Resistance of Non-Small-Cell Lung Cancer to Gefitinib. *N. Engl. J. Med.* **2005,** *352,* 786–792.

21. Yun, C.-H.; Mengwasser, K. E.; Toms, A. V.; Woo, M. S.; Greulich, H.; Wong, K.-K.; Meyerson, M.; Eck, M. J. The T790M Mutation in EGFR Kinase Causes Drug Resistance by Increasing the Affinity for ATP. *Proc. Natl. Acad. Sci. U.S.A.* **2008,** *105,* 2070–2075.

22. Zhou, W.; Ercan, D.; Chen, L.; Yun, C.-H.; Li, D.; Capelletti, M.; Cortot, A. B.; Chirieac, L.; Iacob, R. E.; Padera, R.; Engen, J. R.; Wong, K.-K.; Eck, M. J.; Gray, N. S.; Janne, P. A. Novel Mutant-Selective EGFR Kinase Inhibitors against EGFR T790M. *Nature* **2009,** *462,* 1070–1074.

23. Zhou, W.; Ercan, D.; Jänne, P. A.; Gray, N. S. Discovery of Selective Irreversible Inhibitors for EGFR-T790M. *Bioorg. Med. Chem. Lett.* **2011,** *21,* 638–643.

24. Bokemeyer, C.; Bondarenko, I.; Makhson, A.; Hartmann, J. T.; Aparicio, J.; de Braud, F.; Donea, S.; Ludwig, H.; Schuch, G.; Stroh, C.; Loos, A. H.; Zubel, A.; Koralewski, P. Cetuximab in the First-Line Treatment of Metastatic Colorectal Cancer. *J. Clin. Oncol.* **2009,** *10,* 663–671.

25. http://www.bms.com/news/press_releases/pages/default.aspx (retrieved May 21, 2012).

Targeting HER2 by monoclonal antibodies for cancer therapy

Hasmann

Pharma Research and Early Development (pRED), Roche Diagnostics GmbH, 82377 Penzberg, Germany

CHAPTER OUTLINE

9.1 Introduction 284
9.2 Structure and Function of HER2.. 285
9.3 Trastuzumab (Herceptin™) 286
   9.3.1 Trastuzumab manufacturing and quality control 287
   9.3.2 Trastuzumab's multiple mechanisms of action 288
   9.3.3 Nonclinical pharmacology 289
   9.3.4 Clinical data and therapeutic efficacy of trastuzumab 290
      9.3.4.1 Pharmacokinetic profile in patients 290
      9.3.4.2 Efficacy in metastatic breast cancer 290
      9.3.4.3 Trastuzumab resistance and the treatment beyond progression concept 291
      9.3.4.4 Tolerability and safety of trastuzumab 293
      9.3.4.5 Efficacy of trastuzumab in early breast cancer..............................294
      9.3.4.6 Efficacy of trastuzumab in advanced metastatic gastric cancer 294
      9.3.4.7 Development of a subcutaneous formulation of trastuzumab 296
9.4 Pertuzumab (Perjeta™) 297
   9.4.1 Mechanisms of action and preclinical activity of pertuzumab 297
   9.4.2 Clinical efficacy of pertuzumab..299
9.5 Trastuzumab Emtansine (T-DM1; Kadcyla™) 300
   9.5.1 Characteristics of trastuzumab emtansine and its mechanisms of action 301
   9.5.2 Clinical efficacy..302
9.6 Conclusion... 303
References 303

ABSTRACT

HER2 (ErbB-2) is a member of the human epidermal growth factor receptor tyrosine kinase family which is involved in the regulation of cell proliferation, survival and differentiation. Soon after its discovery, HER2 was shown frequently to be overexpressed in breast cancer and was associated with a worse prognosis. It was identified as a target for drug development and molecular cloning of the gene and expression in cell lines provided a vehicle for the selection of HER2-specific antibodies. The monoclonal antibody trastuzumab is the first HER2-targeting drug approved for cancer treatment. By significantly extending the time to disease progression and overall survival of patients, it has become established in all treatment lines of early and metastatic HER2-positive breast cancer, as well as in HER2-positive advanced metastatic gastric or gastroesophageal junction cancer. Combination of trastuzumab with pertuzumab, a second antibody binding to a distinct epitope on HER2 which implies a different mode of action, took the treatment of HER2-positive metastatic breast cancer to the next level of success. Finally, trastuzumab emtansine is an antibody–drug conjugate that retains all pharmacodynamic activities of trastuzumab and delivers a toxic maytansinoid directly to the tumour cells. Current clinical results indicate that trastuzumab emtansine may be more efficacious and less toxic than trastuzumab plus chemotherapy, and further improvement is expected in combination with pertuzumab.

Keywords/Abbreviations: HER2 dimerization inhibitor (HDI); Human epidermal growth factor receptors (HER1–4 or ErbB-1–4); Epidermal growth factor receptor (EGFR or ErbB-1); c-erbB-2/*c-erbB-2* gene; Tyrosine phosphorylation; Murine antibody 4D5; Human anti-mouse antibody (HAMA); Complementarity determining regions (CDRs); International nonproprietary name (INN); Chinese hamster ovary (CHO) cells; Truncated HER2 (p95HER2); Phospho-inositol-3-kinase (PI3K); Fc-gamma receptors (FcγR); Natural killer cells (NK cells); Antibody-dependent cellular cytotoxicity (ADCC); Progression-free survival (PFS); Pathological complete response (pCR); Mitogen-activated protein kinase (MAPK); Cyclin-dependent kinase (CDK); Immunohistochemistry (IHC); Phosphatase and tensin homologue (PTEN) mutations; Multiple gated acquisition (MUGA) scan; Recombinant human hyaluronidase (rHuPH20); Antidrug antibodies (ADAs); Extracellular domain (ECD) shedding; Antibody–drug conjugates (ADCs); 4-(*N*-maleimidomethyl) cyclohexane-1-carboxylate (MCC); Maximum tolerated dose (MTD).

9.1 INTRODUCTION

The proliferation, differentiation and survival of normal epidermal cells are tightly regulated processes in response to external stimuli like environmental stress, growth factors or hormones. These extracellular factors are often sensed by cell membrane-bound proteins, referred to as growth factor receptors. With the advent of molecular biology techniques, the first such growth factor receptors were cloned and functionally characterized as tyrosine kinases in the early 1980s. (As described in Chapter 8, kinases are enzymes that transfer a phosphate group from ATP to a peptide/protein. Tyrosine kinases are so named because, in the transfer process, a tyrosine residue is phosphorylated). Soon it turned out that the gene encoding the human epidermal growth factor receptor (EGFR, HER1) protein had a high sequence homology to an oncogene (cancer-causing gene), v-erbB, carried by the avian erythroblastosis virus which causes erythroblastosis and fibrosarcomas in susceptible birds. (Erythroblastosis is the abnormal presence of erythroblasts, which are nucleated blood cells and are the immediate precursors of erythrocytes; fibrosarcomas are malignant tumours that arise from fibroblasts, i.e. cells that produce connective tissue). This association suggested that the mammalian equivalents of the avian tumour virus genes would be implicated in cancer and that their normal versions in nonmalignant cells were called proto-oncogenes. Finally, it turned out that there were four members in the family of tyrosine kinase receptors with closely related structure and two different nomenclatures were established, either ErbB-1 (EGFR), ErbB-2, ErbB-3 and ErbB-4, referring to the relationship with the avian erythroblastosis virus, or HER1, HER2, HER3 and HER4, referring to the function as human epidermal growth factor receptors (HER).

HER2 was independently detected by three different research groups. Initially it was described from a chemically induced rat neuroblastoma; the gene was designated *neu* and shown to transform fibroblast cell lines in culture. Subsequently, the rat gene was demonstrated to display close similarity to the *c-erbB-2* gene found in human cancer. Around the same time, another group of investigators found an EGFR-related gene to be amplified in tissue from a human mammary carcinoma which was then found to be identical with the *c-erbB-2* gene. Amplification of the HER2/*neu* oncogene was then described in 1987 to occur with a frequency of 20–30% in breast cancer patients and was a significant predictor of early relapse and short survival. Consequently, HER2 appeared to be an attractive target for cancer treatment.

9.2 STRUCTURE AND FUNCTION OF HER2

The elucidation of major structural features of the HER tyrosine kinase family by X-ray crystallography enabled insight into the activation mechanisms of these growth factor receptors. Like the other HER family members, the HER2 protein consists of a regulatory extracellular domain (ECD) including two cysteine-rich subdomains, a membrane-spanning region of about 22 amino acids, and an intracellular domain that comprises both, the tyrosine kinase function as well as several tyrosine phosphorylation sites on its C-terminal tail. In contrast to HER1, HER3 and HER4, the extracellular domain of HER2 exists in an untethered conformation constitutively exposing its dimerization arm. This particular structure of its extracellular domain resembles the ligand-activated conformation of other HER family members and explains why HER2 neither has nor requires a cognate ligand in order to become activated (*1*), a fact that had been postulated long before its structure was known.

HER2 signalling is strictly dependent on the interaction either with another HER2 molecule (homodimerization) or with a ligand-activated HER family member (heterodimerization). The following mechanisms can lead to the activation of HER2:

Ligand-activated heterodimerization: Upon binding of the respective growth factors, their so-called ligands, the autoinhibitory extracellular domains of HER1, HER3 and HER4, undergo a substantial conformational change which finally results in exposing their dimerization domains (Figure 9.1). This open conformation of the receptors enables their interaction with other HER family members, HER2 being the preferred partner.

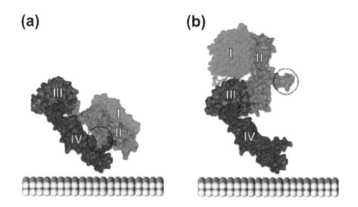

(a)   **(b)**

FIGURE 9.1

Conformation of inactive (a) and EGF (orange-coloured) activated (b) EGFR extracellular domain. (See colour plate.)

Source: Illustrations taken from Ref. (22).

HER2 overexpression: HER2 has been found to be overexpressed in about one of five human breast cancers. In most instances, this overexpression of the protein is caused by amplification of the *c-erbB-2* gene. The resulting high density of about 1–2 million receptors per cell, compared to 20,000–50,000 on normal breast epithelial cells, leads to the formation of HER2 homodimers or oligomers and constitutive activation of the HER2 signalling pathway.

HER2 ECD shedding: The extracellular domain of HER2 can be cleaved by matrix metalloproteinases like ADAM10, released into the circulation and may be found as 'soluble HER2' in tumour patient serum samples. The resulting truncated receptor protein, p95HER2, is retained in the cell membrane. Because the transmembrane domain as well as the intracellular kinase domain are both involved in the formation of dimers, p95HER2 has a tendency to self-associate and therefore represents a constitutively active form of HER2. In addition, p95HER2 may also associate with other HER family members like HER1 or HER3, which in this case would probably not need ligand activation, because there is no interaction between extracellular domains. The role of extracellular domain shedding as an activation mechanism for HER2 is underlined by the prognostic significance of p95HER2 in breast cancer.

Homo- or heterodimerization of HER2 finally leads to the phosphorylation of the six tyrosine residues on its cytoplasmic carboxy-terminal end. The phosphotyrosine sites then recruit docking (adapter) proteins containing Src homology 2 or phosphotyrosine binding domains, like grb2, shc and chk, which eventually initiate the intracellular signalling cascade. HER2 downstream signalling ends up at transcription factors in the cell nucleus and finally regulates physiological processes important for cell survival, proliferation and migration. Quantitative and qualitative modulation of the signalling is implied by the various combinations of receptor homo- and heterodimers, because the number of intracellular tyrosine phosphorylation sites and the type of adapter proteins recruited by them varies within the HER family (2).

9.3 TRASTUZUMAB (HERCEPTIN™)

As soon as the link of HER2 overexpression in breast cancer to a more aggressive course of the disease was established, HER2 became a focus of drug research activities. After monoclonal antibodies directed against the rat version of HER2 had been shown to inhibit the proliferation of cells transfected with the *neu* oncogene, antibodies specifically binding the human HER2 protein were developed. Balb/c mice were immunized with c-erbB-2-transfected NIH 3T3 cells expressing high amounts of human protein. Splenocytes derived from mice generating high antibody titres directed against human HER2 were isolated and fused with myeloma cells in order to create immortal antibody-producing hybridoma cells. The screening strategy for HER2-specific antibodies included high-affinity binding to HER2 as determined by HER2 immunoprecipitation, binding to HER2 on the cell surface as measured by flow cytometry, lack of binding to EGFR and biological characterization by cell proliferation and anchorage-independent growth assays. One of the monoclonal antibodies created and selected by these techniques, called 4D5, strongly inhibited the proliferation of HER2-overexpressing human patient-derived breast cancer cell lines, while HER2 normal expressing cells were unaffected, and it was not cross-reactive to EGFR. As previous experience had shown that the development of human anti-mouse antibody response was a major limitation to the clinical use of murine antibodies, 4D5 was subsequently 'humanized' in order to reduce its immunogenicity. The gene segments encoding the variable regions of the heavy and light chains, V_H and V_L, were cloned and humanized applying a gene conversion mutagenesis approach using long preassembled oligonucleotides (3). The complementarity determining regions of the murine 4D5 antibody were inserted into selected human framework sequences. The heavy chain constant region sequence selected was of the IgG1 subclass; however, two mutations were introduced, into the CH3 domain, to reduce the potential for immunogenicity (3). The optimized antibody bound HER2 with high affinity and specificity and was composed of about 95% human and 5% murine sequences. Each heavy chain was composed of 451 amino acid residues and the kappa light chains of 214 residues. The antibody's international nonproprietary name (INN) issued by the World Health Organization became trastuzumab.

9.3.1 Trastuzumab manufacturing and quality control

Chinese hamster ovary (CHO) cells are the most commonly used mammalian hosts for industrial production of recombinant proteins because of their rapid growth and high productivity. A plasmid containing the trastuzumab-encoding gene construct was transfected into CHO cells in order to establish a master cell bank for its commercial manufacturing in large scale. A series of aliquots of this production cell line is maintained frozen in liquid nitrogen for later reference and restarting of an identical fresh culture, as required.

When a new manufacturing campaign is kicked off, all components of the culture broth have to meet predefined quality criteria including certified origin, purity and being free of microbial contamination. Using a vial of the master cell bank or working cell bank, the cells are thawed, seed train cultures started in spinner flasks and further expanded in bioreactors of increasing volumes. The final manufacturing bioreactors comprise a working volume of about 12,000 l. After about 2 weeks of fermentation with continuous monitoring of many different in-process control parameters, e.g. cell density and viability, pH, oxygen and glucose concentrations, the intact cells and cell debris are removed from the culture broth by centrifugation. The subsequent industrial-scale purification process of the antibody containing cell-free culture supernatant consists of several steps including affinity, anion exchange and hydrophobic interaction chromatography techniques. For safety reasons, a standardized and highly effective virus removal process is then applied on the purified antibody solution.

Before the purified drug substance can be used for manufacturing of the final marketed product, it has to undergo extensive analytical testing procedures. The parameters that need to be assessed relate to the physicochemical (capillary isoelectric focussing, peptide map) and functional properties (binding to HER2, inhibition of cell proliferation) of the protein, as well as to safety aspects (chemical purity confirmed by different technologies, endotoxin content). As complex biological products like antibodies have a certain innate molecular heterogeneity, e.g. in terms of their glycosylation, all these analytical parameters have to meet certain release specification values that are predefined by reference standards.

Interestingly, despite all the extensive protein purification and quality control efforts, the discussion and uncertainties around bovine spongiform encephalitis ('mad cow disease') in the 1990s had an impact on the manufacturing of biological agents like trastuzumab. While bovine peptone was initially used as an amino acid source in the fermentation broth, a new process utilizing porcine peptone was developed and received regulatory approval in 2004.

A drug substance batch that meets all the quality criteria including the above-mentioned ones is finally released for manufacturing of the drug product. The drug substance is prepared at a defined concentration in a standardized buffer, filled in vials under sterile conditions and finally lyophilized.

9.3.2 Trastuzumab's multiple mechanisms of action

Trastuzumab binds with high affinity and specificity to an epitope in the juxtamembrane region of the HER2 extracellular domain (*1*). A variety of different modes of action have been proposed to account for the antitumour activity of trastuzumab. While the majority of these mechanisms described in the literature are in fact (intra)cellular consequences of HER2 signalling inhibition, the following three direct effects mediated by the trastuzumab–HER2 interaction can be considered as trastuzumab's real mechanisms of action:

- Inhibition of ligand-independent HER2 signalling
- Prevention of HER2 shedding
- Engagement of immune effector cells directing them against tumour cells

Inhibition of ligand-independent HER2 signalling was demonstrated using Western blotting and coimmunoprecipitation experiments with HER2-positive breast cancer cell lines. Trastuzumab effectively reduces the phosphorylation of HER2 and various downstream signalling molecules like phospho-inositol-3-kinase (PI3K) and Akt as shown by Western blotting. Furthermore, trastuzumab can disrupt the ligand-independent interaction of HER2 and HER3 that occurs under conditions of HER2 overexpression, while it has little impact on the growth factor-activated association of this potent signalling pair (*4*). Upon trastuzumab treatment, ligandless HER3 dissociates from HER2 in a dose-dependent manner, which is ensued by rapid loss of HER3, PI3K, and Akt phosphorylation, and ultimately results in reduced proliferation.

A second direct effect of trastuzumab is the prevention of proteolytic HER2 extracellular domain cleavage, most likely by sterically blocking enzyme access to the cleavage site. Thus it blocks the formation of the constitutively active truncated version of HER2, called p95HER2. *In vitro*, trastuzumab efficiently inhibits ectodomain shedding from HER2-overexpressing breast tumour cells (*5*), and the decline of the extracellular domain found in patient serum or plasma ('soluble HER2' or 'shed HER2') as early as 1 week after trastuzumab treatment initiation confirms the relevance of this effect. However, although HER2 extracellular domain shedding obviously is an important activation mechanism of HER2, it cannot be exploited for patient selection for the following reasons: (1) there is currently no clinically validated quantitative assay for p95HER2 which could be used on formalin-fixed patient tumour samples and (2) serum HER2 levels are not reliable indicators for the presence of p95HER2, because they are influenced by too many different parameters, like individual tumour load, unpredictable shedding activity in tumours, individual metabolization rate or even metabolic state of patients.

The third mechanism of action of trastuzumab relates to the engagement of immune effector cells directing them against tumour cells. As an IgG1 type of immunoglobulin, trastuzumab can bind with its Fc region to Fc-gamma receptors (FcγR), in particular to the FcγR IIIa (CD16) which is mainly found on natural killer cells (NK cells) and monocytes/macrophages. Thus, trastuzumab bound in large amounts to the surface of HER2-overexpressing cancer cells can recruit NK cells which in turn become activated and secrete various substances, like perforin and granzyme B, which eventually kill the tumour cells. This effect has been termed antibody-dependent cellular cytotoxicity (ADCC) and had already been described *in vitro* and in mouse tumour xenograft models for the murine version of trastuzumab, 4D5 (*3,6*). In patients, a role of this antitumour mechanism is supported by the enhanced accumulation of leukocytes and cytotoxic molecules at the tumour after administration of trastuzumab-containing neoadjuvant therapy. Furthermore, another clinical study investigating the influence of FcγR genotypes on treatment outcome demonstrated that patients carrying the high-affinity variant of FcγR IIIa had a higher overall response rate (ORR) than patients with the low-affinity variant of FcγR IIIa (82% vs. 40%) and a significantly prolonged median progression-free survival (PFS, 72 months vs. 12.9 months). Although this is no direct evidence that ADCC occurs in patient tumours, these results strongly argue for the contribution of Fc-mediated immune effector functions to the overall antitumour effect of trastuzumab. Generally, however, it is not possible to predict or estimate in quantitative terms the contribution of ADCC to trastuzumab's therapeutic efficacy, because it depends on multiple factors that may vary from patient to patient, like the number of NK cells, the activation state of NK cells, the patient's individual immune status, chemotherapy regimens applied (which may compromise the immune system), presence and role of alternative signalling pathways in the tumour cells, etc.

The above-described three direct mechanisms of action entailed by trastuzumab binding to HER2-overexpressing cells should be distinguished from various downstream effects arising as consequences of the primary modes of action. Among the secondary effects of HER2 signalling inhibition by trastuzumab that have been described as mechanisms of action are findings such as reduced PI3K/Akt and mitogen-activated protein kinase signalling, enhanced nuclear import and stabilization of the cyclin-dependent kinase inhibitor p27^{Kip1}, G_0/G_1 cell cycle arrest, diminished secretion of angiogenic factors, and impaired DNA damage response. The occurrence of these downstream effects largely depends on the individual tumour characteristics including the cellular equipment with growth factor receptors and signalling pathways activated.

Of note, an initially postulated mechanism of action of trastuzumab could not be verified. During its early development, it was believed that trastuzumab would downregulate HER2 from the cell surface, thereby eliminating the growth-promoting effects of HER2 overexpression. However, later it was realized that HER2 continued to be overexpressed on tumours after trastuzumab therapy. Using various techniques independent of changes in sample protein resulting from growth inhibition, it could be shown that most of the HER2 protein internalized by endocytosis is recycled to the cell membrane even with trastuzumab bound to it (7). Consequently, the cell surface expression of HER2 does not significantly change during treatment, but its persistent presence may rather enable continued recruitment of immune effector cells and supports the treatment beyond progression concept (see below).

9.3.3 Nonclinical pharmacology

Trastuzumab binds to human HER2 with about threefold higher affinity than the murine parent 4D5 and inhibits the proliferation of HER2-positive tumour cells more efficiently than HER2 normal expressing cells (3). Using various human tumour cell lines, a clear relationship could be established between the extent of HER2 expression and the inhibition of proliferation in vitro. The antitumour effect of trastuzumab was also demonstrated in soft agar colony formation assays and the dose-dependent inhibition of human tumour xenograft growth in vivo was shown in athymic nude mice. In these experiments, growth of the BT-474 xenograft tumour model was completely stopped at dosages above 1 mg/kg, while at lower doses of the drug, tumour growth was slowed dose dependently. Furthermore, trastuzumab proved to enhance the antitumour effect in combination with various cytotoxic agents commonly used for cancer chemotherapy, such as paclitaxel, docetaxel, doxorubicin, epirubicin, cisplatin, carboplatin and vinorelbine. The synergy seen in these in vivo preclinical models eventually encouraged the exploration of trastuzumab combination treatment with chemotherapy in

HER2-positive breast cancer patients. Trastuzumab was also shown to sensitize tumour cells to radiation-induced apoptosis in a HER2-expression-dependent manner, and radiosensitization could be confirmed in a clinical trial. The enhancement of cytotoxic treatment approaches like chemotherapy and radiotherapy by trastuzumab very likely is a result of the shutdown of HER2-associated cell survival signals.

9.3.4 Clinical data and therapeutic efficacy of trastuzumab

Clinical development of trastuzumab single-agent treatment started in 1992, followed by several additional phase I and phase II trials investigating various dosages, treatment schedules and combinations with chemotherapy drugs. A large multinational phase II study enrolled 222 patients from April 1995 to September 1996 and supported trastuzumab's first approval for the treatment of HER2-positive metastatic breast cancer by the US Food and Drug Administration (FDA) in 1998. A phase III trial in metastatic breast cancer started in 1995 investigated the efficacy of trastuzumab in combination with chemotherapy and became the basis for approval in first-line treatment. Clinical development for the adjuvant treatment of HER2-positive early breast cancer was initiated in 2002, resulting in its first marketing authorization approval for this use in 2006. A phase III trial in the second tumour indication for which trastuzumab won approval in 2010, advanced HER2-positive gastric or gastroesophageal junction cancer, was started in 2005. The main clinical results are summarized in the following chapters.

9.3.4.1 Pharmacokinetic profile in patients

Trastuzumab is administered to patients by intravenous (IV) infusion which should last 90 min at first treatment and can be reduced to 30 min subsequently. The dose regimen developed initially involved a 4 mg/kg loading dose followed by a weekly 2 mg/kg maintenance dose. It was reported in Cobleigh et al. (8) that the minimum plasma concentration before the next dosing (C_{min}) was shown to increase from 25 µg/ml after the first administration to about 65 µg/ml at week 20 and the mean concentration at steady state was 59.7 µg/ml.

Later, another dosing regimen was developed, starting with a dose of 8 mg/kg, followed by 6 mg/kg every 3 weeks. Various pharmacokinetic analyses have been reported, including that mean C_{min} values were determined to be 27.3 µg/ml at cycle 2 and 52.7 µg/ml at cycle 15 and the mean steady-state concentration was calculated at 65.47 µg/ml (9).

Early pharmacokinetic studies of trastuzumab underestimated the serum half-life in patients. However, subsequent pharmacokinetic analyses using a two-compartment model estimated the terminal elimination half-life of trastuzumab at 28.5 days.

Trastuzumab has no effect on the pharmacokinetics of various chemotherapeutic drugs. On the other hand, paclitaxel was shown to have no clinically significant influence on the pharmacokinetics of trastuzumab.

9.3.4.2 Efficacy in metastatic breast cancer

The clinical proof of concept for trastuzumab efficacy in HER2-positive breast cancer was provided by a phase II trial in patients who had previously received one or two chemotherapy regimens for their metastatic disease. A total of 222 patients whose tumours showed at least 10% cells with a 2+ or 3+ score using an immunohistochemistry (IHC) assay for HER2 staining were enrolled. The ORR turned out to be 15% as determined by an independent committee and the median duration of response was 9.1 months. The trial confirmed that trastuzumab was generally well tolerated and active as a single agent in heavily pretreated patients.

The clinical efficacy of trastuzumab in first-line treatment of HER2-positive breast cancer was tested in combination with chemotherapy in a phase III trial. A total of 469 patients with HER2-positive (IHC 3+ or 2+) metastatic breast cancer were randomized to receive either chemotherapy alone or chemotherapy in combination with trastuzumab. Chemotherapy, either paclitaxel or an anthracycline like doxorubicin, was administered every 3 weeks for at least six cycles; continuation was allowed according to the discretion of the investigator. Trastuzumab was given at a loading dose of 4 mg/kg, followed by weekly 2 mg/kg maintenance dose until disease progression. After a median follow-up period of 30 months, the time to progression for patients receiving chemotherapy was only calculated at 4.6 months, while the addition of trastuzumab to chemotherapy extended the time to progression to 7.4 months. Response rate (17% vs. 41% with paclitaxel, 42% vs. 56% with anthracyclines) and duration of response (4.5 months vs. 10.5 months with paclitaxel, 6.7 months vs. 9.1 months with anthracyclines) confirmed the improved efficacy achieved by the combination therapy. Median overall survival was also significantly improved by the addition of trastuzumab, from 20.3 to 25.1 months, when both chemotherapy treatment groups were taken together. The magnitude of the effect on overall survival may even be underestimated in this study, because patients who had initially been randomized to the chemotherapy-alone group subsequently received trastuzumab. In summary, this trial convincingly demonstrated enhanced therapeutic efficacy of combination treatment compared to chemotherapy alone and based on this result trastuzumab was rewarded marketing approval for the use as first-line treatment of HER2-positive metastatic breast cancer.

Subsequent to the pivotal studies that lead to health authority approval, two different dosages (4 → 2 mg/kg vs. 8 → 4 mg/kg) of trastuzumab were also tested for first-line treatment of HER2-positive breast cancer without chemotherapy. Although there was significant therapeutic efficacy in monotherapy, with response rates of 24% in the standard dose and 28% in the higher dose, it seemed to be less than in combination with chemotherapy. Therefore, trastuzumab (4 mg/kg loading dose followed by weekly 2 mg/kg) in combination with initial chemotherapy became the standard-of-care first-line treatment for HER2-positive metastatic breast cancer. Table 9.1 provides an overview of various clinical trials investigating trastuzumab in metastatic breast cancer.

9.3.4.3 Trastuzumab resistance and the treatment beyond progression concept

Despite the above-described clinical activity, many of the patients initially responding to trastuzumab-based therapy of metastatic breast cancer experience disease progression sooner or later. Therefore 'trastuzumab resistance' has become a topic of extensive research. Preclinical studies have characterized a wide range of potential mechanisms by which the trastuzumab sensitivity of HER2-positive breast tumour cells can be reduced. Certain mechanisms often associated with resistance to small molecule drugs, e.g. enhanced metabolization or active drug efflux by cell membrane-located multidrug resistance pumps like P-glycoprotein, do not play a role for antibodies. Rather, most of the trastuzumab resistance mechanisms identified in preclinical systems relate to downstream changes in the intracellular HER2 signalling cascade, or to the activation of alternative signalling pathways. Several of these factors, like mutation or loss of phosphatase and tensin homologue, mutations in the

Table 9.1 Trastuzumab Plus Chemotherapy as First-Line Treatment: Results from Key Published Clinical Trials

Study (year)	Study Phase	Regimen	Patients (n)	ORR (%)	Median PFS/TTP (months)	Median OS (months)
Slamon et al. (2001)	Phase III	Chemotherapy + T vs. Chemotherapy alone	469	50 vs. 32*	7.4 vs. 4.6*	25.1 vs. 20.3*
Marty et al. (2005)	Phase II	T + docetaxel vs. docetaxel alone	186	61 vs. 34*	11.7 vs. 6.1*	31.2 vs. 22.7*
Andersson et al. (2011)	Phase III	Docetaxel + T vs. vinorelbine + T	143	59.3 in both arms	12.4 vs. 15.3	35.7 vs. 38.8
Robert et al. (2006)	Phase III	T + paclitaxel vs. T + paclitaxel + carboplatin	196	36 vs. 52*	7.1 vs. 10.7*	32.2 vs. 35.7
Perez et al. (2005)	Phase II	T + paclitaxel + carboplatin 3-week schedule vs. weekly schedule	91	65 vs. 81	9.9 vs. 13.8	27 vs. 38
Valero et al. (2001)	Phase III	T + docetaxel vs. T + docetaxel + carboplatin	263	72 in both arms	11.1 vs. 10.4	37.1 vs. 37.4
Wardley	Phase II	T + docetaxel + capecitabine vs. T + docetaxel	222	70.5 vs. 72.7	17.9 vs. 12.8*	Not available

ORR, overall response rate; OS, overall survival; PFS, progression-free survival; T, trastuzumab; TTP, time to disease progression.
*$p < 0.05$.
Source: Modified from Ref. (21).

PI3K gene (PIK3CA), and insulin-like growth factor-1 receptor expression, have extensively been studied in clinical trials. Although there were certain trends with some parameters which confirmed their role in the HER2 signalling cascade, none of these factors were strongly correlated with clinical efficacy of trastuzumab. Consequently, HER2 protein overexpression or gene amplification in the tumour sample remains the only criterion for patient selection for trastuzumab treatment.

Several explanations may account for the discrepancies between the preclinical and clinical data on trastuzumab sensitivity: whereas *in vitro* studies often focus on a single resistance mechanism, multiple mechanisms of action determine the efficacy of trastuzumab in patients. The impact of any single biomarker is therefore likely to be diluted, which is consistent with the observation of 'trends' rather than 'all-or-nothing' effects in the clinical data. Furthermore, the contribution of ADCC to the efficacy of trastuzumab is not reflected in preclinical systems based on cultured cell lines or xenograft models in immunocompromized mice.

The apparent differences to small molecule resistance mechanisms eventually suggested the concept of 'treatment beyond progression' with trastuzumab. Meanwhile, several clinical studies have confirmed that trastuzumab continues to contribute therapeutic efficacy even after progression on trastuzumab-based therapy (*10*). This is unusual compared to the experience with standard chemotherapeutic agents and may be explained as follows: HER2 overexpression is usually caused by gene amplification and as there is no physiological mechanism to eliminate gene amplification, HER2 remains overexpressed throughout trastuzumab therapy unless there is a selective outgrowth of HER2-negative cells in a heterogeneous tumour. Therefore, binding of trastuzumab can still recruit immune cells to the tumour and contribute to the overall therapeutic effect even if HER2 signalling is no longer essential in a progressing tumour. On the other hand, if HER2 signalling is not persistently blocked, it could contribute to an even more aggressive growth of the cancer.

9.3.4.4 Tolerability and safety of trastuzumab

Compared to other cancer treatments, intravenously administered trastuzumab is tolerated very well by patients. The most frequently reported adverse events associated with trastuzumab are infusion-related symptoms, such as fever and chills, typically occurring only with the first dose (*11*). Toxicity is usually driven by the chemotherapeutic agent used in combination treatment. Consistent with this experience, no maximum tolerated dose (MTD) could be established in the early clinical trials of trastuzumab single-agent treatment.

A lot of discussion and clinical investigation on trastuzumab's safety has centred on its potential to cause symptomatic or asymptomatic cardiac dysfunction. Depending on the study and chemotherapy combination regimen used, the reported incidence of congestive heart failure varies from 0.5% to 8.8% (*11*). On the basis of more recent data on the use of trastuzumab in the adjuvant setting, both asymptomatic and symptomatic side effects seem to be treatable and mostly reversible. The risk of severe chronic heart failure or cardiac events in patients who received anthracyclines as treatment for breast cancer before receiving trastuzumab ranges from 0.6% to 3.9%. In addition, a meta-analysis including 10,955 patients from adjuvant trials showed that the risk of clinically significant cardiac events (grade 3 or 4) related to chemotherapy and 1 year of trastuzumab therapy was 1.9% vs. 0.3% in patients who did not receive the antibody.

With baseline cardiac assessment including history, physical examination, electrocardiogram, echocardiogram and/or multiple gated acquisition scan, and cardiac function monitored about every

3 months during trastuzumab therapy, the risk is quite low. Furthermore, the cardiac side effects of trastuzumab seem to be reversible. Although different hypotheses have been suggested and many questions remain unanswered, there seems to be a close association of cardiac toxicity and predamage to the heart. A preferred view is that trastuzumab may not have a direct toxic effect on cardiomyocytes, but it may inhibit HER2-dependent repair processes after any kind of cardiac predamage that could be caused by chemotherapy, or radiotherapy or that could be age related. This hypothesis is in line with the finding that cardiac events are most frequently observed in patients treated with anthracyclines, chemotherapeutic drugs that are known for their relatively high cardiotoxicity.

9.3.4.5 Efficacy of trastuzumab in early breast cancer

Early breast cancer is defined as a state of disease in which the tumour and any affected regional lymph nodes can be surgically removed, with no residual tumour tissue left at the primary site of disease and absence of any distant or organ metastases. In this situation, prophylactic or 'adjuvant' treatment with chemotherapy and/or endocrine therapy had been shown to prolong both, disease-free and overall survival. Considering the higher molecular tumour heterogeneity at an advanced stage of disease, the impressive efficacy of trastuzumab in metastatic breast cancer suggested that it could even be more useful in early breast cancer.

As various chemotherapeutic agents and treatment regimens had been clinically established, the efficacy of trastuzumab in adjuvant treatment of early breast cancer was investigated in several large trials which allowed comparison of concurrent chemotherapy administration with sequential regimens (Table 9.2). Irrespective of the chemotherapy drugs and regimens used, 1 year of trastuzumab treatment (6 mg/kg every 3 weeks) consistently reduced the rate of disease recurrence at 2 years by nearly 50%. Based on these exciting results, trastuzumab was approved for adjuvant treatment of early breast cancer in 2006. After longer observation times, a meta-analysis of five clinical trials that included over 13,000 HER2-positive breast cancer patients published in 2008 concluded that adjuvant treatment with trastuzumab lowered the disease recurrence rate by 38% and reduced the risk of death by 34%. Finally, updates on three of the large adjuvant trials were presented in December 2012 at the San Antonio Breast Cancer Symposium. The results confirmed that the addition of trastuzumab to adjuvant treatment in patients with HER2-positive early breast cancer significantly improved overall survival, e.g. reducing the risk of death by 37% after a median follow-up of 8.4 years, despite a considerable number of patients having crossed over from the control arms. Thus, long-term survival benefit of 1 year trastuzumab adjuvant treatment has unequivocally been confirmed.

9.3.4.6 Efficacy of trastuzumab in advanced metastatic gastric cancer

Gastric cancer is the fourth most commonly diagnosed and the second most common cancer-related cause of death worldwide. About 10–20% of gastric and gastroesophageal junction cancer overexpress HER2 and most patients present with inoperable advanced or metastatic disease. The median survival time with chemotherapy is less than a year. Because of this high medical need, trastuzumab treatment in combination with chemotherapy was studied in advanced gastric cancer patients. In a phase III trial, 594 patients were randomized to either chemotherapy alone or chemotherapy plus trastuzumab treatment every 3 weeks. The addition of trastuzumab prolonged the median overall survival from 11.8 months in the chemotherapy-only group to 16.0 months in the combination arm. The unprecedented therapeutic improvement in this aggressive disease earned

Table 9.2 Summary of Randomized, Multicentre Phase III Trials in Patients (pts) with HER2-Positive Early-Stage Breast Cancer

Study	Regimen	No. of Pts	Key Inclusion Criteria	Reduction in Risk of Cancer Recurrence vs. Control
NSABP B-31	AC 60/600 mg/m^2 q3wk × 4, then Pac 175 mg/m^2 q3wk × 4 + Tra q1wk × 52	864	Node positive; chemotherapy naïve	52** after 3 years; combined analysis of NSABP B-31 and NCCTG N9831
	AC 60/600 mg/m^2 q3wk × 4, then Pac 175 mg/m^2 q3wk × 4 (control group)	872		
NCCTG N9831	AC 60/600 mg/m^2 q3wk × 4, then Pac 80 mg/m^2 q1wk × 12 + Tra q1wk × 52	808	Node positive and high risk node negative; chemotherapy naïve	
	AC 60/600 mg/m^2 q3wk × 4, then Pac 80 mg/m^2 q1wk × 12 (control group)	807		
HERA (BIG 01-01)	Tra q3wk × 17	1694	Any nodal status; prior treatment with adjuvant (or neoadjuvant) chemotherapy ± radiation	46** after 2 years
	No Tra (control group)	1693		
BCIRG 006	AC 60/600 mg/m^2 q3wk × 4, then Doc 100 mg/m^2 q3wk × 4 + Tra q1wk × 12, then Tra q3wk × 13	1074	Node positive and high risk node negative; chemotherapy naïve	51**
	Doc 75 mg/m^2 + Carbo (AUC = 6) q3wk × 6 + Tra q1wk × 18, then Tra q3wk × 11	1075		39*
	AC 60/600 mg/m^2 q3wk × 4, then Doc 100 mg/m^2 q3wk × 4 (control group)	1073		

AC, doxorubicin plus cyclophosphamide; AUC, area under the concentration–time curve; BCIRG, Breast Cancer International Research Group; BIG, Breast International Group; Carbo, carboplatin; Doc, docetaxel; HERA, herceptin adjuvant trial; NCCTG, North Central Cancer Treatment Group; NSABP, National Surgical Adjuvant Breast and Bowel Project; Pac, paclitaxel; qxwk, every x wk.
*p < 0.001.
**p < 0.0001 vs. comparator.
Source: Modified from Ref. (11).

trastuzumab regulatory approval for the treatment of advanced HER2-positive gastric or gastro-esophageal junction cancer in 2010.

9.3.4.7 Development of a subcutaneous formulation of trastuzumab

Trastuzumab is administered by IV infusion, which means that patients have to plan hospital visits at least every 3 weeks for drug treatment. As trastuzumab is provided as a lyophilized powder, it needs to be reconstituted under sterile conditions and prepared for patient treatment. After the first administration, the actual infusion time can be reduced from 90 min to as short as 30 min. In order to offer more treatment convenience to patients, shorten pharmacy preparation times and reduce overall administration costs, a subcutaneous (s.c.) formulation of trastuzumab has been developed (*12*).

The following major changes had to be introduced: as the volume that can be injected subcutaneously is limited, the trastuzumab solution was concentrated from approximately 21 mg/ml to 120 mg/ml. As this highly concentrated formulation still results in a comparatively large s.c. injection volume of about 5 ml, recombinant human hyaluronidase (rHuPH20), an enzyme that temporarily degrades hyaluronan in the interstitial space, was added as an excipient which facilitates the immediate dispersion of injected volumes over a greater tissue area, enhances systemic absorption of the therapeutic protein, and avoids injection volume-related pain. Finally, a fixed dose per patient instead of weight-based dosing was introduced in order to facilitate the application procedure. This was enabled by the realization that drug clearance of trastuzumab is less dependent on body weight than that of small molecule drugs and was further supported by population pharmacokinetics data as well as drug modelling and simulation efforts.

The development of the trastuzumab s.c. formulation required additional nonclinical and clinical studies, although safety information for rHuPH20 had already been available from clinical experience. First, comparable efficacy at equivalent dosages of the IV and s.c. formulations were demonstrated in a mouse tumour xenograft study. A dye dispersion study helped to determine the amount of hyaluronidase required for optimal drug dispersion when used as an excipient in a liquid formulation. Enhanced drug absorption by addition of rHuPH20 and its recommended concentration when used as an excipient were confirmed by a pharmacokinetic study on minipigs. The minipig was chosen as an appropriate animal model because its skin and the texture of the s.c. tissue are considered to be similar to those of humans. Finally, clinical studies were performed to confirm the s.c. dose required in patients to achieve noninferior trastuzumab serum trough concentrations compared to IV administration and to provide supportive efficacy data. In addition to potential systemic side effects, special focus was on injection reactions related to s.c. administration of 5 ml drug, as well as formation of antidrug antibodies and their potential impact on the trastuzumab exposure and correlation with adverse events. Overall, the clinical studies confirmed the feasibility of an s.c. trastuzumab formulation for breast cancer treatment. A phase III trial involving 596 patients showed noninferior trastuzumab serum trough concentrations achieved on s.c. dosing that resulted in comparable efficacy to the IV regimen. No new safety signals were observed and the apparent s.c. safety profile was consistent with the known safety profile of IV trastuzumab. The successful development of the s.c. formulation will help to minimize patients' hospital visits and drug preparation time, and it will enable trastuzumab administration to be performed at the general practitioner or even self-administration with a single-injection device that is also being developed.

9.4 **PERTUZUMAB (PERJETA™)**

Among a series of HER2-binding monoclonal antibodies that were created and characterized in the late 1980s, the murine parent of trastuzumab, 4D5, was selected for development, because it showed stronger antiproliferative effects on HER2-positive cell lines *in vitro*. However, by continued basic research, Mark X. Sliwkowski at Genentech discovered many years later that one of the discarded antibodies derived from the initial HER2 immunization effort, called 2C4, bound to a different epitope on HER2 which implied differential cell growth inhibition characteristics and eventually a distinct mode of action compared to 4D5 or trastuzumab (*13*). As this molecule suggested a different activity profile, its CDR sequences were subsequently transferred onto a similar IgG1 framework as trastuzumab and a new drug development program was started. The recombinant humanized monoclonal antibody rhuMab 2C4 then became known under the generic name pertuzumab.

9.4.1 **Mechanisms of action and preclinical activity of pertuzumab**

Pertuzumab binds with high affinity and specificity to subdomain II of HER2's extracellular domain, the region which is crucial for the interaction with ligand-activated HER3 or HER1 when HER2 forms heterodimers with them. As mentioned in Section 9.2, ligand-activated heterodimer formation is a major mechanism for activation of HER2 signalling cascades and this cannot be prevented by trastuzumab. However, when pertuzumab is bound onto the dimerization domain (subdomain II, Figure 9.2), it constitutes a steric blockade for dimer formation. Therefore, pertuzumab became known as the first in a new class of HER2 dimerization inhibitors (HDI). As a result, pertuzumab shuts down the HER2-dimer-mediated activation of intracellular signalling pathways associated with cancer cell survival and proliferation. On the other hand, the site of pertuzumab binding on HER2 implies that it cannot inhibit proteolytic extracellular domain shedding.

Because pertuzumab is built on the same IgG1 framework as trastuzumab, it binds FCγR in a similar way, recruits immune effector cells to the tumour and elicits ADCC with the same potency (*14*).

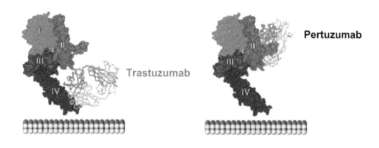

FIGURE 9.2

Trastuzumab and pertuzumab bind to distinct epitopes on HER2 extracellular domain. (See colour plate.)

Source: Illustrations taken from Ref. (22), according to data from Ref. (23).

Consistent with the above-described mechanism, pertuzumab inhibits a number of different human cancer cell lines whose proliferation is driven by HER2 heterodimers. Preclinical data also showed that pertuzumab inhibits growth of a large number of human tumour xenograft models of different origin, including lung, breast, ovarian, and both hormone-dependent as well as androgen-independent prostate cancer. In contrast to trastuzumab, pertuzumab is active on human cancer cell lines and xenograft models that do not overexpress HER2.

With increased understanding of the mechanistic implications of pertuzumab and trastuzumab binding to HER2 on its function it became clear that each of the two antibodies can achieve a biological effect which the other cannot: while trastuzumab inhibits HER2 activation by ECD shedding, pertuzumab prevents ligand-activated HER2 dimerization. These complementary activities suggested the investigation of trastuzumab and pertuzumab in combination. Indeed, combination treatment with the two HER2-binding monoclonal antibodies, pertuzumab and trastuzumab, turned out to result in synergistic antitumour effects *in vitro* and *in vivo* (Figure 9.3) (*14*). In addition, pertuzumab was shown to enhance antitumour activity in combination with cytotoxic chemotherapy drugs representing different modes of action, like paclitaxel (tubulin inhibitor), cisplatin (alkylating agent), gemcitabine (antimetabolite) and irinotecan (topoisomerase inhibitor). These results may be explained by pertuzumab-mediated deprivation of HER2 signalling-related cell survival signals, rendering tumour cells more sensitive to the toxic insult by chemotherapeutic agents.

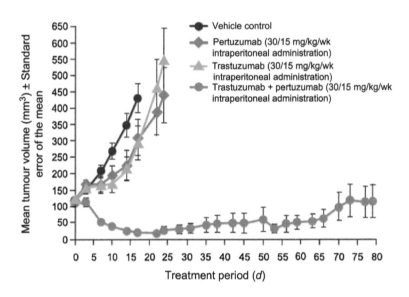

FIGURE 9.3

Synergistic efficacy of pertuzumab + trastuzumab combination therapy in a HER2-positive breast cancer xenograft model. (See colour plate.)

Source: From Ref. (14).

9.4.2 **Clinical efficacy of pertuzumab**

The phase I dose escalation study with pertuzumab included 21 patients with incurable, locally advanced, recurrent or metastatic solid tumours that had progressed on standard therapy. Surprisingly, two partial responses in patients with ovarian cancer and pancreatic islet cell carcinoma were observed. The pharmacokinetic behaviour of pertuzumab supported the use of a fixed dose per patient, independent of body weight, in the subsequent phase II trials.

The preclinical activity in the absence of HER2 overexpression and two responders in the phase I trial prompted clinical efficacy in tumours with normal HER2 expression levels. Consequently, five phase II trials with pertuzumab single-agent therapy were performed in four different cancers, including HER2-negative breast, ovarian, lung, and hormone-refractory prostate cancer, either in chemotherapy-naïve patients or after progression on taxane-based chemotherapy.

Overall, the results of the phase II studies with pertuzumab single-agent studies were disappointing. Although there were several partial responses (five in ovarian, two in breast cancer) and a number of disease stabilizations, the efficacy was considered not to justify embarking on a phase III pivotal trial.

However, pertuzumab had not yet been tested in patients with HER2-positive breast cancer and preclinical data strongly supporting a synergistic antitumour effect of pertuzumab and trastuzumab combination treatment were emerging. Therefore, instead of terminating the programme, another phase II trial was conducted. As trastuzumab was already part of standard-of-care treatment in HER2-positive disease, it was decided to perform a study in patients whose advanced metastatic breast cancer had progressed under trastuzumab treatment. Pertuzumab was added to continued trastuzumab treatment, with no chemotherapy in place. And the result was clearly beyond expectations: In 66 patients treated, a response rate of 24% and a clinical benefit rate (complete response + partial response + stable disease for at least 6 months) of 50% were observed (*15*). Furthermore, five patients experienced complete tumour remission without chemotherapy; just treatment with two antibodies after progression on one of them, in an advanced stage of an aggressive form of breast cancer – a truly unprecedented finding.

When the exciting results of the phase II trial in patients whose advanced breast cancer had progressed under trastuzumab therapy became apparent, a phase III pivotal study for first-line treatment of metastatic breast cancer was initiated. A standard-of-care treatment in this setting, the chemotherapeutic agent docetaxel combined with trastuzumab and placebo added (control group), was compared to the docetaxel plus trastuzumab plus pertuzumab combination (pertuzumab group). A total of 808 patients with HER2-positive metastatic breast cancer were randomized to the two treatment arms. The standard dose of trastuzumab was used, 8 mg/kg body weight, followed by 6 mg/kg every 3 weeks, and pertuzumab was administered at the dosing regimen derived from the phase II trials, 840 mg per patient, followed by 420 mg per patient every 3 weeks. The outcome after first analysis in 2011 was impressive again: while the safety profile was similar in both groups, the median PFS was 12.4 months in the control group but 18.5 months in the pertuzumab group (Figure 9.4) (*16*). This result was corroborated by survival data that became available about a year later: After a median follow-up time of 30 months in both arms, the risk of death was reduced by 34% for patients in the pertuzumab group, compared to those in the control group (hazard ratio = 0.66; $p = 0.0008$). At the time of the analysis, median overall survival had not yet been reached in the pertuzumab group, as more than half of these patients continued to survive. The extraordinary efficacy improvement by the addition of pertuzumab to a standard treatment regimen led to accelerated review by the US FDA and marketing approval in the United States was granted on 8 June 2012.

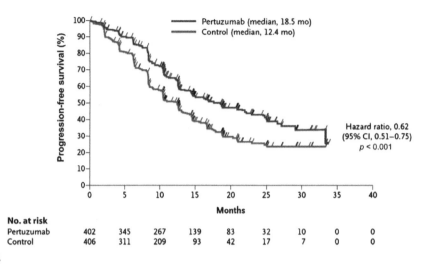

FIGURE 9.4

Prolongation of progression-free survival by addition of pertuzumab to trastuzumab plus docetaxel (control) in first-line metastatic breast cancer. CI, confidence interval. (For colour version of this figure, the reader is referred to the online version of this book.)

Source: From Ref. (16).

Pertuzumab has also been studied in the neoadjuvant setting, which is treatment before surgery. Patients with HER2-positive early breast cancer (nonmetastatic) were randomized to four groups and treated with different drug combinations. After four 3-week treatment cycles, the remaining tumour was removed and the tissue at the site of initial tumour biopsy was analysed for remaining tumour cells. Efficacy was evaluated by the rate of pathological complete response (pCR), which indicates the proportion of patients whose tumour had completely disappeared during presurgical drug treatment. Again, the outcome of this study demonstrated that the addition of pertuzumab to trastuzumab plus docetaxel significantly improved the therapeutic effect, increasing the pCR rate from 29% to 45.8% (17). Interestingly, one patient group in this trial did not receive any chemotherapy but only trastuzumab plus pertuzumab combination treatment and still about 17% of these patients experienced pCR. This result suggests that there may be a group of patients who could be spared cytotoxic chemotherapy. Efforts to identify such patients by molecular characterization for respective biomarkers on their tumours are ongoing. Furthermore, a large phase III trial to study the efficacy of pertuzumab in combination with trastuzumab in the adjuvant treatment of early breast cancer has been initiated.

9.5 TRASTUZUMAB EMTANSINE (T-DM1; KADCYLA™)

Soon after monoclonal antibodies became available as research tools, their high target specificity and physiologic properties prompted the idea to utilize them as carriers for pharmacologic agents to their desired site of activity. However, despite the seemingly simple concept, a multiplicity of difficulties turned up during several decades of relentless efforts to develop antibody–drug conjugates (ADCs). In the case of anticancer drugs, several critical requirements for an effective and safe ADC emerged, including inactive and nontoxic properties of the intact antibody–toxin conjugate, high systemic stability, high tumour specificity and/or selectivity, a good understanding of the target binding on the cell surface and its dynamic properties, and activation/release of the toxin at the tumour or inside the tumour cell. Eventually, all the molecular, pharmacologic and therapeutic knowledge gained over the years with trastuzumab and its target HER2 were exploited to create an ADC that is now in late-stage development and may constitute the next step in the continued progress of HER2-targeted therapies, T-DM1, which received the INN name trastuzumab emtansine (*18*).

9.5.1 Characteristics of trastuzumab emtansine and its mechanisms of action

As its name suggests, trastuzumab emtansine is based on the monoclonal HER2-specific antibody trastuzumab. Via a stable chemical linker, 4-(*N*-maleimidomethyl) cyclohexane-1-carboxylate (MCC), the cytotoxic agent DM1, a maytansine derivative, is conjugated to lysine residues of trastuzumab (Figure 9.5). The ADC trastuzumab emtansine is optimized for its target in the following ways.

Extraordinary tumour selectivity is provided by targeting HER2-positive tumours. As described earlier, HER2 overexpression is usually caused by amplification of the *c-erbB-2* gene leading to orders of magnitude higher HER2 protein levels compared to normal cells. Relative to the closely related EGFR, HER2 undergoes a slow internalization and turnover rate. It has been described that most of the HER2 molecules internalized by endocytosis are recycled to the cell membrane via early endosomes, and both, the HER2 internalization kinetics and its intracellular trafficking, are obviously not changed by trastuzumab binding (*7*). Accordingly, most of the trastuzumab emtansine molecules are either found in the circulation or bound to cell surface HER2, where the active toxin is not supposed to be

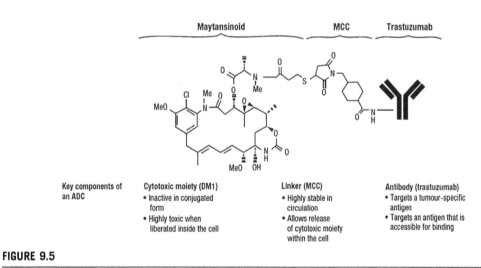

FIGURE 9.5

Schematic structure of trastuzumab emtansine.

Source: From Ref. (18).

released. Therefore, an extremely stable chemical linker was required for this ADC in order to keep systemic toxicity at a minimum. On the other hand, as only a small fraction of internalized HER2 enters the lysosomal degradation pathway eventually releasing the active metabolite lysine-N^{ε}-4-(N-maleimidomethyl)cyclohexane-1-carboxylate-DM1 (MCC-DM1), a highly potent toxin was required. The combination of trastuzumab with the stable linker MCC and the highly toxic tubulin polymerization inhibiting maytansinoid DM1 fulfilled all these criteria best. In addition, the intact conjugate trastuzumab-MCC-DM1 proved not to be toxic until cleaved, but retained all pharmacodynamic properties of trastuzumab: high-affinity binding to HER2, inhibition of ligand-independent HER2 signalling, prevention of HER2 ECD shedding, and activation of ADCC.

In preclinical experiments, trastuzumab emtansine demonstrated enhanced potency compared to trastuzumab on various trastuzumab-sensitive and trastuzumab-resistant cell lines. However, HER2 overexpression was still required for activity. Trastuzumab emtansine was also more potent in various mouse models, including human tumour xenografts derived from HER2-positive breast, lung, ovarian and gastric cancers. Nevertheless, the antitumour activity of trastuzumab emtansine in preclinical models could be further enhanced by the addition of pertuzumab. This may be explained by the fact that a more complete inhibition of HER2-associated antiapoptotic signalling as achieved by trastuzumab plus pertuzumab may render the cells more susceptible to the cytotoxic component DM1.

9.5.2 Clinical efficacy

Trastuzumab emtansine turned out to be well tolerated in patients, with an MTD at 3.6 mg/kg body weight administered every 3 weeks. Two phase II single-arm trials of patients with HER2-positive metastatic breast cancer were conducted with trastuzumab emtansine single-agent therapy. In a study of 112 patients who had received previous chemotherapy and had tumour progression after HER2-directed therapy, trastuzumab emtansine was associated with an objective response rate of 25.9% based on independent review. Another phase II study included 110 patients with HER2-positive metastatic breast cancer who had received previous anthracycline, trastuzumab, taxane, capecitabine, and lapatinib therapy and had evidence of progressive disease on their last regimen. Trastuzumab emtansine single-agent therapy led to an objective response rate of 34.5%, an impressive result in this heavily pretreated patient population.

In a phase III clinical trial involving 991 patients with advanced metastatic HER2-positive breast cancer whose disease had progressed after previous treatment with trastuzumab and a taxane, trastuzumab emtansine single-agent treatment was compared to combination treatment with the chemotherapeutic drug capecitabine plus lapatinib, a small molecule dual HER1 and HER2 tyrosine kinase inhibitor. Trastuzumab emtansine proved to be clearly superior in every clinical end point evaluated: median PFS was 9.6 months vs. 6.4 months, median overall survival at the second interim analysis was 30.9 months vs. 25.1 months, and the objective response rate was 43.6% vs. 30.8% (*19*). After 24 months, 64.7% of patients in the trastuzumab emtansine group but only 51.8% in the capecitabine plus lapatinib group were alive. In addition, patients treated with trastuzumab emtansine experienced significantly less toxicity than those who received capecitabine plus lapatinib. Based on these results, which demonstrate that targeted intracellular delivery of a cytotoxic agent can enhance efficacy and improve safety, trastuzumab emtansine gained U.S. FDA approval for second-line treatment of metastatic HER2-positive breast cancer in February 2013.

Trastuzumab emtansine is still under investigation in a phase III trial for first-line treatment of HER2-positive metastatic breast cancer. The efficacy and safety of single-agent trastuzumab emtansine or trastuzumab emtansine plus pertuzumab will be compared with the combination of trastuzumab with a taxane (paclitaxel or docetaxel).

9.6 CONCLUSION

The introduction of HER2-targeting monoclonal antibodies founded the basis for one of the most exciting success stories in cancer treatment over the past few decades. Trastuzumab is now established in all treatment lines of metastatic breast cancer and has clearly changed the natural course of the disease. While HER2-positivity previously indicated early relapse and short overall survival compared to HER2-negative breast cancer, it is now a positive prognostic factor (20). Small molecule HER2-targeting drugs like lapatinib could not enter early treatment lines and came out with inferior efficacy despite increased toxicity in clinical head-to-head comparisons. Among the fundamental differences between antibody and small molecule drugs, the following properties may explain the success of trastuzumab: antibodies are highly specific for the respective antigen target which avoids toxicity when it is selectively (over)expressed at the tumour. Unlike small molecule tyrosine kinase inhibitors, trastuzumab does not face binding competition by high concentrations of a physiological substrate, is not subject to multidrug resistance causing efflux pumps like P-glycoprotein, and is not inactivated by cytochrome P450 enzyme metabolization. In addition, trastuzumab has multiple modes of action including HER2 signalling inhibition, prevention of extracellular domain shedding, and engaging immune effector cells to turn the patient's immune system against the tumour.

Further progress in HER2-positive cancer therapy is conceivable with the introduction of pertuzumab and trastuzumab emtansine. The HDI pertuzumab has shown synergistic efficacy in combination with trastuzumab and has obtained regulatory approval for use in first-line treatment of HER2-positive metastatic breast cancer. Superior efficacy at reduced toxicity compared to the combination of trastuzumab plus chemotherapy in a clinical trial indicates that trastuzumab emtansine may spare systemic chemotherapy. While further progress is expected with the combination of trastuzumab emtansine plus pertuzumab in HER2-positive breast cancer, more studies are required to move these efficient drugs into other cancer indications in which HER2 overexpression occurs less frequently.

References

1. Cho, H. S.; Mason, K.; Ramyar, K. X.; Stanley, A. M.; Gabelli, S. B.; Denney, D. W., Jr., et al. Structure of the Extracellular Region of HER2 Alone and in Complex with the Herceptin Fab. *Nature* **2003,** *421* (6924), 756–760.
2. Jones, R. B.; Gordus, A.; Krall, J. A.; MacBeath, G. A Quantitative Protein Interaction Network for the ErbB Receptors Using Protein Microarrays. *Nature* **2006,** *439* (7073), 168–174.
3. Carter, P.; Presta, L.; Gorman, C. M.; Ridgway, J. B.; Henner, D.; Wong, W. L., et al. Humanization of an Anti-p185HER2 Antibody for Human Cancer Therapy. *Proc. Natl. Acad. Sci. U S A* **1992,** *89* (10), 4285–4289.

4. Junttila, T. T.; Akita, R. W.; Parsons, K.; Fields, C.; Lewis Phillips, G. D.; Friedman, L. S., et al. Ligand-Independent HER2/HER3/PI3K Complex is Disrupted by Trastuzumab and is Effectively Inhibited by the PI3K Inhibitor GDC-0941. *Cancer Cell* **2009 May 5,** *15* (5), 429–440.

5. Molina, M. A.; Codony-Servat, J.; Albanell, J.; Rojo, F.; Arribas, J.; Baselga, J. Trastuzumab (Herceptin), a Humanized Anti-Her2 Receptor Monoclonal Antibody, Inhibits Basal and Activated Her2 Ectodomain Cleavage in Breast Cancer Cells. *Cancer Res.* **2001,** *61* (12), 4744–4749.

6. Clynes, R. A.; Towers, T. L.; Presta, L. G.; Ravetch, J. V. Inhibitory Fc Receptors Modulate *In vivo* Cytoxicity against Tumor Targets. *Nat. Med.* **2000,** *6* (4), 443–446.

7. Austin, C. D.; De Maziere, A. M.; Pisacane, P. I.; van Dijk, S. M.; Eigenbrot, C.; Sliwkowski, M. X., et al. Endocytosis and Sorting of ErbB2 and the Site of Action of Cancer Therapeutics Trastuzumab and Geldanamycin. *Mol. Biol. Cell.* **2004,** *15* (12), 5268–5282.

8. Cobleigh, M. A.; Vogel, C. L.; Tripathy, D.; Robert, N. J.; Scholl, S.; Fehrenbacher, L., et al. Multinational study of the efficacy and safety of humanized anti-HER2 monoclonal antibody in women who have HER2-overexpressing metastatic breast cancer that has progressed after chemotherapy for metastatic disease. *J Clin Oncol* **1999 Sep,** *17* (9), 2639–2648.

9. Baselga, J.; Carbonell, X.; Castaneda-Soto, N. J.; Clemens, M.; Green, M.; Harvey, V., et al. Phase II study of efficacy, safety, and pharmacokinetics of trastuzumab monotherapy administered on a 3-weekly schedule. *J Clin Oncol* **2005,** *23* (10), 2162–2171.

10. Pegram, M.; Liao, J. Trastuzumab Treatment in Multiple Lines: Current Data and Future Directions. *Clin. Breast Cancer* **2012 Feb,** *12* (1), 10–18.

11. Plosker, G. L.; Keam, S. J. Trastuzumab: A Review of Its Use in the Management of HER2-Positive Metastatic and Early-Stage Breast Cancer. *Drugs* **2006,** *66* (4), 449–475.

12. Bittner, B.; Richter, W. F.; Hourcade-Potelleret, F.; McIntyre, C.; Herting, F.; Zepeda, M. L., et al. Development of a Subcutaneous Formulation for Trastuzumab – Nonclinical and Clinical Bridging Approach to the Approved Intravenous Dosing Regimen. *Arzneimittelforschung* **2012 Sep,** *62* (9), 401–409.

13. Agus, D. B.; Akita, R. W.; Fox, W. D.; Lewis, G. D.; Higgins, B.; Pisacane, P. I., et al. Targeting Ligand-Activated ErbB2 Signaling Inhibits Breast and Prostate Tumor Growth. *Cancer Cell* **2002,** *2* (2), 127–137.

14. Scheuer, W.; Friess, T.; Burtscher, H.; Bossenmaier, B.; Endl, J.; Hasmann, M. Strongly Enhanced Antitumor Activity of Trastuzumab and Pertuzumab Combination Treatment on HER2-Positive Human Xenograft Tumor Models. *Cancer Res.* **2009 Dec 15,** *69* (24), 9330–9336.

15. Baselga, J.; Gelmon, K. A.; Verma, S.; Wardley, A.; Conte, P.; Miles, D., et al. Phase II Trial of Pertuzumab and Trastuzumab in Patients with Human Epidermal Growth Factor Receptor 2-Positive Metastatic Breast Cancer that Progressed during Prior Trastuzumab Therapy. *J. Clin. Oncol.* **2010 Mar 1,** *28* (7), 1138–1144.

16. Baselga, J.; Cortes, J.; Kim, S. B.; Im, S. A.; Hegg, R.; Im, Y. H., et al. Pertuzumab Plus Trastuzumab Plus Docetaxel for Metastatic Breast Cancer. *N. Engl. J. Med.* **2012 Jan 12,** *366* (2), 109–119.

17. Gianni, L.; Pienkowski, T.; Im, Y. H.; Roman, L.; Tseng, L. M.; Liu, M. C., et al. Efficacy and Safety of Neoadjuvant Pertuzumab and Trastuzumab in Women with Locally Advanced, Inflammatory, or Early HER2-Positive Breast Cancer (NeoSphere): A Randomised Multicentre, Open-Label, Phase 2 Trial. *Lancet Oncol.* **2012 Jan,** *13* (1), 25–32.

18. Burris, H. A., III; Tibbitts, J.; Holden, S. N.; Sliwkowski, M. X.; Lewis Phillips, G. D. Trastuzumab Emtansine (T-DM1): A Novel Agent for Targeting HER2+ Breast Cancer. *Clin. Breast Cancer* **2011 Oct,** *11* (5), 275–282.

19. Verma, S.; Miles, D.; Gianni, L.; Krop, I. E.; Welslau, M.; Baselga, J., et al. Trastuzumab Emtansine for HER2-Positive Advanced Breast Cancer. *N. Engl. J. Med.* **2012 Nov 8,** *367* (19), 1783–1791.

20. Dawood, S.; Broglio, K.; Buzdar, A. U.; Hortobagyi, G. N.; Giordano, S. H. Prognosis of Women with Metastatic Breast Cancer by HER2 Status and Trastuzumab Treatment: An Institutional-Based Review. *J. Clin. Oncol.* **2010 Jan 1,** *28* (1), 92–98.

21. Mastro, L. D.; Lambertini, M.; Bighin, C.; Levaggi, A.; D'Alonzo, A.; Giraudi, S., et al. Trastuzumab as First-Line Therapy in HER2-Positive Metastatic Breast Cancer Patients. *Expert Rev. Anticancer Ther.* **2012 Nov,** *12* (11), 1391–1405.

22. Hubbard, S. R. EGF Receptor Inhibition: Attacks on Multiple Fronts. *Cancer Cell* **2005,** *7* (4), 287–288.

23. Franklin, M. C.; Carey, K. D.; Vajdos, F. F.; Leahy, D. J.; de Vos, A. M.; Sliwkowski, M. X. Insights into ErbB Signaling from the Structure of the ErbB2-Pertuzumab Complex. *Cancer Cell* **2004,** *5* (4), 317–328.

Recombinant human erythropoietin and its analogues

10

Wolfgang Jelkmann

Institute of Physiology, University of Luebeck, Luebeck, Germany

CHAPTER OUTLINE

10.1 Introduction 308
10.2 Discovery of Erythropoietin (Epo) 309
10.3 Physiology of Epo...310
    10.3.1 Sites and control of production 310
    10.3.2 Structure of Epo..310
    10.3.3 *EPO* expression 313
    10.3.4 Action of Epo..313
    10.3.5 Assay of Epo...316
10.4 Manufacture of Erythropoiesis-Stimulating Agents (ESAs) 316
    10.4.1 Innovator rhEpo preparations 316
    10.4.2 Biosimilar epoetins 318
    10.4.3 Second-generation ESAs ('biobetter') 319
10.5 Clinical Application of ESAs..320
    10.5.1 Indications for ESA therapy 320
        10.5.1.1 Anaemia due to CKD..320
        10.5.1.2 Anaemia associated with cancer chemotherapy 320
        10.5.1.3 Other applications 321
    10.5.2 Safety...321
        10.5.2.1 General concerns 321
        10.5.2.2 Immunogenicity of ESAs..321
    10.5.3 Non-haematopoietic actions of Epo 322
10.6 Perspectives 323
References 325

ABSTRACT

The renal hormone erythropoietin (Epo) is essential for the viability and proliferation of the erythrocytic progenitors generating reticulocytes in the bone marrow. Human Epo consists of a chain of 165 amino acid residues and four glycans. The synthesis of Epo is strongly stimulated by hypoxia. Epo deficiency is the primary cause of the anaemia in chronic kidney disease (CKD). For anaemia treatment recombinant human Epo (rhEpo/epoetin and epoetin) is very useful. The drug substance is generally manufactured in *EPO* complementary DNA (cDNA) transfected Chinese hamster ovary cell cultures. rhEpo therapy is beneficial not only for CKD patients but also for anaemic cancer patients receiving chemotherapy. The expiry of the patents for the first original rhEpo formulations has initiated the production of similar biological medicinal products ('biosimilars'). The term 'biosimilar' should only be used for a medicinal product approved on a strict regulatory pathway (i.e. European Medicines Agency, Food and Drug Administration, etc.). Unfortunately, many claimed 'biosimilars' have not been through an approval authority. Furthermore, rhEpo analogues with prolonged survival in circulation ('biobetter') have been developed and synthetic Epo mimetic peptides have been developed for therapy. The biological role of Epo outside the bone marrow is the focus of current research.

Keywords/Abbreviations: Anaemia; Chronic kidney disease; Erythrocytic progenitors; Erythropoietin-mimetic peptides; Hypoxia; Immunogenicity; Recombinant human erythropoietin; Red blood cells; Anti-erythropoietin antibodies (Anti-Epo Abs); CREB-binding protein (CBP); Cluster of differentiation (CD); Chinese hamster ovary (CHO); Chronic kidney disease (CKD); Colony-forming unit-erythroid (CFU-E); cAMP response element-binding (CREB); Erythropoietin (Epo); Epo gene (*EPO*); Erythropoietin receptor (Epo-R); Erythropoiesis-stimulating agent (ESA); Haemoglobin (Hb); Haematocrit (Hct); Hypoxia-inducible factor (HIF); Hypoxia-response element (HRE); International Nonproprietary Name (INN); International unit (IU); Intravenous (IV); Janus kinase 2 (JAK-2); Mitogen-activated protein kinase (MAPK); Nuclear factor κB (NF-κB); O_2 pressure (pO_2); Pure red cell aplasia (PRCA); von Hippel–Lindau tumour suppressor protein (pVHL); Red blood cell (RBC); Recombinant human erythropoietin (rhEpo); Subcutaneous (SC); Signal transducer and activator of transcription (STAT).

10.1 INTRODUCTION

Blood cells are continuously produced in the bone marrow. There, the myeloid haematopoietic cells reside, namely pluripotent stem cells and their offspring. The rate of the proliferation and differentiation of the haematopoietic cells is controlled by blood cell lineage-specific hormones (erythropoietin (Epo) is one of these) and locally acting cytokines. Based on *in vitro* observations, some of the haematopoietic growth factors are also called 'colony-stimulating factors', because in their presence clusters of blood cell precursors grow from single cells ('colony-forming units') in bone marrow cultures, appearing as erythroid, nonerythroid or mixed colonies under the microscope. Just like *in vitro*, the haematopoietic growth factors stimulate the production of blood cells in the organism. As can be seen in Table 10.1, some of these proteins are engineered by recombinant DNA technologies for the treatment of patients suffering from the lack of red blood cells (RBCs) (anaemia), white blood cells (neutropenia) or blood platelets (thrombopenia).

Epo is a glycoprotein hormone that regulates the production of the RBCs in a process called *erythropoiesis* (from ancient Greek 'erythrós' for 'red' and 'poietin' for 'making'). RBCs develop in haematopoietic tissues (the bone marrow in adult humans) and circulate for 100–120 days until they are engulfed by mononuclear phagocytes in the bone marrow, spleen and liver. Balancing this loss in healthy humans, the bone marrow produces about 2.5 million young RBCs, 'reticulocytes', every second. Mature human RBCs ('erythrocytes') are anucleated disks full of haemoglobin (Hb), the oxygen (O_2)-binding Fe^{2+}-containing haeme protein that brings about the blood's red colour. The mean Hb concentration ([Hb]) is normally about 140 g/l blood in females and 160 g/l in males, with 1 g Hb binding up to 1.34 ml O_2 ('O_2 transport capacity').

Epo acts to maintain [Hb] and haematocrit (Hct; the volume fraction of RBCs in blood) constant under normoxic conditions, and to hasten erythropoiesis under hypoxic conditions. Epo deficiency

TABLE 10.1 Therapeutically Relevant Haematopoietic Growth Factors

Factor	Molecular Mass (kDa)	Main Cells of Origin	Main Target Cells	Therapeutic Indication*
Erythropoietin	30	Renal fibroblasts and hepatocytes	CFU-E	Anaemia
GM-CSF	14–35	T-lymphocytes, monocytes, endothelial cells, and fibroblasts	CFU-GM, CFU-G, CFU-M, CFU-Eo, and CFU-Meg	Neutropenia
G-CSF	20	Monocytes, fibroblasts, and endothelial cells	CFU-G	Neutropenia
Interleukin-11	23	Myeloid stroma cells	Megakaryocytes and lymphocytes	Thrombopenia
Thrombopoietin	60	Hepatocytes and renal tubular cells	CFU-Meg, megakaryocytes, and thrombocytes	Thrombopenia[†]

BFU, burst-forming unit; CFU, colony-forming unit; CSF, colony-stimulating factor; E, erythroid; G, granulocyte; M, macrophage (monocyte); Eo, eosinophilic granulocyte; Meg, megakaryocyte (large cell releasing thrombocytes).
[*]Approval status, marketing, brand names, indications and dosage forms differ by countries.
[†]Clinically unapproved.

results in anaemia, meaning a decrease in [Hb] and Hct. Endogenous Epo is mainly of renal origin. About 25% of patients with chronic kidney disease (CKD) needed regular RBC transfusions before recombinant human erythropoietin (rhEpo) became available for therapy. Although allogeneic RBC transfusion is a very beneficial manoeuvre, it bears risks such as acute lung injury, haemolytic reactions and transmission of pathogens. Further, repeated RBC transfusions lead to iron overload. Therefore, the introduction of rhEpo as an anti-anaemic drug for stimulation of erythropoiesis has been a major progress in medicine (1,2). This chapter describes the discovery of Epo, its physiological function, the genetic engineering and properties of rhEpo and other erythropoiesis-stimulating agents (ESAs), and aspects of the clinical use of the medicines.

10.2 DISCOVERY OF ERYTHROPOIETIN (EPO)

The discovery that erythropoiesis depends on the body's O_2 supply was made by French high-altitude researchers in the second half of the nineteenth century (reviewed in Refs (3,4)). The O_2 pressure (pO_2) of the air is reduced at altitude. Paul Bert showed that animals living at 4000 m altitude in Bolivia had higher [Hb] than lowlanders. Denis Jourdanet noted that humans living on the Mexican highland suffered from chronic mountain disease. The high-altitude residents had thickened blood, but their symptoms (breathlessness and heart hurry) resembled those of anaemic patients. Jourdanet coined the term 'anoxyhémie' for the lack of O_2 in arterial blood. It was initially believed that the thick blood ('polycythaemia') of the highlanders was inherited. However, in 1890 Francois-Gilbert Viault observed that his RBC number increased from 5 to $8 \times 10^6/\mu l$ within 3 weeks during a trip from Bordeaux in France to Morococha, a mining village at 4500 m altitude in Peru. This led Viault to conclude that erythropoiesis is acutely stimulated when the O_2 content of the blood is reduced.

The hormonal regulation of erythropoiesis was postulated by Carnot and Deflandre from Paris in 1906. These investigators subjected rabbits to a blood-letting, took another blood sample on the next day and injected the serum into normal rabbits. The fact that the RBC concentration increased in the recipients indicated that the donor serum contained a haematopoietic factor ('hémopoiétine'). The specific name 'erythropoietin' was introduced by Eva Bonsdorff and Eeva Jalavisto from Helsinki in 1948. The haematologist Allan Erslev first predicted the potential therapeutic value of the erythropoietic factor in 1953: 'Conceivably isolation and purification of this factor would provide an agent useful in the treatment of conditions associated with erythropoietic depression, such as chronic infection and chronic renal disease' (5).

10.3 PHYSIOLOGY OF EPO

10.3.1 Sites and control of production

The important role of the kidneys in Epo production became clear in 1957 when Leon Jacobson et al. showed that nephrectomized rats fail to respond with the normal increase in the plasma Epo levels under hypoxic conditions. Subsequently, Epo bioactivity and Epo mRNA were extracted from the kidneys of hypoxic rodents. *In situ* hybridization studies identified fibroblasts in the kidney cortex as the primary site of Epo mRNA expression. In addition, liver, spleen, lung, bone marrow and brain cells express Epo mRNA in small amounts; the liver is the main site of Epo synthesis in the prenatal period.

The concentration of circulating Epo increases exponentially with decreasing [Hb] in uncomplicated anaemia (absence of renal disease or inflammation). The underlying feedback regulation, shown in Figure 10.1, is based on the tissue pO_2, which depends mainly on [Hb], arterial pO_2, and the O_2 affinity of the Hb. Compared to other organs, where the tissue pO_2 greatly decreases with reduced blood flow and increased O_2 consumption, the kidneys are most appropriate for regulating Epo production. The pO_2 in the renal cortex is little affected by the rate of blood flow, as the renal O_2 consumption changes in proportion with the blood flow and, thus, the rate of fluid filtration from the renal capillaries (glomeruli) into the tubules (glomerular filtration rate, GFR).

10.3.2 Structure of Epo

The purification of human Epo was a difficult task due to technical limitations and the low Epo levels in biological fluids. The major breakthrough came in 1975 when Takaji Miyake took a concentrate of 2550 l of urine, from patients with severe anaemia, from Japan to Eugene Goldwasser's Institute of Biochemistry in Chicago. Human Epo was purified to homogeneity from this urine and a partial amino acid sequence was obtained. The sequence data were used to identify and isolate the human Epo gene (*EPO*), in 1985, and express it, *in vitro* (6).

Human Epo is an acidic glycoprotein with a molecular mass of 30.4 kDa. The peptide core (60% of the molecule) is a single chain of 165 amino acid residues forming four anti-parallel α-helices, two β-sheets and two intra-chain disulphide bridges (Cys^7–Cys^{161} and Cys^{29}–Cys^{33}). The carbohydrate

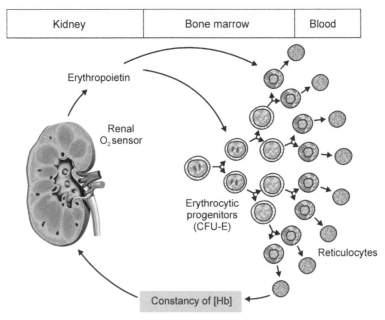

FIGURE 10.1 Scheme of the feedback regulation of erythropoiesis.

Lack of O_2 due to anaemia or hypoxaemia is the primary stimulus for the synthesis of erythropoietin (Epo) in the kidneys. Epo is a survival, proliferation and differentiation factor for the myeloid erythrocytic progenitors, particularly the colony-forming units-erythroid (CFU-Es). The O_2 capacity of the blood increases with the enhanced release of reticulocytes.

portion (40% of the molecule) comprises three complex tetra-antennary N-linked glycans (at Asn^{24}, Asn^{38} and Asn^{83}) and one small O-linked glycan (at Ser^{126}). The N-glycans serve a variety of functions, including the protection of the glycoprotein from proteases and the modulation of its receptor-binding affinity. The *in vivo* erythropoietic activity increases and receptor binding decreases with increasing N-glycan number. The O-glycan is probably not essential. Figure 10.2 shows the predicted structure of glycosylated human Epo.

As can be seen in Figure 10.3, endogenous human Epo as well as rhEpo and its analogues present with several glycosylation isoforms, which can be separated by isoelectric focussing and detected by immunoblotting of human urine concentrates. These methods are also used for proof of rhEpo doping in sports. Immunoblots of urine from untreated humans show several Epo isoform bands in the isoelectric point (pI) range 3.77–4.70. Blots from subjects treated with rhEpo contain bands more in the basic range (pI 4.42–5.21). For *in vivo* activity, the terminal sialic acid (syn. N-acetylneuraminic acid) residues of the N-glycans are very important. Asialo-Epo is rapidly removed via asialoglyco-protein receptors (ASGPRs) expressed on hepatocytes; the loss of terminal sialic acid exposes a terminal galactose sugar residue that is bound by the ASGPR; the introduction of additional sialy-lated N-glycans into rhEpo, following site-directed mutagenesis, results in products with a prolonged survival in circulation (7).

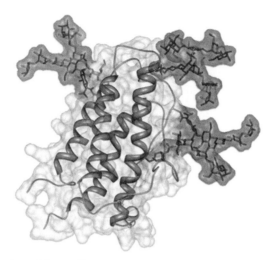

FIGURE 10.2 Predicted structure of human Epo.

The three complex *N*-linked glycans are shown in purple and the single *O*-linked glycan in pink. (See colour plate.)

Source: The figure is reproduced with permission from the Woods Group using Chimera. [Woods Group. (2005–2012) GLYCAM Web. Complex Carbohydrate Research Center, University of Georgia, Athens, GA, USA (http://www.glycam.org)].

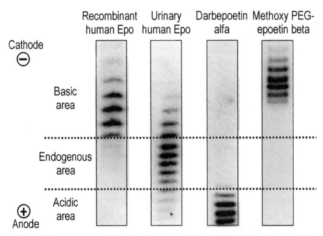

FIGURE 10.3 Demonstration of Epo glycosylation isoforms by isoelectric focusing and immunoblotting of human urine concentrates.

Endogenous Epo isoforms are more acidic than rhEpo isoforms. Hyperglycosylated rhEpo (darbepoetin alfa) migrates more in the acidic range and pegylated rhEpo (methoxy PEG-epoetin beta) more in the basic range than endogenous Epo or regular rhEpo.

Source: With permission modified from the world anti-doping Agency (WADA; www.wada-ama.org).

10.3.3 *EPO* expression

The human *EPO* (2.9 kb; www.genenames.org/data/hgnc_data.php?hgnc_id=3415), on the long arm of chromosome 7 (q11–q22), contains five exons, which encode a 193 amino acid residues prohormone, and four introns. The amino acid leader sequence of 27 residues and the *C*-terminal arginine are cleaved prior to Epo's secretion. *EPO* expression is controlled by several transcription factors (*8*). The *EPO* promoter possesses binding sites for GATA-2 (bonding with the nucleic acid sequence GATA) and nuclear factor κB (NF-κB), both inhibiting *EPO* transcription. GATA-2 and NF-κB are likely responsible for the impaired *EPO* expression in anaemic patients suffering from inflammatory diseases. Of major biological importance is the *EPO* enhancer, which possesses a hypoxia-response element (HRE) that is activated on binding specific hypoxia-inducible transcription factors (HIFs). The HIFs are heterodimeric proteins composed of an α- and a β-subunit (*9*). There are isoforms of the α-subunits (α-1 and -2), which control both shared and unique target genes. The main factor inducing *EPO* expression is HIF-2, which comprises the subunits HIF-2α and HIF-1β (*10*).

As illustrated in Figure 10.4, the HIF α-subunits are O_2 labile, rendering them active only under hypoxic conditions. Their *C*-terminus contains two proline residues (Pro^{405} and Pro^{531} in HIF-2α) that are hydroxylated in the presence of O_2 by means of HIF-α-specific α-ketoglutarate-requiring dioxygenases (*11*). Prolyl hydroxylated HIF-α binds to the von Hippel–Lindau tumour suppressor protein (pVHL), the substrate recognition subunit of an E3 ubiquitin ligase (Ub), and undergoes immediate proteasomal degradation. Further, an asparagine residue (Asn^{847} in HIF-2α) is hydroxylated in the presence of O_2 by means of the so-called factor inhibiting HIF. Thereby, the binding of the transcriptional coactivator p300/cAMP response element-binding (CREB)-binding protein is prevented. In the absence of O_2, HIF-α combines with HIF-1β in the nucleus and the dimeric transcription complex activates the expression of genes possessing HREs.

Divalent metals such as cobalt (Co^{2+}) stabilize HIF-α and induce Epo production under normoxic conditions. Also, α-ketoglutarate competitors (clinical term: 'HIF stabilizers') stimulate Epo production and erythropoiesis, as α-ketoglutarate is a cofactor in HIF-α hydroxylation.

HREs and HIFs are not only relevant for Epo-producing cells but also expressed by almost all tissues (*9*). More than 200 genes have been identified that are influenced by HIF binding to HREs, including those encoding the vascular endothelial growth factor, the glucose transporters 1 and 3 and several glycolytic enzymes.

10.3.4 Action of Epo

Erythrocytes are the offspring of a small pool of self-perpetuating haematopoietic stem cells called colony-forming units generating granulocytic, erythrocytic, megakaryocytic, and monocytic progeny (CFU-GEMMs). CFU-GEMMs are also known as $CD34^+$ cells because they express the membrane glycoprotein CD34 (CD means 'cluster of differentiation', with the number indicating specific membrane marker proteins) (*12*). The earliest cell of the erythrocytic compartment (the 'erythron') is called burst-forming unit-erythroid (BFU-E). Its progeny goes through about 12 divisions, eventually giving rise to several hundred nucleated red cells (erythroblasts) within 10–20 days. The first daughter cell of the BFU-E is the colony-forming unit-erythroid (CFU-E). Unlike the BFU-Es, CFU-Es express many Epo receptor (Epo-R) molecules. In the presence of Epo, CFU-Es and their descendants (first

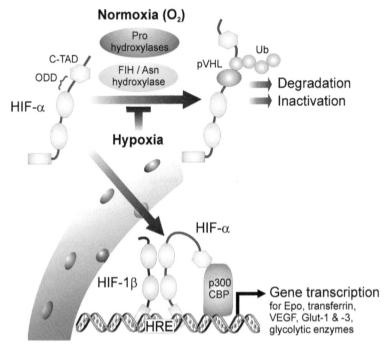

FIGURE 10.4 Scheme of the O_2-dependent transcriptional activity of the HIFs (hypoxia-inducible factors).

In the presence of O_2 (normoxia), HIF-α is hydroxylated by prolyl (Pro) hydroxylases at specific prolyl residues in its oxygen-dependent degradation domain (ODD) and targeted for proteasomal degradation by the pVHL-E3 ubiquitin ligase (Ub) complex. Further, HIF-α is hydroxylated by FIH (factor inhibiting HIF) at an asparaginyl (Asn) residue in its *C*-terminal transactivation domain (*C*-TAD). In the absence of O_2 (hypoxia), HIF-α hydroxylation does not occur and HIF-α heterodimerizes with HIF-1β in the nucleus. Supported by co-activators (p300/CBP) the transcriptional complex binds to the HREs (hypoxia-response elements) in the DNA. Apart from *EPO*, HIF target genes include those encoding transferrin, vascular endothelial growth factor (VEGF), glucose transporters (Glut-1 and -3) and several glycolytic enzymes.

follow proerythroblasts) divide 3–5 times generating 8–64 erythroblasts within 7–8 days. This process can be investigated *in vitro* utilizing cultures of haematopoietic tissue. The staining with diaminobenzidine, a reagent for Hb, allows one to visualize and count the colonies. Each CFU-E colony contains between 8 and 64 haemoglobinized cells. As proliferation progresses, the nuclei of the descendants become smaller and their cytoplasm more basophilic ('basophilic erythroblasts'), which is due to the presence of ribosomes. When the cells begin to synthesize Hb, they are called 'polychromatic erythroblasts'. Once the level of 'orthochromatic erythroblasts' (syn. 'normoblasts') is reached, the cells do not divide any more but extrude their nuclei and become reticulocytes. Over a few days, these complete the synthesis of Hb, start to degrade their organelles and eventually enter the blood vessels. Circulating reticulocytes lose their filamentous structures (mainly polyribosomes), diminish in size and become mature erythrocytes within 1–2 days.

CFU-Es and proerythroblasts are the primary Epo targets. Actually, most CFU-Es undergo apoptosis in the absence of Epo, whereas they proliferate in large number at high Epo concentrations. At later stages of erythropoiesis the expression of the Epo-R decreases, and the cells become Epo-insensitive. Reticulocytes lack Epo-R molecules.

The human Epo-R is a membrane-spanning ∼59 kDa glycoprotein (484 amino acid residues and one N-glycan) that is a member of the cytokine receptor superfamily I. As shown in Figure 10.5, two Epo-R molecules form a homodimer that binds a single Epo molecule. Most of the ligand/Epo-R interaction occurs in a hydrophobic flat region of the Epo-R (Phe^{93}, Met^{150}, and Phe^{205}). Binding of Epo induces a conformational change of the Epo-R and activation of Janus kinase 2, a cytoplasmic tyrosine kinase coupling to the Epo-R. On phosphorylation, the Epo-R provides docking sites for proteins containing Src homology 2 (SH2) domains (for example, the SH2-containing adapter protein, SHC). The major Epo-R signalling pathways involve signal transducer and activator of transcription 5

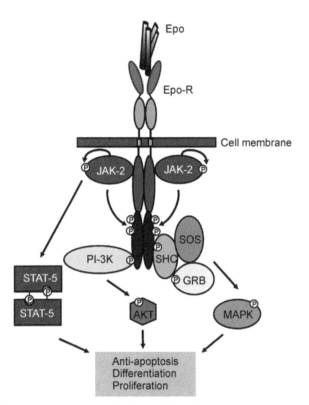

FIGURE 10.5 Model of the Epo receptor (Epo-R) and signal transduction.

On binding of Epo (4 antiparallel α-helices) the Epo-R homodimer undergoes a conformational change inducing the autophosphorylation of Epo-R-associated Janus kinases (JAK-2). In turn, this causes a tyrosine phosphorylation of the Epo-R and of STAT-5 (signal transducer and activator of transcription 5), and the activation of the PI-3K (phosphatidyl-inositol 3-kinase)/AKT and MAPK (mitogen-activated protein kinase) pathways. The action of Epo is terminated when the Epo/Epo-R-complex is internalized following dephosphorylation of the Epo-R.

(STAT-5), phosphatidyl-inositol 3-kinase (PI-3K)/AKT (an important serine/threonine-specific protein kinase), and SHC/mitogen-activated protein kinase (MAPK) (*13*). STAT-5 induces the expression of anti-apoptotic proteins (for example Bcl-x_L). PI-3K and MAPK prevent pro-apoptotic proteins (Bax, Bad) from activating caspases, specific cysteine proteases otherwise executing apoptosis. In addition, MAPK activates genes promoting cell proliferation.

The action of Epo is terminated when the Epo/Epo-R-complex is internalized following Epo-R dephosphorylation by the tyrosine phosphatase, Src homology phosphatase-1. After internalization about 40% of Epo is degraded by the proteasome and 60% is resecreted. Epo-R-mediated uptake by the target cells is considered a major mechanism of the degradation of circulating Epo.

10.3.5 Assay of Epo

The earliest tests were bioassays in which radioactive iron (^{59}Fe) was injected into rodents followed by measurements of the incorporation of the iron into the RBCs (*1,3,4*). One international unit (IU) of Epo is defined as the dose producing the same erythropoiesis-stimulating response in the animals as 5 µmol cobaltous chloride. Thus, similar to the insulin unit, the Epo unit is based on biological activity. In 1972 the current International Reference Preparation (second IRP) of impure human urinary Epo (specific activity 2 IU/mg protein) and in 1992 the first purified recombinant DNA-derived human Epo (IS 87/684; 130,000 IU/mg fully glycosylated protein) were established (*3*).

In vivo bioassays for Epo show high variability. The same holds true for *in vitro* bioassays, i.e. the measurement of ^{59}Fe incorporation into haeme in foetal mouse liver cell cultures. Therefore, immunoassays are commonly used, including radioimmunoassays, where the Epo in the sample competes with radiolabeled Epo for anti-Epo antibody (anti-Epo Ab) binding, and enzyme-linked immunosorbent assays, where the Epo of the sample is immobilized on a microtiter plate followed by the addition of anti-Epo Ab linked to an enzyme and, finally, the enzyme's substrate producing a colour change. The normal concentration of immunoreactive Epo in human plasma is about 15 U/l, equivalent to approximately 5 pmol/l. Immunoassays cannot distinguish between fully glycosylated Epo and asialo-Epo which is not active *in vivo*. Therefore, *in vivo* bioassays are still required to calibrate rhEpo for therapeutic purposes.

10.4 MANUFACTURE OF ERYTHROPOIESIS-STIMULATING AGENTS (ESAs)

10.4.1 Innovator rhEpo preparations

rhEpo is produced with the use of cells transfected with human *EPO* complementary DNA (cDNA) (the coding sequence of the gene) linked to an expression vector ('recombinant DNA'). The therapeutic rhEpo drug substances must be manufactured in mammalian host cells because Epo is a complex glycoprotein (*14,15*). Transformed bacteria such as *Escherichia coli* are primarily useful for the production of non-glycosylated recombinant proteins like insulin or somatotropin. In mammalian cells, the synthesis of *N*-glycans starts in the cytosol with the anchoring of sugar molecules to dolichol, which is then transferred to the growing polypeptide in the endoplasmic reticulum. There, several sugars are removed, and the protein is folded and moved to the Golgi complex, where further mannose elimination occurs followed by the addition of *N*-acetylglucosamine, galactose and sialic acid.

Chinese hamster ovary (CHO) cells deficient in the dihydrofolate reductase gene are most commonly used for the large-scale manufacture of rhEpo or other therapeutic glycoproteins. This is because gene amplification can be achieved in these cells by co-selection in the presence of methotrexate (8). The structures of the sialylated N- and O-glycans of rhEpo expressed in CHO cells have been studied by means of fast protein and high-performance liquid chromatography, high-pH anion-exchange chromatography and nuclear magnetic resonance spectroscopy (15). There are di-, tri-, tri'- and tetra-antennary N-acetyllactosamine-type oligosaccharides, which can be completely or partially sialylated. Three different types of α2-3-linked sialic acids are present, namely N-acetylneuraminic (95%), N-glycolylneuraminic (2%) and N-acetyl-9-O-acetylneuraminic (3%) acid. CHO cell lines express the α2,3-sialyltransferase but not a functional α2,6-sialyltransferase. This explains why CHO cells add sialic acid exclusively in an α2-3-linkage, whereas the α2-6-linkage seen in human oligosaccharides is missing. On the other hand, CHO cells add N-glycolylneuraminic acid residues (Neu5Gc) to the glycans, which human cells cannot synthesize.

According to the World Health Organization, eukaryotic cell-derived rhEpo, whose peptide core is identical with that of the native human Epo, is given the international non-proprietary drug name (INN) 'epoetin'. Differences in the amino acid residues chain are indicated by a random prefix (e.g. 'darbepoetin'). Differences in the glycosylation pattern were supposed to be indicated by a specific Greek letter added to the name (e.g. 'epoetin alfa'), but this was not always applied (see 10.4.2). Two innovator CHO cell-derived epoetins have been used for over two decades in clinical routine, namely epoetin alfa and epoetin beta. Epoetin alfa is marketed primarily under the trade names Epogen® (Amgen), Procrit® (Johnson & Johnson) and—outside the USA—Eprex® or Erypo® (Ortho Biotech). Epoetin beta is marketed under the brand names NeoRecormon® (F. Hoffmann-LaRoche) and Epogin® (Chugai). Epoetin alfa and epoetin beta are alike in their four glycosylation sites and their secondary structure. However, they exhibit distinct differences in their glycosylation patterns. For example, there are less N-glycans with non-sialylated outer Galβ1–4GlcNAc moieties and O-glycans with Galβ1–3GalNAc in epoetin alfa compared to epoetin beta. Even so, clinically, the two epoetins are considered equivalent. In 2009, another original CHO cell-derived rhEpo (epoetin theta) has been launched in the European Union (EU), which is marketed as Biopoin® (CT Arzneimittel) and Eporatio® (Ratiopharm).

In some East European, Asian and Latin American countries, CKD patients have been treated with epoetin omega, which is expressed in *EPO* cDNA-transfected baby hamster kidney (from Syrian hamster) cells. Epoetin omega has been traded as Repotin® (Bioclones) and Epomax® (Lek; Cryopharma Pizzard y Salud). Although epoetin omega is probably non-inferior to epoetin alfa in correcting renal anaemia in CKD patients on dialysis, this medicine is not widely used. In contrast to CHO cell-derived rhEpo, epoetin omega has an N-glycan with phosphorylated oligomannoside chains and it possesses less O-glycans (8).

Another recombinant product earlier marketed in the EU was epoetin delta (Dynepo®, Shire Pharmaceutical). This unique rhEpo was homologously expressed on gene activation in a human fibrosarcoma cell line (HT-1080) into which a DNA fragment was transfected that activated the native *EPO* promoter (8). Contrasting the other epoetins, epoetin delta lacked Neu5Gc residues as it was human cell-derived. In 2009 Shire Pharmaceutical decided to withdraw the marketing authorization for commercial reasons (not affording profits).

The epoetins are purified by a number of chromatographic procedures to remove xenogenic proteins, oncogenes, pyrogens and microorganisms. The medicinal products are formulated as dry powders or as isotonic sodium chloride/sodium citrate buffered solutions for intravenous (IV) or subcutaneous (SC) administration. They are stabilized by the addition of either polyoxyethylene sorbitan (Tween-20 or Tween-80) or human albumin (2.5 mg/ml) and benzyl alcohol (1%).

10.4.2 Biosimilar epoetins

Since the patents for the innovator epoetins have expired, manufacturers other than the originators have developed biosimilars (short for 'similar biological medicinal products' at full length) of epoetin alfa in the EU and in other parts of the world (*16*). Biosimilars are approved copy versions of an already authorized biological medicinal product with demonstrated similarity in physicochemical characteristics, efficacy and safety, based on a comprehensive comparability exercise. Since developers of biosimilars usually do not have access to the innovators' proprietary data they engineer their own manufacturing process to obtain a product as similar as possible to the original. It is evident from Table 10.2 that the manufacturing processes required for recombinant biopharmaceuticals are so complex that they cannot be exactly copied by companies other than the innovator (see 10.4.1). This is an important difference between biosimilars and generics, the latter being copied small, chemically synthesized, drugs. Particularly glycoproteins exhibit differences compared to the original with respect to the microstructure of their glycans.

In the EU, the Committee for Medicinal Products for Human Use of the European Medicines Agency, which is responsible for the assessment of novel medicines, has provided overarching guidelines regards the approval of biosimilars and product-specific requirements for epoetins. In the

---

TABLE 10.2 Production Steps Influencing Recombinant Medicinal Products

Establishment of 'Master Cell Bank' (MCB)
  Sequence of cDNA
  Type of vector/plasmid
  Accessory DNA elements (promoter, etc.)
  Type of host cell
  Technique of transfection
  Propagation of host cell clone
  Establishment of 'Working cell bank' (WCB, derived from the MCB to initiate a production batch)
Maintenance of production cultures
  Composition of culture medium (pH; serum, vitamins, etc.)
  Type of culture vials/bottles
  Type of fermentor/bioreactor
Extraction and purification of recombinant product from culture medium
  Removal of process-related impurities (pathogens, host-cell proteins, DNA, toxins, heavy metals, etc.)
  Removal of product-related impurities (e.g. aggregates, breakdown products, etc.)
Analysis of product with respect to safety, purity, identity, potency and protein content)
Formulation (stabilizer, solvent, etc.)

Source: Modified from Jelkmann (14).

USA the Food and Drug Administration has been developing similar regulatory pathways for bio-similars (previously called 'follow-on biologics').

The primary rationale for the use of biosimilars is cost saving. The two biosimilar epoetins available in the EU are used at the same dose(s) and dosing regimen(s) for indications of the reference product, which is the original epoetin alfa marketed in the EU (Eprex®/Erypo®, Ortho Biotech) marketed outside the USA. One of the biosimilars has received the INN epoetin alfa (Binocrit®, Sandoz; Epoetin alfa Hexal®, Hexal Biotech; Abseamed®, Medice Arzneimittel Putter) and the other, the INN epoetin zeta (Silapo®, Stada; Retacrit®, Hospira).

The term 'biosimilar' should only be used for copied biopharmaceuticals approved under a defined regulatory pathway, as described. Physicochemical and functional investigations of purported epoetin alfa copies manufactured and used in some Asian and Latin American countries have revealed major isoform differences, batch-to-batch variations in biological activities as well as endotoxin contamination of some of the products (*16*).

10.4.3 Second-generation ESAs ('biobetter')

Attempts have been successful to develop recombinant Epo molecules with prolonged survival in circulation. First, the rhEpo mutein (a product with altered amino acid residues sequence) darbepoetin alfa was approved as an anti-anaemic drug early in 2000 (Aranesp®; Amgen). Darbepoetin alfa is a hyperglycosylated long-acting analogue of rhEpo, containing two additional *N*-linked glycans in positions 30 and 88 (*7*). The extra carbohydrates are added in association with the exchange of five amino acids through site-directed mutagenesis (*7*). The carbohydrate portion of darbepoetin alfa amounts to 51% of the total mass, which is 37.1 kDa. Darbepoetin alfa stimulates erythropoiesis through binding to the Epo-R. However, compared to Epo, darbepoetin alfa has a lower affinity for the Epo-R; receptor-binding affinity is defined by the association and dissociation rates (k_{on}/k_{off}). Darbepoetin alfa binds more slowly to the Epo-R and dissociates faster than Epo. Because the ligand/Epo-R complex is internalized following dephosphorylation, the slower association and faster dissociation explains why darbepoetin alfa has a longer survival in circulation than Epo. Compared to the short terminal half-life of IV administered epoetins (6–9 h), the half-life of darbepoetin alfa is three- to fourfold longer (25 h), which allows for less frequent clinical application.

Another biobetter is methoxy polyethylene glycol-epoetin beta (methoxy PEG-epoetin beta; Mircera®; F. Hoffmann-LaRoche, Chugai) which has an even longer half-life (130–140 h) on IV injection. The drug substance contains a methoxy PEG polymer of approximately 30 kDa integrated via amide bonds between the amino groups of either Ala[1], Lys[45] or Lys[52] of epoetin beta (total molecular mass about 60 kDa). The prolonged *in vivo* survival of methoxy PEG-epoetin beta is also due to a reduced Epo-R binding affinity. Compared to Epo, methoxy PEG-epoetin beta has a very slow association rate (k_{on}) for the Epo-R and a slightly faster dissociation rate (k_{off}).

In contrast to native human Epo and the epoetins, which are traditionally calibrated in IU, concentrations and doses of darbepoetin alfa and methoxy PEG-epoetin beta are expressed in micrograms. None of the conventional *in vivo* bioassays is suitable for valid quantification of the activities of the long-acting, second-generation ESAs. As the specific activity of rhEpo amounts to about 200,000 IU/mg peptide, 1 µg of darbepoetin alfa or of methoxy PEG-epoetin beta peptide corresponds biophysically to 200 IU rhEpo peptide. Clinically, however, the long-acting biobetter products enable reduction of the doses to below the predicted 1:200 ratio.

10.5 CLINICAL APPLICATION OF ESAs

The therapy with ESAs can be indicated for the alleviation of chronic anaemia, but it is not indicated as a substitute for RBC transfusions in patients who have life-threatening low [Hb] values and require immediate correction of anaemia.

10.5.1 Indications for ESA therapy

10.5.1.1 Anaemia due to CKD

Lack of Epo is the primary cause of the anaemia of CKD. Other pathogenetic factors include shortening of RBC survival, bleeding, malnutrition, and inhibition of the growth of erythrocytic progenitors by inflammatory cytokines and uraemia toxins. The anaemia in CKD is usually normochromic and normocytic, meaning the average Hb content and size of the RBCs are within normal limits. [Hb] values decrease with the impairment of renal function. Significant anaemia develops in general when the GFR falls from the normal value of about 120 to below 30 ml/min and 1.73 m^2 body surface area.

As soon as rhEpo became available in 1985, a series of clinical studies were conducted to assess the effectiveness of the medicine in CKD patients. The studies demonstrated that rhEpo could restore [Hb], abrogate the necessity of RBC transfusions, and improve the overall well-being in patients requiring dialysis. The results were so impressive that rhEpo was approved as an anti-anaemic drug for treatment of CKD patients in 1988, only 3 years after its development (3,4). Eschbach (17) has reported a typical case of that era: within 8 weeks of rhEpo therapy anaemia was corrected in a 39-year-old patient with nephrotic syndrome, who had experienced androgen treatment, unsuccessful renal transplantation, transfusion of 313 units of RBCs, and human immunodeficiency virus infection. The patient's improvement was impressive; he was able to meet the physical demands of running his country store and to perform sports again. The availability of rhEpo and its analogues has since transformed the lives of millions of CKD patients. Patients at the pre-dialysis stage as well as patients on maintenance dialysis are today routinely treated with ESAs (1,2). The therapy not only abolishes the need for RBC transfusions but also prevents the hyperdynamic cardiac state, thus minimizing the risk of progression towards left ventricular hypertrophy and its associated mortality. Furthermore, the treatment improves physical performance and brain function. The target Hct in patients on ESA therapy is consensually set at 0.30–0.36 ([Hb] 100–120 g/l).

Causes of the rare cases of hyporesponsiveness to ESAs include iron deficiency, infection and inflammation, bleeding and haemolysis, coexisting medical conditions such as malignancy, secondary hyperparathyroidism, aluminium toxicity, vitamin B_{12} or folate deficiency, and protein malnutrition. The most common cause of resistance to ESA therapy is the reduced iron availability, which is indicated by a low serum ferritin concentration (<100 µg/l), a low transferrin saturation (<20%, calculated from serum iron and total iron-binding capacity) and a high proportion of hypochromic (Hb-deficient) RBCs (>10%). The administration of IV iron decreases the ESA dosage requirements.

10.5.1.2 Anaemia associated with cancer chemotherapy

The primary goals of the therapy with ESAs (epoetins or darbepoetin alfa) in cancer patients receiving chemotherapy are to maintain the [Hb] above the transfusion trigger, to increase exercise tolerance, prevent fatigue and improve quality-of-life parameters (*18*). Of note, ESAs should be administered at the lowest dose possible and the treatment should increase [Hb] to the lowest level possible to avoid RBC transfusions. While ESA therapy does not impact on survival or disease progression, it may be associated with an increased incidence of venous thrombo-embolic events. Whether ESAs stimulate tumour growth is a controversial issue. Although cancer cells express Epo-R mRNA to some extent, evidence suggests that they lack functional Epo-R protein (*13*).

10.5.1.3 Other applications

The anaemias associated with acquired immunodeficiency syndrome (AIDS), hepatitis C infection, bone marrow transplantation, myelodysplastic syndromes, autoimmune diseases and heart failure can be alleviated by ESAs (*1,2*). Currently, this treatment is only approved for AIDS patients, particularly those on zidovudine. Another indication for ESAs can be the anaemia of prematurity to reduce the number of RBC transfusions in newborns. In the surgical setting, rhEpo may be administered preoperatively in order to stimulate erythropoiesis in phlebotomy programmes for autologous RBC re-donation or correction of a pre-existing anaemia, and post-operatively for recovery of RBC mass in certain interventions.

The practice of using ESAs as doping substances is to be condemned. The detection of recombinant ESAs is possible by analysis of urine species from athletes. Details of the doping issue can be found elsewhere (*19*).

10.5.2 Safety

10.5.2.1 General concerns

ESAs are contraindicated in patients with known hypersensitivity to non-human cell-derived products. In addition, patients with a hypersensitivity to human albumin should not be treated with formulations stabilized with this protein. Acute toxic effects of rhEpo have never been reported. Chronic treatment with ESAs may result in erythrocytosis if the Hct is not carefully monitored. In pregnancy, rhEpo should be administered very cautiously, because the risks for the foetus have not been evaluated in humans. Still, to the best of present knowledge rhEpo has no mutagenic potential.

The most common side effect of ESA therapy is flu-like symptoms, such as headache, aches in the joints, fever and feeling of weakness, particularly at the start of treatment. A more serious common side effect (affects 1–10 users in 100) is arterial hypertension. The increase in blood pressure can be explained partly by the elevated blood viscosity and by the abolishment of hypoxia-induced vasodi-latation in association with the [Hb]-related increase in O_2 supply. A very rare unwanted effect of ESA therapy is pure red cell aplasia (PRCA) occurring in CKD patients due to the formation of neutralizing anti-Epo Abs. Patients suffering from PRCA become anaemic because the bone marrow ceases to produce RBCs. The exposure-adjusted incidence of this form of PRCA is at present only <0.03 per 10,000 patient-years (*20*).

10.5.2.2 Immunogenicity of ESAs

Notwithstanding the very low incidence of anti-Epo Abs formation, the question of immunogenicity of biopharmaceuticals is of general interest (*16*). The probability of an immune response towards foreign proteins increases as a consequence of structural differences (primary structure and post-transcriptional modifications), protein aggregation, or impurities in the formulation. Preclinical studies do not provide reliable information on the immunogenicity of biopharmaceuticals in humans. Because native human Epo and rhEpo are almost identical in structure, rhEpo-containing medicines are normally not immunogenic, even on SC administration. However, changes in the manufacturing process or the formulation of an ESA may cause an immune reaction. Here, an instructive lesson was the transiently increased incidence of anti-Epo Ab-induced PRCA in ESA-treated CKD patients in the years 1998–2003, amounting to over 200 cases worldwide (20). Almost all the patients received, SC, an epoetin alfa formulation (Eprex®/Erypo®, Ortho Biotech) marketed outside the USA. In 1998, the distributor had replaced human serum albumin by polysorbate-80 and glycine in order to minimize the risk of transmission of pathogens. In addition, the medicine was newly marketed in pre-filled syringes with uncoated rubber stoppers. After these were replaced by Teflon®-coated stoppers, the incidence of anti-Epo Ab-induced PRCA greatly decreased again. The root cause of the temporary adverse event was never clearly identified, but it may have been related to epoetin-loaded micelles or to leachates released from the rubber stoppers that acted as an adjuvant in the immune response.

Another lesson was learned in a more recent clinical trial with a biosimilar epoetin that also caused neutralizing anti-Epo Abs formation on SC administration (20). Thorough investigations carried out by the manufacturer revealed that the problem was not related to the biosimilar epoetin *per se* but to abnormally high tungsten levels in the pre-filled syringes. The tungsten caused the protein to unfold and, subsequently, to form aggregates. The likely source of the tungsten was heat-resistant pins used in the manufacture of the glass syringes (16).

10.5.3 Non-haematopoietic actions of Epo

It has been hypothesized that the action of Epo is not restricted absolutely to the erythrocytic progenitors. Preclinical studies have suggested pleiotropic effects of ESAs with respect to apoptosis and protection against damage caused by hypoxia, ischaemia/reperfusion, cytotoxic agents or inflammation. Actually, Epo-R mRNA is expressed in a variety of non-haematopoietic tissues and cell types, including endothelial, cardiac, renal and neuronal cells. However, data around the role of Epo outside the bone marrow are conflicting (13). First, the presence of Epo-R mRNA does not prove the translation of functional Epo-R protein. Previous commercial antibodies used for immunological detection of Epo-R protein were not specific and cross-reacted with a number of other proteins (for example, the heat shock proteins Hsp70 and Hsp90), rendering immunohistochemical (study of a whole tissue section) and immunoblotting (study of tissue extracts) findings fraught with problems. Epo binding is detected on the surface of erythrocytic progenitors but not generally on cells from other tissues, such as heart, kidney or brain. Second, preclinical studies often used levels of Epo 10- to 1000-fold greater than the maximum plasma level observed in patients receiving a proof dose.

A recent hypothesis suggests that Epo could exert non-haematopoietic effects via heteromeric receptors involving Epo-R and CD131 (the β-common receptor shared by cytokines such as interleukin-3, IL-3, and granulocyte–macrophage colony-stimulating factor) (21). Data used to support the role of the Epo-R/CD131 heteromer were generated using mice in which the CD131 had been knocked out. This research has led to the engineering and testing of non-haematopoietic Epo derivatives and analogues. In preclinical studies asialo-Epo (Epo lacking terminal sialic acid) and carbamoylated Epo (Epo with homocitrulline due to treatment with cyanate) were assigned cyto-protective actions without stimulating erythropoiesis. Further, Epo-R/CD131-binding regions were suggested within the Epo molecule for the development of tissue-protective peptides (*21*). For example, ARA290 is a synthetic linear peptide composed of only 11 amino acids which lacks ery-thropoietic potential but reportedly binds to the Epo-R/CD131 heteroreceptor. However, there have also been studies showing no evidence of an interaction between Epo-R and CD131 (*13*).

10.6 PERSPECTIVES

Anaemia treatment with recombinant ESAs is cost-intensive. Hence, the question has been raised whether there are alternative therapeutic options (*16,22*). Research has focused on the anaemia due to CKD which affects millions of people worldwide. One strategy has been to replace recombinant ESAs by Epo mimetic peptides (EMPs), synthetic cyclic peptides of about 20 amino acids (two cysteine residues form a disulphide bond). EMPs show no sequence homology to Epo but signal through the Epo-R. Figure 10.6 shows the structure of the most advanced product, peginesatide, a ~45 kDa pegylated dimeric peptide. Peginesatide consists of two identical 21-amino acid residue chains (about 2 kDa each) covalently bonded to a linker derived from iminodiacetic acid and β-alanine to which a lysine-branched bis-methoxy PEG chain is attached in order to prolong systemic circulation (*23*). Peginesatide was approved in the USA in 2012 for SC or IV treatment of anaemic adult CKD patients

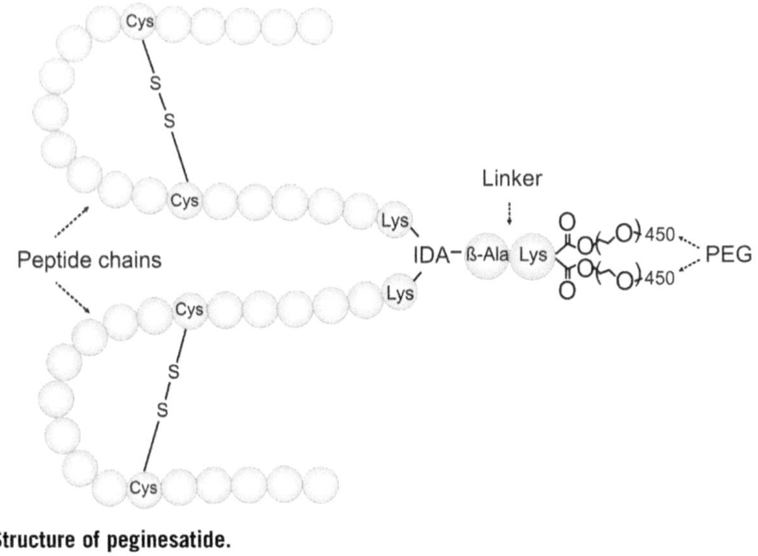

FIGURE 10.6 Structure of peginesatide.

Two identical, synthetic Epo-mimetic peptides of 21 amino acid residues are bonded via a linker derived from iminodiacetic acid (IDA) and β-alanine (β-Ala) to a lysine-branched bis-(methoxypoly(ethylene glycol)) moiety, i.e. two 20 kDa PEG chains. The total mass of the molecule amounts to ~45 kDa.

on dialysis (Omontys®, Affymax/Takeda), but the drug was recalled in 2013 due to serious hypersensitivity reactions, including life-threatening and fatal events in patients receiving Omontys®. In an alternative approach, EMPs have been constructed onto human IgG-based scaffolds by recombinant DNA technology. The seminal compound, CNTO 528 (Centocor), increased [Hb] on IV administration in a phase I study in healthy men. The follow-on product CNTO 530, a dimeric EMP fused to a human IgG4 Fc scaffold, was shown to expand the pool of erythrocytic progenitors *in vitro* and *in vivo*.

Gene therapy could be another treatment option (8). In the first *EPO* therapy trial on CKD patients an autologous *ex vivo* approach was chosen. In brief, individual dermal core samples ('Biopumps', Medgenics) were incubated *in vitro* with *EPO* cDNA inserted into a vector containing the cytomegalovirus promoter and the simian virus-40 polyA site and then reimplanted under the abdominal skin. Clinical trials are in progress with respect to the efficacy, safety and immunogenicity of this treatment.

Other strategies focus on the chemicals that activate the endogenous *EPO*. Since the HIF-α hydroxylases require α-ketoglutarate for action, α-ketoglutarate competitors prevent HIF-α degradation and increase *EPO* expression (*11*). A phase I trial investigating effects of the HIF-α prolyl hydroxylase inhibitor (HIF-PHI) FG-2216 (FibroGen/Astellas) in CKD patients was successful; other HIF-PHIs are in clinical trials. Possible advantages of the use of HIF-PHIs—compared to recombinant ESAs and synthetic EMPs—include the lower production costs and the oral route of administration. However, the safety of the HIF-PHIs must be studied very carefully, before they enter clinical routine. The HIFs induce the expression of >200 genes apart from *EPO* (*9,11*), which may result in serious unwanted effects.

In summary, the evolution of Epo from its discovery to the widespread use of it and its surrogates in the clinic provides one of the success stories of the biopharmaceutical era. The key points in this scientific journey are described in Box 10.1. Research continues apace with the aim to provide even better, more effective medicines in this sector.

BOX 10.1 KEY LEARNING POINTS

- The human Epo gene (*EPO*) is encoded in chromosome 7q11–22.
- Endogenous Epo is mainly of renal origin; its production is stimulated under hypoxic conditions (anaemia and hypoxaemia).
- Human Epo is a 30.4 kDa glycoprotein (165 amino acid residues, three *N*-glycans, and one *O*-glycan); the *N*-glycans confer the *in vivo* stability.
- The human Epo receptor is a membrane-spanning 59 kDa glycoprotein (484 amino acid residues and one *N*-glycan); it is a member of the cytokine receptor superfamily I, forms homodimers and signals through Janus tyrosine kinase 2 (JAK-2).
- Epo is essential for the survival, proliferation and differentiation of erythrocytic progenitors; Epo deficiency results in anaemia.
- Recombinant human Epo (rhEpo, epoetin) has the same polypeptide backbone and similar glycosylation sites as the endogenous form.
- Clinically approved erythropoiesis-stimulating agents (ESAs) presently include original and biosimilar epoetins, a long-acting pegylated rhEpo-derivative, and a long-acting hyperglycosylated mutated rhEpo.
- ESA therapy avoids red blood cell transfusions and improves quality of life in patients with renal anaemia and other anaemias, such as cancer chemotherapy-associated anaemia.
- Non-haematopoietic tissue-protective actions of Epo are the focus of present research.

References

1. Jelkmann, W., Ed. *Erythropoietin: Molecular Biology and Clinical Use;* F.P. Graham Publishing: Johnson City, TN, 2003.
2. Elliott, S. G., Foote, M., Molineux, G., Eds. *Erythropoietins, Erythropoietic Factors, and Erythropoiesis,* 2nd ed.; Birkhäuser: Basel, Boston, Berlin, 2009.
3. Jelkmann, W. Erythropoietin After a Century of Research: Younger than Ever. *Eur. J. Haematol* **2007,** *78,* 183–205.
4. Fisher, J. W. Landmark Advances in the Development of Erythropoietin. *Proc. Soc. Exp. Biol. Med* **2010,** *235,* 1398–1411.
5. Erslev, A. Humoral Regulation of Red Cell Production. *Blood* **1953,** *8,* 349–357.
6. Goldwasser, E. Erythropoietin: A Somewhat Personal History. *Perspect. Biol. Med.* **1996,** *40,* 18–31.
7. Elliott, S.; Lorenzini, T.; Asher, S.; Aoki, K.; Brankow, D.; Buck, L.; Busse, L.; Chang, D.; Fuller, J.; Grant, J.; Hernday, N.; Hokum, M.; Hu, S.; Knudten, A.; Levin, N.; Komorowski, R.; Martin, F.; Navarro, R.; Osslund, T.; Rogers, G.; Rogers, N.; Trail, G.; Egrie, J. Enhancement of Therapeutic Protein In Vivo Activities through Glycoengineering. *Nat Biotechnol* **2003,** *21* (4), 414–421.
8. Jelkmann, W. Control of Erythropoietin Gene Expression and Its Use in Medicine. *Methods Enzymol* **2007,** *435,* 179–197.
9. Semenza, G. L. Involvement of Oxygen-Sensing Pathways in Physiologic and Pathologic Erythropoiesis. *Blood* **2009,** *114,* 2015–2019.
10. Haase, V. H. Hypoxic Regulation of Erythropoiesis and Iron Metabolism. *Am. J. Physiol. Renal. Physiol.* **2010,** *299,* F1–F13.
11. Bruegge, K.; Jelkmann, W.; Metzen, E. Hydroxylation of Hypoxia-Inducible Transcription Factors and Chemical Compounds Targeting the HIF-Hydroxylases. *Curr. Med. Chem.* **2007,** *14,* 1853–1862.
12. Hattangadi, S. M.; Wong, P.; Zhang, L.; Flygare, J.; Lodish, H. F. From Stem Cell to Red Cell: Regulation of Erythropoiesis at Multiple Levels by Multiple Proteins, RNAs, and Chromatin Modifications. *Blood* **2011,** *118* (24), 6258–6268.
13. Jelkmann, W. Recombinant EPO Production – Points the Nephrologist Should Know. *Nephrol. Dial. Transplant.* **2007,** *22,* 2749–2753.
14. Hokke, C. H.; Bergwerff, A. A.; Van Dedem, G. W.; Kamerling, J. P.; Vliegenthart, J. F. Structural Analysis of the Sialylated *N*- and *O*-Linked Carbohydrate Chains of Recombinant Human Erythropoietin Expressed in Chinese Hamster Ovary Cells. Sialylation Patterns and Branch Location of Dimeric *N*-Acetyllactosamine Units. *Eur. J. Biochem.* **1995,** *228* (3), 981–1008.
15. Elliott, S.; Sinclair, A. M. The Effect of Erythropoietin on Normal and Neoplastic Cells. *Biologics* **2012,** *6,* 163–189.
16. Jelkmann, W. Biosimilar Recombinant Human Erythropoietins ("epoetins") and Future Erythropoiesis-Stimulating Treatments. *Expert. Opin. Biol. Ther.* **2012,** *12* (5), 581–592.
17. Eschbach, J. W. The Anemia of Chronic Renal Failure: Pathophysiology and the Effects of Recombinant Erythropoietin. *Kidney. Int.* **1989,** *35,* 134–148.
18. Aapro, M.; Jelkmann, W.; Constantinescu, S. N.; Leyland-Jones, B. Effects of Erythropoietin Receptors and Erythropoiesis-Stimulating Agents on Disease Progression in Cancer. *Br. J. Cancer.* **2012,** *106* (7), 1249–1258.
19. Ghigo, E., Lanfranco, F., Strasburger, C. J., Eds. *Hormone Use and Abuse by Athletes;* Springer: New York, 2011.
20. Macdougall, I. C.; Roger, S. D.; de Francisco, A.; Goldsmith, D. A. J.; Schellekens, H.; Ebbers, H.; Jelkmann, W.; London, G.; Casadevall, N.; Hörl, W. H.; Kemeny, M.; Pollock, C. Antibody-Mediated

Pure Red Cell Aplasia in Chronic Kidney Disease Patients Receiving Erythropoiesis Stimulating Agents: New Insights. *Kidney. Int.* **2012,** *81* (8), 727–732.

21. Brines, M.; Cerami, A. Erythropoietin-Mediated Tissue Protection: Reducing Collateral Damage from the Primary Injury Response. *J. Intern. Med.* **2008,** *264* (5), 405–432.

22. Macdougall, I. C. New Anemia Therapies: Translating Novel Strategies from Bench to Bedside. *Am. J. Kidney. Dis.* **2012,** *59* (3), 444–451.

23. Woodburn, K. W.; Holmes, C. P.; Wilson, S. D.; Fong, K. L.; Press, R. J.; Moriya, Y.; Tagawa, Y. Absorption, Distribution, Metabolism and Excretion of Peginesatide, a Novel Erythropoiesis-Stimulating Agent, in Rats. *Xenobiotica* **2012,** *42* (7), 660–670.

Lysosomal storage disorders: current treatments and future directions

11

Charles W. Richard III

Oxyrane US, Inc., Burlington, MA 01803, USA

CHAPTER OUTLINE

11.1 Introduction 328
11.2 Clinical Aspects of LSDs 329
11.3 Therapies for LSDs 330
    11.3.1 Enzyme replacement therapies 330
        11.3.1.1 Gaucher disease 330
        11.3.1.2 ERT development in other LSDs.. 332
    11.3.2 Improving Efficacy by Enzyme Modifications to Improve Cellular Uptake and Tissue
        Targeting 335
        11.3.2.1 Muscle targeting.. 335
        11.3.2.2 Bone targeting... 335
        11.3.2.3 CNS targeting of neuropathic LSDs 336
    11.3.3 Cross-correction through Stem Cell Therapies 336
        11.3.3.1 Haematopoietic stem cell therapies.. 336
    11.3.4 Enzyme stabilization (chaperone therapy) 337
    11.3.5 Substrate reduction therapy..337
11.4 Conclusions 338
References 338

ABSTRACT

Lysosomal storage disorders (LSDs) represent a group of about 50 genetic disorders caused by deficiencies of lysosomal proteins. The missing lysosomal protein causes a build-up of toxic metabolites in the cells of patients, leading to progressive multisystem disease and premature death. Although individually rare, the combined prevalence of all lysosomal disorders is estimated to be 1 in 8000 births. This chapter describes progress in several different LSD treatment modalities including enzyme replacement therapy, haematopoietic stem cell therapy, chaperone (enzyme stabilization) therapy, and substrate reductions therapy, and highlights new treatment directions for the future.

Keywords/Abbreviations: Lysosomal storage disorders (LSD); Enzyme replacement therapy (ERT); Endoplasmic reticulum (ER); Mannose-6-phosphate (M6P); Glycogen storage disease (GSD); Mucopolysaccharidosis (MPS); Metabolic disease; Enzyme replacement therapy; Biopharmaceuticals; Metabolic cross-correction; Chaperone therapy; Substrate reduction therapy; Haematopoietic stem cell therapies (HSCT); Globotriaosylceramide (Gb3).

Introduction to Drug Research and Development. http://dx.doi.org/10.1016/B978-0-12-397176-0.00011-X

11.1 INTRODUCTION

Lysosomal storage disorders (LSDs) are a heterogeneous group of approximately 50 recessively inherited metabolic diseases (Table 11.1). Most involve genetic mutations causing loss of function of key catabolic enzymes residing in the lysosome that are involved in the degradation of macromolecules. The substrate for the missing enzyme gradually builds up over time, leading to cellular

Table 11.1 Examples of Lysosomal Storage Diseases

Lysosomal Storage Disorder	Defective Protein	Storage Materials
Sphingolipidosis		
Fabry disease	α-galactosidase A	Globotriaosylceramide & globotriaosylsphingosine
Gaucher disease, types I, II, III	β-glucosidase	Glucosylceramide, GM1, GM2, GM3, GD3, glucosylsphingosine
Metachromatic leukodystrophy	Arylsulfatase A	Sulfatide & lysosulfatide
Globoid cell leukodystrophy	Galactocerebroside β-galactosidase	Galactosylceramide, psychosine
Niemann–Pick A & B	Sphingomyelinase	Sphingomyelin, cholesterol
GM1 gangliosidosis	β-galactosidase	GM2
GM2 gangliosidosis (Tay–Sachs)	β-hexosaminidase A	GM2
GM2 gangliosidosis (Sandhoff)	β-hexosaminidase A & β-hexosaminidase B	GM2
Glycogenosis		
Glycogen storage disease II (Pompe)	α-glucosidase	Glycogen
Mucopolysaccharidosis		
MPS I (Hurler, Scheie)	α-iduronidase	Dermatan & heparan sulfate
MPS II (Hunter)	Iduronate sulfatase	Dermatan & heparan sulfate
MPS IIIA (Sanfilippo)	Heparan N-sulfatase	Heparan sulfate
MPS IIIB (Sanfilippo)	N-acetylglucosaminidase	Heparan sulfate
MPS IVA (Morquio A)	n-acetylgalacatosamine 6-sulfatase	Keratan and chondroitin sulfate
MPS IVB (Morquio B)	β-galacatosidase	Keratan sulfate
MPS VI (Maroteaux–Lamy)	Arylsulfatase B	Dermatan sulfate
MPS VII (Sly)	β-glucuronidase	Dermatan, keratin, and chondroitin sulfate
MSD (Austin disease)	Formylglycine generating enzyme	Heparan sulfate, dermatan sulfate, Chondroitin sulfate, sulfolipids

Left column gives the name of the disease, middle column, the deficient protein and the right column, the stored compounds.

dysfunction and clinical symptomatology. Other types of lysosomal dysfunction that can lead to LSDs include defects in proteins involved in posttranslational processing, enzyme trafficking, enzyme activators, and nonenzymatic lysosomal transmembrane proteins. Continually updated reviews of individual LSDs are available in two excellent online textbooks (*1,2*).

The mechanisms by which increased substrate storage triggers cell dysfunction and ultimately cell death is incompletely understood, but hypotheses put forth include mechanical disruption by expansion of lysosomal volume and number, alterations in calcium homeostasis, impairment of autophagy, activation of downstream signal transduction pathways, alterations in sphingolipid trafficking, build-up of storage material in organelle membranes outside the lysosome, disruption of endocytic raft membrane signalling, synaptic dysfunction in neurodegenerative LSDs and downstream secondary central nervous system (CNS) inflammation typified by microglial or macrophage proliferation (*3–6*).

11.2 CLINICAL ASPECTS OF LSDs

The majority of LSDs are caused by mutations in acid hydrolases involved in the degradation of macromolecules within the lysosomes (Table 11.1). The type of accumulated macromolecule substrate is predicated on the place of the enzyme in the metabolic pathway, but can include mucopolysaccharides, glycogen, glycoproteins, sphingolipids and various other lipid classes. Although lysosomal enzymes are uniformly expressed in most cells, the substrates they act upon are less uniformly distributed. This, in turn, determines which tissues and organs are more affected in any particular LSD: excess deposition of gangliosides in the CNS can lead to brain disease while high expression of keratin and dermatan sulfate in skeletal tissue can lead to skeletal and cartilage abnormalities.

There is considerable variation of clinical presentation within most LSDs, ranging from rapid-progressing and severe infantile-onset forms to attenuated late-onset forms in juveniles and adults. Much of the variation can be explained by residual activity of the enzyme in the metabolic pathway involved. Complete or near-complete loss of enzymatic activity causes rapidly progressive infantile-onset forms (e.g. type II Gaucher disease, neuronopathic forms of mucopolysaccharidosis (MPS) types I, II, II, glycogen storage disease type II (GSDII), and cholesterol ester storage disease) whereas partially enzymatic activity of just a few percentage can be enough to prevent neurodegeneration and attenuate the phenotype. This hypothesis is at least partially supported by genotype/phenotype correlations within human genotypes in which truncating mutations, with complete or near-complete loss of enzyme cause severe disease, whereas missense mutations that are often associated with partially functioning enzyme cause milder disease. Within the milder forms, however, there must be other environmental, epigenetic, or polygenic influences since siblings and twins with identical mutations may exhibit highly divergent clinical presentations as regards to age of onset and severity of disease manifestation.

Despite the legislative and regulatory mandates encouraging increased commercial interest in rare and orphan drug development (*7*), proving the therapeutic benefit of therapies for LSDs is not without challenges. The very rarity of these ultra-orphan diseases often means a systematic understanding of disease progression and natural history is not well understood. As most paediatricians do not commonly encounter these rare disorders in their daily practice, many of the attenuated forms are only properly diagnosed after a tortuous diagnostic odyssey, consequently at a time point where some of the organ pathology may be irreversible. Although cross-sectional studies of putative disease-associated biomarkers may have been examined in typically small sample sizes, longitudinal studies tracking potential surrogate biomarkers 'reasonably likely to predict clinical disease' over time have been lacking. Because clinical symptomatology and disease progression is different between early-onset and late-onset forms, the already-rare patient population must be segregated into different clinical trial cohorts with different outcome measures. Patients are best treated in regional Centres of Excellence with multidisciplinary speciality treatment teams, although standards of care guidelines vary even among these centres, and many centres are too widely dispersed for easy clinical trial recruitment and execution. The development of clinical rating scales from expert panels that would be useful at all disease stages is increasingly available for some of the more common rare diseases, but is still in its infancy in ultra-orphan LSDs. Placebo-controlled trials remain the gold standard; however, the rapid downhill course in infantile forms precludes placebo controls on ethical grounds, even when subjects on placebo have the option of continuing on to open-label drug at the conclusion of the initial investigative phase. Although the regulatory agencies allow for some 'flexibility' when considering marketing approval, the burden of proof for safety and efficacy remains, essentially, as statistically rigorous as for the more common diseases. Despite these challenges, an increasing number of larger pharmaceutical companies have now jumped into the rare and orphan drug-development mix along with the smaller 'rare and orphan disease' speciality biotechnology companies.

11.3 THERAPIES FOR LSDs

11.3.1 Enzyme replacement therapies

Lysosomal enzymes are glycoproteins that are synthesized in the endoplasmic reticulum and transported via Golgi to the lysosomes. The posttranslational addition of mannose-6-phosphate (M6P) residues in the Golgi targets most of the enzyme to the lysosome via M6P receptors. A smaller proportion escapes this pathway and is secreted from the cell. The soluble secreted enzyme can be recaptured by cells (M6P) by cell-surface M6P receptors and (re) trafficked back to the lysosome (Figure 11.1). This recapture or 'cross-correction' mechanism has been exploited therapeutically by the intravenous administration of exogenous enzyme which is taken up by tissues and organs missing the enzyme and delivered to the lysosome, where it then degrades accumulated substrate and restores cellular homeostasis. We know from both human and genetically engineered mouse models that small increases in residual enzyme can have a dramatic impact on the clinical phenotype; uptake of only small amounts of catalytic enzyme can have profound therapeutic effects in lysosomal storage diseases.

11.3.1.1 Gaucher disease

The first enzyme replacement therapy (ERT) trials were developed by Brady et al. for autosomal recessive β-glucocerebrosidase deficiency (Gaucher disease), first from native enzyme extracted from human placenta and later from recombinant human enzyme derived from mammalian (Chinese hamster ovary) cell culture (*8–10*). Gaucher is an autosomal recessive disorder caused by a deficiency of the catalytic enzyme β-glucocerebrosidase. The primary cellular site of pathology in Gaucher disease is the macrophage/monocyte system; the bone marrows of reticuloendothelial organs of affected individuals become infiltrated with lipid-laden 'foam' cells known as Gaucher cells. Type I Gaucher disease is characterized by hepatosplenomegaly, anaemia, thrombocytopaenia, and severe skeletal disease resulting in pain and fractures. Although most lysosomal glycoproteins are trafficked to the lysosome by M6P receptors, β-glucocerebrosidase is trafficked through binding to the LIMP2

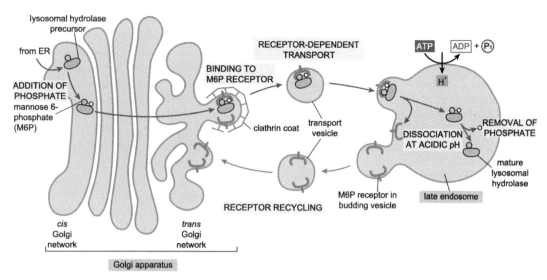

FIGURE 11.1

The transport of newly synthesized lysosomal hydrolases to lysosomes. The precursors of lysosomal hydrolases are covalently modified by the addition of mannose-6-phosphate (M6P) groups in the *cis* Golgi network. They then become segregated from all other types of proteins in the *trans* Golgi network because adaptins in the clathrin coat bind the M6P receptors, which, in turn, bind the modified lysosomal hydrolases. The clathrin-coated vesicles produced bud off from the trans Golgi network and fuse with late endosomes. At the low pH of the late endosome, the hydrolases dissociate from the M6P receptors, and the empty receptors are recycled to the Golgi apparatus for further rounds of transport. It is not known which type of coat mediates vesicle budding in the M6P receptor recycling pathway. In the late endosomes, the phosphate is removed from the mannose sugars attached to the hydrolases, further ensuring that the hydrolases do not return to the Golgi apparatus with the receptor. (For colour version of this figure, the reader is referred to the online version of this book.)

Source: Reproduced with permission from B Alberts; et al. Molecular Biology of the Cell, *4th Ed.; Garland Science: NYC, 2002.*

mannose receptors. To enhance macrophage uptake, the *N*-linked oligosaccharide is modified by the sequential removal of sialic acid, β-galactosyl-, and β-*N*-acetylglucosylamino residues. These modifications expose terminal mannose residues necessary for enhanced uptake through mannose receptors on macrophage membranes and subsequent targeting to macrophage lysosomes. The pivotal clinical trial was conducted in only 12 nonneuropathic type I Gaucher patients in an open-label uncontrolled study which showed a decrease in organomegaly and improvements in blood indices over a 6 month period (*8*). Long-term extension trials and patient registries have showed sustained reductions of organomegaly and improvements in haematocrit and platelets, increases in bone density, and overall improvement in quality of life (*11*). Post-marketing assessment of treatment outcomes has progressed sufficiently over time to allow for establishment of therapeutic goals and treatment guidelines by expert panels (*12*). ERT in Gaucher has proven remarkably safe, with infusion-related reactions treated conservatively by premedication and slowing infusion rates. Antibodies can develop over time but effects do not seem to be clinically significant or neutralizing and with many Gaucher patients the antibody titres decrease (tolerize) over time.

As outlined by Desnick and Schuchman in a recent review (*10*), the general principles learnt from treating more than 5500 type I Gaucher patients worldwide with three β-glucocerebrosidase products have included the following: (1) chronic biweekly intravenous dosing is safe and well-tolerated and results in significant and sustained clinical benefit by reduction of organomegaly and improvement in haemoglobin and platelets; (2) enzyme delivery is receptor-mediated and dose-dependent, with reversal in certain cells and organs over time; (3) stopping or reducing the dose can result in substrate re-accumulation and worsening of disease; (4) ERT does not cross the blood–brain barrier (BBB) and thus does not alter the progressive neurologic manifestations of neuropathic types; and (5) early initiation of ERT can prevent some of the irreversible damage (e.g. bone disease and fibrosis) that is hard to alter in advanced disease.

11.3.1.2 ERT development in other LSDs

Buoyed by the initial success in Gaucher disease, there has been increasing confidence that catalytic enzyme can be efficacious once they have entered the preferred low-pH environment of the lysosome of target cells with some degree of residual 'health'. Current efforts are focused on efficient delivery and biodistribution of drugs to the most severely affected tissues and organs as well as starting therapy early enough to salvage function. A more sobering realization of the potential benefits of cross-correction has been the realization that the pathological storage in type I Gaucher disease is essentially confined to the readily accessible macrophages of the reticuloendothelial cells. In other LSDs, the cells of interest are the principal structural and functional components of the affected organs, such as renal glomerular/podocyte cells in Fabry disease, myocytes with low-copy number receptors in Pompe disease and neurons sitting behind the BBB in the MPS. The half-life of most intravenously administered enzyme in blood is of the order of minutes to hours. Biodistribution to target organs is hampered by this short plasma half-life and sequestration of enzyme in the huge reticuloendothelial cell reserves of the liver.

11.3.1.2.1 Fabry disease

Anderson-Fabry disease is an X-linked multisystem disorder caused by a deficiency of α-galactosidase. Fabry disease is characterized by progressive cardiomyopathy, renal insufficiency, keratopathy, vascular disease causing neuropathy, paraesthesias and strokes due to lysosomal build-up of globotriaosylceramide (Gb3) and potentially more pathologic derivative lyso-Gb3 (*13*). Males are more severely affected, but heterozygous females have milder disease thought to be due to skewing of X-inactivation.

The safety and effectiveness of ERT with two forms of intravenously administered recombinant α-galactosidase was demonstrated by reductions of Gb3 storage in kidney and heart biopsy (*14,15*), though the true value of the surrogate biomarker Gb3/lyso-GB3 in plasma, urine, and tissues as a predictor of clinical benefit has been controversial. Demonstrating robust improvements in glomerular filtration rate have been difficult to show in the typically short duration of most controlled ERT clinical trials, though a larger placebo-controlled study has shown of reduced incidence of renal and vascular events in Fabry patients with pre-existing significant renal disease (*16*). Post-marketing studies in registries and small cohort studies have shown improvement in stabilizing renal disease, improving cardiac involvement and deceasing extremity pain and gastrointestinal manifestations and generally improving quality of life (*17*). As for many LSDs, there probably exist irreversible points that may be different for different tissues/organs where ERT will have a minimal effect on slowing disease progression. Treatment guidelines by expert panels have been developed (*18*), with current trends are towards early treatment even in minimally symptomatic patients, extending also to female patients.

Unlike Gaucher disease, about three-fourth of all classic male Fabry patients inevitably raise IgG antibodies to exogenous enzyme, whereas most later-onset males and heterozygous females do not. Insufficient treatment or development of sustained high-titre antibodies can cause a rise in plasma and urinary Gb3/lyso-Gb3, though the precise clinical significance as to whether this portends inevitable clinical deterioration and ERT failure remains uncertain (*19*).

11.3.1.2.2 Pompe disease

GSDII (Pompe disease, acid maltose deficiency) is an autosomal recessive neuromuscular disorder caused by deficiency of the enzyme acid α-glucosidase and marked by progressive muscle weakness due to lysosomal build-up of glycogen in skeletal, cardiac, bulbar, diaphragmatic and smooth muscle (*20,21*). Classical infantile Pompe disease is characterized by hypertrophic cardiomyopathy, hypotonia due to skeletal myopathy, respiratory insufficiency and death secondary to cardiorespiratory failure by the age of 2 years. Adult-onset forms do not have significant cardiomyopathy, but show a slower progressive proximal muscle weakness which is invariably associated with respiratory insufficiency that causes significant morbidity and early mortality. Predominantly a muscle disease, ERT should be more straightforward, but targeting of uptake into skeletal muscle has proven difficult as large amounts of enzyme are needed. The factors responsible include the paucity of M6P receptors in muscle as compared to heart, the formation of neutralizing sustained high-titre antibodies, and the relative resistance of type II muscle fibres to correction.

ERT in classical infantile-onset Pompe was shown to improve (1) growth and weight, (2) muscle function and (3) to rapidly reverse the cardiomyopathy as measured by left ventricular mass index. All 18 Pompe infants survived 18 months (15 not requiring invasive ventilation) whereas all infants in an age-matched, retrospective natural history cohort would have predicted to have died (*22*). Treatment was not as effective in a second study of 21 patients who began ERT after 6 months of age. When followed over time, infants in the pivotal ERT trial had variable clinical outcomes; with ~50% of the patients being dependent on invasive ventilation when followed up to the age of 41½ months (*23*). In late-onset patients, clinical trials demonstrated improved walking capacity, and stabilized pulmonary and neuromuscular function (*24*). Treatment guidelines by expert panels have been developed (*25*). The marked variability in clinical outcomes with ERT is believed to result from multiple factors, including age and extent of disease-related pathology when starting ERT, underlying genotype, residual enzyme activity, and the presence of high sustained antibody titres. Perhaps more so than any other LSD, sustained high-titre antibodies leads to neutralization of enzyme, and regression of motor milestones and ultimately decreased survival (*26*). Different strategies to induce tolerance have met with some success, including a combination therapy of methotrexate, anti-CD-20 monoclonal antibody (rituximab), and intravenous gamma globulin (*27*). Risk–benefit considerations and Timing (preventative vs reactive) of various combination immunosuppressive regiments is the focus of very active current research.

11.3.1.2.3 Mucopolysaccharidosis

MPS encompasses seven clinical disorders caused by 11 enzyme deficiencies (MPS I, II, II, III, IV, VII, and IX) (Table 11.1). ERTs are now approved for three of these disorders (MPS I, II, and VI) and are in development for MPS IV, MPS VII as well as neuropathic forms of MPS II and IIIA. MPS enzyme deficiencies cause accumulation of sulfated carbohydrate polymers composed of a central core protein attached to disaccharide branches deriving from sulfate monosaccharides or glycosaminoglycans (GAGs). Keratan sulfate and chondroitin sulfate build-up in cartilage and cornea, dermatin sulfate in connective tissue, and heparan sulfate in somatic organs and the brain (28,29).

11.3.1.2.3.1 Mucopolysaccharidosis type I.

MPS I (Hurler, Hurler–Scheie and Scheie syndrome) is an autosomal recessive disorder caused by deficiency of catalytic enzyme α-L-iduronidase and subsequent GAG build-up in multiple tissues and organs. The most common manifestations of MPS I include a characteristic coarse facies, corneal clouding, large tongue, hearing loss, hydrocephaly, cardiomyopathy, respiratory problems, hepatosplenomegaly, inguinal and umbilical hernia, dysostosis multiplex, and limited joint mobility. The most severe Hurler patients have inexorable neurocognitive decline, leading to inanition and early death, whereas attenuated Scheie patients are spared CNS disease. Four clinical trials of recombinant α-L-iduronidase have been conducted, encompassing patients of all phenotypes and age ranges. In placebo-controlled trials and the open-label extensions, ERT decreased urinary GAGs, improved physical capacity as measured by increased walking distance, improved respiratory function, improved myocardial function by decreasing left ventricular hypertrophy, decreased hepatosplenomegaly, stabilized or improved joint range of motion and generally improved the quality of life (30–32). The beneficial effects of iduronidase ERT were sustained over time (33). Although half the patients experienced at least one infusion-related reaction and more than 90% develop antibodies to iduronidase, overt clinical deterioration due to sustained, high-titre antibodies is rare.

11.3.1.2.3.2 Mucopolysaccharidosis type II.

MPS II (Hunter syndrome) is an X-linked disorder caused by a deficiency of the catalytic enzyme iduronate-2-sulfatase causing characteristic facial features, severe restrictive and obstructive airway disease, cardiomyopathy, hepatosplenomegaly, inguinal hernias, joint stiffness, skeletal deformities including pelvic dysplasia and scoliosis, hearing loss due to recurrent otitis and sensorineural deficit, sleep apnoea, and developmental delay and neurological regression in the most severe neuropathic forms. Four clinical trials of idursulfase have been conducted in patients with MPS II, encompassing patients over an age of 5 years and older. ERT has been shown to improve walking capacity, increase respiratory function as measured by increase in forced vital capacity, and decrease urinary GAG levels (34), with clinical effects sustained over time (35), and no safety concerns when treating juvenile patients under the age of 6 years (36). Consensus treatment guidelines have been published (37,38). Immediate and/or biphasic infusion-related reactions occurred in over half of the clinical trial participants. Antibodies develop in about half the patients, but may tolerize over time with no overt clinical consequences.

11.3.1.2.3.3 Mucopolysaccharidosis type VI.

MPS VI (Maroteaux–Lamy syndrome) is an autosomal recessive disease caused by deficiency of the enzyme N-acetylgalactosamine-4-sulfatase or arylsulfatase B. Patients with MPS VI exhibit a wide variability of progressive multisystemic disease. Systems affected include hypertrophic cardiomyopathy leading to heart failure, chronic respiratory obstruction, cornea clouding, glaucoma, skeletal abnormalities (short truck and gibbus), and obstructive and sensorineural hearing loss. A short neck, elevated epiglottis and other defects contribute to the respiratory problems and sleep apnoea (39). Three clinical trials of recombinant human arylsulfatase B have been conducted in adolescent and adult patients over the age of 5 years with severe disease manifestations. ERT improved growth, increased mobility and lessened fatigue as measured by improvements in walking capacity and stair-climbing capacity; and improved pulmonary function (40,41). Expert treatment guidelines have been published (42). About half the patients experience infusion-related reactions. Mostly all patients develop antibodies to arylsulfatase B, but like many MPSs the clinical significance is not clear.

11.3.2 Improving Efficacy by Enzyme Modifications to Improve Cellular Uptake and Tissue Targeting

Because some tissues and organs are mostly refractory to ERT treatment (cartilage, bone, heart valve, skeletal muscle, and brain) attempts have been made to modify lysosomal proteins to improve penetration and uptake in directed target tissues (*43*).

11.3.2.1 Muscle targeting

Current α-glucosidase used in Pompe disease ERT has only one M6P per mole of protein and therefore comparatively little enzyme is taken up in skeletal tissue. To maximize enzyme uptake through the sparsely populated M6P receptors on skeletal muscle, two glyco-engineered forms of α-glucosidase have been developed. In the first instance, α-glucosidase conjugated to synthetic oligosaccharide chains containing *bis*-M6P greatly increased the affinity of α-glucosidase for the M6P receptor, resulted in significantly higher glycogen clearance, and evidenced greater improvements in muscle strength in mouse models of Pompe. This preparation is currently being considered for clinical trials (*44*). In the second instance, α-glucosidase expressed in the yeast *Yarrowia lipolytica* was glyco-engineered to hyper-mannosyl phosphorylate the terminal *N*-glycans. A novel mannose-uncapping enzyme together with α-mannosidase was used to generate a highly M6P-modified form of α-glucosidase. Uptake of this α-glucosidase by Pompe disease patient fibroblasts was 18 times more efficient than the current therapeutic enzyme, and resulted in more effectively reduced cardiac muscular glycogen storage in a mouse model of the disease (*45*). This molecule is advancing to clinical trials. To improve muscle uptake a modified α-glucosidase coupled with an insulin growth factor II (IGFII) moiety that targets the IGFII binding site of the M6P receptor has been developed. Now not dependent on posttranslational performance of proper glycosylation in cell culture, this α-glucosidase fusion protein demonstrated increased cellular uptake in vitro and increased tissue glycogen depletion (*45a*). This molecule has advanced to exploratory clinical trials in adult Pompe, with good preliminary tolerability but no reported outcome data.

Enhanced targeting to skeletal muscle using an antibody fragment (3E10Fv) directed at the muscle-expressed surface protein ENT2 has shown promise in early preclinical work in myotubular myopathy, myotonic dystrophy, and Pompe disease (*46*). Whether enhanced uptake of enzyme through ENT2 muscle receptors leads to effective trafficking to the lysosome (as with cross-correction via M6P receptors) or not remains to be demonstrated. Adjunctive therapies such as B2 agonists with ERT that can increase cell-surface M6P expression through enhanced receptor-mediated uptake of ERT could reduce the dosage requirements for ERT and allow for better delivery to skeletal muscle (*47*).

11.3.2.2 Bone targeting

Although not a lysosomal disorder, hypophosphatasia is an autosomal recessive disorder caused by the deficiency of tissue nonspecific alkaline phosphatase (TNSALP). Hypophosphatasia causes incomplete development of the axial skeleton and rib cage and an early death in infantile forms due to respiratory insufficiency as well as rickets and brittle bones in juvenile and adult forms. A TNSALP–IgG1 fusion molecule tagged to a length of acidic oligopeptide effectively targeted the enzyme to the bone and resulted in dramatic improvements in bone mineralization, respiratory function and survival in the infantile-onset form of disease (*48*). Whether this approach could improve outcomes in any LSDs with skeletal manifestations or not remains to be evaluated.

11.3.2.3 CNS targeting of neuropathic LSDs

Since lysosomal enzymes do not readily cross the BBB, one direct approach has been to bypass the BBB altogether by direct injection of recombinant enzyme into the intrathecal space using an indwelling drug delivery device. In rodent and monkey models, good spreading was seen from the thoracolumbar site of administration via cerebrospinal fluid to spinal cord to brain. Brain parenchymal penetration was also observed from superficial convexities into the deeper structures of the brain (*49*). This product has advanced into exploratory clinical trials for MPS II and MPSIII A patients with neurocognitive disability, but outcome data have not been reported.

Two MPS I exploratory clinical trials involving intrathecal delivery of enzyme are currently underway in children with cognitive symptoms and in adults with spinal cord compression. Two case reports describe successful delivery of recombinant enzyme intrathecally in adult patients with MPS I and MPS VI with spinal cord compression (*50,51*).

Alternative approaches have focused on modification of the enzyme with peptide leader sequences that bind to specific receptors on the brain capillary endothelium. Enzymes are transported from the blood stream to the brain tissue using receptor-mediated transcytosis utilizing transferrin receptor, insulin receptor and receptors of the low-density lipoprotein receptor super-family (*52–55*).

11.3.3 Cross-correction through Stem Cell Therapies

A full discussion of cross-correction through stem cell or gene therapies is outside of the scope of this medical chemistry text, but some brief examples are noted here.

11.3.3.1 Haematopoietic stem cell therapies

Haematopoietic stem cell therapy (HSCT) has mainly been used in patients with the severe form of MPS I (Hurler syndrome). Administered stem cells differentiate in the brain to microglia, and provide missing enzyme through cross-correction to CNS tissue. If performed before the age of 2 years, HSCT preserves neurocognition and prolongs overall survival due to reduced heart and liver disease (*56*). HSCT improves hepatosplenomegaly and upper limb movement, and often slows general disease progression. But much of musculoskeletal disease continues to progress, vision usually worsens, cardiac valve disease persists and often progresses, and growth is stunted (*57*). HSCT for other MPS disorders has been mixed.

The chief drawback hindering more rapid adoption of HSCT is the difficulty in finding compatible stem cell donors and the high morbidity and mortality associated with the initial toxic ablation procedure and subsequent immunosuppression. Better tissue matching techniques especially through umbilical cord stem cells, improved graft vs host prophylaxis and more targeted partial conditioning regimens have improved survival and decreased transplant-related morbidity. Even with these technical advancements, there may never be enough engraftment achieved in many LSDs (especially brain) to provide proper levels of cross-correction.

11.3.4 **Enzyme stabilization (chaperone therapy)**

Lysosomal enzymes are synthesized and secreted into the endoplasmic reticulum initially in an unfolded state. Specific molecules called chaperones (e.g. heat shock proteins and calnexin) support proper conformational folding of proteins during their transport to the Golgi, where they undergo further post-translational glycosylation before final transit to the lysosome. In some of the attenuated forms of disease, LSDs are due to missense mutations which partially preserve the catalytic activity of enzymes, but decrease the stability of the enzyme, caused by an unfolded protein response and consequent degradation in proteasomes. Certain competitive inhibitors (chaperones) at the active site of LSD enzymes can increase the thermal stability of these mutant enzymes enough to permit transiting to lysosomes (58).

The best studied is the effect of the chaperone 1-deoxygalactonojirimycin (migalastat hydrochloride) on α-galactosidase in Fabry disease. A cell-based assay to test the effect of migalastat on stability was devised to identify those patients most likely to respond to therapy. In phase II clinical trials, migalastat was generally safe and well-tolerated, with increases in α-galactosidase levels in most patients tested and reductions of Gb3 in kidney biopsies from patients who showed the greatest increase in α-galactosidase activity (59). Two large phase III trials in both naïve patients and those switching from existing ERT are nearing completion.

General advantages of pharmacological chaperones are (1) better biodistribution due to their small molecular weight in comparison to ERT and (2) the potential for brain penetration. Disadvantages include the fact that achieving the proper dosing regimen is difficult since the competitive inhibitors that stabilize the enzyme also prevent the enzyme from degrading substrate in the lysosome. Enough time must lapse between doses for the chaperone molecule to dissociate from the enzyme. In one early trial of an experimental chaperone for Pompe disease, the chaperone actually worsened neuromuscular symptoms. Another disadvantage is that treatment is restricted to patients with addressable missense mutations – molecular analysis of a large number of Fabry patients demonstrated that only 10–15% are amenable to chaperone therapy (60).

11.3.5 **Substrate reduction therapy**

An alternative to providing the missing enzyme through ERT is, instead, to reduce the amount of substrate available to degrading enzyme (substrate reduction or deprivation therapy). There are two major advantages of this approach; first, a single drug can be used to treat diseases that involve storage of any glucosylceramide-derived sphingolipid disease (Gaucher, Fabry and the GM1 and GM2 gangliosidoses and diseases where glycosphingolipids are stored secondarily to the primary defect). The second advantage is that the small molecular size permits better biodistribution to less accessible organs such as brain and bone.

The first drugs to be developed were orally active inhibitors of uridine diphosphate glucosylceramide synthetase, the enzyme that catalyses the first committed step in the biosynthesis of glycosphingolipids that are derived from β-glucosylceramide (glucocerebrosidase). Many of the early ceramide inhibitors of glucocerebrosidase synthetase also inhibited 1-O-acylceramide synthetase causing increased intracellular ceramide and cell toxicity. Medical chemistry efforts led to increased potency of the ceramide analogues as glucosylceramide synthetase inhibitors while showing decreased potency towards 1-O-acylceramide synthetase (61).

Miglustat was the first glucosylceramide synthetase inhibitor approved for the treatment of Gaucher disease (62). Minor gastrointestinal symptoms and tremors in a few patients have limited wide acceptance of its use. An improved glucosylceramide inhibitor (eliglustat) is now progressing through phase III clinical trials for Gaucher disease (*63*).

11.4 CONCLUSIONS

ERT, chaperone enhancement therapy and substrate reduction therapy continue to be active areas of research, expanding to even the most ultra-orphan of rare genetic disorders. Combination therapies of these newly approved products promise potentially better outcomes for LSD patients. As stem cell and gene therapy matures, the very real potential for cures in early-diagnosis patients looms tantalizingly large. The wider acceptance of newborn screening programmes will lead to the identification of pre-symptomatic LSD patients and holds the promise of getting therapy to patients before irreversible damage has occurred. Continued expansion of long-term outcome studies, focusing on improved quality of life for both patients and families as well as economic benefit to health care services, will help assure continued access to these expensive life-saving medicines for LSD patients who so deservedly need them.

References

1. Rimoin, D.L.; Connor, J.M.; Pyeritz, R.E.; Korf, B.R. *Emery and Rimoin's Principles and Practice of Medical Genetics*, 5th Edition plus e-edition: Continually Updated Online Reference, 3-Volume Set, **2006**.
2. Valle, D.; Beaudet, A.L.; Vogelstein, B.; Kinzler, K.W.; et al. Eds. *Online Metabolic and Molecular Bases of Inherited Disease*. http://www.ommbid.com/ Published January 2006. Updated March 28, 2011.
3. Vitner, E. B.; Platt, F. M.; Futerman, A. H. Common and Uncommon Pathogenic Cascades in Lysosomal Storage Diseases. *J. Biol. Chem.* **2010**, *285*, 20423–20427.
4. Ballabio, A.; Gieselmann, V. Lysosomal Disorders: From Storage to Cellular Damage. *Biochim. Biophys. Acta* **2009**, *1793* (4), 684–696.
5. Jeyakumar, M.; Dwek, R. A.; Butters, T. D.; Platt, F. M. Storage Solutions: Treating Lysosomal Disorders of the Brain. *Nat. Rev. Neurosci.* **2005**, *6* (9), 713–725.
6. Platt, F. M.; Lachmann, R. H. Treating Lysosomal Disorders: Current Practice and Future Prospects. *Biochim. Biophys. Acta* **2009**, *1793*, 737–745.
7. Cote, T. R.; Xu, K.; Pariser, A. R. Accelerating Orphan Drug Development. *Nat. Rev. Drug Discovery* **2010**, *12*, 901–902.
8. Barton, N. W.; Brady, R. O.; Dambrosia, J. M., et al. Replacement Therapy for Inherited Enzyme Deficiency – Macrophage-Targeted Glucocerebrosidase for Gaucher Disease. *N. Engl. J. Med.* **1991**, *324*, 1464–1470.
9. Deegan, P. B.; Cox, T. M. Imiglucerase in the Treatment of Gaucher Disease: A History and Perspective. *Drug Des. Devel. Ther.* **2012**, *6*, 81–106.
10. Desnick, R. J.; Schuchman, E. H. Enzyme Replacement Therapy for Lysosomal Diseases: Lessons from 20 Years of Experience and Remaining Challenges. *Annu. Rev. Genomics Hum. Genet.* **2012**, *13*, 307–355.
11. Weinreb, N. J.; Charrow, H. C.; Andersson, P., et al. Effectiveness of Enzyme Replacement Therapy in 1028 Patients with Type 1 Gaucher Disease after 2 to 5 Years of Treatment; A Report from the Gaucher Registry. *Am. J. Med.* **2002**, *113*, 112–119.

12. Cox, T. M.; Aerts, J. M.; Belmatoug, N. Management of Non-Neuronopathic Gaucher Disease with Special Reference to Pregnancy, Splenectomy, Bisphosphonate Therapy, Use of Biomarkers and Bone Disease Monitoring. *J. Inherited Metab. Dis.* **2008,** *3,* 319–336.
13. Germain, D. P. Fabry Disease. *Orphanet J. Rare Dis.* **2010,** *5,* 30.
14. Desnick, R. J. Enzyme Replacement for Fabry Disease; Lessons from Two alpha-Galactosidase A Orphan Products and One FDA Approval. *Expert Opin. Biol. Ther.* **2004,** *4,* 1167–1176.
15. Beck, M. Agalsidase Alfa for the Treatment of Fabry disease: New Data on Clinical Efficacy and Safety. *Expert Opin. Biol. Ther.* **2009,** *9,* 255–261.
16. Banikazemi, M.; Bultas, J.; Waldek, S., et al. Agalsidase-Beta Therapy for Advanced Fabry Disease: A Randomized Trial. *Ann. Intern. Med.* **2007,** *146* (2), 77–86.
17. Schaefer, R. M.; Tylki-Szymanska, A.; Hilz, M. J. Enzyme Replacement Therapy for Fabry Disease. A Systematic Review of available Evidence. *Drugs* **2009,** *69,* 2179–2205.
18. Eng, C. M.; Germain, D. P.; Banikazemi, M., et al. Fabry Disease: Guidelines for the Evaluation and Management of Multi-Organ System Involvement. *Genet. Med.* **2006,** *8* (9), 539–548.
19. Rombach, S. M.; Aerts, J. M.; Poorthuis, B. J., et al. Long-term Effect of Antibodies against Infused Alpha-Galactosidase A in Fabry Disease on Plasma and Urinary (Lyso) Gb3 Reduction and Treatment Outcome. *PLoS One* **2012,** *7* (10), e47805.
20. van der Ploeg, A.; Reuser, A. J. J. Pompe's Disease. *Lancet* **2008,** *372,* 1342–1353.
21. Kishnani, P. S.; Beckemeyer, A. A.; Mendelsohn, N. J. The New Era of Pompe Disease: Advances in the Detection, Understanding of the Phenotypic Spectrum, Pathophysiology, and Management. *Am. J. Med.* **2012,** *160,* 1–7.
22. Kishnani, p. S.; Corzo, D.; Nicolino, M., et al. Recombinant Human Acid [Alpha]-Glucosidase: Major Clinical Benefits in Infantile-Onset Pompe Disease. *Neurology* **2007,** *68,* 99–109.
23. Nicolino, M.; Byrne, B.; Wraith, J. E., et al. Clinical Outcomes after Long-term Treatment with Alglucosidase Alfa in Infants and Children with Advanced Pompe Disease. *Genet. Med.* **2009,** *11,* 210–219.
24. van der Ploeg, A.; Clemens, P. R.; Corzo, D., et al. A Randomized Study of Alglucosidase Alfa in Late Onset Pompe's Disease. *N. Engl. J. Med.* **2010,** *362,* 1396–1406.
25. Culper, E. J.; Berger, K. I.; Leshner, R. T., et al. Consensus Treatment Recommendations for Late-Onset Pompe Disease. *Muscle Nerve* **2012,** *45* (3), 319–333.
26. Banugaria, S. G.; Prater, S. N.; Ng, Y. K. The Impact of Antibodies on Clinical Outcomes in Diseases Treated with Therapeutic Protein: Lessons Learned from Infantile Pompe Disease. *Genet. Med.* **2011,** *13* (8), 729–736.
27. Messinger, Y. H.; Mendelsohn, N. J.; Rhead, W., et al. Successful Immune Tolerance Induction to Enzyme Replacement Therapy in CRIM-Negative Infantile Pompe Disease. *Genet. Med.* **2012,** *14,* 135–142.
28. Valayannopoulos, V.; Wijburg, F. A. Therapy for the Mucopolysaccharidoses. *Rheumatology* **2011,** *50* (Suppl. 5), v49–v59.
29. Muenzer, J. Overview of the Mucopolysaccharidoses. *Rheumatology* **2011,** *50* (Suppl. 5), v4–12.
30. Clarke, L. A.; Wraith, J. E.; Beck, M., et al. Long-term Efficacy and Safety of Laronidase in the Treatment of Mucopolysaccharidosis I. *Pediatrics* **2009,** *123,* 229–240.
31. Wraith, J. E.; Clarke, L. A.; Beck, M., et al. Enzyme Replacement Therapy for Mucopolysaccharidosis I: A Ranadomized, Double-Blinded, Placebo-Controlled Multinational Study of Recombinant Human alpha-L-iduronidase (laronidase). *J. Pediatr.* **2004,** *144* (5), 151–157.
32. Harada, H.; Uchiwa, H.; Nakamura, M. Laronidase Replacement Therapy Improves Myocardial Function in Mucopolysaccharidosis I. *Mol. Genet. Metab.* **2011,** *103* (3), 215–219.
33. Sifuentes, J.; Doroshow, R.; Hoft, R. A Follow-up Study of MPS I Patients Treated with Laronidase Enzyme Replacement Therapy for 6 Years. *Mol. Genet. Metab.* **2007,** *90,* 171–180.
34. Muenzer, J.; Wraith, J. E.; Beck, M., et al. A Phase II/III Clinical Study of Enzyme Replacement Therapy with Idursulfase in Mucopolysaccharidosis II (Hunter Syndrome). *Genet. Med.* **2006,** *8* (8), 465–473.

35. Muenzer, J.; Beck, M.; Eng, C. M., et al. Long-term, Open-Labeled Extension Study of Idursulfase in the Treatment of Hunter Syndrome. *Genet. Med.* **2011**, *13* (2), 95–101.

36. Muenzer, J.; Beck, M.; Giugliani, R., et al. Idursulfase Treatment of Hunter Syndrome in Children Younger than 6 Years: Results from the Hunter Outcome Survey. *Genet. Med.* **2011**, *13* (2), 102–109.

37. Muenzer, J.; Bodamer, O.; Burton, B. The Role of Enzyme Replacement Therapy in Severe Hunter Syndrome—An Expert Panel Consensus. *Eur. J. Pediatr.* **2012**, *171*, 181–188.

38. Wraith, J. E.; Scarpa, M.; Beck, M., et al. Mucopolysaccharidosis Type II (Hunter Syndrome): A Clinical Review and Recommendations for Treatment in the Era of Enzyme Replacement Therapy. *Eur. J. Pediatr.* **2012**, *167*, 267–277.

39. Valayannopoulos, V.; Nicely, H.; Harmatz, P., et al. Mucopolysaccharidosis VI. Orphanet. *J. Rare. Dis.* **2010**, *5* (5).

40. Harmatz, P.; Giugliani, R.; Schwartz, I., et al. Enzyme Replacement Therapy for Mucopolysaccharidosis VI: A Phase 3, Randomized, Double-Blind, Placebo-Controlled, Multinational Study of Recombinant Human N-Acetylgalactosamine 4-Sulfatase (Recombinant Human Arylsulfatase B or rhASB) and Follow-on, Open-Label Extension Study. *J. Pediatr.* **2006**, *148*, 533–539.

41. Harmatz, P.; Giugliani, R. D.; Schwartz, I. V., et al. Long-term Follow-up of Endurance and Safety Outcomes During Enzyme Replacement Therapy for Mucopolysaccharidosis VI: Final Results of Three Clinical Studies of Recombinant Human N-Acetylgalactosamine 4-Sulfatase. *Mol. Genet. Metab.* **2008**, *94*, 469–475.

42. Giugliani, R.; Harmatz, P.; Wraith, J. E. Management Guidelines for Mucopolysaccharidosis VI. *Pediatrics* **2007**, *120*, 405–418.

43. Lachmann, R. H. Enzyme Replacement Therapy for Lysosomal Diseases. *Curr. Opin. Pediatr.* **2011**, *23*, 588–593.

44. Zhou, Q.; Stefano, J. E.; Harrahy, J., et al. Strategies for Neoglycan Conjugation to Human Acid α-Glucosidase (2009). *Bioconjugate Chem.* **2011**, *22* (4), 741–751.

45. Tiels, P.; Baranova, E.; K., Piens., et al. Tiels, P.; Baranova, E.; Piens. K.; et al. A Novel Glycoside Hydrolase Enables Modification of Glyco-Engineered Yeast-Produced Lysosomal Enzymes to Contain High Levels of Mannose-6-Phosphate for Improved Therapeutic Efficacy. *Nat. Biotechnol.* **2012**, *30*, 1225–1231.

45a. Maga, J. A.; Zhou, J.; Kambampati., R., et al. Glycosylation-independent Lysosomal Targeting of Acid α-Glucosidase Enhances Muscle Glycogen Clearance in Pompe Mice. *J. Biol. Chem.* **2013**, *288* (3), 1428–1438.

46. Hansen., et al. Antibody Mediated Transduction of Therapeutic Proteins into Living Cells. *Sci. World J.* **2005**, *5*, 782–788.

47. Koeberl, D. D.; Luo, X.; Sun, B., et al. Enhanced Efficacy of Enzyme Replacement Therapy in Pompe Disease through Mannose-6-Phosphate Receptor Expression in Skeletal Muscle. *Mol. Genet. Metab.* **2011**, *103* (2), 107–112.

48. Whyte, M. P.; Greenberg, C. R.; Salman, N. J., et al. Enzyme-Replacement Therapy in Life-Threatening Hypophosphatasia. *N. Engl. J. Med.* **2012**, *366* (10), 904–913.

49. Calias, P.; Papisov, M.; Pan, J., et al. CNS Penetration of Intrathecal-Lumbar Idursulfase in the Monkey, Dog and Mouse: Implications for Neurological Outcomes of Lysosomal Storage Disorder. *PLoS One* **2012**, *7* (1), e30341.

50. Munoz-Rojas, M. V.; Vieira, T.; Costa, R., et al. Intrathecal Enzyme Replacement Therapy in a Patient with Mucopolysacchasridosis Type I and Symptomatic Spinal Cord Compression. *Am. J. Med. Genet.* **2008**, *146A*, 2538–2544.

51. Munoz-Rojas, M. V.; Horovitz, D. D.; Jardim, L. B., et al. Intrathecal Administration of Recombinant Human N-acetylgalactosamine 4-sulfatase to a MPS VI Patient with Pachymeningitis Cervicalis. *Mol. Genet. Metab.* **2010**, *99*, 346–350.

52. Cecchelli, R.; Berezowski, V.; Lundquist, S., et al. Modelling of the Blood-Brain Barrier in Drug Discovery and Development. *Nat. Rev. Drug Discovery* **2007**, *6* (8), 650–661.

53. Pardridge, W. M. Biopharmaceutical Drug Targeting to the Brain. *J. Drug Targeting* **2010,** *18* (3), 157–167.

54. Gabathuler, R. Approaches to Transport Therapeutic Drugs across the Blood–Brain Barrier to Treat Brain Diseases. *Neurobiol. Dis.* **2010,** *37,* 48–57.

55. Boado, R. J.; Hui, E. K.; Lu, J. Z., et al. AGT-181: Expression in CHO cells and Pharmacokinetics, Safety, and Plasma Iduronidase Enzyme Activity in Rhesus Monkeys. *J. Biotechnol.* **2009,** *144,* 135–141.

56. Moore, D.; Connock, M. J.; Wraith, E., et al. The Prevalence of and Survival in Mucopolysaccharidosis I: Hurler, Hurler-Scheie and Scheie Syndromes in the UK. *Orphanet J. Rare Dis.* **2008,** *3,* 24.

57. Aldenhoven, M.; Boelens, J. J.; de Koning, T. J. The Clinical Outcome of Hurler Syndrome after Stem Cell Transplantation. *Biol. Blood Marrow Transplant.* **2008,** *14,* 485–498.

58. Parenti, G. Treating Lysosomal Storage Diseases with Pharmacological Chaperones: From Concept to Clinics. *EMBO Mol. Med.* **2009,** *1* (5), 268–279.

59. Benjamin, E. R.; Flanagan, J. J.; Schilling, A., et al. The Pharmacological Chaperone 1-Deoxygalactonojirimycin Increases Alpha-Galactosidase A Levels in Fabry Patient Cell Lines. *J. Inherited Metab. Dis.* **2009,** *32,* 424–440.

60. Schiffmann, R.; Germain, D.P.; Castelli, J.; et al. Phase 2 Clinical Trials of the Pharmacological Chaperone AT1001 for the Treatment of Fabry Disease. *58th Annual Meeting American Society of Human Genetics*; 11–15 November 2008, Philadelphia, PA.

61. Platt, F. M.; Butters, T. D. In *Lysosomal Disorders of the Brain;* Platt, F. M., Walkley, S. U., Eds.; Oxford Univ. Press: Oxford, 2004; pp. 381–408.

62. Cox, T.; Lachmann, R.; Hollak, C., et al. Novel Oral Treatment of Gaucher's Disease with N-Butyldeoxynojirimycin (OGT 918) to Decrease Substrate Biosynthesis. *Lancet* **2000,** *355,* 1481–1485.

63. Lukina, E.; Watman, N.; Arreguin, E. A., et al. Improvement in Hematological, Visceral, and Skeletal Manifestations of Gaucher Disease Type 1 With Oral Eliglustat Tartrate (Genz-112638) Treatment: 2-Year Results of a Phase 2 Study. *Blood* **2010,** *116* (20), 4095–4098.

Hormone replacement therapy

12

Stanley M. Roberts

School of Chemistry (MIB), The University of Manchester, Manchester M1 7DN, UK

CHAPTER OUTLINE

12.1 Background and Overview 344
12.2 History of Oestrogen Therapy 344
    12.2.1 Premarin® 344
    12.2.2 Alternatives to Premarin® 345
12.3 HRT and Cancer 346
12.4 HRT and Osteoporosis 348
12.5 HRT and Heart Disease 348
12.6 The Women's Health Initiative 348
12.7 Improved Protocols for HRT 352
12.8 Conclusion 352
References 353

ABSTRACT

Menopause affects the quality of life for women, sometimes severely. The symptoms can be addressed by treatment with steroids, namely, an oestrogen alone or an oestrogen with a progestin. The treatment is called hormone replacement therapy (HRT). However, for years, HRT has been associated with side effects, for instance, an increased risk of breast or endometrial cancer and stroke. The potentially positive effect of reducing some forms of heart disease is still a matter of debate, while the HRT-related decrease in bone fracture/osteoporosis is more certain. The rise, fall and current status of HRT is summarized in this chapter.

Keywords/Abbreviations: Oestrogen/Estrogen; Progesterone/Progestins; Conjugated equine oestrogen (CEE); Selective oestrogen receptor modulator (SERM); Combined hormone replacement therapy (cHRT); Heart and oestrogen/progesterone replacement therapy (HERS) initiative; Women's health initiative (WHI); Selective oestrogens, menopause and response to therapy (SMART) initiative; Diethylstilbestrol (DES); US Food and Drug Administration (FDA); Low-density lipoprotein (LDL); UK Medical Research Council (MRC); University College London (UCL).

Introduction to Drug Research and Development. http://dx.doi.org/10.1016/B978-0-12-397176-0.00012-1

12.1 BACKGROUND AND OVERVIEW

Menopause is the result of a gradual change that human females experience in midlife (usually 45–55 years of age). It is marked by the end of the menstrual cycle and signals the end of the reproductive period.

Menopause is not an illness or a disorder but a natural progression in females (both human and some animals). Menopause is caused by a change in the natural balance of hormones involved in the reproductive cycle. This cycle involves the actions of steroids (oestrogens and progesterone) leading to the establishment of the uterine lining and subsequent shedding (in the absence of pregnancy) at the end of about 4 weeks. In other words, menopause is due to a change in oestrogen and progesterone production, principally in the ovaries. Oral administration of the 'missing' steroids essentially delays the onset of the menopause and/or ameliorates the symptoms.

Oestrogen products were first approved by the US Food and Drug Administration (FDA) in 1941 for alleviation of the symptoms of menopause, which are, principally, night sweats, vaginal dryness and atrophy. It is estimated that about 10% of the menopausal population finds the symptoms particularly problematic. Thereafter, the perceived advantages of oestrogen supplements embraced other conditions associated with ageing in women, for example, osteoporosis and even some forms of heart disease. It is also clear that menopause was (and is) often linked to other conditions such as depression, forgetfulness and apnoea (sleeplessness), although at least some of these trends are more likely to be associated with 'midlife' circumstances that apply equally to both sexes.

However, over time, the risks associated with the daily intake of these steroids became more firmly established. In particular, a study in 2002 provided evidence for the increased risk of development of breast and uterine cancer. In this chapter the history of hormone replacement therapy (HRT) will be analyzed, as an example of the ebb and flow of the fortunes of a basket of drugs that have been marketed for a long time. The story does not have an ending; the 'risk–benefit' analysis of HRT is still a matter of debate.

12.2 HISTORY OF OESTROGEN THERAPY

Oestrogen and the other sex hormones were identified in the 1920s and the 1930s and their therapeutic potential in terms of birth control, infertility and alleviating the symptoms of menopause became evident when their individual biological functions were established. In the context of this chapter, the key issues are that oestrogen is the hormone that allows the lining of the uterus to build up in preparation for pregnancy, while progesterone is the hormone that causes the lining to shed when no fertilization occurs. After child-bearing age, the natural supply of oestrogen in women diminishes.

The actions of oestrogen are mediated by oestrogen receptors, which are expressed in the ovary, uterus and breast. This receptor is a dimeric nuclear protein that binds specific DNA sequences called hormone response element which can activate the transcription of over 100 oestrogen receptor-regulated genes.

12.2.1 Premarin®

The first source of oestrogen was the urine of pregnant mares and this material was introduced by the company Ayerst in the United States under the name Premarin®. It is commonly referred to as conjugated equine oestrogen (CEE). Contemporaneously, Schering marketed a product obtained from human pregnancy urine for the amelioration of night sweats and hot flushes in menopausal women. However, this material was less readily available than CEE.

Oestrogen therapy was immediately perceived by some health practitioners as a useful tool to replace oestrogen deficiency that started in women at midlife. Indeed, over time it became common for medical practitioners to prescribe oestrogens for practically all menopausal women for an indefinite period.

Premarin® contains up to 10 oestrogens, principally oestrone sulfate (ca. 55%) and equilin sulfate (ca. 27%), with smaller amounts of 17-alpha-/17-beta-oestradiol sulfate, equilenin sulfate, 17-alpha-/17-beta-dihydroequilenin sulfate and 17-alpha-/17-beta-dihydroequilin sulfate. Such sulfates are known as conjugated oestrogens. In the body, inter-conversion of these materials take place; for example, oestrone is converted into oestradiol *in vivo*.

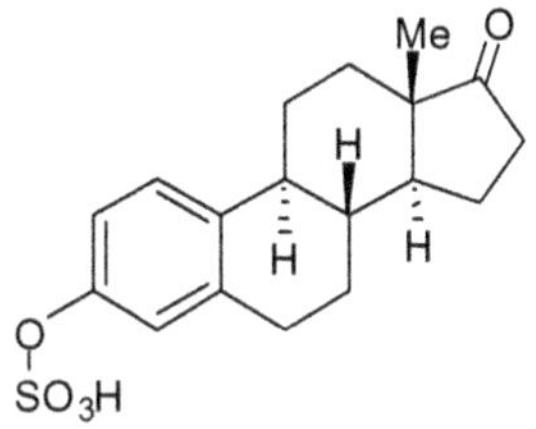

Estrone sulfate

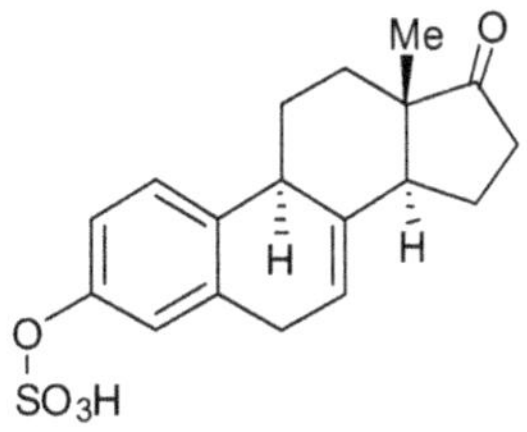

Equilin sulfate

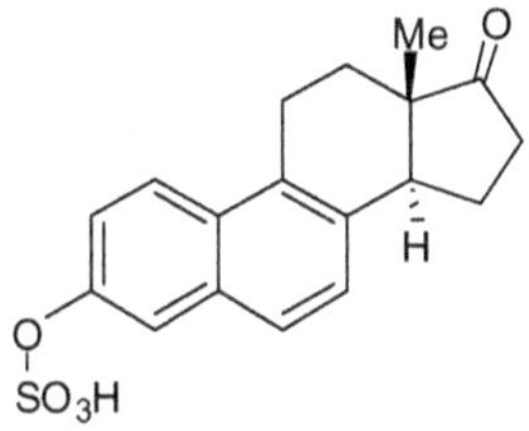

Equilenin sulfate

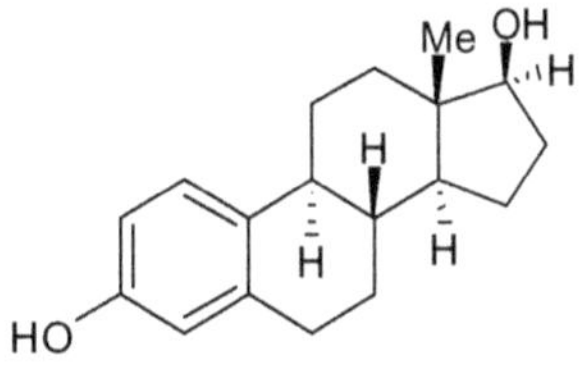

17β-estradiol

The more widespread use of hormone interventions continued in the 1960s and into the 1970s, not only with HRT drugs such as Premarin® but also with long-acting progestin contraceptives such as medroxyprogesterone acetate, also known as Provera® (discovered simultaneously by Searle and Upjohn in 1956), other oral 'progestogen-only' contraceptives such as the synthetic levonorgestrel and combined hormonal contraceptive agents such as mestranol/norethisterone. Note that the term progestogen refers to the group of compounds with progesterone-type activity, while progestin refers more specifically to synthetic progesterone analogues.

12.2.2 Alternatives to Premarin®

The subtle mixture of steroids that comprises Premarin or CEE and the fact that it is a 'natural product' that is relatively easy to obtain meant that it has remained the therapy of choice. Indeed, it has been argued that some of the minor components of the mixture of steroids contributed to the therapeutic effect by metabolism to active species. However, other synthetic materials were tried over the years, a fact welcomed by animal rights groups who were concerned about the restraint of mares for periods of up to 6 months in order to efficiently collect the urine for the CEE product.

One of the earliest possible alternatives was diethylstilbestrol (DES), a synthetic non-steroidal oestrogen, which was approved for menopausal symptoms and vaginitis in the early 1940s. It was also used 'off label' for the prevention of miscarriage. However, in 1971, DES was shown to cause a rare cancer, vaginal adenocarcinoma, and, over the course of time, was shown to cause a variety of adverse medical complications. The major use of DES thereafter was reserved for advanced prostate cancer and advanced breast cancer in post-menopausal women. The discovery of tamoxifen, a selective oestrogen receptor modulator (SERM – see later), reduced the use of DES which was last marketed in 1997. Interestingly, DES was designed, synthesized and tested by Sir Robert Robinson (Oxford) and Sir Edward Charles Dodds (Middlesex Hospital, UK), with funding from the United Kingdom's Medical Research Council (MRC). Since the MRC had the policy not to patent inventions, various companies such as Upjohn, Eli Lilly, Merck and Wyeth Labs were able to make and sell DES throughout the period of its medical usage.

An alternative approach was to use only one of the more active components of the CEE mixture, i.e. 17-beta-oestradiol, as an ester, for example, 17-beta-oestradiol acetate (Estrace®) and oestradiol valerate (Estrofem®). Cenestin® contains a mixture of synthetic oestrogen hormones derived from plants (soya and yam) and was approved for the treatment of menopausal symptoms in 1999.

While beyond the scope of this chapter, it is noteworthy that there are several complementary medicines available for the treatment of the symptoms of menopause, for instance, Femarelle® which comprises material obtained by extraction of tofu combined with powdered flaxseed.

Despite the stricter safety requirements following the US Food Drug and Cosmetics Act of 1962, only sporadically did safety issues regarding HRT come to the fore, including a possible link between oestrogen usage and uterine cancer. It is noteworthy, in this context, that such adverse effects may only become evident after long-term exposure to the drugs. Another adverse effect that only became evident over time was that some women became psychologically addicted to the oestrogen supplement.

12.3 HRT AND CANCER

By 1975, Premarin® was the fifth leading prescription drug in the United States, while, contemporaneously, research into the connection between oestrogen therapy and cancers continued. As a result of this work, it was found that oestrogens caused both breast and cervical cancer in animals. It was mooted that the evidence for onset of cervical cancer had been masked in earlier studies due to animals succumbing to mammary cancers before the development of the more slowly developing cervical cancer.

Slowly, the consensus of opinion regarding the risk/benefit of HRT began to shift, with more studies concluding that oestrogens could be influential in the development of cancer in those organs and tissues which are normally oestrogen dependant, e.g. the genital tract and breasts. In one of the early studies, extreme hyperplasia and, particularly, endometrial adenocarcinoma were observed in women who had prolonged oestrogen therapy (after 7 years of HRT, the risk increased severalfold) (*1*).

As a result, in 1976, the FDA required that the risks of oestrogen intake should be indicated by a written warning to be given out with the drug. The result of the release of this new information was that the number of oestrogen prescriptions fell by 50% during the later half of the 1970s.

After the uterine cancer scare, it was proposed that the risk factor would be reduced by the addition of progesterone (the hormone that causes shedding of the uterine lining) or similarly acting progestogens such as medroxyprogesterone acetate or levonorgestrel. The addition of progesterone to the oestrogen induced a regular bleed, reducing the potentially dangerous build-up of the uterine lining. It is noteworthy that post-menopausal women do not produce progesterone and the addition of the steroid in the combined HRT (cHRT) prescriptions caused 25% of women to experience pre-menstrual tension and some to start bleeding again.

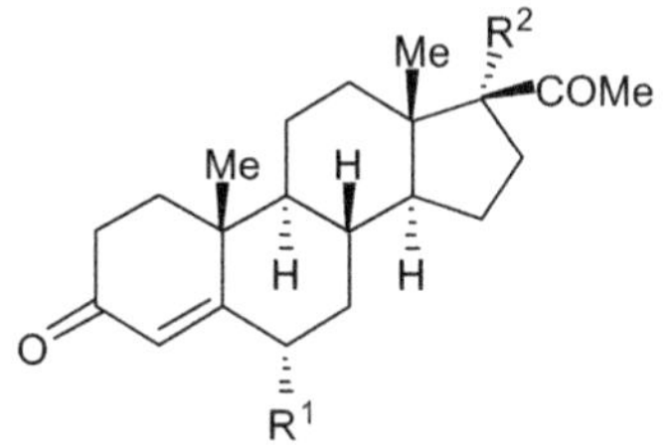

R^1 = H, R^2 = H progesterone
R^1 = Me, R^2 = OCOMe medroxyprogesterone acetate

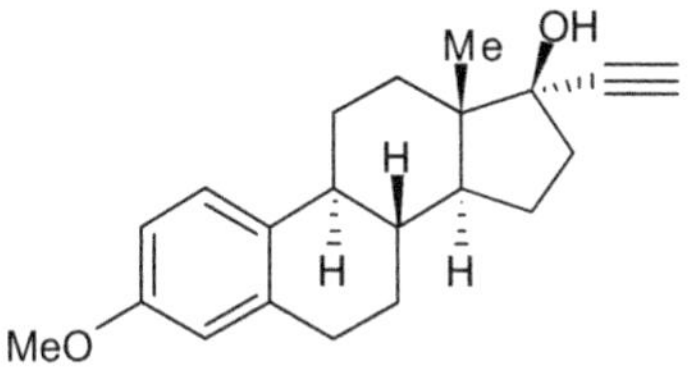

Mestranol

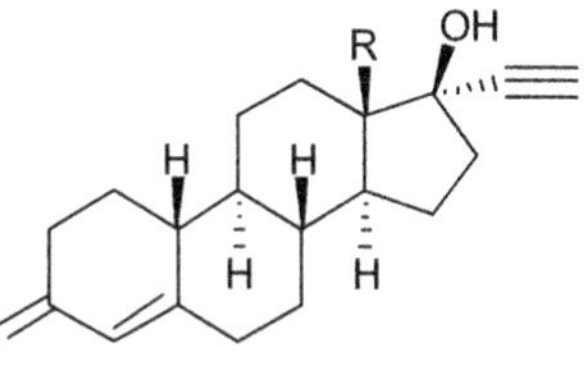

R = Me norethisterone
R = Et levonorgestrel(-)
R = Et norgestrel(±)

A typical combination of steroids for HRT comprises CEE and medroxyprogesterone acetate in the ratio ca. 1:5 as one tablet (marketed as Premique®). Alternatively, the equine oestrogen conjugate can be taken alone for 14–16 days and then in combination with progestin norgestrel or levonorgestrel for the following 12–14 days (a therapy marketed as Prempak® or Prempro®).

By the late 1980s, cHRT was being widely prescribed to women who retained their uterus (i.e. had not undergone a hysterectomy). While the risk of breast cancer was known to increase on oestrogen therapy (2), only in the late 1980s and early 1990s were studies published on the increased risk of breast cancer with the combined oestrogen/progesterone therapy (i.e. cHRT). Indeed, the risk of breast cancer was shown to be greater with cHRT than with oestrogen alone.

A detailed analysis of small-scale studies showed that the increased risk of breast cancer amongst HRT users did not begin until after about 5 years of treatment. However, the study showed that after 15 years of use there was a 30% increase in the risk of breast cancer (3). Other studies confirmed a 40% increase in the risk of breast cancer for users of cHRT; the induction period (that is, the period before the advent of risk) was less certain.

12.4 HRT AND OSTEOPOROSIS

Balanced against the increased risk of developing some cancers, there were signs of positive benefits of HRT, particularly with the long-term usage of oestrogen for the prophylaxis of osteoporosis, particularly for women who had had ovaries removed (oophorectomy). A National Institute of Health report in 1984 estimated that one-quarter of women over 65 years and half of women over 75 years would suffer from bone resorption (the root cause of osteoporosis), making oestrogen prophylaxis of osteoporosis an attractive opportunity. Consequently, in 1986, the FDA approved oestrogens for the treatment of post-menopausal osteoporosis, together with a high calcium diet and exercise.

12.5 HRT AND HEART DISEASE

Interest in the use of oestrogen therapy to counter osteoporosis was complemented by studies, including one called the Nurses' Health Study (4), which showed that post-menopausal women taking oestrogen were roughly 50% less likely to develop coronary heart disease than those who were not.

Thus the consensus view in the early to mid 1990s was that the benefits of HRT for overcoming the disconcerting effects of the menopause, the debilitating effects of osteoporosis and life-threatening heart disease outweighed the potential risks from the increased likelihood of contracting some cancers. Major professional medical organizations such as the American College of Physicians recommended that all women should consider 10–20 years of 'preventive hormone therapy' for maximum benefit. This and similar recommendations led to a boost in sales for products such as Premarin® by as much as 40%.

However, more detailed studies in the late 1990s (e.g. the Post-Menopausal Oestrogen/Progestin Interventions Trial) on the effect of HRT on heart disease were less positive. Thus, while HRT reduced the levels of low-density lipoprotein (LDL or 'bad' cholesterol), it increased triglyceride (fat) levels (5). Further work, for example, the Heart and Oestrogen/Progestin Replacement Study (HERS) showed that, while cholesterol LDL levels did improve, there was no difference in women taking HRT with respect to heart attacks over a period of 4 years. Oestrogen neither alone nor in combination with medroxyprogesterone stopped the further deterioration of the arteries in women with pre-existing atherosclerosis (6).

In 1996, an association between HRT and the increased incidence of blood clots was made (7). A higher incidence of venous embolism was found reflecting a risk of pulmonary embolism (blood clots in the lung).

12.6 THE WOMEN'S HEALTH INITIATIVE

The Women's Health Initiative (WHI) was a comprehensive study, started in the late 1990s, designed to investigate heart disease, breast and colon cancer, bone fracture and the role of HRT (as well as diet, vitamin and calcium supplements) in preventing these conditions and diseases. Over 26,000 women were involved in the whole study: of them 16,000 healthy women aged 50–79 years received Premique® (Prempro®) (0.625 mg Premarin®/2.5 mg Provera®) or placebo. Another cohort of 10,739 women who had had a hysterectomy was prescribed 0.625 mg Premarin® or placebo. Premarin® and Prempro® were chosen for study since they were commonly prescribed forms of oestrogen alone and cHRT, respectively, at that time. In 2002, the accumulated data revealed that there was clear evidence that the combined therapy led to an increased risk of invasive breast cancer (8). After 5 years of use, there was a 26% increase in the risk of breast cancer (i.e. eight more women would contract breast cancer for every 10,000 women). The Prempro® study was curtailed in the light of the findings. Interestingly, studies in animal models suggest that other steroid mixtures such as oestradiol and progesterone are potentially less harmful (9).

In the wider context, the WHI study showed a 29% increase in heart attacks for the cHRT group compared to placebo and 41% more strokes in the cHRT group compared to placebo. Onset of blood clots was four times higher in the first 2 years of use, decreasing thereafter to about two times. On the positive side, there was a reduction in hip fractures (34%) and a reduction in colorectal cancer (37%) (10).

In contrast, oestrogen alone did not appear to dramatically alter the risk of heart disease or breast cancer; however, oestrogen therapy was shown to cause an increased risk of stroke and a decreased the risk of hip fracture. The oestrogen-alone study was terminated in 2003/2004 with the conclusion that while HRT is effective for prevention of post-menopausal osteoporosis, oestrogen therapy should be considered 'only for women at significant risk from osteoporosis who cannot take non-oestrogen medications', for example, vitamin D/calcium or a bis-phosphonate such as zoledronic acid.

The commercial effect of the WHI study was profound. In 2001, in the United States, there were 45 million prescriptions for Premarin® (generating $2 billion) and 21.4 million prescriptions for Prempro®. Within a short span of time the use of these two top-selling drugs fell by 40% and 75%, respectively. The number of women using HRT in the United States fell from 18.5 million in 2002 to 7.6 million in 2004. Net sales suffered for the companies involved but, still, Wyeth's sales of Premarin®, Prempak®/Prempro® and related hormones topped at $1 billion in 2006. Indeed, there are a considerable number of products available in the marketplace produced by a variety of drug companies (Table 12.1). Most of the products are based on the traditional combinations of oestrogen and a well-known progestogen. Hence, it is noteworthy that the therapies involving Femapak and Angeliq involve relatively new progestins, namely, dydrogesterone and drospirenone, respectively. The former compound is called a 'retro-progestogen' due to the inversion of stereochemistry at the C9/C10 ring junction, while the cyclopropane groups in drospirenone provide structural features that have stimulated new synthetic methodology. Early studies suggested that dydrogesterone may be less of a risk than medroxyprogesterone in terms of potentiating breast cancer (11).

Dydrogesterone

Drospirenone

Table 12.1 Some of the Products Available for HRT*

Name	Oestrogen	Progestogen	Delivery	Source
Conjugated oestrogens only				
Premarin®	CEE	None	Tablet	Wyeth
Oestradiol (ester) only				
Bedol®	Oestradiol	None	Tablet	ReSource Medica
Zumenon®	Oestradiol	None	Tablet	Solvay
Elleste-Solo®	Oestradiol	None	Tablet and patch	Meda
Estraderm®	Oestradiol	None	Patch	Novartis
Estradot®	Oestradiol	None	Patch	Novartis
Evoral®	Oestradiol	None	Patch	Janssen-Cilag
Fematrix®	Oestradiol	None	Patch	Solvay
FemSeven®	Oestradiol	None	Patch	Merck
Oestrogel®	Oestradiol	None	Gel	Ferring
Sandrena®	Oestradiol	None	Gel	Organon
Climaval®	Oestradiol valerate	None	Tablet	Novartis
Progynova®	Oestradiol Valerate	None	Tablet	Organon
Conjugated oestrogen/progestogen				
Premique®	CEE	Medroxyprogesterone acetate	Tablet	Wyeth
Prempak® (Prempro®)	CEE	(Levo)norgesterol	Tablet	Wyeth
Oestradiol (esters) with progestogen				
Angeliq®	Oestradiol	Drospirenone	Tablet	Schering
Clinique®	Oestradiol	Norethisterone	Tablet	ReSource Medica
Elleste-Duet®	Oestradiol	Norethisterone acetate	Tablet	Meda
Kliovance®	Oestradiol	Norethisterone acetate	Tablet	Novo Nordisk[†]
Estracombi®	Oestradiol	Norethisterone acetate	Patch	Novartis
Evoral®	Oestradiol	Norethisterone acetate	Patch	Janssen-Cilag
Femapak®/Femoston® oestradiol	Dydrogesterone	Patch	Solvay	—
FemSeven Conti®	Oestradiol	Levonorgestrel	Patch	Merck
Climagest®	Oestradiol valerate	Norethisterone	Tablet	Novartis**
Cyclo-Progynova®	Oestradiol valerate	Norgestrel	Tablet	Viatris

Table 12.1 Some of the Products Available for HRT* *(continued)*

Name	Oestrogen	Progestogen	Delivery	Source
Indivina®	Oestradiol valerate	Medroxyprogesterone acetate	Tablet	Orion[‡]
Nuvella®	Oestradiol valerate	Levonorgestrel	Tablet	Schering

*U.K. nomenclature. Product names may vary from country to country.
[†]Also available from Novo Nordisk as Kliofem®, Trisequens®, and Novofem®.
[**]Also available from Novartis as Climesse®.
[‡]Also available from Orion as Tridestra®.

In recent years, organizations such as the UK Committee on the Safety of Medicines advise that for HRT, the minimum effective dose should be used for the shortest duration, with treatment being stopped altogether prior to trauma (e.g. surgery) or immobilization (e.g. long-distance travel). This advice reflects the data set out in the HRT risk table *(12)*. Particularly striking numbers from Table 12.2 relate to the risk of increased breast cancer or endometrial cancer using the standard cHRT or oestrogen-alone treatments over an extended period. The increased risk of venous thromboembolism is also evident on using HRT over a 5-year period. In comparison, the risk of increased coronary heart disease is relatively small. Indeed, it continues to be suggested that perimenopausal women may even benefit from HRT through 30–50% protection from heart disease *(13)*.

Table 12.2 Hormone Replacement Therapy Risk Table

Disease	Age Range (years)	Background Incidence*		Additional Cases (Oestrogen Alone)[†]		Additional Cases (Combined Oe/Pg)[†]	
		5 years	10 years	5 years	10 years	5 years	10 years
Breast cancer	50–59	10	20	2	6	6	24
	60–69	15	30	3	9	9	36
Endometrial cancer	50–59	2	4	4	32	–	–
	60–69	3	6	6	48	–	–
Ovarian cancer	50–59	2	4	<1	1	<1	1
	60–69	3	6	<1	2	<1	2
Venous thromboembolism	50–59	5	–	2	–	7	–
	60–69	8	–	2	–	10	–
Stroke	50–59	4	–	1	–	1	–
	60–69	9	–	3	–	3	–
Coronary heart disease	70–79	29–44	–	–	–	15	–

*Per 100 women in Europe.
[†]Per 100 women using oestrogen alone or combined oestrogen/progestogen.

12.7 IMPROVED PROTOCOLS FOR HRT

Of course work has been continuing to accentuate the benefits of HRT over the risks. Different methods of delivery have been explored because, when oestrogen is taken orally, it is first processed by the liver to stimulate the production of proteins associated with heart disease and stroke, such as C-reactive protein, activated protein C and clotting factors. When delivered by trans-dermal patch, the oestrogen is not processed by the liver immediately (i.e. avoids first-pass metabolism) and this lessens the risk of heart disease and stroke.

The Selective Oestrogens, Menopause and Response to Therapy (SMART) initiative (2011) is investigating the employment of a SERM together with conjugated oestrogen for moderate to severe menopausal symptoms and post-menopausal osteoporosis. Bazedoxifene is a third-generation SERM (N.B. tamoxifen was first generation) in that it can be distinguished from other agonists and antagonists in that it modulates oestrogen activity differently in different tissues, e.g. it acts as an agonist at the oestrogen receptors in bone and as an antagonist at oestrogen receptors in breast and the uterus (*14*). The mechanism of SERMs such as bazedoxifene is related to (1) the ratio of co-activator and co-repressor proteins in the different cell types and (2) the conformation of the receptor on drug binding which determines whether the complex preferentially recruits coactivators (giving the agonist response) relative to the recruitment of corepressors (giving the antagonist response).

Bazedoxifene   Tamoxifen

Initial results from the study are promising, showing no cases of endometrial hyperplasia and a reduction in bone turnover (*15*). However, the trial is in its early stages and long-term benefits/risks will have to be assessed in due course.

12.8 CONCLUSION

In a typical medical doctor's surgery in the United Kingdom in 2012, the general practitioner has no problem in prescribing drugs such as Premarin® or the cHRT Kliovance® for women experiencing symptoms of menopause. However, the medical strategy is to prescribe the drug at as low a dose as possible for as short a time as is sensible. This pathway seeks to ameliorate the more distressing effects of menopause while minimizing the most overt risks of HRT, namely, the onset of breast cancer or endometrial cancer later in life.

References

1. (a) Smith, D. C., et al. *New Engl. J. Med.* **1975**, *293*, 1164–1167. (b) Ziel, H. K.; Finkle, W. D. *New Engl. J. Med.* **1975**, *293*, 1167–1170. (c) Ryan, K. *New Engl. J. Med.* **1975**, *293*, 1199.
2. Hoover, R., et al. *New Engl. J. Med.* **1976,** *295*, 401–405.
3. Steinberg, J., et al. *J. Am. Med. Assoc.* **1991,** *265*, 1985–1990.
4. Stampfer, M. J., et al. *New Engl. J. Med.* **1991,** *325*, 756. Barrett-Connor, E., et al. *J. Am. Med. Assoc.* **1991,** *265*, 1861–1867. It should be said that a contemporaneous (Framingham) study came to the opposite conclusion.
5. Miller, V. T., et al. *J. Am. Med. Assoc.* **1995,** *273*, 199–208.
6. Hulley, S., et al. *J. Am. Med. Assoc.* **1998,** *280*, 605. Herrington, D.M.; Howard, T.D. *New Engl. J. Med.* **2003**, *349*, 519–521.
7. Daly, E., et al. *Lancet* **1996**, *348*, 977–980. and associated articles.
8. Rossouw, J. E., et al. *J. Am. Med. Assoc.* **2002,** *288*, 321–333. Later it was suggested that the hazard of HRT causing heart disease may have been over-estimated, see Shapiro, S. *J. Am. Med. Assoc.* **2007**, *298*, 623 and the riposte by Shapiro, J.E. *J. Am. Med. Assoc.* **2007**, *298*, 624.
9. Wood, C. E., et al. *Breast Cancer Res. Treat.* **2007**, *101*, 125–134.
10. WHI Investigators Writing Group. *J Am. Med. Assoc.* **2002,** *288*, 321–333. Nelson, H.D., et al. *J. Am. Med. Assoc.* **2002,** *288*, 872; Billeci, A.M. et al. *Curr. Vasc. Pharmacol.* **2008**, *6*, 112–123.
11. Gadducci, A., et al. *Gynacol. Endocrinol.* **2009**, *25*, 807–815.
12. *British National Formulary*, September 2008. p. 391.
13. Van der Schouw, Y. T.; Grobbee, D. E. *Eur. Heart J.* **2005**, *26*, 1358–1361.
14. Gennari, L., et al. *Ther. Clin. Risk Manage.* **2008**, *4*, 1229–41242.
15. Levine, J. P. *Gend. Med.* **2011,** *8*, 57–68.

Design of the anti-HIV protease inhibitor darunavir

13

Arun K. Ghosh*,†, Bruno D. Chapsal*

*Department of Chemistry, Purdue University, West Lafayette, IN 47907, USA,
†Department of Medicinal Chemistry, Purdue University, West Lafayette, IN 47907, USA

CHAPTER OUTLINE

13.1 Introduction 356
13.2 The Target Enzyme: HIV-1 Aspartic Acid Protease 357
13.3 Advent of Protease Inhibitors, HAART, and Structural Insights from Saquinavir............. 358
13.4 Cyclic Ethers to Mimic Peptide Bonds: Inspiration from Natural Products 360
13.5 Design of Conceptually New Cyclic Ether-derived Ligands and Corresponding Protease Inhibitors ... 361
    13.5.1 Exploration of ligands from cyclic ethers to cyclic sulfones....................................364
    13.5.2 Design and development of bicyclic bis-tetrahydrofuran (bis-THF) ligand....................366
13.6 Design Strategy to Combat Drug Resistance by Targeting Protein Backbone: 'Backbone Binding Concept' 367
13.7 Design of PIs Promoting Strong Backbone Interactions from S2 to S2′ Subsites 369
13.8 The 'Backbone Binding Concept' and Its Relevance to Combat Drug Resistance 370
13.9 Selection of Darunavir as a Promising Drug Candidate... 372
    13.9.1 Thermodynamic and kinetic effects behind darunavir's high binding affinity for the protease 374
    13.9.2 Darunavir inhibits HIV-1 protease dimerization: a unique dual mode of action 375
13.10 Convenient Syntheses of the bis-THF Ligand 376
13.11 Clinical Use of Darunavir in the Management of HIV-1 Infection................................ 377
13.12 Design of Potent Inhibitors Targeting the Protease Backbone 377
13.13 Conclusion.. 378
References 379

ABSTRACT

The development of HIV protease inhibitors (PIs) and their inclusion in highly active antiretroviral therapies (HAARTs) marked the beginning of a treatment breakthrough in the management of human immunodeficiency virus (HIV)/acquired immunodeficiency syndrome (AIDS). The HAART treatment regimen can cut HIV viral load to undetectable levels. Nonetheless, the rapid emergence of HIV drug resistance has continued to seriously compromise long-term treatment options for HIV-infected patients. Our structure-based design strategy to develop PIs that specifically target the enzyme's backbone atoms has resulted in a number of very potent inhibitors with superior drug resistance profiles. Of particular note, our development of stereochemically defined bis(tetrahydrofuranyl) urethane as a high-affinity P2 ligand

Introduction to Drug Research and Development. http://dx.doi.org/10.1016/B978-0-12-397176-0.00013-3

355

has led to the development of exceedingly potent inhibitors. One of these inhibitors, darunavir, has shown exceptional potency against the HIV-1 virus and superior activity against multi-PI-resistant viral strains. Our backbone binding strategy was corroborated with detailed crystal structure analyses of darunavir-bound protease complexes which revealed a series of conserved interactions between the inhibitor and key backbone atoms of HIV-1 protease. Darunavir first received accelerated US Food and Drug Administration approval in 2006 for highly treatment-experienced patients with little therapeutic options. It has now become a leading PI in the fight against HIV infection and drug resistance.

Keywords/Abbreviations: Acquired immunodeficiency syndrome (AIDS); Amprenavir (APV); bis in die (twice a day) (b.i.d.); Cell culture inhibitor concentration at >95% inhibition (CIC_{95}); Darunavir (DRV); Highly active antiretroviral therapy (HAART); Human immunodeficiency virus type 1 (HIV-1); Half maximal inhibitory concentration (IC_{50}); Indinavir (IDV); Constant of inhibition (K_i); Mutant (mut); Nelfinavir (NFV); HIV-1 protease inhibitor (PI); *quaque die* (once daily) (q.d.); Ritonavir (RTV); Saquinavir (SQV); Tipranavir (TPV); Wild type (*wt*); Coadministration of a PI with low-dose ritonavir (/r).

13.1 INTRODUCTION

The human immunodeficiency virus (HIV)/acquired immunodeficiency syndrome (AIDS) epidemic continues to plague the world on a pandemic scale. According to the Joint United Nations Programme on HIV/AIDS (UNAIDS)/World Health Organization's 2011 progress report on HIV/AIDS, an estimated 34 million people are infected with the HIV-1 virus and 1.8 million died as a result of AIDS (*1*). With the understanding of the HIV-1 virus life cycle and identification of key viral enzymes as therapeutic targets, several classes of drugs, that can effectively block HIV-1 replication, have been developed. To date, about 25 clinically approved antiretrovirals are available for the treatment of HIV/AIDS (*2*).

The approval of the first HIV-1 protease inhibitors (PIs) in the mid-1990s and their combination with reverse transcriptase inhibitors initiated the first highly active antiretroviral therapy (HAART) treatments. The advent of HAART led to durable suppression of HIV-1 infection and improvement of $CD4^+$ lymphocyte cell counts in HIV-1-infected patients. HAART has significantly decreased the morbidity and mortality associated with HIV-1/AIDS and improved the quality of life of patients in industrialized nations (*3,4*). Despite these major clinical benefits, PI-based HAART therapies are faced with many challenges that seriously complicate long-term management of HIV-1 infection. Many drawbacks such as stringent dosing regimens, high pill burden, ever-escalating treatment cost, drug-related debilitating side effects, and toxicities compromise treatment adherence among patients (*5*).

Perhaps, the most alarming problem is the rapid emergence of drug resistance that quickly renders a selected treatment ineffective (*6,7*). An estimated 10–25% of newly infected individuals in the United States and Europe harbour drug-resistant HIV-1 variants, including those with PI-specific resistance (*8*). In addition, first-generation PIs exhibited poor oral absorption, high serum protein binding, low metabolic stability and poor pharmacokinetic properties. Lipodistrophy, hyperlipidaemia, insulin resistance, and premature ageing are common side effects that typically affect patients under PI-based regimens (*9,10*).

For some years, we have been interested in designing a conceptually new generation of nonpeptidic PIs that would maintain activity against drug-resistant HIV-1 variants and possess improved

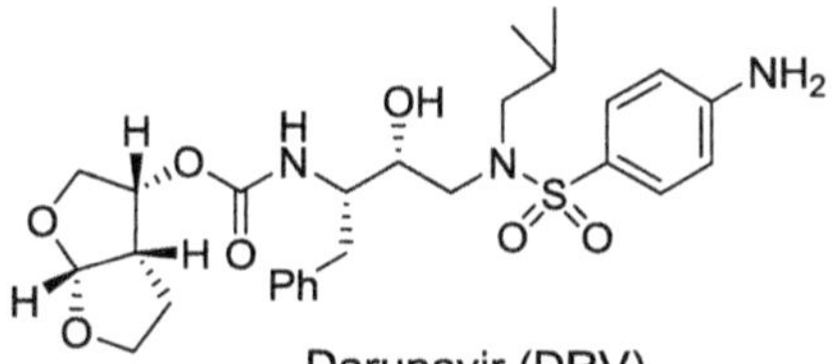

Darunavir (DRV),
Prezista® (1)

FIGURE 13.1

Structure of Darunavir.

pharmacological properties. We have been particularly interested in designing novel inhibitor scaffolds and ligands by drawing our inspiration from nature (*11,12*). These research efforts led to the design and discovery of novel ligand templates and a series of exceedingly potent inhibitor leads. Further optimization of these lead structures culminated in the development of darunavir (DRV) (**1**, Figure 13.1). DRV displayed superb activity against a wide range of drug-resistant HIV-1 variants (*13*). X-ray structural studies of DRV-bound HIV-1 proteases revealed a network of close hydrogen bonding interactions with the protease backbone atoms. These hydrogen bonding interactions with backbone residues may be responsible for DRV's superior resistance profile compared to other PIs (*14,15*). Our main design strategy to create a network of hydrogen bonds with backbone atoms of the enzyme's active site, designated as our 'backbone binding concept', has now emerged as a useful approach to combat drug resistance (*15*). Due to the unique ability of DRV to inhibit multidrug-resistant HIV-1 viruses, and as a result of its favourable pharmacological properties, this inhibitor was selected for clinical development. Ultimately, DRV received US Food and Drug Administration (FDA) approval in 2006, under the trade name Prezista®, for the treatment of patients harbouring highly drug-resistant viruses (*16,17*). In 2008, its approval was extended to all HIV/AIDS patients including paediatrics (*16*). The unique structural design of DRV, its molecular interactions with the HIV-1 protease, and its resilience against multidrug-resistant HIV-1 variants have subsequently stimulated immense research interest (*18,19*). In the present chapter, we will describe the strategy, concepts, and inspiration behind the design of DRV and its subsequent clinical development.

13.2 THE TARGET ENZYME: HIV-1 ASPARTIC ACID PROTEASE

The HIV-1 protease is an aspartic acid protease that is critical for proper assembly and final maturation of HIV viral particles at the late stage of the viral replication. It catalyses the hydrolysis of peptide bonds at the various polypeptide cleavage sites of the virally encoded *Gag* and *Gag-Pol* polyproteins and produces all essential viral enzymes and structural proteins, including the reverse transcriptase, the protease itself, and the integrase (*20*). HIV-1 protease is a relatively small enzyme that consists of a dimer of two identical subunits, each having 99 amino acid residues. Two aspartic acid residues, Asp25 and Asp25′, form the catalytic dyad at the dimer interface (*20*). The catalytic site of the protease is covered on the top by two flexible β-strand sheets, which form flaps that open in the absence of a substrate and fold down on substrate binding, locking the peptide in an appropriate local environment that favours catalysis (*20*).

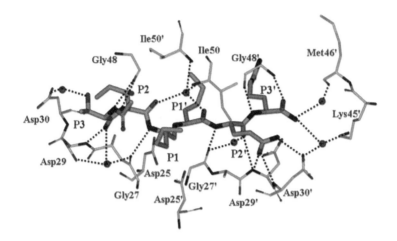

FIGURE 13.2

X-ray crystal structure of the HIV-1 protease in complex with a substrate analogue (green) showing closed conformation flaps. (See colour plate.)

The HIV protease is a highly specific enzyme. The natural substrates of the protease bind with a minimum sequence of seven amino acid residues that standard nomenclature designates as P4 to P1 and P1′ to P3′. Each of these residues bind to enzymatic subsites which are designated as S4 to S3′, respectively. The peptide bond cleavage takes place between the P1 and P1′ residues as they bind into the S1 and S1′ enzyme subsites surrounding the protease's catalytic dyad, respectively. The S1 and S1′ subsites accommodate highly hydrophobic amino acid residues. Both S2 and S2′ subsites accept polar or hydrophobic residues, while S3 and S3′ are mostly hydrophobic. The peptide bond cleavage occurs via insertion of a conserved water molecule originally bound to the catalytic dyad into the substrate peptide bond. It forms a *gem*-diol tetrahedral transition state that dissociates to form two cleavage products (*20*). Design of nonhydrolysable peptide analogues based on isosteres of the transition state have permitted extensive studies including X-ray crystal structure analysis of the HIV protease bound to these analogues. Structures of the protease bound with various hexapeptide analogues unveiled a series of hydrophobic and hydrogen bond interactions that connect the substrate with the protease's binding site (*21*). A series of highly conserved hydrogen bond interactions between the substrate and the backbone atoms of protease residues 25–29 in the catalytic site, and 48–50 in the flap region of the protease, could clearly be highlighted. Figure 13.2, which shows the crystal structure of the protease bound to a peptide analogue of the polyprotein p2/NC cleavage site (PDB code: 2AOD) (*22*), provides an outlook on these interactions.

13.3 ADVENT OF PROTEASE INHIBITORS, HAART, AND STRUCTURAL INSIGHTS FROM SAQUINAVIR

Saquinavir (SQV, **2**, Figure 13.3) was developed by Roche and became the first FDA-approved HIV-1 PI in 1995 (*23*). Thereafter, several other PIs followed through clinical development and received

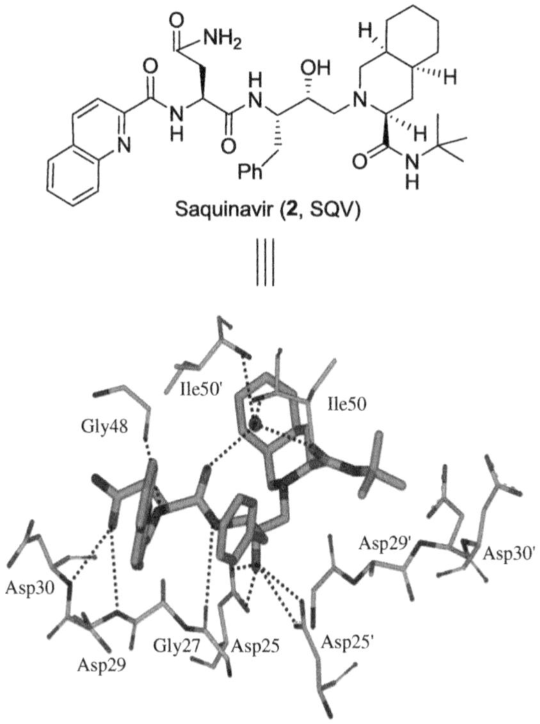

Saquinavir (**2**, SQV)

FIGURE 13.3

Saquinavir (2) and X-ray structure of SQV–HIV-1 protease complex. (See colour plate.)

approval. First-generation PIs were designed based on noncleavable hydroxyethylene or hydroxyethylamine-based isosteres (*20*). These PIs typically possessed multiple amide/peptide-like bonds and have been initially developed to extensively interact with all subsites of the protease.

As reversible competitive inhibitors, PIs were tested for their *in vitro* enzyme inhibitory potency. An inhibitor's potency is indicated by its inhibition constant (K_i value). K_i characterizes enzyme's inhibition by an inhibitor. It is measured by observing the change in enzyme's activity at different concentrations of inhibitor and natural substrate. The PI's antiviral activity was determined on HIV-infected cell lines, and the minimum concentration for 50% inhibition (ID_{50}) of viral replication was measured by measurement of p24 viral antigen production. SQV showed very impressive enzyme inhibitory and antiviral activity ($K_i = 0.23$ nM, antiviral $ID_{50} = 12$–25 nM). SQV's FDA approval marked the important beginning of HAART treatment regimens. In combination with reverse transcriptase inhibitors, SQV-based regimens significantly reduced viral loads, improved CD4$^+$ cell counts in patients, and halted the progression of HIV/AIDS. However, one of SQV's major limitations has been its poor bioavailability, possibly owing to its high peptide-like structural features. The X-ray crystal structure of SQV-bound HIV-1 protease provided important insights into its interactions in the enzyme's active site (*24*). As shown in Figure 13.3, SQV is involved in numerous interactions with the protease. Of particular interest, the P2 asparagine side chain's carbonyl forms a hydrogen bond with the protease's backbone NH of the Asp30 residue, while the quinaldic acid's carbonyl interacts with the Asp29 NH. All these molecular interactions in the active site are presumably responsible for SQV's impressive HIV-1 protease inhibitory and antiviral activity.

We became interested in SQV because of its clinical efficacy and its early X-ray crystallography and structure–activity relationship studies. Based on these key available information, we sought to design new templates that could effectively replace the amide or peptide bonds of SQV and mimic their specific binding interactions in the protease's active site. We were particularly interested in designing cyclic ether or sulfone-derived templates wherein conformationally restricted and suitably positioned ether oxygen atoms could induce the same hydrogen bonding interactions as the carbonyl oxygens of SQV's P2 or P3 amide bond/peptide bonds. Our choice of cyclic ethers was motivated by the intriguing occurrence of these five- or six-membered structural templates in a vast majority of bioactive natural products. Also, the cyclic carbon backbone could offer exquisite structural complementarity and effectively fit in an enzyme's hydrophobic pocket. Due to their limited degrees of rotational freedom, cyclic ethers might favourably contribute to binding entropy and the inhibitor's free binding energy. Furthermore, unlike peptide-based compounds, ethers are not subject to degradation by peptidases and may therefore offer improved metabolic stability.

13.4 CYCLIC ETHERS TO MIMIC PEPTIDE BONDS: INSPIRATION FROM NATURAL PRODUCTS

Natural products have historically been a great source of inspiration for the development of new therapeutic agents. Numerous clinically approved molecules are either natural products or natural product-derived analogues (25–27). Honed by millions of years of evolution, organisms have developed a variety of fine-tuned molecules and molecular templates that are custom-made to interact with their biological microenvironment (28). The diversity of structural templates, scaffolds, and functionalities of natural products and consideration of their evolutionary fitness motivated us to harness nature's molecular probes for the design of new therapeutics. We have been particularly interested in conformationally constrained heterocyclic scaffolds, especially cyclic ethers and acetals (11,12). The natural products laulimalide (**3**) (29), ginkgolide (**4**) (30), platensimycin (**5**) (31), and monensin A (**6**) (32) depicted in Figure 13.4 have shown very potent biological properties. These molecules do not have any peptidic features, yet they bind to their respective biological target with high affinity. Halichondrin B's analogue eribulin mesylate (**7**) (33) is a potent anticancer agent and contains numerous cyclic ethers.

The cyclic ether templates in the depicted molecules very likely form polar hydrogen bonding interactions with peptides or protein residues. In essence, these molecules are peptide mimics. We then speculated that the design and incorporation of such templates could alleviate problems inherent to peptide and peptidomimetic-derived drugs. Studies of these molecules revealed that cyclic ether scaffolds not only contribute to the overall geometry of the molecule but also may serve as important pharmacophores that specifically bind to their respective biological target. An X-ray crystal structure (PDB Code:3HNZ) (34) of the recently discovered natural antibiotic, platensimycin (**5**) bound to gram-positive bacterial enzyme beta-ketoacyl synthase II (FabF), shows the cyclic ether oxygen creating hydrogen bond interactions with the Thr270 residue of the enzyme (Figure 13.5). The unique polycyclic scaffold of platensimycin defined a structurally new class of

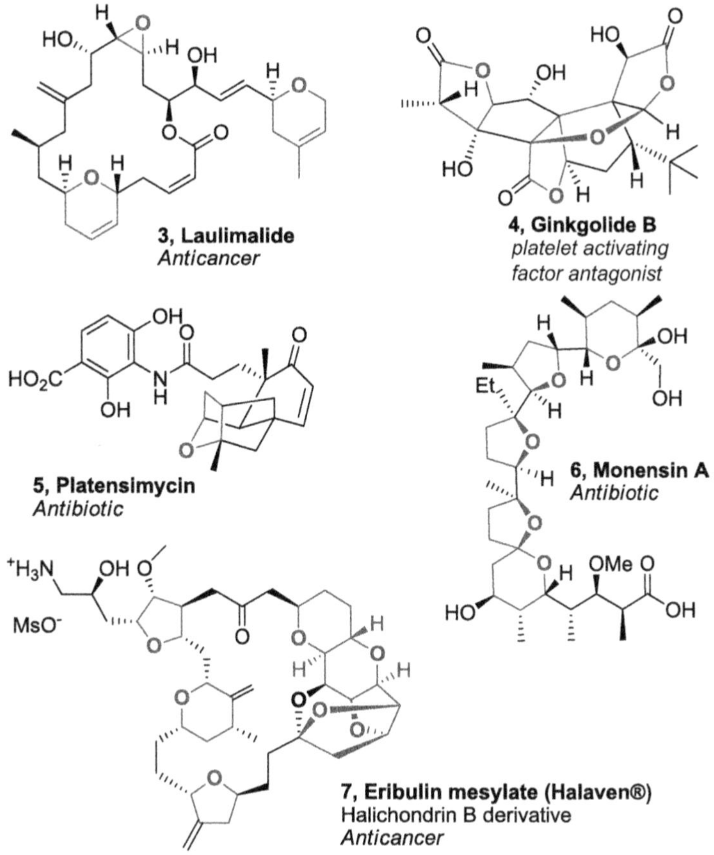

FIGURE 13.4

Cyclic ether-containing natural products and analogues. (For colour version of this figure, the reader is referred to the online version of this book.)

antibiotics that may provide new opportunities in the fight against drug-resistant infections (*31*). It is therefore relevant that these cyclic ether-containing natural products being the result of millions of years of evolutionary selection possess adequate biological fitness. This inspiration from nature has fuelled our design of novel PIs with improved antiviral activity and better pharmacological properties.

13.5 DESIGN OF CONCEPTUALLY NEW CYCLIC ETHER-DERIVED LIGANDS AND CORRESPONDING PROTEASE INHIBITORS

The X-ray crystal structure of SQV-bound HIV-1 protease showed that the P2 asparagine side chain carbonyl appeared to form a hydrogen bond with the backbone NH of Asp30 (*24*). Based on this

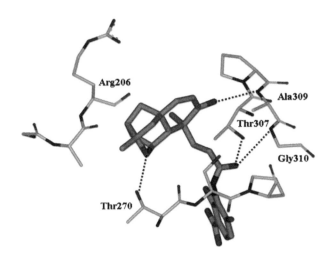

FIGURE 13.5

X-ray crystal structure of platensimycin (5) bound to bacterial FabF enzyme. (See colour plate.)

observation, our plan was to introduce a stereochemically defined tetrahydrofuran oxygen that could mimic the binding interactions of the P2 asparagine's carbonyl of SQV. To test this design hypothesis, we synthesized a (2R,3′R)-tetrahydrofuranyl (THF) glycine and incorporated it in inhibitor **8** (Figure 13.6). Indeed, this inhibitor exhibited much improved enzyme inhibitory potency and antiviral activity (IC$_{50}$ (half maximal inhibitory concentration) = 0.05 nM, CIC$_{95}$ = 8 nM) compared to SQV. The enzyme inhibitory IC$_{50}$ was determined *in vitro* as the concentration of inhibitor needed to inhibit by 50% the cleavage of a fluorescent natural substrate by the protease. Tests were run at fixed protease's and substrate's concentrations, temperature and pH, and substrate cleavage was quantified by high-performance liquid chromatography with ultraviolet detection or by measurement of fluorescence. Antiviral activity was determined *in vitro* by the enzyme-linked immunosorbent assay p24 which measures the inhibition of HIV infection on a cell culture. The resulting CIC$_{95}$ (cell culture inhibitory concentration) represents the inhibitor's concentration which inhibited by greater than 95% the spread of infection on the cells, as measured by a greater than 95% reduction in viral p24 antigen production.

When we tested inhibitor **8**, the ring stereochemistry and position of the oxygen appeared critical to the observed potency (*35*). The corresponding (2S,3′S)-THF glycine with 3′-(S)-ring stereochemistry (inhibitor **9**) resulted in a 23-fold loss of enzyme inhibitory potency compared to SQV. The removal of the ring oxygen also resulted in compound **10** with nearly 10-fold loss of potency. We presumed that the cyclic ether oxygen in **8** was suitably positioned to mimic the asparagine carbonyl group and form a hydrogen bond with the Asp30 NH. In an effort to reduce PI's molecular weight, the P3 quinaldic moiety was removed and inhibitor **11** with a simple 3-(S)-tetrahydrofuranyl urethane ligand was designed (*36*). The enzyme inhibitory potency of **11** was reduced significantly (IC$_{50}$ = 132 nM, CIC$_{95}$ = 800 nM) compared to SQV; however, the molecular weight of **11** was much reduced (515 Da vs 670 Da). The corresponding 3-(R)-furanyl urethane derivative **12** was fourfold less potent than the 3-(S)-isomer. Again, importance of ring stereochemistry was apparent as the 3-(R)-isomer was

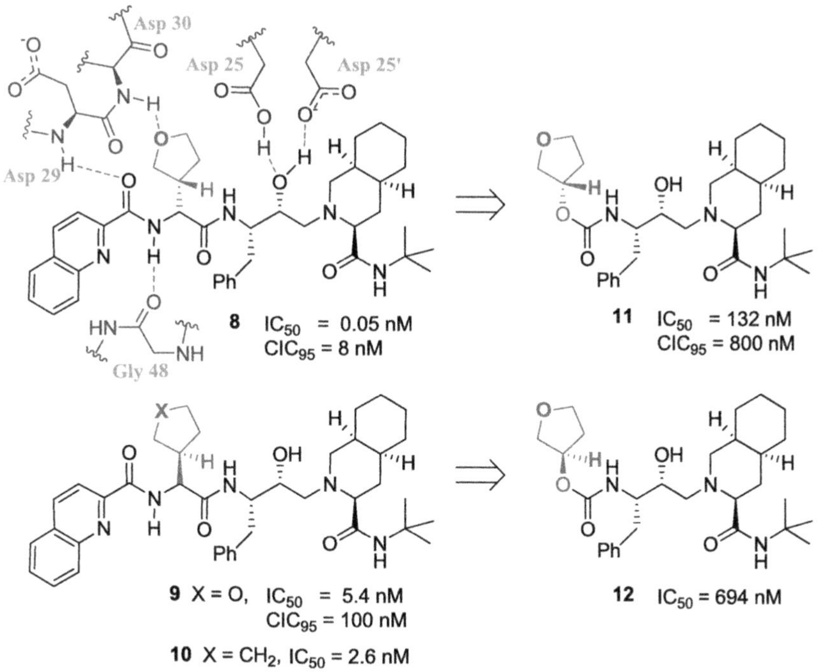

FIGURE 13.6

Introducing stereochemically defined cyclic ether-based ligands into PIs. (See colour plate.)

significantly less potent. X-ray crystal structure of **11**-bound HIV-1 protease revealed that the THF ring oxygen in **11** was within proximity to form hydrogen bonds with the backbone NHs of both Asp29 and Asp30 residues.

To further probe the potency-enhancing effect of the 3-(S)-THF ligand and its specific preference for the residues of the S2 subsite of the protease, we investigated the effect with other dipeptide isosteres. As shown in Figure 13.7, combination of the 3-(S)-THF P2 ligand with a hydroxyethylene isostere containing a chiral aminoindanol as a P2′ ligand provided inhibitor **13**, with marked enhancement of enzyme inhibitory potency and antiviral activity (*36*). The corresponding 'all-carbon' cyclopentanyl P2 ligand (compound **14**) showed significant loss of activity and this demonstrated the importance of the cyclic oxygen in the 3-(S)-THF ligand in inhibitor **13**. Subsequently, investigators at Vertex laboratory utilized this 3-(S)-THF urethane in the design of VX-478 (**15**). As can be seen, combination of the 3-(S)-THF as a P2 ligand with an (R)-hydroxyethyl(sulfonamide) isostere developed by Vazquez et al. (*37*) and Tung et al. (*38*) provided the very potent inhibitor **15**. This inhibitor exhibited favourable pharmacological properties and further clinical development resulted in its FDA approval under the name amprenavir (APV) (*38*). The X-ray crystal structure of **15**-bound HIV-1 protease showed the THF ligand fitting in the enzyme's S2 subsite with the cyclic oxygen of this THF ligand within hydrogen bonding distance to the Asp29 and Asp30 NH bonds (Asp29 NH: 3.4 Å, Asp30

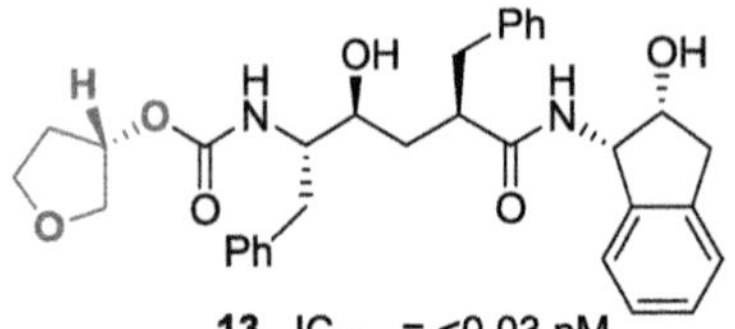

13 IC_{50} = <0.03 nM
CIC_{95} = 3 nM

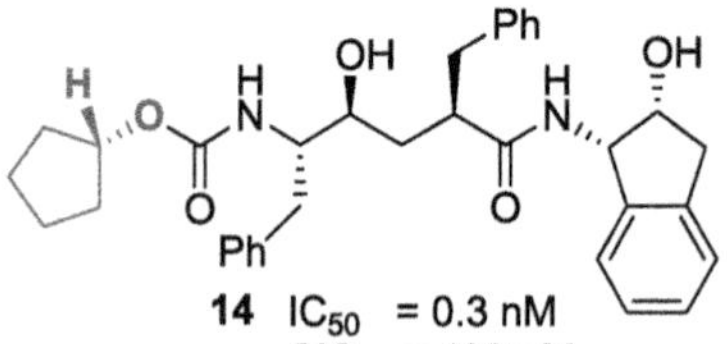

14 IC_{50} = 0.3 nM
CIC_{95} = 400 nM

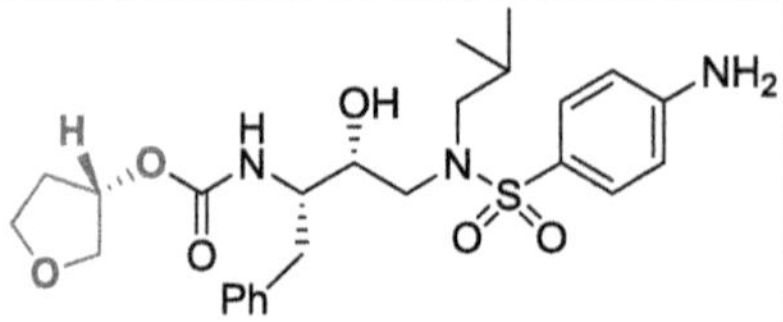

15 Amprenavir (APV, VX-478)
K_i = 0.6 nM
IC_{90} = 40 nM

FIGURE 13.7

Tetrahydrofuranyl P2 ligand-derived protease inhibitors. (For colour version of this figure, the reader is referred to the online version of this book.)

NH: 3.5 Å) (*38*). However, these distances corresponded to rather weak hydrogen bonding interactions.

13.5.1 Exploration of ligands from cyclic ethers to cyclic sulfones

Following the development of cyclic ethers as effective mimics of amide carbonyl oxygen, we explored cyclic thiolane- and sulfolane-derived P2 ligands so as to potentially create closer hydrogen bonding with the Asp29 and Asp30 residues (*39–41*). Inhibitor **16** (Figure 13.8), containing a 3-(*S*)-sulfolane ligand, displayed comparable enzyme inhibitory potency and antiviral activity to 3-(*S*)-THF-based inhibitor **11**. Introduction of an isopropyl substituent on the 3-(*R*)-sulfolane ring further improved potency (inhibitor **17**). Incorporation of these ligands into the (*R*)-hydroxyethyl(sulfonamide) provided very potent inhibitors **18** and **19**. An energy-minimized model structure of inhibitor **16** (Figure 13.9) showed that the sulfone oxygen *cis* to the 3-(*S*)-hydroxy is within proximity to form hydrogen bonds with Asp29's and Asp30's NH bonds. The increase in potency observed with isopropyl-substituted sulfolanes may originate from optimized van der Waals interactions between the ligand and the enzyme's S2 hydrophobic pocket.

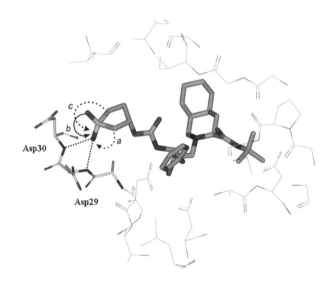

FIGURE 13.8

Potent sulfolane-based protease inhibitors. (For colour version of this figure, the reader is referred to the online version of this book.)

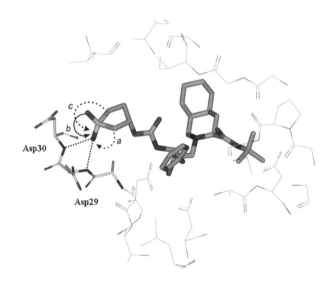

FIGURE 13.9

Design strategy based on modelling of 3-(S)-sulfolane-based inhibitor **16**. (See colour plate.)

13.5.2 Design and development of bicyclic bis-tetrahydrofuran (bis-THF) ligand

Based on the possible ligand-binding site interactions of the sulfolane-based ligand, we speculated that the P2 ligand could be further optimized to improve potency and interactions with backbone residues in the S2 subsite. Considering that a sulfone oxygen is a weak 'hydrogen bond acceptor', we designed bicyclic P2 ligands wherein a suitably positioned cyclic ether oxygen would mimic the sulfolane's oxygen in **16** while the bicyclic structure would improve interactions with the S2 hydrophobic pocket. A cyclic ether oxygen would act as a better hydrogen bond acceptor and thus more efficiently interact with Asp29 and Asp30 NH bonds. As shown in Figure 13.9, various bicyclic rings could be formed (modes: a–c) through the *cis* oxygen and α-carbons of the sulfolane ring in **16**. Several bicyclic ligands were explored in combination with the SQV- or hydroxyethylsulfonamide-based isosteres.

The structures and potency of the designed inhibitors are shown in Figure 13.10. Following design **a**, inhibitor **20** with a (4S)-hexahydro-2*H*-cyclopenta[*b*]furan-4-ol ligand was prepared. It was more potent than the 3-(S)-THF-based inhibitor **11** or sulfolane-based inhibitor **16** (*42*). To induce multiple interactions with both Asp29 and Asp30 NH backbone bonds, we decided to incorporate another cyclic ether oxygen on the top cyclopentane ring. Inhibitor **21** that contained the corresponding fused *bis*-tetrahydrofuran (*bis*-THF) P2 ligand exhibited substantially improved enzyme inhibitory potency and antiviral activity compared to other cyclic ether and cyclic sulfone-based inhibitors (*42,43*). Our design of an oxaspiro ligand (design **b**) and its incorporation into the hydroxyethyl(arylsulfonamide) isostere resulted in inhibitor **22** with good enzyme inhibitory potency; however, antiviral activity was modest. Further design of bicyclic ligands via design **c** resulted in cyclopentanyl tetrahydrofuran ligand. Incorporation of this ligand in the sulfonamide isostere provided the incredibly potent inhibitor **23**.

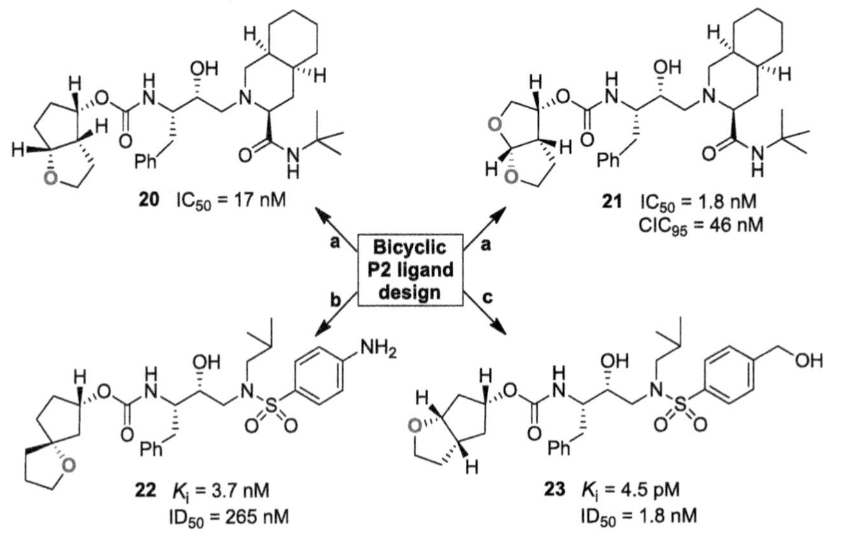

FIGURE 13.10

Design of bicyclic P2 ligands. (For colour version of this figure, the reader is referred to the online version of this book.)

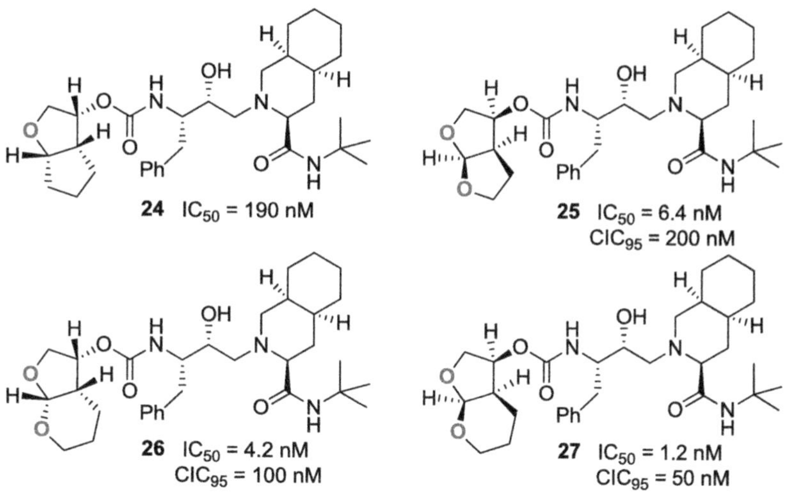

FIGURE 13.11

PIs with various bicyclic ligands. (For colour version of this figure, the reader is referred to the online version of this book.)

Of particular interest, incorporation of the *bis*-THF ligand into these sulfonamide-based isosteres also resulted in exceedingly potent inhibitors, which will be described later. To establish the importance of both oxygens on the *bis*-THF rings, we have prepared a ligand analogue that only contained the top cyclic ether oxygen on the bicyclic structure. As shown in Figure 13.11, the corresponding inhibitor **24** with SQV-based isostere showed substantial loss of potency (*42*). To ascertain the importance of ring stereochemistry, we have prepared the enantiomeric *bis*-THF ligand and synthesized the corresponding inhibitor **25**. This inhibitor showed reduced enzyme inhibitory and antiviral potency compared to **21**. Similarly, inhibitors **26** and **27** with hexahydrofuropyran ligands showed reduced potencies compared to **21**. Thus, it was apparent that both ether oxygens on the *bis*-THF rings in **21** were critical to the observed potency.

Indeed, the X-ray crystal structure of the **21**-bound HIV-1 protease revealed that the *bis*-THF ligand was optimally located in the protease active site (*42,43*). Both *bis*-THF oxygens were nicely lined up within hydrogen bonding distance to the main-chain NH amide bonds of the Asp29 and Asp30 residues (*42*).

13.6 DESIGN STRATEGY TO COMBAT DRUG RESISTANCE BY TARGETING PROTEIN BACKBONE: 'BACKBONE BINDING CONCEPT'

Despite the major clinical benefits obtained with PIs, drug resistance has quickly emerged and compromised long-term effectiveness of PI-based regimens. Among all antiretroviral enzyme targets, the HIV-1 protease exhibits the highest rate of mutations (*44*). Nearly 45 of the 99 residues in HIV protease monomer can undergo mutations that are relevant to clinical resistance to current

PIs (*45–47*). Nearly 20 years of research on mutant proteases and PI-specific resistance has provided critical insights into the acquisition and structural implications of drug resistance-associated mutations (*46*). Protease mutations can be distinguished into several categories. Active site mutations are generally primary resistance mutations and involve protease residues that stand in the direct vicinity of the PI's binding site. These mutations disrupt the PI's interactions with the protease and decrease its binding affinity. Emergence of active site mutations is typically specific to a PI, although many induce cross-resistance with other PIs (*48*). Numerous X-ray crystallographic studies on mutant protease complexes provided critical insights into how mutations can evolve at residues that are not well conserved. At the same time, they revealed domains that are highly conserved. Mutations of 37 protease residues out of 99 are considered very rare. Residues Asp25, Gly27, Ala28, Asp29, and Gly49 are particularly highly conserved within the protease-binding site (*47,49*). In regard to these observations, we considered that targeting conserved residues of the protease as well as domains that are least prone to structural alteration may provide a viable strategy for developing PIs that can effectively evade drug resistance.

The development of the first-generation HIV-1 PIs has provided an incredible wealth of information and knowledge regarding structure–activity relationships. The X-ray crystallographic information showing binding interactions in the protease active site has guided new drug design strategies to combat drug resistance. Our initial structure-based design strategy was to identify and target the conserved areas shared by mutant and wild-type proteases. We have notably compared the crystal structures of various mutant protease–inhibitor complexes with those of the wild-type protease. Superimposition of these structures consistently revealed minimal distortion of the protease backbone around the enzyme's active site (*49–52*). Despite structural rearrangements, conformational shifts, or loss of dimer stability observed with highly mutated proteases, superimposition of these structures consistently demonstrated conservation of the protein backbone conformation at the base of the enzyme's binding site, extending from the S2 to S2′ subsites around the catalytic dyad. The crystal structure of a multi-PI-resistant protease containing 20 mutations revealed a substantial deformation of its external structure, and its flaps adopted a larger than usual open conformation in its native state (*53*). Yet, superimposition of its crystal structure with that of the wild-type protease still indicated a near-identical match of Cα backbone carbon atoms for both enzymes at the base of the binding site. Figure 13.12 illustrates how other superimposed structures of several drug-resistant HIV-1 protease variants containing 10–14 mutations and even HIV-2 protease, that differs by 40 residues, show minimal deviation of the backbone atoms in the active site (*52*). This observation is relevant to the fact that the protease cannot extensively alter its overall structure around the active site without compromising its catalytic fitness necessary for viral replication (*15,45,46*).

Based on these observations, we hypothesized that an inhibitor that maximizes interactions, particularly hydrogen bonds, with backbone atoms in the protease's active site would likely conserve these interactions with mutant proteases (*15*). This 'backbone binding concept' would result in inhibitors that would maintain high binding affinity with mutant enzymes and thereby retain their inhibitory potency. Using this design concept, we have created a variety of novel inhibitors that form extensive interactions with the backbone atoms of highly conserved protease backbone residues. The resulting inhibitors have been shown to maintain near-full antiviral activity against a panel of multidrug-resistant HIV-1. The following section will describe the culmination of these designs with the discovery of DRV (*19*).

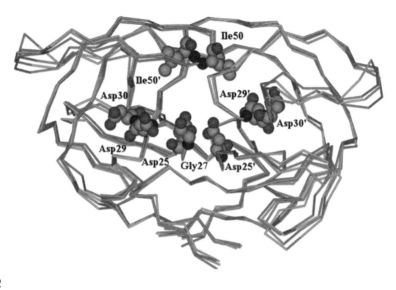

FIGURE 13.12

Superimposition of X-ray crystal structures of mutant proteases' backbones. (See colour plate.)

13.7 DESIGN OF PIs PROMOTING STRONG BACKBONE INTERACTIONS FROM S2 TO S2′ SUBSITES

Our research objective remained to address the critical issues of HIV-1 drug resistance by developing inhibitors that could maintain potency against various mutant HIV-1 proteases. Our initial designs of PIs containing cyclic ether P2 ligands were based on the SQV-derived hydroxyethylene isostere. As described earlier, X-ray structure of **21**-bound HIV-1 protease revealed that the *bis*-THF ligand was involved in the formation of robust hydrogen bonding interactions with the Asp29 and Asp30 backbone NHs in the S2 subsite (*42*). However, the X-ray structure also showed that the SQV isostere would not interact with protease backbone atoms in the S2′ subsite (*24*). We logically envisioned structural modification of the SQV isostere so as to introduce additional backbone binding interactions in the S2′ subsite and we explored suitability of the hydroxyethyl (arylsulfonamide) isostere for the introduction of new functionalities that could create such inter-actions. Similar to cyclic ether ligands, we first considered alkyl ether substituents on the aryl ring as these could similarly induce hydrogen bonding interactions with the backbone NH bonds of both Asp29′ and Asp30′ residues. We first incorporated the *bis*-THF ligand in a hydroxyethyl (4-methoxyphenylsulfonamide)isostere which contained a *p*-methoxyphenylsulfonamide as the P2′ ligand (*54*). As shown in Figure 13.13, the resulting inhibitor **28** displayed impressive enzyme inhibitory potency ($K_i = 14$ pM vs $K_i = 0.12$ nM for SQV, in-house enzyme inhibitory assay) and superb antiviral activity ($IC_{50} = 1.3$ nM in MT-4 cells). The corresponding *p*-toluenesulfonamide analogue **29** showed significant reduction of potency (a, $K_i = 1.2$ nM, $ID_{50} = 3.5$ nM; SQV showed a $K_i = 1.2$ nM in this assay) compared to the methoxy derivative **28** (*54*). This result suggested that the methoxy group in **28** may have been involved in additional interactions in the S2′ subsite of the

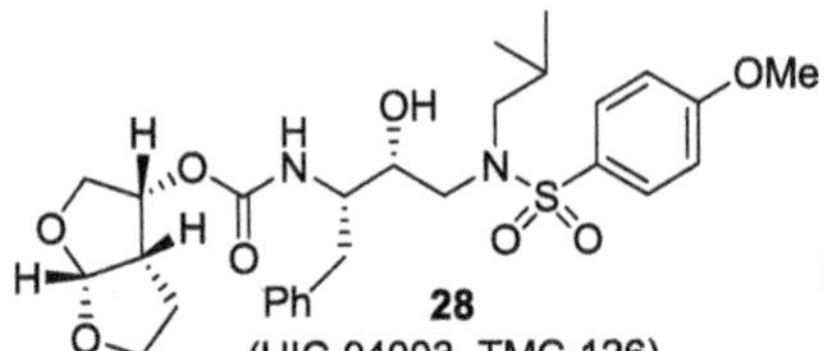

28
(UIC-94003, TMC-126)
K_i = 14 pM, IC_{50} = 1.4 nM

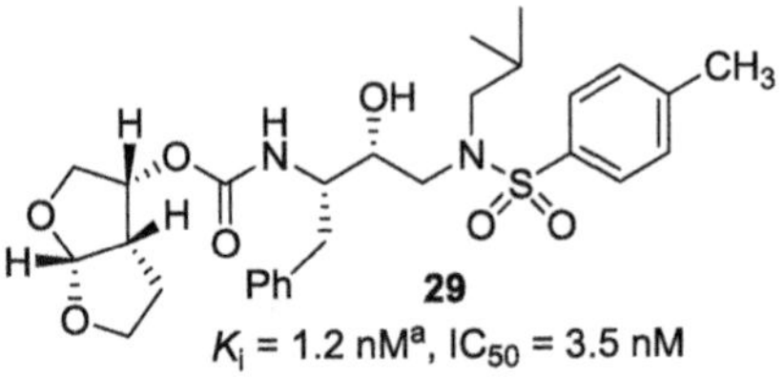

29
K_i = 1.2 nM[a], IC_{50} = 3.5 nM

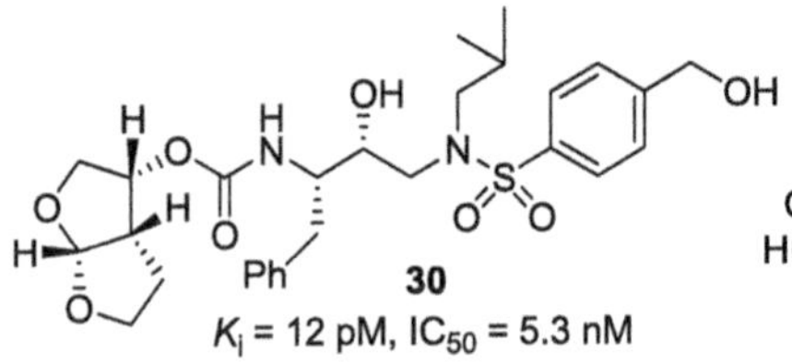

30
K_i = 12 pM, IC_{50} = 5.3 nM

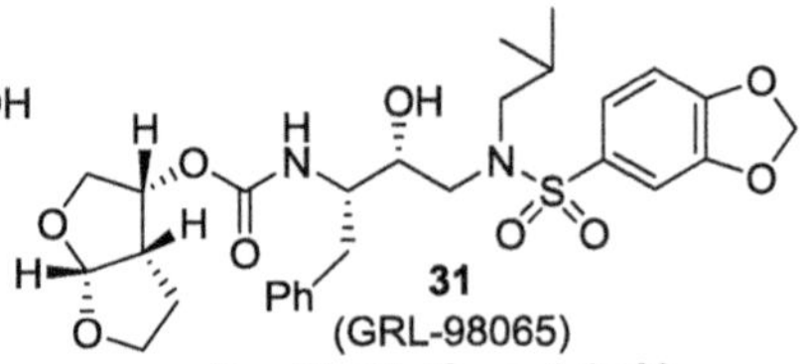

31
(GRL-98065)
K_i = 14 pM, IC_{50} = 1.4 nM

FIGURE 13.13

Structures of novel PIs based on hydroxyethyl(arylsulfonamide) isosteres. (For colour version of this figure, the reader is referred to the online version of this book.)

HIV-1 protease (*54*). We also speculated that a variety of other functionalities such as benzodioxalane-, benzimidazole-, benzodioxane-, benzyl alcohol-, or aniline-derived sulfonamides could form strong hydrogen bonds with backbone atoms in the S2′ subsite. Indeed, inhibitors **30**, **31** and **1** containing *p*-(hydroxymethyl)phenyl, benzo[*d*][1,3]dioxole, and *p*-aminophenyl sulfonamides, respectively, as the P2′ ligand were all exceedingly potent (*55*).

13.8 THE 'BACKBONE BINDING CONCEPT' AND ITS RELEVANCE TO COMBAT DRUG RESISTANCE

Inhibitor **28**, later renamed TMC-126, was tested against a panel of HIV-1 clinical isolates including mutant variants and its activity was compared with those of clinically available PIs (*56,57*). Against wild-type HIV, its antiviral activity was very impressive (IC_{50} range 0.3–0.5 nM) and it surpassed all previously approved PIs (*57*). In an effort to gain molecular insights into the ligand-binding site interactions, we determined a high-resolution X-ray crystal structure of the **28**-bound HIV-1 protease complex (*58*). As shown in Figure 13.14, inhibitor **28** made extensive interactions throughout the HIV-1 protease active site. Most significantly, both oxygen atoms of the *bis*-THF P2 ligand formed strong hydrogen bonds with the Asp29 and Asp30 backbone NHs in the S2 subsite. The *p*-methoxy oxygen on the P2′ sulfonamide ligand also formed strong hydrogen bonds with the backbone NH of Asp30′ as well as with the Asp30′side chain carboxylate in the S2′ subsite (*58*). Furthermore, the P1 and P1′ benzyl and isobutyl groups appeared to ideally pack the S1 and S1′ hydrophobic subsites effectively.

In essence, inhibitor **28** has been shown to make strong backbone interactions with conserved residues from the S2 to S2′ subsites of the protease-binding site. Based on our stated assumptions, such an inhibitor should retain activity against mutant proteases. Indeed, as shown in Table 13.1, enzyme

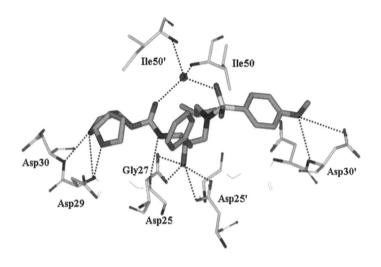

FIGURE 13.14

X-ray crystal structure of inhibitor **28**–HIV-1 protease complex. (See colour plate.)

inhibitory potency of **28** remained less than 100 pM and fold changes in potency ($K_{i\ mut}/K_{i\ wt}$) were no greater than five against mutant proteases known to be resistant to several first-generation PIs (*59,60*). As shown in Table 13.2, in subsequent antiviral studies involving multi-PI-resistant HIV-1 strains isolated from patients, the activity of **28** did not vary substantially (IC$_{50}$ = 0.5–5 nM), whereas all other PIs exhibited a high level of drug resistance (from 33 nM to >1 μM) (*57*). As it turned out,

Table 13.1 Inhibitory Potency of **28** against Wild-Type and Mutant Proteases

Enzyme	K_i (pM)	$K_{i\ wt}/K_{i\ mut}$	Vitality*
WT	14	1	1
D30N	<5	0.33	0.3
V32I	8	0.57	0.5
I84V	40	2.85	1
V32I/I84V	70	5	0.7
M46F/V82A	<5	0.33	0.1
G48V/L90M	<5	0.33	0.1
V82F/I84V	7	0.5	0.1
V82T/I84V	22	1.57	0.1
V32I/K45I/F53L/A71V/I84V/L89M	31	2.2	0.1
V32I/L33F/K45I/F53L/A71V/I84V	46	3.3	0.1
20R/36I/54V/71V/82T	31	2.2	0.1

*The vitality is a measure of the enzymatic fitness of a mutant protease in the presence of an inhibitor, here **28**, (determined as $(K_i{}^*k_{cat}/K_m)_{mut}/(K_i{}^*k_{cat}/K_m)_{wild}$).

Table 13.2 Sensitivities of **28** against HIV-1 Isolated From Heavily Drug-experienced Patients

Virus*	IC$_{50}$ µM (fold change)					
	RTV	**IDV**	**SQV**	**NFV**	**APV**	**28**
WT	0.044 (1)	0.013 (1)	0.010 (1)	0.023 (1)	0.025 (1)	0.0007 (1)
1	>1 (>23)	>1 (>77)	0.27 (27)	>1 (>43)	0.27 (11)	0.004 (6)
2	>1 (>23)	0.49 (38)	0.037 (4)	0.33 (14)	0.28 (11)	0.0013 (2)
3	>1 (>23)	0.49 (38)	0.036 (4)	>1 (>43)	0.26 (10)	0.001 (1)
4	>1 (>23)	0.21 (16)	0.033 (3)	0.09 (4)	0.31 (12)	0.0016 (2)
5	>1 (>23)	>1 (>77)	0.31 (31)	0.41 (18)	0.67 (27)	0.0024 (3)
6	>1 (>23)	0.30 (23)	0.19 (19)	>1 (>43)	0.16 (6)	0.0005 (1)
7	>1 (>23)	>1 (>77)	0.12 (12)	>1 (>43)	0.49 (20)	0.0055 (8)
8	>1 (>23)	0.55 (42)	0.042 (4)	>1 (>43)	0.15 (6)	0.001 (1)

IDV, indinavir; NFV, nelfinavir; RTV, ritonavir.
Amino acid substitutions identified in the protease-encoding regions of viruses compared to the consensus sequence cited from the Los Alamos database. See Ref. (57) for details.

inhibitor **28** (later renamed TMC-126) displayed a high genetic barrier to drug resistance acquisition and *in vitro* selection of HIV-1 strains resistant to this inhibitor proved difficult to attain (*57*). We attributed this resilience towards drug resistance and impressive activities against a wide spectrum of drug-resistant HIV variants to the extensive binding interactions that **28** makes with the protease, particularly through a network of hydrogen bonding interactions with the backbone atoms in the S2 to S2′ subsites.

13.9 SELECTION OF DARUNAVIR AS A PROMISING DRUG CANDIDATE

As described previously, the combination of the *bis*-THF ligand with (*R*)-(hydroxyethyl) sulfonamide isosteres containing a variety of P2′ sulfonamide functionalities capable of interacting with the backbone atoms in the S2′ subsite led to exceptionally potent PIs. PIs **30**, **31** and **1** have all shown marked drug resistance properties comparable to **28**. However, inhibitor **1** (Figure 13.15, later named TMC-114, then DRV) with a *p*-aminophenylsulfonamide P2′ ligand offered superior pharmacokinetic properties compared to **28**. DRV also displayed favourable pharmacological properties and drug resistance profiles over other PIs. Subsequently, this inhibitor was selected for clinical development. *In vitro* assays showed that DRV is extensively metabolized by cytochrome P450 isozyme CYP3A4. Ritonavir (RTV) is a first-generation PI that has also long been known to be an excellent ligand of CYP3A4 enzyme. Its coadministration at low doses with other PIs has been exploited to improve and maintain therapeutic levels of a PI in patients and therefore promote 'pharmacoenhancement' (*61*). A twice-a-day (b.i.d.) 600/100 mg DRV/RTV (DRV/r) dose increased the oral bioavailability of DRV in healthy human volunteers by up to 82% as opposed to 37% when administered alone, and systemic exposure increased by 14-fold (*62*).

DRV exhibited impressive antiviral activity against a broad range of HIV-1 viral strains. This PI effectively inhibited the replication and infectivity of various wild-type HIV-1 and HIV-2 isolates at

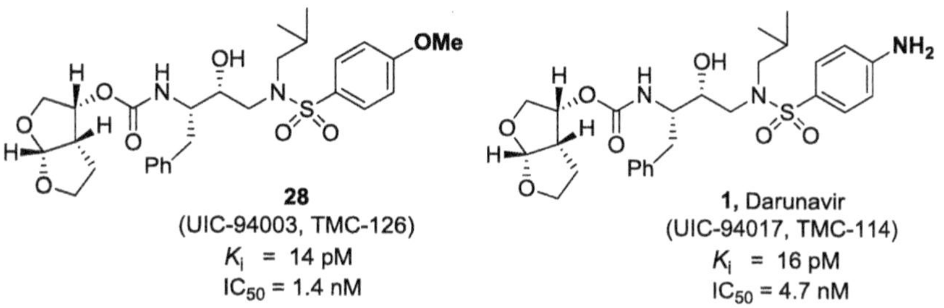

FIGURE 13.15

Structures of TMC-126 (**28**) and TMC-114 (**1**).

very low concentrations (IC$_{50}$ = 0.003 ± 0.0001 μM). In comparison, six clinically approved PIs tested showed 6- to 16-fold higher IC$_{50}$ values (*13*). Although DRV showed high potency, its cytotoxicity proved to be relatively low with a Concentration of inhibitor inducing 50% cytotoxicity(CC$_{50}$)/half maximal effective concentration(EC$_{50}$) selectivity index > 20,000 (*13*). More importantly, DRV consistently exhibited superb antiviral activity against a broad range of HIV-1 variants with resistance-associated protease mutations. As shown in Table 13.3, DRV consistently maintained excellent antiviral activity (IC$_{50}$ = 0.003–0.029 μM) when tested against a series of HIV-1 strains, which were selected for their resistance against SQV, APV, indinavir (IDV), nelfinavir (NFV), or RTV after exposure of the virus to increasing concentrations of the corresponding PI. Only the APV-resistant strain induced a more substantial loss of activity of DRV (fold change in IC$_{50}$ = 73), probably due to the structural similarities between these two inhibitors (*13*). DRV also exhibited excellent antiviral activity against a panel of highly multi-PI-resistant HIV-1 variants

Table 13.3 Activity of Darunavir against HIV-1 Clinical Isolates in Phytohemagglutinin-activated Peripheral Blood Mononuclear Cells (PHA-PBMC) Infected Cells*

Virus	IC$_{50}$ Values (μM)					
	SQV	APV	IDV	NFV	RTV	DRV (1)
Wild Type (WT)	0.010	0.023	0.018	0.019	0.027	0.003
1	0.004	0.011	0.018	0.033	0.032	0.003
2	0.23 (23)	0.39	>1 (>56)	0.54 (28)	>1 (>37)	0.004 (1)
3	0.30 (30)	0.34	>1 (>56)	>1 (>53)	>1 (>37)	0.02 (7)
4	0.35 (35)	0.75 (33)	>1 (>56)	>1 (>53)	>1 (>37)	0.029 (10)
5	0.14 (14)	0.16 (7)	>1 (>56)	0.36 (19)	>1 (>37)	0.004 (1)
6	0.31 (31)	0.34 (15)	>1 (>56)	>1 (>53)	>1 (>37)	0.013 (4)
7	0.037 (4)	0.28 (12)	>1 (>56)	0.44 (23)	>1 (>37)	0.003 (1)
9	0.029 (3)	0.25 (11)	0.39 (22)	0.32 (17)	0.44 (16)	0.004 (1)

*Amino acid substitutions identified in the protease-encoding regions of viruses compared to the consensus sequence cited from the Los Alamos database. See Ref. (13) for details.

isolated from AIDS patients that previously failed anti-HIV regimens after receiving 9–11 anti-HIV drugs (*13*). All variants showed 9–14 resistance-associated mutations of the protease (PR) at baseline. Against all viruses, DRV displayed IC_{50}s of 3–30 nM and fold changes in activity of less than 10, whereas all other PIs (SQV, APV, IDV, NFV, and RTV) showed substantially reduced or even no activity at all. In subsequent studies involving 19 recombinant HIV-1 clinical isolates that displayed multidrug resistance to an average of five PIs, fold changes in DRV's IC_{50} remained less than 4 (*63*).

The impressive ability of DRV to effectively suppress multidrug-resistant HIV-1 variants was further evaluated in more elaborate studies. Using high-throughput cell-based assays, DRV was tested against 1501 recombinant clinical HIV-1 isolates that possessed resistance to at least one PI (EC_{50} fold change > 4). From these assays, 80% of viral samples retained susceptibility to DRV (EC_{50} fold change < 4), while only 50% remained sensitive to other PIs (*63*). These results further illustrated the superior resistance profile of DRV compared to other PIs and emphasized its excellent clinical potential to treat patients harbouring multidrug-resistant HIV-1 strains.

In addition to treating drug-resistant HIV-1 infection, preventing the onset of drug resistance during PI treatment represents one of the most compelling challenges of HIV-1 therapies. Compared to other PIs, DRV exhibited an exceptionally high genetic barrier to the development of drug resistance. *In vitro* selection of HIV-1 variants resistant to DRV by exposing a wild-type laboratory HIV-1 strain to increasing concentrations of the PI proved difficult to achieve and the virus failed to acquire resistance to DRV. Multiple passages at low concentration of DRV (<200 nM) were necessary to select viruses with some level of resistance (*63*). In subsequent studies, only when *in vitro* selection was initiated with multidrug-resistant HIV isolates at baseline as opposed to wild-type strains, the viruses could effectively acquire *in vitro* resistance to DRV (*64*).

The X-ray crystal structure of DRV-bound HIV-1 protease was determined at high resolution. As shown in Figure 13.16, the structure revealed a number of important interactions between DRV and backbone atoms of the HIV-1 protease (*65*). The *bis*-THF P2 ligand forms strong hydrogen bonds with the Asp29 and Asp30 backbone amide NHs, anchoring DRV in the S2 subsite. The *p*-aminophenylsulfonamide functionality, on the other side, interacts with the backbone amide of Asp30′ and its carboxylic acid side chain through the amine functionality.

The central alcohol functionality of DRV serves as a transition state mimic and forms hydrogen bonds with the catalytic residues Asp25 and Asp25′. A tetra-coordinated water molecule mediates hydrogen bonding interactions between Ile50 and Ile50′ in the flaps and the urethane carbonyl and sulfonamide oxygen of DRV. The urethane NH of DRV forms a strong hydrogen bond with the Gly27 backbone carbonyl, while the P1′ isobutyl and P1 benzyl groups further enhance binding through hydrophobic interactions (*65*). These extensive interactions effectively allowed DRV to tightly clutch the protease backbone and may be a main contributing factor for DRV's impressive antiviral profile against multidrug-resistant HIV-1 variants.

13.9.1 Thermodynamic and kinetic effects behind darunavir's high binding affinity for the protease

The impressive antiviral activity and resistance profile of DRV has attracted a lot of research interest to elucidate its mode of binding at the molecular level. Thermodynamic studies have shown that, unlike most other PIs, DRV's binding to the wild-type protease is largely enthalpically driven

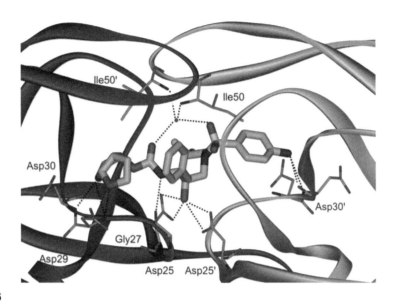

FIGURE 13.16

X-ray structure of DRV-bound HIV-1 protease complex. (See colour plate.)

($\Delta H = -12.1$ kcal mol^{-1}) (*66*). This unique feature likely originates from the numerous hydrogen bonding interactions that DRV creates in the protease active site, especially via its high-affinity *bis*-THF ligand. A study of the energetic impact of protease mutations on DRV's binding affinity showed that the *bis*-THF mostly retained strong interactions with the S2 subsite, while the isobutyl and arylsulfonamide groups most likely sustained the strongest loss of binding affinity (*66*). Kinetic studies have also demonstrated that DRV's dissociation rate is extremely small, several orders of magnitude lower than that of most other PIs, and provided additional evidences to explain DRV's superior antiviral activity (*67*). DRV's inherent binding domain being entrenched into the natural substrate's binding envelope has also been considered to likely explain the inhibitor's exceptional ability to escape the onset of drug resistance (*18*). Interestingly, an ultrahigh-resolution X-ray crystal structure analysis of the DRV-bound protease and kinetic studies have also suggested a possible second, allosteric binding site of DRV situated on the outer surface of the protease flap (*65,68*).

13.9.2 Darunavir inhibits HIV-1 protease dimerization: a unique dual mode of action

As mentioned in Section 13.2, the HIV protease is a homodimeric enzyme that results from dimerization of two identical 99-amino acid peptide subunits. Dimerization is an essential step to the formation of catalytically viable protease enzyme and inhibition of this process has been considered as a possible therapeutic option to ward off viral replication (*69,70*). DRV's impressive antiviral activity, superior resistance profile, and high genetic barrier to emergence of drug resistance partly stem from its unique binding mode and strong affinity for the HIV-1 protease. We also discovered that DRV inhibits dimerization of the protease. A set of HIV-1 expression assay experiments utilizing fluorescent protein-tagged HIV-1 protease monomers and intermolecular fluorescence resonance energy transfer (FRET) analysis showed that DRV inhibits the protease dimerization at concentrations as low as 0.01 μM (*71*). In comparison, no other clinically available PI except tipranavir (TPV) exhibited the same ability. Interestingly, this DRV-mediated inhibition mechanism only seemed to involve the protease at its nascent stage since DRV could not promote dissociation of the dimer once the enzyme fully formed. A subsequent study established that HIV-1 clonal variants highly resistant to DRV, which contained a combination of four DRV resistance-associated protease mutations, led to loss of DRV's ability to inhibit protease dimerization. These results strongly suggested that DRV's antiviral activity in a clinical context may in fact also rely on its inhibition of protease dimerization (*71,72*). In order to get insights into the origin of this mechanism, molecular dynamic simulations of DRV's binding to the protease monomer have recently been conducted. Results suggested that DRV's binding location was close to the original protease binding site and that it inhibits dimerization by disrupting the flap's interactions between the two monomers (*73*). This unique dual mode of action may explain DRV's impressive genetic barrier to drug resistance.

13.10 CONVENIENT SYNTHESES OF THE *BIS*-THF LIGAND

Our first synthesis of the chiral *bis*-THF ligand starting from (S)- or (R)-malic acid was not suitable for large-scale preparation of the ligand for the exploration of PI analogues (*43*). We subsequently developed a practical and easily scalable three-step synthesis of the racemic *bis*-THF ligand (Figure 13.17) (*74*). A subsequent Lipase-PS Amano-mediated optical resolution of the racemic ligand alcohol provided gram quantities of the optically active *bis*-THF ligand **38** in high

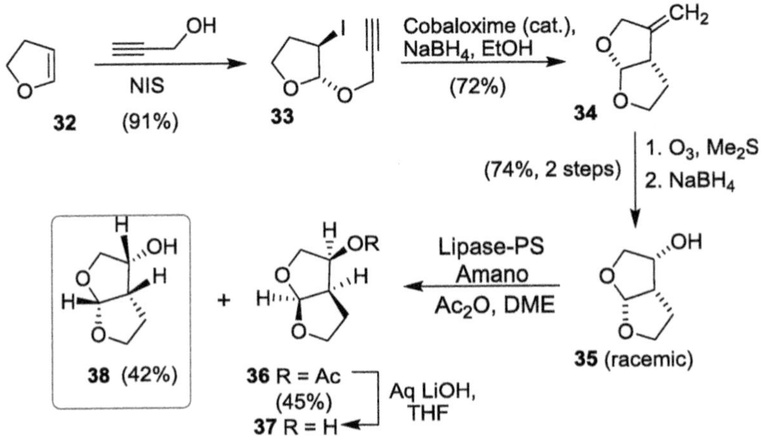

FIGURE 13.17

Practical synthesis of optically active bis-THF ligand. (For colour version of this figure, the reader is referred to the online version of this book.)

enantiomeric purity (enantiomeric excess (ee) > 96%). We have subsequently developed a number of other asymmetric synthetic routes to get optically active *bis*-THF (*75–77*).

13.11 CLINICAL USE OF DARUNAVIR IN THE MANAGEMENT OF HIV-1 INFECTION

Since its accelerated approval in 2006, DRV has become the PI of choice in the treatment of highly drug-experienced patients. In combination with other novel investigational drugs, it has even provided successful salvage therapies to patients with little or no treatment options left. Together with the second-generation PI, TPV, DRV provides a new means to combat HIV-1 drug resistance. However, DRV has become a much preferred therapeutic option with respect to its superior tolerability and fewer adverse effects. Many HIV-1 treatment guidelines now widely recommend DRV. Since the expansion of DRV's approval to treatment-naïve adult and paediatric populations, use of this PI has become increasingly widespread. With its convenient once-a-day RTV-boosted 800/100 mg dosing, DRV is now offering durable treatment for first-line therapy with good tolerability and low probability for the development of drug resistance. Although DRV-based treatment still comes at higher cost compared to first-generation PIs, its superior efficacy and ability to delay the onset of drug resistance offers a clear long-term advantage in terms of cost-effectiveness. Recent studies concluded that, compared to other PIs, DRV/r 600/100 mg b.i.d. treatment may actually offer a more favourable economic profile (*78*).

13.12 DESIGN OF POTENT INHIBITORS TARGETING THE PROTEASE BACKBONE

Based on inhibitor **28** (TMC-126) framework featuring the *bis*-THF P2 ligand, numerous other PIs have been developed. As shown in Figure 13.18, GlaxoSmithKline developed a new PI, brecanavir (**39**, GW0385) that combined the *bis*-THF ligand with a P1′-P2′-modified hydroxyethyl(sulfonamide) isostere. This PI exhibited exceptional potency ($K_i = 15$ fM) and very low nanomolar antiviral activities against both wild-type and mutant HIV-1 strains (*79*). This inhibitor underwent phase III clinical evaluation but was subsequently discontinued due to 'unsurmountable formulation issues'. Gilead Sciences developed TMC-126 analogue **40**, that contained a 4-diethylphosphonylmethoxy substituent on the phenyl P1 ligand (Figure 13.18) (*80*). The phosphonate moiety was introduced in order to promote better intracellular retention. This PI displayed excellent antiviral activity, higher resilience to drug resistance, and lower cross-resistance susceptibility compared to other PIs, including DRV. To explain this unexpected mechanism of drug resistance evasion, it was proposed that the phosphonate group would act as a solvent-anchoring point that triggers more favourable binding entropy compensation when binding to mutant proteases (*81*).

Employing our backbone binding concept to combat drug resistance, we have continued to design new PIs that could create additional interactions with the protease backbone (*14,15*). Inhibitors **41** and **42** were designed with a novel hexahydrofuropyran P2 ligand in an attempt to create closer hydrogen bonds with the Asp30 residue (*82,83*). Both inhibitors exhibited excellent antiviral potencies against a panel of multidrug-resistant HIV-1 viruses that matched and even surpassed those of DRV. We have

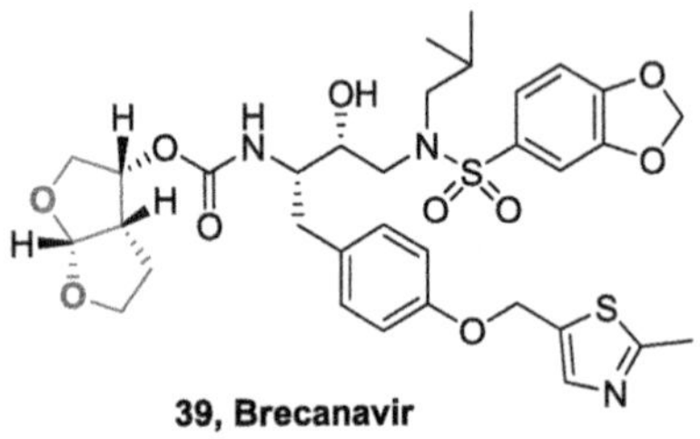

39, Brecanavir
$K_i = 15$ fM, $IC_{50} = 0.7$ nM

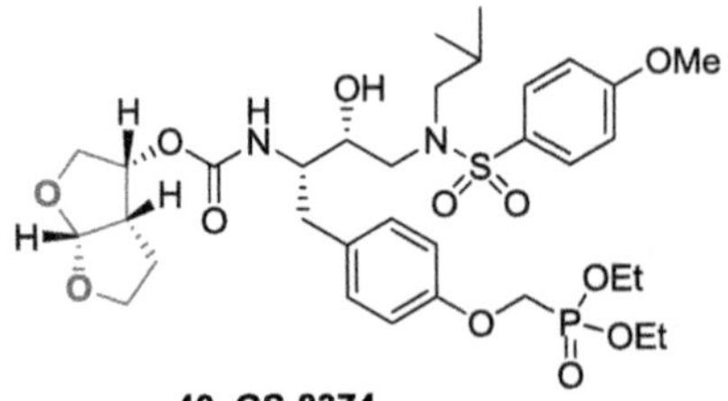

40, GS-8374
$K_i = 8.1$ pM, $EC_{50} = 3.5$ nM

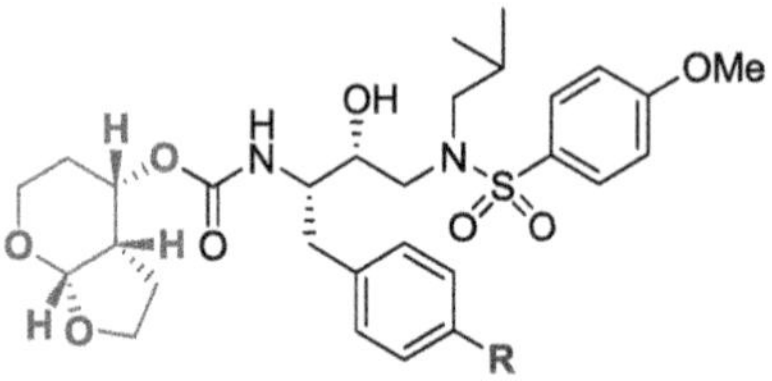

41 GRL-0476 (R = H), $IC_{50} = 0.5$ nM
42 GRL-1388 (R = OMe), $IC_{50} = 0.2$ nM

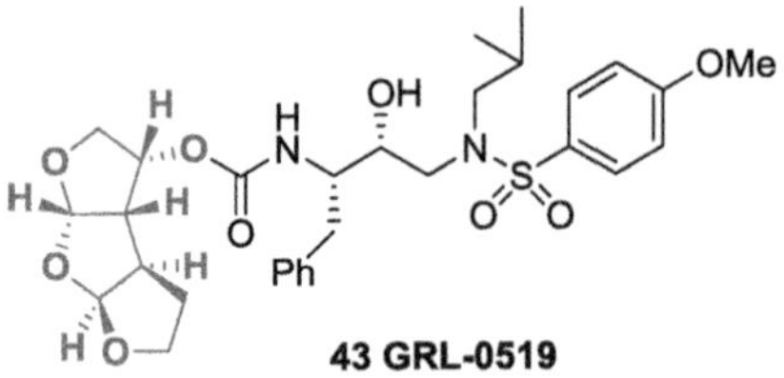

43 GRL-0519
$K_i = 5.9$ pM, $IC_{50} = 1.8$ nM

FIGURE 13.18

Structures of novel protease inhibitors. (For colour version of this figure, the reader is referred to the online version of this book.)

also designed tris-THF-based inhibitor **43** in order to promote additional interactions in the S2 subsite. Most notably, this PI exhibited almost 10-fold improvement of potency against highly resistant clinical HIV-1 strains compared to DRV. It also inhibited protease dimerization at least 10-fold more effectively compared to DRV (*84*).

13.13 CONCLUSION

The advent of HAART therapies has dramatically improved the treatment of HIV-1 infection/AIDS and markedly decreased morbidity and mortality, especially in industrialized nations. However, current treatment regimens have been hampered by numerous issues. Among those, the rapid emergence of drug resistance and cross-resistance has seriously complicated the long-term management of HIV/AIDS. Our motivation to address this urgent concern has propelled our research on designing novel protease inhibitors with improved resistance profiles and enhanced pharmacological properties. Our design strategy to target the protein backbone to combat drug resistance stemmed from our observations that the protease backbone conformation remains relatively uniform despite the occurrence of multiple viral mutations. Our 'backbone binding' approach has resulted in conceptually new classes of exceedingly potent PIs that maintained potency against a broad spectrum of multidrug-resistant HIV-1 strains. For our structure-based design, we continually draw inspiration from bioactive natural products and their seemingly limitless structural features. Our research endeavours led to the development of unprecedented cyclic/polycyclic ether-derived molecular templates that could effectively mimic peptide bonds and function as high-affinity ligands and scaffolds for a new generation of PIs. Incorporation of these privileged ligands resulted in much improved pharmacological properties.

These design strategies culminated in the discovery of DRV, a new-generation protease inhibitor that provides fresh hope in the combat against HIV-1 drug resistance. In addition to its superb antiviral potency, DRV displays a unique dual mode of action, inhibiting protease dimerization as well as the catalytic activity of the HIV-1 protease. The excellent resistance profile of DRV has generated immense interest regarding the origin of its exceptionally high affinity for the protease and its unique mechanism to evade drug resistance. The accelerated FDA approval of this inhibitor has been a milestone in the combat against multidrug-resistant HIV-1 infections. DRV has raised the standard for the development of new antiretroviral treatments and their clinical applications. Our 'backbone binding' concept has fuelled our continuing efforts to develop more potent inhibitors and may prove to be a useful strategy in the development of new and improved treatments in other areas of medicine.

Acknowledgements

Financial support by the National Institutes of Health (GM53386) is gratefully acknowledged. We thank Dr K.V. Rao (Purdue University) for helpful discussions.

References

The most important references are indicated with an asterisk as*

1. UNAIDS/WHO. *Global HIV/AIDS Response – Progress Report 2011*, November 2011. http://www.unaids.org/en/resources/publications/2011/ (accessed May 2012).
2. *De Clercq, E. Anti-HIV Drugs: 25 Compounds Approved within 25 Years after the Discovery of HIV. *Int. J. Antimicrob. Agents* **2009**, *33*, 307–320.
3. Palella, F. J.; Delaney, K. M.; Moorman, A. C.; Loveless, M. O.; Fuhrer, J.; Satten, G. A.; Aschman, D. J.; Holmberg, S. D. Declining Morbidity and Mortality among Patients with Advanced Human Immunodeficiency Virus Infection. *New Engl. J. Med.* **1998**, *338*, 853–860.
4. Sepkowitz, K. A. AIDS – The First 20 Years. *N. Engl. J. Med.* **2001**, *344*, 1764–1772.
5. Waters, L.; Nelson, M. Why Do Patients Fail HIV Therapy? *Int. J. Clin. Pract.* **2007**, *61*, 983–990.
6. Menendez-Arias, L. Molecular Basis of Human Immunodeficiency Virus Drug Resistance: An Update. *Antiviral Res.* **2010**, *85*, 210–231.
7. Wensing, A. M. J.; van Maarseveen, N. M.; Nijhuis, M. Fifteen Years of HIV Protease Inhibitors: Raising the Barrier to Resistance. *Antiviral Res.* **2010**, *85*, 59–74.
8. Tang, J. W.; Pillay, D. Transmission of HIV-1 Drug Resistance. *J. Clin. Virol.* **2004**, *30*, 1–10.
9. Barbaro, G. Long-term Effects of Protease-Inhibitor-Based Combination Therapy. *Lancet* **2004**, *363*, 900–901.
10. Capeau, J. Premature Aging and Premature Age-Related Comorbidities in HIV-Infected Patients: Facts and Hypotheses. *Clin. Infect. Dis.* **2011**, *53*, 1127–1129.
11. *Ghosh, A. K. Harnessing Nature's Insight: Design of Aspartyl Protease Inhibitors from Treatment of Drug-Resistant HIV to Alzheimer's Disease. *J. Med. Chem.* **2009**, *52*, 2163–2176.
12. *Ghosh, A. K. Capturing the Essence of Organic Synthesis: From Bioactive Natural Products to Designed Molecules in Today's Medicine. *J. Org. Chem.* **2010**, *75*, 7967–7989.
13. *Koh, Y.; Nakata, H.; Maeda, K.; Ogata, H.; Bilcer, G.; Devasamudram, T.; Kincaid, J. F.; Boross, P.; Wang, Y.-F.; Tie, Y.; Volarath, P.; Gaddis, L.; Harrison, R. W.; Weber, I. T.; Ghosh, A. K.; Mitsuya, H. Novel *bis*-Tetrahydrofuranylurethane-Containing Nonpeptidic Protease Inhibitor PI UIC-94017 (TMC114) with

Potent Activity against Multi-PI-Resistant Human Immunodeficiency Virus in vitro. *Antimicrob. Agents Chemother.* **2003**, *47*, 3123–3129.

14. *Ghosh, A. K.; Chapsal, B. D.; Weber, I. T.; Mitsuya, H. Design of HIV Protease Inhibitors Targeting Protein Backbone: An Effective Strategy for Combating Drug Resistance. *Acc. Chem. Res.* **2008**, *41*, 78–86.

15. *Ghosh, A. K.; Anderson, D. D.; Weber, I. T.; Mitsuya, H. Enhancing Protein Backbone Binding – A Fruitful Concept for Combating Drug-Resistant HIV. *Angew. Chem. Int. Ed.* **2012**, *51*, 1778–1802.

16. On June 23, 2006, *FDA Approves Darunavir for Treatment-Experienced HIV-Infected Patients, and in December 13, 2008, For Treatment-Naive Patients and Pediatrics.* http://www.accessdata.fda.gov/drugsatfda_docs/label/2008/021976s009lbl.pdf (accessed May 2012).

17. Ghosh, A. K.; Dawson, Z. L.; Mitsuya, H. Darunavir, a Conceptually New HIV-1 Protease Inhibitor for the Treatment of Drug-Resistant HIV. *Bioorg. Med. Chem.* **2007**, *15*, 7576–7580.

18. Lefebvre, E.; Schiffer, C. A. Resilience to Resistance of HIV-1 Protease Inhibitors: Profile of Darunavir. *AIDS Rev.* **2008**, *10*, 131–142.

19. *Ghosh, A. K.; Chapsal, B. D.; Mitsuya, H. Darunavir, a New PI with Dual Mechanism: From a Novel Drug Design Concept to New Hope against Drug-Resistant HIV. *Aspartic Acid Proteases as Therapeutic Targets* In Ghosh, A. K., Ed. , Vol. 45; Wiley-VCH Verlag GmbH & Co. KGaA: Weiheim, Germany, 2010; pp 205–243.

20. Brik, A.; Wong, C.-H. HIV-1 Protease: Mechanism and Drug Discovery. *Org. Biomol. Chem.* **2003**, *1*, 5–14.

21. Gustchina, A.; Sansom, C.; Prevost, M.; Richelle, J.; Wodak, S. Y.; Wlodawer, A.; Weber, I. T. Energy Calculations and Analysis of HIV-1 Protease-Inhibitor Crystal Structures. *Protein Eng.* **1994**, *7*, 309–317.

22. Tie, Y.; Boross, P. I.; Wang, Y.-F.; Gaddis, L.; Liu, F.; Chen, X.; Tozser, J.; Harrison, R. W.; Weber, I. T. Molecular Basis for Substrate Recognition and Drug Resistance from 1.1 to 1.6 Å Resolution Crystal Structures of HIV-1 Protease Mutants with Substrate Analogs. *FEBS J.* **2005**, *272*, 5265–5277.

23. Roberts, N. A.; Martin, J. A.; Kinchington, D.; Broadhurst, A. V.; Craig, J. C.; Duncan, I. B.; Galpin, S. A.; Handa, B. K.; Kay, J.; Krohn, A. Rational Design of Peptide-Based HIV Proteinase Inhibitors. *Science* **1990**, *248*, 358–361.

24. Krohn, A.; Redshaw, S.; Ritchie, J. C.; Graves, B. J.; Hatada, M. H. Novel Binding Mode of Highly Potent HIV-Proteinase Inhibitors Incorporating the (R)-Hydroxyethylamine Isostere. *J. Med. Chem.* **1991**, *34*, 3340–3342.

25. Newman, D. J.; Cragg, G. M. Natural Products as Sources of New Drugs over the 30 Years from 1981 to 2010. *J. Nat. Prod.* **2012**, *75*, 311–335.

26. Li, J. W.-H.; Vederas, J. C. Drug Discovery and Natural Products: End of an Era or an Endless Frontier? *Science* **2009**, *325*, 161–165.

27. Ganesan, A. The Impact of Natural Products upon Modern Drug Discovery. *Curr. Opin. Chem. Biol.* **2008**, *12*, 306–317.

28. Clardy, J.; Walsh, C. Lessons from Natural Molecules. *Nature* **2004**, *432*, 829–837.

29. (a) Quinoa, E.; Kakou, Y.; Crews, P. Fijianolides, Polyketide Heterocycles from a Marine Sponge. *J. Org. Chem.* **1988**, *53*, 3642–3644. (b) Corley, D. G.; Herb, R.; Moore, R. E.; Scheuer, P. J.; Paul, V. J. Laulimalides. New Potent Cytotoxic Macrolides from a Marine Sponge and a Nudibranch Predator. *J. Org. Chem.* **1988**, *53*, 3644–3646. (c) Jefford, C. W.; Bernardinelli, G.; Tanaka, J.-I.; Higa, T. Structures and Absolute Configurations of the Marine Toxins, Latrunculin A and Laulimalide. *Tetrahedron Lett.* **1996**, *37*, 159–162. (d) Tanaka, J.-I.; Higa, T.; Bernardinelli, G.; Jefford, C. W. New Cytotoxic Macrolides from the Sponge. *Fasciospongia Rimosa. Chem. Lett.* **1996**, 255–256.

30. Nakanishi, K. Terpene Trilactones from *Gingko biloba*: From Ancient Times to the 21st Century. *Bioorg. Med. Chem.* **2005**, *13*, 4987–5000.

31. Palanichamy, K.; Kaliappan, K. P. Discovery and Syntheses of "Superbug Challengers" – Platensimycin and Platencin. *Chem. Asian J.* **2010**, *5*, 668–703.

32. Kevin, D. A.; Meujo, D. A. F.; Hamann, M. T. Polyether Ionophores: Broad-Spectrum and Promising Biologically Active Molecules for the Control of Drug-Resistant Bacteria and Parasites. Expert Opin. *Drug Discovery* **2009**, *4*, 109–146.

33. Towle, M. J.; Salvato, K. A.; Budrow, J.; Wels, B. F.; Kuznetsov, G.; Aalfs, K. K.; Welsh, S.; Zheng, W.; Seletsky, B. M.; Palme, M. H.; Habgood, G. J.; Singer, L. A.; DiPietro, L. V.; Wang, Y.; Chen, J. J.; Quincy, D. A.; Davis, A.; Yoshimatsu, K.; Kishi, Y.; Yu, M. J.; Littlefield, B. A. In Vitro and in Vivo Anticancer Activities of Synthetic Macrocyclic Ketone Analogues of Halichondrin B. *Cancer Res.* **2001**, *61*, 1013–1021.

34. Singh, S. B.; Ondeyka, J. G.; Herath, K. B.; Zhang, C.; Jayasuriya, H.; Zink, D. L.; Parthasarathy, G.; Becker, J. W.; Wang, J.; Soisson, S. M. Isolation, Enzyme-Bound Structure and Antibacterial Activity of Platencin A1 from *Streptomyces platensis. Bioorg. Med. Chem. Lett.* **2009**, *19*, 4756–4759.

35. *Ghosh, A. K.; Thompson, W. J.; Holloway, M. K.; McKee, S. P.; Duong, T. T.; Lee, H. Y.; Munson, P. M.; Smith, A. M.; Wai, J. M.; Darke, P. L.; Zugay, J. A.; Emini, E. A.; Schleif, W. A.; Huff, J. R.; Anderson, P. S. Potent HIV Protease Inhibitors: The Development of Tetrahydrofuranylglycines as Novel P$_2$-Ligands and Pyrazine Amides as P$_3$-Ligands. *J. Med. Chem.* **1993**, *36*, 2300–2310.

36. *Ghosh, A. K.; Thompson, W. J.; McKee, S. P.; Duong, T. T.; Lyle, T. A.; Chen, J. C.; Darke, P. L.; Zugay, J. A.; Emini, E. A.; Schleif, W. A.; Huff, J. R.; Anderson, P. S. 3-Tetrahydrofuran and Pyran Urethanes as High-Affinity P2-Ligands for HIV-1 Protease Inhibitors. *J. Med. Chem.* **1993**, *36*, 292–294.

37. Vazquez, M. L.; Bryant, M. L.; Clare, M.; Decrescenzo, G. A.; Doherty, E. M.; Freskos, J. N.; Getman, D. P.; Houseman, K. A.; Julien, J. A.; Kocan, G. P.; Mueller, R. A.; Shieh, H. S.; Stallings, W. C.; Stegeman, R. A.; Talley, J. J. Inhibitors of HIV-1-Protease Containing the Novel and Potent (*R*)-(Hydroxyethyl)sulfonamide Isostere. *J. Med. Chem.* **1995**, *38*, 581–584.

38. Tung, R. D.; Livingston, D. J.; Rao, B. G.; Kim, E. E.; Baker, C. T.; Boger, J. S.; Chambers, S. P.; Deininger, D. D.; Dwyer, M. D.; Elsayed, L.; Fulghum, J.; Li, B.; Murcko, M. A.; Navia, M. A.; Novak, P.; Pazhanisamy, S.; Stuber, C.; Thompson, J. A. Design and Synthesis of Amprenavir, a Novel HIV Protease Inhibitor. In *Protease Inhibitors in AIDS Therapy;* Ogden, R. C., Flexner, C. W., Eds.; Marcel Dekker, Inc.: New York, 2001; pp 101–137.

39. Ghosh, A. K.; Thompson, W. J.; Lee, H. Y.; McKee, S. P.; Munson, P. M.; Duong, T. T.; Darke, P. L.; Zugay, J. A.; Emini, E. A.; Schleif, W. A.; Huff, J. R.; Anderson, P. S. Cyclic Sulfolanes as Novel and High-Affinity P2-Ligands for HIV-1 Protease Inhibitors. *J. Med. Chem.* **1993**, *36*, 924–927.

40. *Ghosh, A. K.; Lee, H. Y.; Thompson, W. J.; Culberson, C.; Holloway, M. K.; McKee, S. P.; Munson, P. M.; Duong, T. T.; Smith, A. M.; Darke, P. L.; Zugay, J. A.; Emini, E. A.; Schleif, W. A.; Huff, J. R.; Anderson, P. S. The Development of Cyclic Sulfolanes as Novel and High-Affinity P2-Ligands for HIV-1 Protease Inhibitors. *J. Med. Chem.* **1994**, *37*, 1177–1188.

41. Ghosh, A. K.; Thompson, W. J.; Munson, P. M.; Liu, W.; Huff, J. R. Cyclic Sulfone-3-Carboxamides as Novel P2-Ligands for Ro 31-8959 Based HIV-1 Protease Inhibitors. *Bioorg. Med. Chem. Lett.* **1995**, *5*, 83–88.

42. *Ghosh, A. K.; Kincaid, J. F.; Walters, D. E.; Chen, Y.; Chaudhuri, N. C.; Thompson, W. J.; Culberson, C.; Fitzgerald, P. M. D.; Lee, H. Y.; McKee, S. P.; Munson, P. M.; Duong, T. T.; Darke, P. L.; Zugay, J. A.; Schleif, W. A.; Axel, M. G.; Lin, J.; Huff, J. R. Nonpeptidal P2 Ligands for HIV Protease Inhibitors: Structure-Based Design, Synthesis, and Biological Evaluation. *J. Med. Chem.* **1996**, *39*, 3278–3290.

43. Ghosh, A. K.; Thompson, W. J.; Fitzgerald, P. M. D.; Culberson, J. C.; Axel, M. G.; McKee, S. P.; Huff, J. R.; Anderson, P. S. Structure-Based Design of HIV-1 Protease Inhibitors: Replacement of Two Amides and a 10π−Aromatic System by a Fused *bis*-Tetrahydrofuran. *J. Med. Chem.* **1994**, *37*, 2506–2508.

44. Vergne, L.; Peeters, M.; Mpoudi-Ngole, E.; Bourgeois, A.; Liegeois, F.; Toure-Kane, C.; Mboup, S.; Mulanga-Kabeya, C.; Saman, E.; Jourdan, J.; Reynes, J.; Delaporte, E. Genetic Diversity of Protease and Reverse Transcriptase Sequences in Non-Subtype-B Human Immunodeficiency Virus Type 1 Strains:

Evidence of Many Minor Drug Resistance Mutations in Treatment-Naive Patients. *J. Clin. Microbiol.* **2000,** *38,* 3919–3925.

45. Weber, I.; Agniswamy, J. HIV-1 Protease: Structural Perspectives on Drug Resistance. *Viruses* **2009,** *1,* 1110–1136.

46. Ali, A.; Bandaranayake, R. M.; Cai, Y.; King, N. M.; Kolli, M.; Mittal, S.; Murzycki, J. F.; Nalam, M. N. L.; Nalivaika, E. A.; Ozen, A.; Prabu-Jeyabalan, M. M.; Thayer, K.; Schiffer, C. A. Molecular Basis for Drug Resistance in HIV-1 Protease. *Viruses* **2010,** *2,* 2509–2535.

47. *Stanford HIV drug resistance database.* http://HIVdb.stanford.edu/ (accessed Oct 2012).

48. Wu, T. D.; Schiffer, C. A.; Gonzales, M. J.; Taylor, J.; Kantor, R.; Chou, S.; Israelski, D.; Zolopa, A. R.; Fessel, W. J.; Shafer, R. W. Mutation Patterns and Structural Correlates in Human Immunodeficiency Virus Type 1 Protease Following Different Protease Inhibitor Treatments. *J. Virol.* **2003,** *77,* 4836–4847.

49. Hong, L.; Zhang, X. C.; Hartsuck, J. A.; Tang, J. Crystal Structure of an *in vivo* HIV-1 Protease Mutant in Complex with Saquinavir: Insights into the Mechanisms of Drug Resistance. *Protein Sci.* **2000,** *9,* 1898–1904.

50. Clemente, J. C.; Moose, R. E.; Hemrajani, R.; Whitford, L. R. S.; Govindasamy, L.; Reutzel, R.; McKenna, R.; Agbandje-McKenna, M.; Goodenow, M. M.; Dunn, B. M. Comparing the Accumulation of Active- and Nonactive-Site Mutations in the HIV-1 Protease. *Biochemistry* **2004,** *43,* 12141–12151.

51. Wang, W.; Kollman, P. A. Computational Study of Protein Specificity: The Molecular Basis of HIV-1 Protease Drug Resistance. *PNAS* **2001,** *98,* 14937–14942.

52. *Ghosh, A. K.; Sridhar, P. R.; Leshchenko, S.; Hussain, A. K.; Li, J.; Kovalevsky, A. Y.; Walters, D. E.; Wedekind, J. E.; Grum-Tokars, V.; Das, D.; Koh, Y.; Maeda, K.; Gatanaga, H.; Weber, I. T.; Mitsuya, H. Structure-Based Design of Novel HIV-1 Protease Inhibitors to Combat Drug Resistance. *J. Med. Chem.* **2006,** *49,* 5252–5261.

53. Agniswamy, J.; Shen, C. H.; Aniana, A.; Sayer, J. M.; Louis, J. M.; Weber, I. T. HIV-1 Protease with 20 Mutations Exhibits Extreme Resistance to Clinical Inhibitors through Coordinated Structural Rearrangements. *Biochemistry* **2012,** *51,* 2819–2828.

54. *Ghosh, A. K.; Kincaid, J. F.; Cho, W.; Walters, D. E.; Krishnan, K.; Hussain, K. A.; Koo, Y.; Cho, H.; Rudall, C.; Holland, L.; Buthod, J. Potent HIV Protease Inhibitors Incorporating High-Affinity P2-Ligands and (*R*)-(Hydroxyethylamino)Sulfonamide Isostere. *Bioorg. Med. Chem. Lett.* **1998,** *8,* 687–690.

55. *Ghosh, A. K.; Sridhar, P. R.; Kumaragurubaran, N.; Koh, Y.; Weber, I. T.; Mitsuya, H. Bis-Tetrahydrofuran: A Privileged Ligand for Darunavir and a New Generation of HIV Protease Inhibitors That Combat Drug Resistance. *ChemMedChem* **2006,** *1,* 939–950.

56. Ghosh, A. K.; Pretzer, E.; Cho, H.; Hussain, K. A.; Duzgunes, N. Antiviral Activity of UIC-PI, a Novel Inhibitor of the Human Immunodeficiency Virus Type 1 Protease. *Antivir. Res.* **2002,** *54,* 29–36.

57. *Yoshimura, K.; Kato, R.; Kavlick, M. F.; Nguyen, A.; Maroun, V.; Maeda, K.; Hussain, K. A.; Ghosh, A. K.; Gulnik, S. V.; Erickson, J. W.; Mitsuya, H. A Potent Human Immunodeficiency Virus Type 1 Protease Inhibitor, UIC-94003 (TMC-126), and Selection of a Novel (A28S) Mutation in the Protease Active Site. *J. Virol.* **2002,** *76,* 1349–1358.

58. Ghosh, A. K.; Kulkarni, S.; Anderson, D. D.; Hong, L.; Baldridge, A.; Wang, Y.-F.; Chumanevich, A. A.; Kovalevsky, A. Y.; Tojo, Y.; Amano, M.; Koh, Y.; Tang, J.; Weber, I. T.; Mitsuya, H. Design, Synthesis, Protein-Ligand X-Ray Structure, and Biological Evaluation of a Series of Novel Macrocyclic Human Immunodeficiency Virus-1 Protease Inhibitors to Combat Drug Resistance. *J. Med. Chem.* **2009,** *52,* 7689–7705.

59. Erickson, J. W.; Gulnik, S. V.; Mitsuya, H.; Ghosh, A. K. Fitness Assay and Associated Methods. US 07470506, 2008.

60. Gulnik, S. V.; Suvorov, L. I.; Liu, B. S.; Yu, B.; Anderson, B.; Mitsuya, H.; Erickson, J. W. Kinetic Characterization and Cross-Resistance Patterns of HIV-1 Protease Mutants Selected under Drug Pressure. *Biochemistry* **1995,** *34,* 9282–9287.

61. Kempf, D. J.; Marsh, K. C.; Kumar, G. N.; Rodrigues, A. D.; Denissen, J. F.; McDonald, E.; Kukulka, M. J.; Hsu, A.; Granneman, G. R.; Baroldi, P. A.; Sun, E.; Pizzuti, D.; Plattner, J. J.; Norbeck, D. W.; Leonard, J. M. Pharmacokinetic Enhancement of Inhibitors of the Human Immunodeficiency Virus Protease by Co-administration with Ritonavir. *Antimicrob. Agents Chemother.* **1997,** *41,* 654–660.

62. *Latest Label Information for Prezista® (Darunavir Ethanolate).* http://www.accessdata.fda.gov/drugsatfda_docs/label/2012/021976s026, 202895s002lbl.pdf (accessed June 2012).

63. *De Meyer, S.; Azijn, H.; Surleraux, D.; Jochmans, D.; Tahri, A.; Pauwels, R.; Wigerinck, P.; de Bethune, M.-P. TMC114, a Novel Human Immunodeficiency Virus Type 1 Protease Inhibitor Active against Protease Inhibitor-Resistant Viruses, Including a Broad Range of Clinical Isolates. *Antimicrob. Agents Chemother.* **2005,** *49,* 2314–2321.

64. Koh, Y.; Amano, M.; Towata, T.; Danish, M.; Leshchenko-Yashchuk, S.; Das, D.; Nakayama, M.; Tojo, Y.; Ghosh, A. K.; Mitsuya, H. In Vitro Selection of Highly Darunavir-Resistant and Replication-Competent HIV-1 Variants by Using a Mixture of Clinical HIV-1 Isolates Resistant to Multiple Conventional Protease Inhibitors. *J. Virol.* **2010,** *84,* 11961–11969.

65. *Kovalevsky, A. Y.; Liu, F.; Leshchenko, S.; Ghosh, A. K.; Louis, J. M.; Harrison, R. W.; Weber, I. T. Ultra-High Resolution Crystal Structure of HIV-1 Protease Mutant Reveals Two Binding Sites for Clinical Inhibitor TMC114. *J. Mol. Biol.* **2006,** *363,* 161–173.

66. Cai, Y.; Schiffer, C. A. Decomposing the Energetic Impact of Drug Resistant Mutations in HIV-1 Protease on Binding DRV. *J. Chem. Theory Comput.* **2010,** *6,* 1358–1368.

67. Dierynck, I.; De Wit, M.; Gustin, E.; Keuleers, I.; Vandersmissen, J.; Hallenberger, S.; Hertogs, K. Binding Kinetics of Darunavir to Human Immunodeficiency Virus Type 1 Protease Explain the Potent Antiviral Activity and High Genetic Barrier. *J. Virol.* **2007,** *81,* 13845–13851.

68. Kovalevsky, A. Y.; Ghosh, A. K.; Weber, I. T. Solution Kinetic Measurements Suggests HIV-1 Protease Has Two Binding Sites for Darunavir and Amprenavir. *J. Med. Chem.* **2008,** *51,* 6599–6603.

69. Kohl, N. E.; Emini, E. A.; Schleif, W. A.; Davis, L. J.; Heimbach, J. C.; Dixon, R. A. F.; Scolnick, E. M.; Sigal, I. S. Active Human Immunodeficiency Virus Protease is Required for Viral Infectivity. *PNAS* **1988,** *85,* 4686–4690.

70. Wlodawer, A.; Miller, M.; Jaskólski, M.; Sathyanarayana, B. K.; Baldwin, E.; Weber, I. T.; Selk, L. M.; Clawson, L.; Schneider, J.; Kent, S. B. H. Conserved Folding in Retroviral Proteases: Crystal Structure of a Synthetic HIV-1 Protease. *Science* **1989,** *245,* 616–621.

71. *Koh, Y.; Aoki, M.; Danish, M. L.; Aoki-Ogata, H.; Amano, M.; Das, D.; Shafer, R. W.; Ghosh, A. K.; Mitsuya, H. Loss of Protease Dimerization Inhibition Activity of Darunavir is Associated with the Acquisition of Resistance to Darunavir by HIV-1. *J. Virol.* **2011,** *85,* 10079–10089.

72. Mitsuya, H.; Ghosh, A. K. Development of HIV-1 Protease Inhibitors, Antiretroviral Resistance, and Current Challenges of HIV/AIDS Management. In *Aspartic Acid Proteases as Therapeutic Targets;* Ghosh, A. K., Ed.; Wiley-VCH Verlag GmbH & Co. KGaA: Weinheim, Germany, 2010; pp 245–262; Chapter 9.

73. Huang, D.; Caflisch, A. How Does Darunavir Prevent HIV-1 Protease Dimerization? *J. Chem. Theory Comput.* **2012,** *8,* 1786–1794.

74. Ghosh, A. K.; Chen, Y. Synthesis of Optical Resolution of High-Affinity P$_2$-Ligands for HIV-1 Protease Inhibitors. *Tetrahedron Lett.* **1995,** *36,* 505–508.

75. Ghosh, A. K.; Leschenko, S.; Noetzel, M. Stereoselective Photochemical 1,3-Dioxolane Addition to 5-Alkoxymethyl-2(5H)-Furanone: Synthesis of *bis*-Tetrahydrofuranyl Ligand for HIV Protease Inhibitor UIC-94017 (TMC-114). *J. Org. Chem.* **2004,** *69,* 7822–7829.

76. Ghosh, A. K.; Li, J.; Sridhar, P. R. A Stereoselective Anti-Aldol Route to (3R,3aS,6aR)-Hexahydrofuro[2,3-b]furan-3-ol: A Key Ligand for a New Generation of HIV Protease Inhibitors. *Synthesis* **2006,** *18,* 3015–3018.

77. *Ghosh, A. K.; Martyr, C. D. Darunavir (Prezista®): A HIV-1 Protease Inhibitor for Treatment of Multidrug-Resistant HIV. In *Modern Drug Synthesis;* Li, J. J., Jonhson, D. S., Eds.; John Wiley & Sons, Inc.: Hoboken, NJ, 2010; pp 29–44.

78. Mauskopf, J.; Annemans, L.; Hill, A. M.; Smets, E. A Review of Economic Evaluations of Darunavir Boosted by Low-Dose Ritonavir in Treatment-Experienced Persons Living with HIV Infection. *Pharmacoeconomics* **2010,** *28,* 1–16.

79. Hazen, R.; Harvey, R.; Ferris, R.; Craig, C.; Yates, P.; Griffin, P.; Miller, J.; Kaldor, I.; Ray, J.; Samano, V.; Furfine, E.; Spaltenstein, A.; Hale, M.; Tung, R.; St Clair, M.; Hanlon, M.; Boone, L. In Vitro Antiviral Activity of the Novel, Tyrosyl-Based Human Immunodeficiency Virus (HIV) Type 1 Protease Inhibitor Brecanavir (GW640385) in Combination with Other Antiretrovirals and against a Panel of Protease Inhibitor-Resistant HIV. *Antimicrob. Agents Chemother.* **2007,** *51,* 3147–3154.

80. He, G. X.; Yang, Z. Y.; Williams, M.; Callebaut, C.; Cihlar, T.; Murray, B. P.; Yang, C.; Mitchell, M. L.; Liu, H. T.; Wang, J. Y.; Arimilli, M.; Eisenberg, E.; Stray, K. M.; Tsai, L. K.; Hatada, M.; Chen, X. W.; Chen, J. M.; Wang, Y. J.; Lee, M. S.; Strickley, R. G.; Iwata, Q.; Zheng, X. B.; Kim, C. U.; Swaminathan, S.; Desai, M. C.; Lee, W. A.; Xu, L. H. Discovery of GS-8374, a Potent Human Immunodeficiency Virus Type 1 Protease Inhibitor with a Superior Resistance Profile. *MedChemComm* **2011,** *2,* 1093–1098.

81. Cihlar, T.; He, G.-X.; Liu, X.; Chen, J. M.; Hatada, M.; Swaminathan, S.; McDermott, M. J.; Yang, Z.-Y.; Mulato, A. S.; Chen, X.; Leavitt, S. A.; Stray, K. M.; Lee, W. A. Suppression of HIV-1 Protease Inhibitor Resistance by Phosphonate-Mediated Solvent Anchoring. *J. Mol. Biol.* **2006,** *363,* 635–647.

82. Ghosh, A. K.; Chapsal, B. D.; Baldridge, A.; Steffey, M. P.; Walters, D. E.; Koh, Y.; Amano, M.; Mitsuya, H. Design and Synthesis of Potent HIV-1 Protease Inhibitors Incorporating Hexahydrofuropyranol-Derived High Affinity P2 Ligands: Structure-Activity Studies and Biological Evaluation. *J. Med. Chem.* **2011,** *54,* 622–634.

83. Ide, K.; Aoki, M.; Amano, M.; Koh, Y.; Yedidi, R. S.; Das, D.; Leschenko, S.; Chapsal, B.; Ghosh, A. K.; Mitsuya, H. Novel HIV-1 Protease Inhibitors (PIs) Containing a Bicyclic P2 Functional Moiety, Tetrahydropyrano-Tetrahydrofuran, That Are Potent against Multi-PI-Resistant HIV-1 Variants. *Antimicrob. Agents Chemother.* **2011,** *55,* 1717–1727.

84. Ghosh, A. K.; Xu, C. X.; Rao, K. V.; Baldridge, A.; Agniswamy, J.; Wang, Y. F.; Weber, I. T.; Aoki, M.; Miguel, S. G. P.; Amano, M.; Mitsuya, H. Probing Multidrug-Resistance and Protein-Ligand Interactions with Oxatricyclic Designed Ligands in HIV-1 Protease Inhibitors. *ChemMedChem* **2010,** *5,* 1850–1854.

The case of anti-TNF agents

14

Denis Mulleman*,†, Marc Ohresser*, Hervé Watier*,†

* Centre National de la Recherche Scientifique - Unité Mixte de Recherche 7292,
Université François Rabelais de Tours, Tours, France,
† Centre Hospitalier Régional et Universitaire de Tours, Tours, France
This work has been done in the frame work of the Laboratory of Excellence MAb
Improve and the MAb Mapping project

CHAPTER OUTLINE

14.1 The Story of anti-TNF: From Septic Shock to Inflammatory Diseases 386
14.2 From a Clinically Validated Target to the Design of anti-TNF Biopharmaceuticals 387
    14.2.1 Development of anti-TNF-α mAbs 387
    14.2.2 Development of soluble receptors..391
14.3 Marketing Approvals and Post-Marketing Clinical Experience 392
    14.3.1 Efficacy (biomarkers and sequential treatment) 392
    14.3.2 Tolerance..393
14.4 Structure–Clinical Activities Relationships Drawn from the Clinical Experience............................. 393
    14.4.1 Immunogenicity 394
    14.4.2 mAbs, granulomas, and membrane TNF 394
14.5 Conclusions 395
References 395

ABSTRACT

Parallel studies in septic shock and cancer led to the discovery of the endogenous factors named cachectin and tumour necrosis factor (TNF), respectively, that were shown to be structurally and functionally similar. Further studies identified two forms of TNF, TNF-α and TNF-β. The anticipation that specific anti-TNF antibody might have therapeutic potential in the resolution of sepsis was not realized; however, anti-TNF-α agents have been shown to have dramatic therapeutic efficacy in a number of inflammatory diseases. Currently, there are five anti-TNF agents approved that are equally effective in the treatment of rheumatoid arthritis but exhibit differing efficacy in other inflammatory conditions. Three are full-length immunoglobulin G (IgG) anti-TNF-α antibodies, one is an anti-TNF-α Fab fragment and another, a TNF receptor–Fc fusion protein. Their different structures reflect recent advances in our ability to apply genetic engineering for patient benefit.

Keywords/Abbreviations: Tumour necrosis factor (receptor) (TNF(R)); Monoclonal antibody (mAb); Rheumatoid arthritis (RA); Cytokines; Lymphotoxin (LT); Pharmacokinetic (PK); TNF-alpha-converting enzyme (TACE); Paratope; Idiotype; Human anti-murine antibody (HAMA); Crohn's disease (CD); Fraction, crystallizable (Fc); Fragment antigen binding (Fab); Immunoglobulin G (IgG); Complementarity-determining region (CDR); Polyethylene glycol (PEG); PEG-ylation; Antidrug antibody (ADA); Variable heavy (VH)/variable light (VL) domain; Natural killer (NK) cell.

14.1 THE STORY OF anti-TNF: FROM SEPTIC SHOCK TO INFLAMMATORY DISEASES

The discovery of tumour necrosis factor (TNF) was reported in 1975, coincidently the same year that the hybridoma technology for the production of monoclonal antibodies (mAbs) of pre-selected specificity was published. It was described as a human macrophage-derived substance inducing haemorrhagic necrosis of solid tumours. Concurrently, cachectin was identified, in a rabbit model of infection, as a product deriving from macrophages and provoking the breakdown of energy stores from adipocytes and muscle cells. Biochemical studies demonstrated that these factors were structurally and functionally similar and TNF was recognized as a prototypical multifunctional cytokine, operating in immunity and inflammation and regulating endocrine, cardiovascular, and metabolic systems.

The presence of TNF was closely associated with adverse outcome from sepsis in animal models and, in the clinical setting, patients with elevated TNF levels exhibited higher incidence and severity of adult respiratory distress syndromes and a higher mortality rate than did patients without raised levels of TNF (1). Reports of TNF involvement in sepsis dominated in the literature in the late 1980s and it was considered to be an endogenous 'toxin' and an ideal target for sepsis therapy. As such it appeared to be an ideal target for the generation of TNF neutralizing mAbs. Preclinical studies in the early 1990s showed anti-TNF mAbs to exert a protective effect in different models of gram-negative shocks, in rodents, pigs and nonhuman primates. These preclinical data encouraged several companies to initiate clinical trials. The first open-label study in 80 patients with severe sepsis or septic shock revealed that administration of a murine anti-TNF mAb was well tolerated but did not provide clinical benefit (2). In a large randomized control trial involving 971 patients, another murine monoclonal anti-TNF mAb also failed to improve survival beyond 28 days (3).

While most people interested in TNF were focused on infection and sepsis, Feldmann et al. found that TNF was implicated in the prolonged proinflammatory condition observed in the synovial tissue of rheumatoid arthritis (RA) patients (4,5). At that time cytokines were perceived as functioning within a complex network of mediators acting in cascade, with redundancy and parallel pathways of action, suggesting that neutralizing just one would not be sufficient to control inflammation. However, the feasibility of controlling inflammation by neutralizing TNF alone was clearly demonstrated for a collagen-induced arthritis model (6) and TNF was recognized as a potentially interesting therapeutic target in inflammatory diseases. Since the application of anti-TNF mAbs in sepsis had not lived up to its early promise companies decided to switch and investigate possible clinical benefits for anti-TNF mAbs in RA.

14.2 FROM A CLINICALLY VALIDATED TARGET TO THE DESIGN OF anti-TNF BIOPHARMACEUTICALS

Initially TNF was thought to be a unique soluble mediator; however, in 1984 it was shown that a cytokine previously described as lymphotoxin (LT, or LT-α) (Figure 14.1(a)) was closely similar to TNF; thus two TNFs were defined as TNF-α and TNF-β. Both cytokines exhibited similar biological activities, now explained by the fact that both are binding to TNFR1 (p55 or p60, CD120a, *TNFRSF1A* gene product) and TNFR2 (p75 or p80, CD120b, *TNFRSF1B* gene product) (Figure 14.1(b)). TNF-β is secreted as a soluble factor of 40 kDa (a homotrimer of 17 kDa subunits, Figure 14.1(a)), whilst TNF-α is a membrane protein of 26 kDa that assembles in trimers that are rapidly cleaved by a TNF-α-converting enzyme (TACE) to release a soluble factor of 40 kDa (Fig. 14.1(b)) (7). Although exhibiting similar biological profiles soluble TNF-α and TNF-β show only ~30% identity, at the amino acid sequence level and even less for surface-exposed residues (Figure 14.1(a)). These structural differences are sufficient to be distinguished by mAbs and anti-TNF-α-specific mAbs were generated that neutralized TNF-α (i.e. to prevent its binding to TNFR1 and TNFR2), and vice versa for anti-TNF-β-specific antibodies. Experimental and clinical data suggested TNF-α to be the relevant target to address inflammation in RA and resulted in a focus on generating anti-TNF-α neutralizing agents.

For indications of acute sepsis rapid neutralization of TNF-α was essential and the use of a murine mAb was justified, even if the drug was rapidly neutralized by the humoural immune response of the patients (human anti-murine antibodies). Similarly, as for anti-venom serotherapy, the administration of fragment antigen binding (Fab) or F(ab′)$_2$ fragments of anti-TNF-α immunoglobulin G (IgG), such as afelimomab (suffix momab for mouse mAb), could be effective, despite their short plasma half-life (Figure 14.2). By contrast, chronic inflammatory diseases like RA and Crohn's disease (CD) required repeated administrations of the product to achieve sustained and long-lasting TNF-α neutralization. This need was partially met by the development of humanization techniques that reduced immunogenicity (Figure 14.2) and the use of a full-length IgG to retain the required long plasma half-life (~21–15 days).

The discovery of membrane TNF receptors (TNFR1 and TNFR2) was rapidly followed by the discovery of soluble TNF receptors (sTNFR1 and sTNFR2), physiologically produced by TNFR shedding or by alternative splicing (Figure 14.1(a) and (c)). These soluble forms maintained their TNF-binding properties, behaving as natural TNF-α and TNF-β antagonists. These findings led pharmaceutical companies to explore the development of soluble TNF receptors as therapeutics to neutralize TNF. In addition to neutralization of TNF it was essential that these biologics should exhibit extended plasma half-lives, as for antibodies, in order to maintain sustained TNF neutralization.

14.2.1 Development of anti-TNF-α mAbs

Two murine mAbs with specificity and high affinity for TNF-α (suffix momab) were engineered to create chimeric antibodies with therapeutic potential. The variable heavy (VH) and variable light (VL) regions of the murine antibody A2 were fused to the constant regions of human IgG1 heavy chain and kappa light chains, respectively, to create a chimeric antibody (suffix ximab) now known as infliximab or Remicade® (Figure14.2). An additional level of humanization was achieved for the antibody ner-elimomab by grafting the murine complementarity-determining regions (CDRs) into human VH and VL framework domains to generate the IgG4 antibody CDP-571 (Humicade®) (suffix zumab).

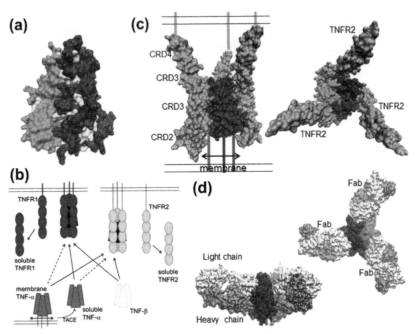

FIGURE 14.1 Mode of binding of TNF-α and TNF-β to their receptors and TNF-α neutralization by monoclonal antibodies.

(a) Tridimensional structure of TNF-α drawn from pdb 1tnf, showing the surfaces of the three trimers in green, red and blue (behind). TNF-β, which has the same trimeric structure as TNF-α, shares with it a limited number of residues in surface; these residues are shown in yellow. This limited surface homology explains why anti-TNF-α mAbs do not bind to TNF-β, and vice versa. (b) Although different, TNF-α (in blue) and TNF-β (in yellow) bind to the same receptors, TNFR1 (in red) and TNFR2 (in orange). TNFR1 and TNFR2 have four cysteine-rich domains (CRDs); TNFR2 (p75) has a higher molecular weight than TNFR1 (p55), mainly due to the presence of glycans in the membrane proximal part of the receptor. Binding of TNF-α or TNF-β to TNFR1 or TNFR2 leads to the trimerization of the receptor, triggering the signalling cascade. TNF-α also differs from TNF-β in that it is a membrane protein secondarily cleaved by the metalloprotease TACE to generate soluble TNF-α, while TNF-β is secreted as a soluble cytokine. Both membrane and soluble TNF-α bind to TNFR1 and TNFR2, although with subtle differences (dotted lines indicate a slightly lower affinity). TNFR1 and TNFR2 can also be cleaved by metalloproteases (or be subjected to alternative spicing), generating soluble receptors that maintain affinity for TNF-α and TNF-β. (c) Tridimensional structure of TNF-α bound to three truncated forms of TNFR2 (limited to the four CRDs), drawn from pdb 3alq. The three monomers of TNF-α are represented with the same colours as in (a); TNFR2 is in orange. The left and right figures are distinguished by a 90° rotation, showing the cytokine laterally and from the top, respectively. CRD2 and CRD3 of TNFR2 bind laterally to two monomers of TNF-α, in a furrow which parallels the main axis of the trimer. This mode of binding, which characterizes the TNF-TNFR family, is at the origin of the receptor trimerization. (d) Representation of what could be the binding of anti-TNF Fab fragments to TNF-α. In the absence of available tridimensional structures of anti-TNF-α mAbs complexed with TNF-α, this representation is that of the binding of Fab fragments of ruplizumab, a neutralizing mAb binding to CD40L, another member of the TNF-α family (pdb 1i9r). VH and CH1 domains from the ruplizumab heavy chain are shown in dark grey, and the VL and CL domains of the

A similarly humanized anti-TNF-α(Fab′)2 fragment (hTNF40) was covalently linked to polyethylene glycol (PEG) to generate the therapeutic CDP-870 or certolizumab (Cimzia®) (Figure 14.2). The ultimate level of humanization was achieved with the development of fully human mAbs (suffix mumab), either from phage display libraries (adalimumab) or from mice expressing human immunoglobulin genes (golimumab) (Figure 14.2).

Whilst these technologies could maximize the extent of humanization it is not possible to entirely eliminate the potential for immunogenicity since the antibody specificity depends on the unique structure of the antigen-binding site (paratope) which is the product of unique gene recombination and hypermutation events. The potentially immunogenic non-self-epitopes generated due to the unique structure of the paratope is referred to as the *idiotype*. The currently available choice of recombinant anti-TNF agents was generated, at least in part, due to competition between pharmaceutical companies for a presence in a lucrative market, rather than in response to clinical necessity. A key factor in the success of anti-TNF-α therapeutics has been their long serum half-lives, achieved either by the use of full-length mAbs, as an Fc (fragment crystallisable) fusion protein or by 'pegylation', i.e. the covalent attachment of PEG chain(s).

There is a potential downside to the presence of the Fc region in an antibody therapeutic because the immune complex formed, between target and antibody, may be taken up by an Fc receptor expressing antigen-presenting cells, leading to a humoural antidrug antibody (ADA) response. Extension of plasma half-lives by 'pegylation' has been achieved for certolizumab (Figure 14.3) and other drugs, e.g. the interferons. In addition to the prolongation of half-lives pegylation can contribute to a reduction in immunogenicity (Figure 14.3).

The focus on neutralization of TNF-α and providing for long plasma half-lives has resulted in little attention being paid to other immune mechanisms that can be mediated by intact IgG and, particularly, the immune complexes that are formed. Full-length IgG1 subclass anti-TNF-α mAbs (infliximab, adalimumab and golimumab) are approved and have gained 'blockbuster' status; however, the IgG4 anti-TNF-α mAbs (Humicade®) never reached the market due to lack of efficacy. This does not necessarily mean that the IgG4 format was a wrong choice. Each human IgG subclass exhibits a unique profile of Fc effector functions (see Chapter 3), and the IgG1 and IgG4 subclasses are distinguished on the basis of their ability to recruit effector mechanisms (complement, FcγR expressing cytotoxic cells, … etc.), which can be of critical importance for targets expressed on cell membranes, e.g. trastuzumab. The principle mode of action of anti-TNF agents has frequently been considered only with respect to neutralization of soluble TNF and the fact that TNF-α is both a soluble antigen and a cell membrane protein has been too frequently overlooked. Thus it remains possible that the choice of the subclass for an anti-TNF antibody is not irrelevant because anti-TNF-α IgG1 mAbs have the potentiality to kill TNF-α-expressing cells, whereas IgG4 may not. In fact, elimination of inflammatory cells and

---

ruplizumab light chain are in light grey. The left and right figures are distinguished by a 90° rotation on the Fab axis, showing the cytokine from the top and laterally, respectively. In the present example, the main axis of the Fab is orthogonal to the main axis of the cytokine. The epitope is straddling two monomers and the trimeric symmetry allows the binding of three Fabs to a single cytokine (3:1 ratio). Taking into consideration the mode of binding of TNF-α to its receptors (c), it is obvious that the binding of anti-TNF-α mAbs like that depicted here will prevent the binding of TNF-α to its receptor, which characterizes the neutralization function. (See colour plate.)

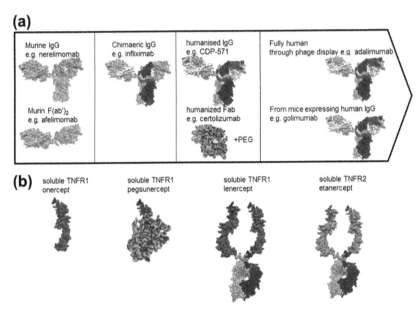

FIGURE 14.2 Biopharmaceutical agents developed to neutralize TNF-α.

(a) Technological improvements in the design of monoclonal antibodies against TNF-α. Starting from murine mAbs (green) produced by the hybridoma technology, full-length IgG such as nerelimomab or F(ab′)$_2$ fragments such as afelimomab were developed and tested in humans. The next technological advance was the development of a chimeric mAb through recombining cDNAs coding for VH and VL domains deriving from an anti-TNF-α murine mAb (green) with cDNAs coding for a human heavy (γ) chain and a kappa (κ) light chain, respectively (the κ-chains are shown in light grey, one γ-chain is in dark grey and the other in black). Infliximab was generated with a γ1-heavy chain and is thus is an IgG1 subclass protein. The next step was the development of humanized mAbs, always starting from a murine mAb with a given antigenic specificity, by using refinements of DNA recombination technologies. CDRs from the murine mAbs (in green) are grafted onto frameworks derived from human VH and VL domains. Development of the CDP-571 was discontinued due to lack of efficacy, may be because it was generated as an IgG4 subclass protein (γ4 heavy chain); certolizumab, was developed as an IgG1 Fab fragment complexed with polyethylene glycol chains (in blue and red; see details in Figure 14.3). The last step was the development of fully human mAb, produced directly from human immunoglobulin genes, expressed either in phage libraries (top) or in transgenic mice (bottom). Note that the paratope (in blue) remains the product of both recombinations between human genes and hypermutations (artificial in phage display and natural in transgenic mice), and will have a unique structure, i.e. non-self. The two mAbs currently approved are both IgG1. (b) Initial grouping in the development of anti-TNF soluble receptors. Initially TNFR1 (in red) was developed as a soluble single-chain receptor (onercept), as a single chain complexed to polyethylene glycol chains (schematized in blue and red) (pegsunercept), or as a dimeric receptor fused with an IgG1 Fc fragment. TNFR2 was only developed as a fusion protein with human IgG1 Fc fragment, and is the only soluble TNFR approved (etanercept); its detailed structure is shown in Figure 14.4. (See colour plate.)

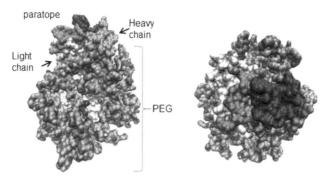

FIGURE 14.3 Model of pegylated Fab.

The Fab fragment of an anti-HIV-1 gp120 IgG1κmAb (clone b12) was used as a model mAb. Its three-dimensional structure (pdb #1hzh) was drawn using surface representation software (VMD software, University of Illinois), with the light and heavy chains in light and dark grey, respectively. The three CDRs from the light chain and the three CDRs from the heavy chain are shown in orange and red, respectively, and together form the paratope (antigen-binding site), better seen on the right. It is known that two 20 kDa molecules of polyethylene glycol (PEG) are bound to the C-terminal part of the heavy chain. The exact conformation of the two PEG chains is still debated. Starting from small PEG chains co-crystallized with proteins (pdb 1to2 and 3b6j), PEG chains of the correct molecular weight were built by juxtaposing existing structures, and were freely represented covering the Fab. Carbon and oxygen atoms of the PEG are shown in blue and red, respectively. In this ideal representation, PEG seems to protect the Fab from degradation and capture by antigen-presenting cells whilst not occluding the antigen-binding site of the mAb. However, PEG chains are very mobile structures and could adopt many different configurations. Precisely how they protect the Fab is not known. (See colour plate.)

TNF-α-expressing immune effectors could account for some therapeutic properties of IgG1 subclass anti-TNF antibodies, as well as some adverse effects.

14.2.2 Development of soluble receptors

The potential ability for soluble TNF receptors to neutralize both TNF-α and TNF-β suggested a possible therapeutic advantage over mAbs. The initial development of soluble TNF receptors focused on TNFR1 (p55), the most abundantly expressed receptor (Figure 14.2(b)). The drugs developed have the suffix ercept, for 'receptor'. The soluble form of TNFR1 (onercept) had a too short half-life to be clinically useful in chronic disease, therefore, a pegylated form was generated (pegsunercept), as for certolizumab (*cf. supra*); however, the affinity proved to be too low for clinically efficacy. The next development was the generation of the sTNFR1–Fc fusion protein lenercept. The presence of the Fc provided for the associated prolonged half-life and since the Fc is a homodimer, lenercept is composed of two sTNFR1 protein chains, conferring the potential for increased avidity of binding to multimers of sTNF-α protein. The soluble TNF molecule forms trimers with the result that three identical epitopes may be available to engage with lenercept or divalent anti-TNF antibodies to form immune complexes. In spite of these improvements in sTNF binding, lenercept proved not to be efficacious in the treatment of TNF-driven inflammatory diseases. The parallel development of a TFNR2–Fc (p75–Fc) fusion

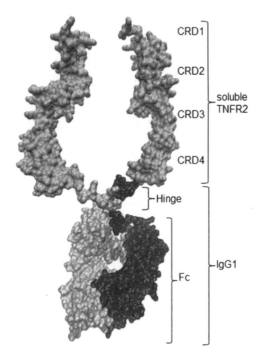

FIGURE 14.4 Model of etanercept.

Although etanercept is TNFR2-Fc, this representation was made by juxtaposing the structure of TNFR1 (pdb #1ext) and the hinge and the Fc portion of an anti-HIV-1 gp120 IgG1 mAb (clone b12) (pdb #1hzh). Both are drawn as surface representations (VMD software, University of Illinois), in orange for TNFR, and dark grey and black for the two γ1-heavy chains. The N-glycans of the Fc are shown in red and blue. TNFR1 was preferred to TNFR2 because TNFR2 (pdb #3alq) is only known in association with TNF-α (Figure 14.1(c)) while TNFR1 alone has been crystallized. The four CRDs of TFR1 and TNFR2 are very similar in their structure. (See colour plate.)

protein led to the generation of a therapeutically effective drug that was approved and marketed as Enbrel® (etanercept). The explanation for the therapeutic efficacy of etanercept, but not lenercept, is not resolved. Comparisons of the clinical profile of etanercept vs anti-TNF-α mAbs (*cf. infra*) suggest that neutralization of both TNF-β and TNF-α do not provide additional therapeutic benefit.

14.3 MARKETING APPROVALS AND POST-MARKETING CLINICAL EXPERIENCE

14.3.1 Efficacy (biomarkers and sequential treatment)

In inflammatory bowel diseases, anti-TNF mAbs, but not etanercept, are superior to placebo to achieve initial and long-term maintenance in patients having insufficient response to conventional immunosuppressive drugs such as methotrexate and azathioprine. (*8–10*). By contrast, all anti-TNF agents are effective in RA and other rheumatic diseases, such as psoriatic arthritis (*11,12*). Although they are effective in a broad spectrum of inflammatory diseases no individual therapeutic is clearly more efficacious than another (*10,12–14*). Given the apparently similar results, irrespective to the drug administered (*11*), no comparative trial has been undertaken to evaluate the relative efficacy of anti-TNF agents.

Whilst a majority of patients achieve a clinical response, nonresponders are encountered and secondary loss of response may occur in approximately one-third of the patients; this may be due to patients becoming refractory, e.g. through the development of ADA. Interestingly, in cases of primary or secondary absence of response to one of the therapeutics, clinicians may switch to an alternative drug and achieve clinical response. Analysis of sequential treatments in RA suggests that the probability of achieving a response is lower with the second anti-TNF agent than with the first one (*11*).

14.3.2 Tolerance

There is no evidence that anti-TNF agents increase the risk of RA and CD patients developing all-site malignancies in the short term, in particular lymphomas, in comparison to patients receiving conventional immunosuppressive drugs (*15–20*). There is however a clear evidence for an increased risk of non-melanoma skin cancer (and possibly melanoma) in RA patients receiving anti-TNF agents.

A recent meta-analysis in CD showed no significant difference in the risk of serious infection in patients treated with TNF-α inhibitors as compared with controls. Similarly, meta-analysis of studies in which early RA patients received immune modulators (methotrexate) or anti-TNF agents did not show any significant difference in the risk of infections (*19*). This finding concurs with similar meta-analyses in spondyloarthritis and psoriatic arthritis (*16,21*). However, an increased risk of serious infections in RA patients treated with anti-TNF agents versus those not exposed to anti-TNF has been reported in a meta-analysis of observational studies (*22*). The discrepancies between observational registers and randomized controlled trials may be explained by the difference in terms of risk factors of infection between patients enrolled in trials and those treated in the 'real-life' setting.

Importantly, a reduced risk of cardiovascular morbidity has been reported for patients receiving anti-TNF agents since the disease itself is a risk factor of morbidity (*23*). However, some thrombo-embolic observations in patients with few risk factors for venous thrombosis have raised the question of a causative relationship between these events and anti-TNF agents (*24*).

Although anti-TNF agents have a profile of being well-tolerated some adverse events may occur. Identification of individual risk factors and preventive action (screening for cancer and chronic infection cancer and vaccination) are essential. The absolute rate of adverse event should not outweigh the substantial benefits of these biopharmaceuticals.

14.4 STRUCTURE–CLINICAL ACTIVITIES RELATIONSHIPS DRAWN FROM THE CLINICAL EXPERIENCE

With five approved anti-TNF agents on the market, some of them for 15 years, the accumulated clinical data provide a remarkable opportunity to conduct 'case studies'. In addition to binding TNF, the five products display a diversity of functions, dependent on individual parameters, e.g. epitope specificity, heavy-chain isotype, etc. Consequently, they also exhibit different patterns of therapeutic efficacy and adverse reactions. Studies to establish relationships between the structure of a given anti-TNF agent and clinical outcomes could contribute to our understanding of the mechanisms operative *in vivo*. It should be noted that some prospective anti-TNF therapeutics, e.g. Humicade® or lenercept, were 'counter-selected' during their clinical development and too little clinical data is available for case-study analysis.

14.4.1 Immunogenicity

Immunogenicity of biopharmaceuticals results in the formation of ADA that can impact clinical response, lead to accelerated clearance of drug and the cause of adverse events. This risk is encountered with both chimeric (25,26) and human mAbs (27). The incidence and frequency of ADA development is extremely variable across reported studies. The presence of ADA can vary depending on the time since treatment, the disease entity, dose, …, etc. Differences in assay methods and protocols may account for discordance across such studies. Drug interference (ADA–drug interaction) is a major source of false-negative results with most assays and underestimates ADA detection. Concomitant immunosuppressive drug administration reduces the immunogenicity of mAbs.

The production of neutralising ADA has been reported for all full-length anti-TNF antibodies, regardless of the level of humanization. Interestingly, although few studies have been reported, neutralising ADA has not been reported for etanercept or certolizumab pegol. The lack of immunogenicity of etanercept has logic, because, apart from the linker region, the protein is entirely human without any obvious accessible foreign epitopes. By contrast, the apparent lack of immunogenicity of certolizumab pegol is less clear since certolizumab does express an idiotype. Whether this lack of immunogenicity is due to the presence of PEG chains and/or to the absence of an Fc portion is not evident. Monovalency of the Fab could prevent the formation of immunogenic aggregates and the absence of an Fc could prevent FcγR-mediated capture of immune complexes by antigen-presenting cells. However, the fact that abciximab, a marketed unpegylated Fab fragment, is immunogenic suggests that the presence of PEG chains associated with certolizumab may be crucial to prevent its immunogenicity. PEG chains could mask potential epitopes, although probably not the idiotype, or prevent the formation of aggregates and capture by antigen-presenting cells.

14.4.2 mAbs, granulomas, and membrane TNF

Clinical observations and experience suggest a difference in the mode of action between anti-TNF-α mAbs and etanercept. Thus, only anti-TNF-α mAbs are indicated for treatment in CD and exhibit some benefit in vasculitis, diseases which have in common the presence of inflammatory granulomas. Anti-TNF-α mAbs are also characterized by a higher incidence of tuberculosis infection, where granuloma formation is critical to control *Mycobacterium tuberculosis*. Anti-TNF-α mAbs, but not etanercept, thus appear as granuloma-destroying agents. The mechanisms underlying therapeutic benefit in RA and psoriasis would seem to be similar since etanercept and mAbs exhibit comparable efficacy in these indications. It has been suggested that etanercept binds membrane TNF-α more weakly than do mAbs, due to either a higher k_{off} or the fact that membrane TNF is not necessarily trimeric, a requirement if both TNFR2 domains of etanercept are to bind membrane TNF. The higher binding affinity of anti-TNF-α mAbs for membrane TNF-α may result in more efficacious inhibition of juxtacrine signals between TNF-α and TNFR1 and TNFR2 within granulomas. Another possibility, not exclusive of the former, is that mAb binding to membrane TNF-α could lead to antigen bridging and triggering of a pro-apoptotic signal (reverse signalling), resulting in the elimination of inflammatory cells. A further possible explanation, again not exclusive, is that better mAb binding to membrane TNF-α could favour the recruitment of immune effector functions, such as complement-dependent cytotoxicity and antibody-dependent cell-mediated cytotoxicity by FcγR-expressing killer cells. The fact that CD patients homozygous for the 158Val alloform of FcγRIIIA show a better biological response to infliximab than patients expressing the 158Phe alloform, of lower affinity for IgG1, suggests that FcγRIIIA-expressing cells such as NK cells and macrophages contribute to the response to anti-TNF-α mAbs (28,29). If most of these explanations are valid for infliximab, adalimumab and golimumab, which are all full-length IgG1 mAbs, it does not explain why certolizumab exhibits therapeutic benefit in CD (although not approved for this indication in Europe, it is in the US); certolizumab is monomeric, does not aggregate membrane TNF-α or recruit immune effectors. It is suggested that an antagonistic action on juxtacrine signals and, possibly, induction of apoptosis (in the absence of antigen bridging?) may explain its clinical profile.

14.5 CONCLUSIONS

In conclusion, the development of anti-TNF biologics as effective therapeutics for a number of inflammatory diseases may be seen, in retrospect, as the positive outcome of an intellectual, scientific and clinical saga. After the failure of anti-TNF-α mAbs in sepsis, few would have gambled on the future of these drugs! However, there are now five approved biologics on the market that are recognized as major advances in therapeutics. They result from the application of innovative basic and applied science allied to successful industrial and commercial exploitation; several have acquired 'blockbuster' status, i.e. annual sales values $> 1 \times 10^9$ dollars (US). Their structural diversity is associated with some differences in terms of efficacy and safety profiles, but the underlying mechanisms remain to be established. Among the approved anti-TNF agents etanercept is distinguished as being a receptor–Fc fusion protein and certolizumab pegol as a pegylated Fab fragment. The therapeutic efficacy and immunogenicity profile of certolizumab pegol appear to be exceptional; the absence of an Fc means that its pharmacokinetics is independent on FcRn-mediated pathways; its biodistribution profile into inflamed tissues, normal tissues, and its ability to cross the placenta remain to be evaluated. As often happens in biology, the exceptions have contributed to advances in knowledge, and this process is still ongoing for anti-TNF agents.

References

1. Marks, J. D.; Marks, C. B.; Luce, J. M.; Montgomery, A. B.; Turner, J.; Metz, C. A., et al. Plasma Tumor Necrosis Factor in Patients with Septic Shock. Mortality Rate, Incidence of Adult Respiratory Distress Syndrome, and Effects of Methylprednisolone Administration. *Am. Rev. Respir. Dis.* **1990,** *141,* 94–97.
2. Fisher, C. J., Jr.; Opal, S. M.; Dhainaut, J. F.; Stephens, S.; Zimmerman, J. L.; Nightingale, P., et al. Influence of an Anti-tumor Necrosis Factor Monoclonal Antibody on Cytokine Levels in Patients with Sepsis. The CB0006 Sepsis Syndrome Study Group. *Crit. Care Med.* **1993,** *21,* 318–327.

3. Abraham, E.; Wunderink, R.; Silverman, H.; Perl, T. M.; Nasraway, S.; Levy, H., et al. Efficacy and Safety of Monoclonal Antibody to Human Tumor Necrosis Factor Alpha in Patients with Sepsis Syndrome. A Randomized, Controlled, Double-Blind, Multicenter Clinical Trial. TNF-alpha MAb Sepsis Study Group. *J. Am. Med. Assoc.* **1995,** *273,* 934–941.
4. Feldmann, M.; Brennan, F. M.; Maini, R. N. Role of Cytokines in Rheumatoid Arthritis. *Annu. Rev. Immunol.* **1996,** *14,* 397–440.
5. Brennan, F. M.; Chantry, D.; Jackson, A.; Maini, R.; Feldmann, M. Inhibitory Effect of TNF Alpha Antibodies on Synovial Cell Interleukin-1 Production in Rheumatoid Arthritis. *Lancet* **1989,** *2,* 244–247.
6. Williams, R. O.; Feldmann, M.; Maini, R. N. Anti-tumor Necrosis Factor Ameliorates Joint Disease in Murine Collagen-Induced Arthritis. *Proc. Natl. Acad. Sci. U S A* **1992,** *89,* 9784–9788.
7. Lejeune, F. J.; Lienard, D.; Matter, M.; Ruegg, C. Efficiency of Recombinant Human TNF in Human Cancer Therapy. *Cancer Immun.* **2006,** *6,* 6.
8. Shao, L. M.; Chen, M. Y.; Cai, J. T. Meta-analysis: The Efficacy and Safety of Certolizumab Pegol in Crohn's Disease. *Aliment. Pharmacol. Ther.* **2009,** *29,* 605–614.
9. Ford, A.C.; Sandborn, W.J.; Khan, K.J.; Hanauer, S.B.; Talley, N.J.; Moayyedi, P. Efficacy of Biological Therapies in Inflammatory Bowel Disease: Systematic Review and Meta-analysis. *Am. J. Gastroenterol.* *106,* 644–659 [quiz 60].
10. Huang, X.; Lv, B.; Jin, H. F.; Zhang, S. A Meta-analysis of the Therapeutic Effects of Tumor Necrosis Factor-Alpha Blockers on Ulcerative Colitis. *Eur. J. Clin. Pharmacol.* **2011,** *67,* 759–766.
11. Alonso-Ruiz, A.; Pijoan, J. I.; Ansuategui, E.; Urkaregi, A.; Calabozo, M.; Quintana, A. Tumor Necrosis Factor Alpha Drugs in Rheumatoid Arthritis: Systematic Review and Meta-analysis of Efficacy and Safety. *BMC Musculoskelet. Disord.* **2008,** *9,* 52.
12. Saad, A. A.; Symmons, D. P.; Noyce, P. R.; Ashcroft, D. M. Risks and Benefits of Tumor Necrosis Factor-Alpha Inhibitors in the Management of Psoriatic Arthritis: Systematic Review and Metaanalysis of Randomized Controlled Trials. *J. Rheumatol.* **2008,** *35,* 883–890.
13. Aaltonen, K. J.; Virkki, L. M.; Malmivaara, A.; Konttinen, Y. T.; Nordstrom, D. C.; Blom, M. Systematic Review and Meta-analysis of the Efficacy and Safety of Existing TNF Blocking Agents in Treatment of Rheumatoid Arthritis. *PLoS One* **2012,** *7,* e30275.
14. Peyrin-Biroulet, L.; Deltenre, P.; de Suray, N.; Branche, J.; Sandborn, W. J.; Colombel, J. F. Efficacy and Safety of Tumor Necrosis Factor Antagonists in Crohn's Disease: Meta-analysis of Placebo-Controlled Trials. *Clin. Gastroenterol. Hepatol.* **2008,** *6,* 644–653.
15. Askling, J.; Fahrbach, K.; Nordstrom, B.; Ross, S.; Schmid, C. H.; Symmons, D. Cancer Risk with Tumor Necrosis Factor Alpha (TNF) Inhibitors: Meta-analysis of Randomized Controlled Trials of Adalimumab, Etanercept, and Infliximab Using Patient Level Data. *Pharmacoepidemiol. Drug Saf.* **2011,** *20,* 119–130.
16. Dommasch, E. D.; Abuabara, K.; Shin, D. B.; Nguyen, J.; Troxel, A. B.; Gelfand, J. M. The Risk of Infection and Malignancy with Tumor Necrosis Factor Antagonists in Adults with Psoriatic Disease: A Systematic Review and Meta-analysis of Randomized Controlled Trials. *J. Am. Acad. Dermatol.* **2011,** *64,* 1035–1050.
17. Mariette, X.; Matucci-Cerinic, M.; Pavelka, K.; Taylor, P.; van Vollenhoven, R.; Heatley, R., et al. Malignancies Associated with Tumour Necrosis Factor Inhibitors in Registries and Prospective Observational Studies: A Systematic Review and Meta-analysis. *Ann. Rheum. Dis.* **2011,** *70,* 1895–1904.
18. Mariette, X.; Reynolds, A.V.; Emery, P. Updated Meta-analysis of Non-melanoma Skin Cancer Rates Reported from Prospective Observational Studies in Patients Treated with Tumour ·Necrosis Factor Inhibitors. *Ann. Rheum. Dis. 71,* e2.
19. Thompson, A. E.; Rieder, S. W.; Pope, J. E. Tumor Necrosis Factor Therapy and the Risk of Serious Infection and Malignancy in Patients with Early Rheumatoid Arthritis: A Meta-analysis of Randomized Controlled Trials. *Arthritis Rheum.* **2011,** *63,* 1479–1485.

20. Siegel, C. A.; Marden, S. M.; Persing, S. M.; Larson, R. J.; Sands, B. E. Risk of Lymphoma Associated with Combination Anti-tumor Necrosis Factor and Immunomodulator Therapy for the Treatment of Crohn's Disease: A Meta-analysis. *Clin. Gastroenterol. Hepatol.* **2009,** *7,* 874–881.
21. Fouque-Aubert, A.; Jette-Paulin, L.; Combescure, C.; Basch, A.; Tebib, J.; Gossec, L. Serious Infections in Patients with Ankylosing Spondylitis with and without TNF Blockers: A Systematic Review and Meta-analysis of Randomised Placebo-Controlled Trials. *Ann. Rheum. Dis.* **2010,** *69,* 1756–1761.
22. Bernatsky, S.; Habel, Y.; Rahme, E. Observational Studies of Infections in Rheumatoid Arthritis: A Metaanalysis of Tumor Necrosis Factor Antagonists. *J. Rheumatol.* **2010,** *37,* 928–931.
23. Barnabe, C.; Martin, B. J.; Ghali, W. A. Systematic Review and Meta-analysis: Anti-tumor Necrosis Factor Alpha Therapy and Cardiovascular Events in Rheumatoid Arthritis. *Arthritis Care Res.* **2011,** *63,* 522–529.
24. Petitpain, N.; Gambier, N.; Wahl, D.; Chary-Valckenaere, I.; Loeuille, D.; Gillet, P. Arterial and Venous Thromboembolic Events During Anti-TNF Therapy: A Study of 85 Spontaneous Reports in the Period 2000–2006. *Biomed. Mater. Eng.* **2009,** *19,* 355–364.
25. Baert, F.; Noman, M.; Vermeire, S.; Van Assche, G.; Carbonez, A., et al. Influence of Immunogenicity on the Long-term Efficacy of Infliximab in Crohn's Disease. *N. Engl. J. Med.* **2003,** *348,* 601–608.
26. Wolbink, G. J.; Vis, M.; Lems, W.; Voskuyl, A. E.; de Groot, E.; Nurmohamed, M. T., et al. Development of Antiinfliximab Antibodies and Relationship to Clinical Response in Patients with Rheumatoid Arthritis. *Arthritis Rheum.* **2006,** *54,* 711–715.
27. Bartelds, G. M.; Krieckaert, C. L.; Nurmohamed, M. T.; van Schouwenburg, P. A.; Lems, W. F.; Twisk, J. W., et al. Development of Antidrug Antibodies against Adalimumab and Association with Disease Activity and Treatment Failure During Long-term Follow-up. *J. Am. Med. Assoc.* **2011,** *305,* 1460–1468.
28. Louis, E.; El Ghoul, Z.; Vermeire, S.; Dall'Ozzo, S.; Rutgeerts, P.; Paintaud, G., et al. Association between Polymorphism in IgG Fc Receptor IIIa Coding Gene and Biological Response to Infliximab in Crohn's Disease. *Aliment. Pharmacol. Ther.* **2004,** *19,* 511–519.
29. Moroi, R.; Endo, K.; Kinouchi, Y.; Shiga, H.; Kakuta, Y.; Kuroha, M., et al. FCGR3A-158 Polymorphism Influences the Biological Response to Infliximab in Crohn's Disease through Affecting the ADCC Activity. *Immunogenetics* **2013;** http://dx.doi.org/10.1007/s00251-013-0679-8.

Discovery of the cholesterol absorption inhibitor, ezetimibe

15

C. Robin Ganellin

Department of Chemistry, Christopher Ingold Laboratories, University College London, UK

CHAPTER OUTLINE

15.1 Introduction 400
15.2 Past Approaches to Lowering Levels of Circulating Cholesterol 401
    15.2.1 Bile acid sequestrants 401
    15.2.2 Hypocholesterolaemic drugs 402
    15.2.3 Inhibition of cholesterol absorption by fibrates: inhibitors of ACAT 403
    15.2.4 Statins: inhibitors of HMG-CoA reductase 404
    15.2.5 Inhibition of cholesterol absorption by saponins 405
15.3 Further Work on ACAT Inhibitors 406
15.4 Inhibition of a Novel Mechanism for Cholesterol Uptake and the Discovery of SCH 48461 408
15.5 Design of Ezetimibe (SCH 58235) 408
15.6 Ezetimibe in Human Studies 412
15.7 Identification of a New Mechanism for Cholesterol Uptake 413
15.8 Conclusion 414
References 414

ABSTRACT

A brief introduction to cholesterol and lipoproteins is followed by a short account of the types of drugs which reduce cholesterol levels (bile sequestrants, hypocholesterolaemics, fibrates as acyl-CoA:cholesterol acyltransferase (ACAT) inhibitors, statins as 3-hydroxy-3-methylglutaryl-coenzyme A reductase inhibitors, and saponins). This chapter continues with a background to the extraordinary research effort at the Schering-Plough Research Institute on ACAT inhibitors and the structure–activity studies that led to the lead cholesterol absorption inhibitor, the azetidinone SCH 48461. Impressive studies of the metabolism of SCH 48461 and the outstanding application of the results to drug design led to ezetimibe, which was shown to be a potent inhibitor of intestinal cholesterol uptake by a then unknown mechanism. Continued investigation using ezetimibe as a tool culminated in the remarkable discovery of a previously unknown mechanism for cholesterol uptake.

Keywords/Abbreviations: Cholesterol; Ezetimibe; Fibrates; HMG-CoA reductase; Hypocholesterolaemic drugs; LDL cholesterol; SCH 48461; SCH 58235; Statins; Acyl-coA:cholesterol acyltransferase (ACAT); High-density lipoprotein (HDL); 3-Hydroxy-3-methylglutaryl-coenzyme A (HMG-CoA); Intermediate-density lipoprotein (IDL); Low-density lipoprotein (LDL); Niemann–Pick C1-like 1 (NPC1L1); peroxisome proliferator-activated receptor (PPAR-α); Very low-density lipoprotein (VLDL).

Introduction to Drug Research and Development. http://dx.doi.org/10.1016/B978-0-12-397176-0.00015-7

15.1 INTRODUCTION

Cholesterol (Figure 15.1) is essential in the construction and maintenance of membranes in animal cells. It reduces the permeability of the plasma membrane to neutral solute and to ions. In many nerve cells cholesterol is a constituent of the myelin sheath which provides insulation for the conduction of nerve impulses.

Cholesterol is also the precursor for several biochemical pathways. In the liver, cholesterol is sequestered, with a significant portion diverted to bile, which is then stored in the gall bladder. Bile contains bile salts, which solubilize fats in the digestive tracts and aid in their intestinal absorption. Cholesterol is an important precursor for the synthesis of the D vitamins and the steroid hormones.

Cholesterol is so important for animal life that it is synthesized in most cells from simple molecules in a complex 37-step process that starts with the intracellular protein enzyme 3-hydroxy-3-methylglutaryl-coenzyme A (HMG-CoA) reductase that catalyses the production of mevalonic acid from HMG-CoA. Coenzyme A is derived from adenosine triphosphate, pantothenic acid (vitamin B5) and cysteamine. The typical adult total body synthesis of cholesterol is about 1 g/day.

An average adult body contains approximately 140 g of sterols, mainly in the form of cholesterol, derived from two main sources. These are the de novo synthesis of cholesterol by the liver (and extrahepatic sites) and absorption from the intestine. Intestinal absorption is composed of dietary cholesterol (300–500 mg), biliary excretion (800–1200 mg) and desquamated cells (300 mg). A large part of the cholesterol absorbed results from the reabsorption of cholesterol which was initially excreted through the bile. Dietary cholesterol is particularly high in foods such as cheese, egg yolk, beef, pork, poultry, fish and shrimp.

Most dietary cholesterol is esterified, and esterified cholesterol is poorly absorbed and has low solubility in blood, thus it is transported in the circulatory system within lipoproteins. The latter are complex particles that have an exterior of amphiphilic proteins and lipids whose outward-facing surfaces are water soluble, while the inward-facing surfaces are lipid soluble, and the cholesterol esters are carried internally. Cholesterol is recycled. The liver plays a central role in balancing

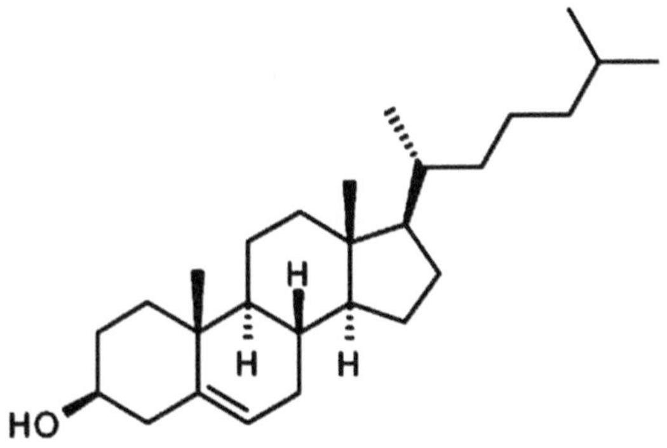

FIGURE 15.1

Cholesterol.

cholesterol from all sources. The liver excretes it in a non-esterified form (via bile) into the digestive tract. About 50% of the excreted cholesterol is reabsorbed by the small bowel back into the bloodstream as a part of the overall process of enterohepatic recirculation.

Lipoprotein particles have cell-targeting signals that direct the lipid they carry to certain tissues, so that there are several types of lipoprotein particle within blood called, in order of increasing density:

Chylomicrons
Very low-density lipoprotein (VLDL)
Intermediate-density lipoprotein (IDL)
Low-density lipoprotein (LDL)
High-density lipoprotein (HDL).

The more lipid and less protein in the particle, the less dense it is. Additional lipoproteins called apolipoproteins act as ligands for specific receptors on cell membranes that determine cholesterol transport. When the LDL receptor system is deregulated, many LDL molecules appear in the blood without receptors on the peripheral tissues. These LDL molecules are oxidized and can contribute to atherosclerotic plaque formation on the walls of blood vessels. The narrowing of blood vessels by plaques is the main cause of heart attacks, stroke and peripheral vascular disease, providing the basis for the association of LDL cholesterol with the concept of 'bad' cholesterol; it is a misnomer because it should be called LDL cholesterol.

In contrast, HDL particles are thought to transport cholesterol back to the liver for excretion or to other tissues that use cholesterol for synthesis of hormones. Having large numbers of large HDL particles has been shown to be cardioprotective, so that HDL cholesterol is associated with the idea of 'good' cholesterol.

The balance between LDL and HDL cholesterol is mostly genetically determined, but can be changed by body build, medication, food choices and other metabolic factors.

Atherosclerotic coronary heart disease is a major cause of death and cardiovascular morbidity in the western world. For this reason, much effort has been expended in seeking to reduce the serum cholesterol levels by drug treatment.

The level of blood plasma cholesterol in the body is affected by the absorption of dietary cholesterol from the intestines, the endogenous biosynthesis of cholesterol, removal of cholesterol from the blood circulation and the reabsorption of cholesterol from the bile.

A high level of dietary saturated fat and cholesterol has been considered to be a major factor contributing to high levels of serum lipids, however, many patients are unable to control their diet or to remain on a strict dietary regimen. A smaller subset of people is genetically predisposed to high serum cholesterol levels despite a strict diet. Hence the basis of medical treatment is the pharmacological help provided by various types of drugs, as follows.

15.2 PAST APPROACHES TO LOWERING LEVELS OF CIRCULATING CHOLESTEROL

15.2.1 Bile acid sequestrants

Bile acids are synthesized in the liver cells by oxidation of cholesterol to give cholic acid and chenodeoxycholic acid, which are then further transformed in the intestine. Removing bile acid from the

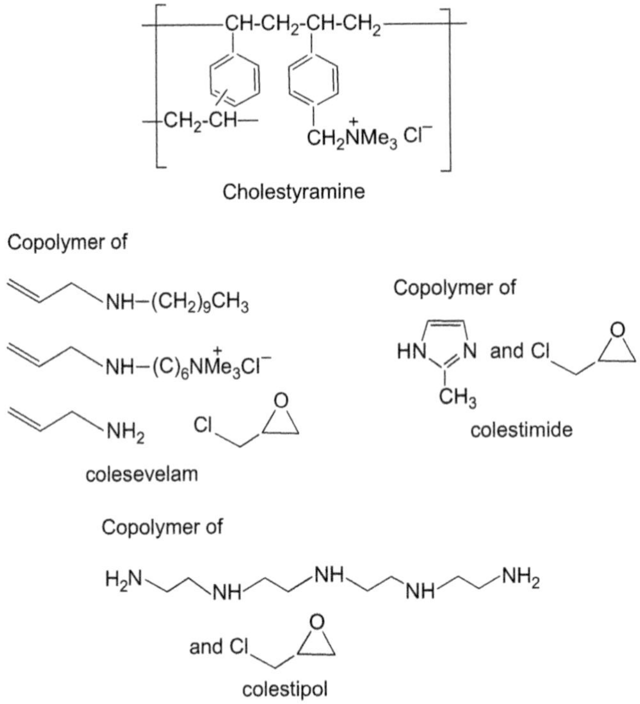

Cholestyramine

Copolymer of

$NH-(CH_2)_9CH_3$

$NH-(C)_6N\overset{+}{M}e_3Cl^-$

NH_2

colesevelam

Copolymer of

$HN \diagdown N$ and Cl
   CH_3

colestimide

Copolymer of

H_2N ... NH ... NH ... NH ... NH_2

and Cl

colestipol

FIGURE 15.2

Bile acid sequestrants.

bile pool leads to upregulation of bile acid synthesis and corresponding increase in transformation of the cholesterol into bile acid and a resulting drop in cholesterol plasma levels.

An early approach to reducing plasma cholesterol levels was the design of bile acid sequestrants that bind bile acids in the gut to prevent their reabsorption. Such sequestrants are cross-linked polymeric cationic gels that serve as ion exchange resins. They exchange anions, such as chloride ions, for bile acids. By doing so, they bind bile acids and sequester them thus preventing reabsorption and enterohepatic circulation. These large polymeric structures are excreted together with the bound bile acids, in faeces. Examples of such drugs are cholestyramine, colesevelam, colestipol and colestimide (Figure 15.2).

The use of bile acid sequestrants is known to inhibit intestinal cholesterol absorption, however, the large dose, limited efficacy, and unpleasant side effects such as constipation, diarrhoea and flatulence have limited the use of these drugs.

15.2.2 Hypocholesterolaemic drugs

The 1950s saw several developments in the discovery of drugs to lower cholesterol levels (hypocholesterolaemic drugs).

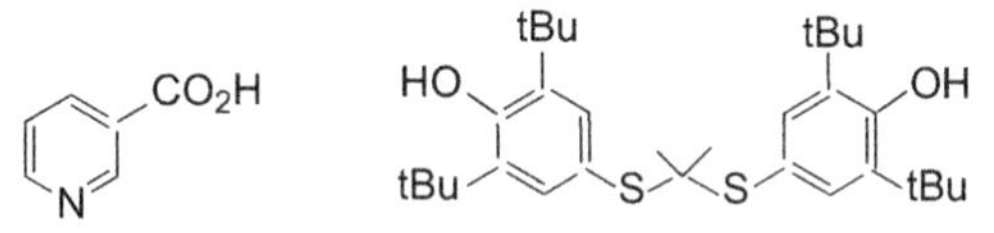

FIGURE 15.3

Niacin and probucol.

Niacin (also known as vitamin B3 and nicotinic acid, Figure 15.3) was discovered at the University of Saskatchewan in 1955 to lower serum cholesterol levels in humans when given in high doses. Niacin binds to a receptor that causes inhibition of fat breakdown in adipose tissue. Lipids that are liberated from adipose tissues are normally used to build VLDL in the liver, which are precursors of the LDL. Because niacin blocks the breakdown of fats, it causes a decrease of free fatty acids in the blood and, consequently, decreases the secretion of VLDL and cholesterol in the liver. However, vasodilatation (skin flushing) is a dose-limiting side effect.

Probucol (Figure 15.3) lowers the level of cholesterol in the bloodstream by increasing the rate of LDL metabolism. Additionally, probucol may inhibit cholesterol synthesis and delay cholesterol absorption. However, it may also lower the level of HDL cholesterol but it has been suggested that it increases HDL function and slows the progression of atherosclerosis.

15.2.3 Inhibition of cholesterol absorption by fibrates: inhibitors of ACAT

Acyl-CoA:cholesterol acyltransferase (ACAT) catalyses the formation of cholesterol esters from cholesterol and long-chain fatty-acyl-coenzyme A. ACAT serves as a regulator of intracellular cholesterol homoeostasis and supplies cholesteryl esters for lipoprotein assembly in the liver and small intestine. Inhibition of this enzyme blocks absorption of intestinal cholesterol and may also inhibit the deposition of cholesteryl esters in the vascular wall (associated with the formation of atherosclerotic plaque).

It was reported in 1953 that clinical trials in France had found that phenylacetic acid and several analogues lowered cholesterol levels. A few years later researchers at Imperial Chemical Industries Laboratories (ICI), England, screened in rats various similar compounds that had previously been synthesized at the company as plant hormone analogues, and discovered the hypocholesterolaemic activity of a series of aryloxyisobutyric acids. The optimum drug selected was the ethyl ester clofibrate, reported in 1962, which was approved in the USA in 1967 and became established as a very effective drug. However, after some years it was found to cause unacceptable side effects and was replaced by safer analogues, such as gemfibrozil, fenofibrate, bezafibrate, ciprofibrate, and etofibrate (Figure 15.4).

In the 1980s the fibrates were shown to act at the peroxisome proliferator-activated receptor (PPAR-α) in muscle, liver and other tissues. PPAR-α belongs to a class of intracellular receptors that modulate carbohydrate and fat metabolism. Activation of PPAR-α signalling results in increased β-oxidation in the liver, decreased hepatic triglyceride secretion, and increased lipoprotein lipase activity resulting in an increase in VLDL clearance, and an increase in HDL levels.

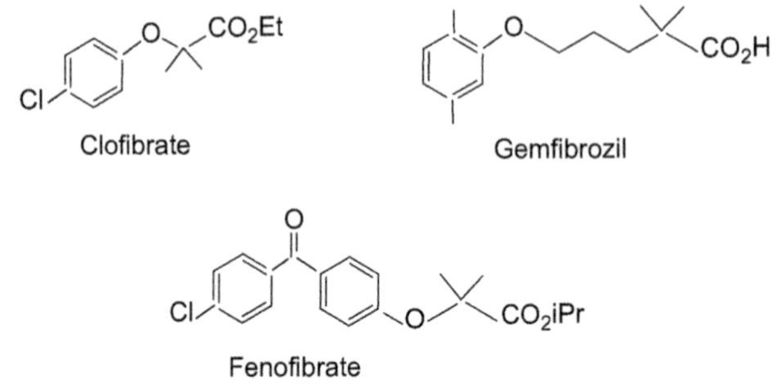

FIGURE 15.4

Some fibrates.

15.2.4 Statins: inhibitors of HMG-CoA reductase

In 1976 two research groups independently isolated mevastatin from fungi. Endo who was with the Sankyo Company in Tokyo screened over 8000 microbial extracts for evidence of inhibition of sterol biosynthesis and isolated mevastatin from *Penicillium citrinum*. Beecham Laboratories in England isolated mevastatin from *Penicillium brevicompactum* and named it compactin. A second, more active compound was isolated in Tokyo from the fungus *Monascus ruber*; the same compound was isolated at Merck Sharp and Dohme (Rahway, NJ, USA) from the fungus *Aspergillus terreus* who named it mevinolin, now known as lovastatin (Figure 15.5).

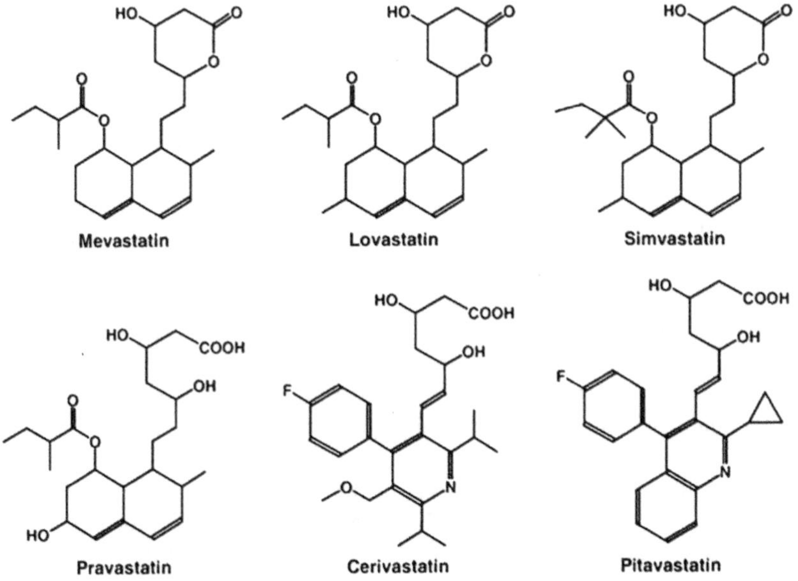

FIGURE 15.5

Some statins.

Mevastatin (Figure 15.5) was shown by Endo to inhibit the enzyme that regulated the hepatic synthesis of cholesterol where at least 60% of the body's cholesterol is synthesized. The enzyme inhibited, as later identified, an HMG-CoA reductase (see Section 15.1) which catalyses the rate-limiting conversion of HMG-CoA to mevalonic acid. Mevastatin lowers cholesterol levels in the liver causing more LDL-cholesterol receptors to be expressed. Circulating LDL-cholesterol levels then fall due to its increased uptake by the greater number of receptors.

Lovastatin was the first statin to be developed and was launched by Merck Sharp and Dohme (USA) in 1987. It is a twofold more potent methyl homologue of mevastatin. Other statins (Figure 15.5) have followed, such as simvastatin (MSD), pravastatin (Sankyo), fluvastatin (Novartis), atorvastatin (Pfizer), rosuvastatin (AstraZeneca), and pitavastatin (Nissan Chemical Industries). By 2003, atorvastatin (Lipitor^{TM}) became the best-selling pharmaceutical in history and, as a class, the most profitable drugs in the pharmaceutical history.

Statins have been effective in decreasing mortality in people with pre-existing cardiovascular diseases. They are also advocated for use in patients at a high risk of developing heart disease. Statins can lower LDL cholesterol by enough to account for a 60% decrease in a number of cardiac events (heart attack and sudden cardiac death) and a 17% reduced risk of stroke after long-term treatment. They have less effect than the fibrates or niacin in reducing triglycerides and raising HDL cholesterol.

The positive results with the statins led to attempts to achieve even better outcomes of treatment by increasing dose levels or combining statins with other drugs such as fibrates. The consequence was the appearance of unwanted side effects such as raised liver enzymes and muscle problems; the statin, cerivastatin (Bayer) (Figure 15.5), was especially toxic resulting in 52 fatal cases of rhabdomyolysis (muscle breakdown) and the resulting kidney failure. Unfortunately these results came to light only during post-marketing surveillance and the drug was voluntarily withdrawn from the market in 2001. Another 385 nonfatal cases of rhabdomyolysis were reported. Rhabdomyolysis was also exacerbated in cases where a fibrate had been co-administered with a statin, e.g. gemfibrozil with cerivastatin.

15.2.5 Inhibition of cholesterol absorption by saponins

Some natural products have been reported to inhibit cholesterol absorption. Saponins, which are steroidal glycosides found in plants, were explored by Pfizer researchers in the 1990s, who published on two synthetic analogues, tiqueside (CP-88818) (Figure 15.6) (a synthetic saponin beta-tigogenin

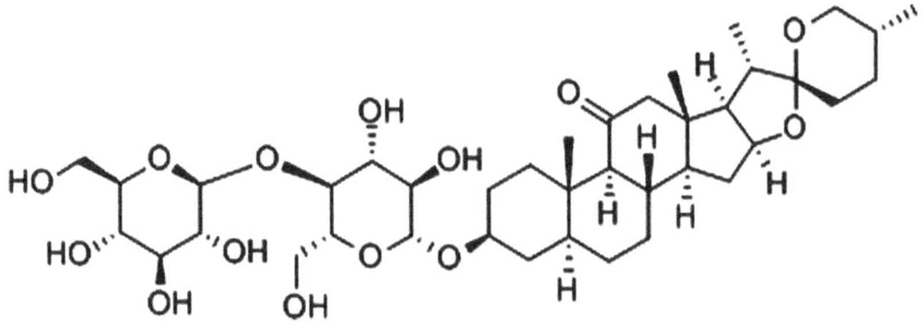

FIGURE 15.6

Synthetic saponin analogue, pamaqueside (tiqueside is the non-carbonyl analogue).

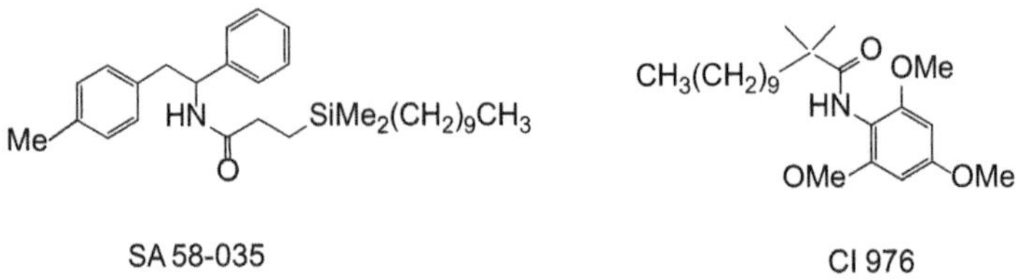

FIGURE 15.7

More ACAT inhibitors.

cellobioside whose structure is based upon naturally occurring saponins) and a ketonic analogue, pamaqueside (Figure 15.6) (CP 148623).

Both these compounds were shown to produce a dose-dependant reduction in plasma LDL-cholesterol levels in hypercholesterolaemic patients. However, although pamaqueside advanced to phase III clinical trials, neither pamaqueside nor tiqueside was marketed.

15.3 FURTHER WORK ON ACAT INHIBITORS

A renewed interest in fibrates appeared in the 1990s when the mode of action of fibrates became understood and the results of several large clinical trials demonstrated the efficacy of fibrates on cardiovascular events (*1*). Although statins had become the first-line hypolipidaemic drugs, fibrates were still a widely prescribed class of hypolipidaemic drugs. Of special interest for the present discussion was the amide compounds identified by research at the companies Sandoz and Parke Davis, in particular, SA 58-035 and CI 976, respectively (Figure 15.7).

A broad chemical programme was underway to discover novel ACAT inhibitors at Schering-Plough Research Institute (Kenilworth, NJ, USA) [e.g. Ref. (*2*)]. They initially explored the conformational preference of the phenylethyl amide class of ACAT inhibitors, exemplified by SA 58-035, by making conformationally constrained analogues (*3*). Of various approaches made, one involved bridging of the two carbon atoms of the ethyl linkage and the nitrogen atom in the simplest ring possible, namely a 2-azetidinone. Their idea was that the four-membered ring system would provide a rigid scaffold to allow investigation of the substitution pattern and the three-dimensional relationship of substituents. An early target was the diphenyl derivative A containing an N-(*p*-methoxyphenyl) group, synthesized by ethyl phenylacetate condensation with a benzylideneaniline, as shown in Scheme 15.1, where Ar = *p*-methoxyphenyl. Two products were isolated from the reaction, A and B, and both were tested.

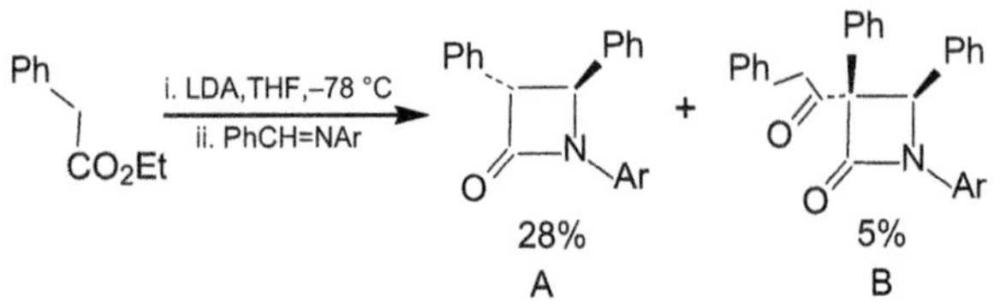

SCHEME 15.1

Synthesis of the lead compound B (*3*).

The compounds were tested in two principal biological screens. The first was an *in vitro* ACAT assay measuring inhibition of the esterification of cholesterol by oleic acid. The second test was *in vivo*, measuring inhibition of the rise in plasma cholesterol levels and hepatic cholesteryl ester accumulation (as a marker for cholesterol absorption) in 7 day cholesterol-fed hamsters (*2*).

Product A was not active @ 100 mg/kg but the second product, B, arising by acylation of the initial adduct (Scheme 15.1) did show some activity, particularly *in vivo*. This lead was followed up by making a series of 2-azetidinones which were mono- or disubstituted at position-3 using H, Et, Ph or Ph(CH$_2$)$_3$–; in addition, H, phenyl or *p*-methoxyphenyl groups were introduced at position-4, and *p*-methoxyphenyl or 2,4,6-trimethoxyphenyl on the N (*3*). These methoxy groups were introduced by analogy with the researchers' previous work on ACAT inhibitors, and the trimethoxy groups were added by analogy with the Parke Davis compound CI 976.

Analogues with the N1 position substituted with the trimethoxyphenyl group were much more active *in vitro* but were less active *in vivo*. Other 2-azetidinone analogues too showed a poor correlation between *in vivo* and *in vitro* activities. This divergence between the two sets of results led the researchers to follow the *in vivo* results. The compound with *trans* substituents was of sufficient interest to be resolved into its two enantiomers and the (+) enantiomer was found to be a twofold better ACAT inhibitor (IC$_{50}$ = 11 μM). The (−) enantiomer (SCH 48461, Scheme 15.2) had much greater *in vivo* activity, however (ED$_{50}$ approximately 2 mg/kg/day for inhibition of the rise in hepatic cholesteryl esters in the 7 day hamster cholesterol absorption assay), whereas the (+) enantiomer had no

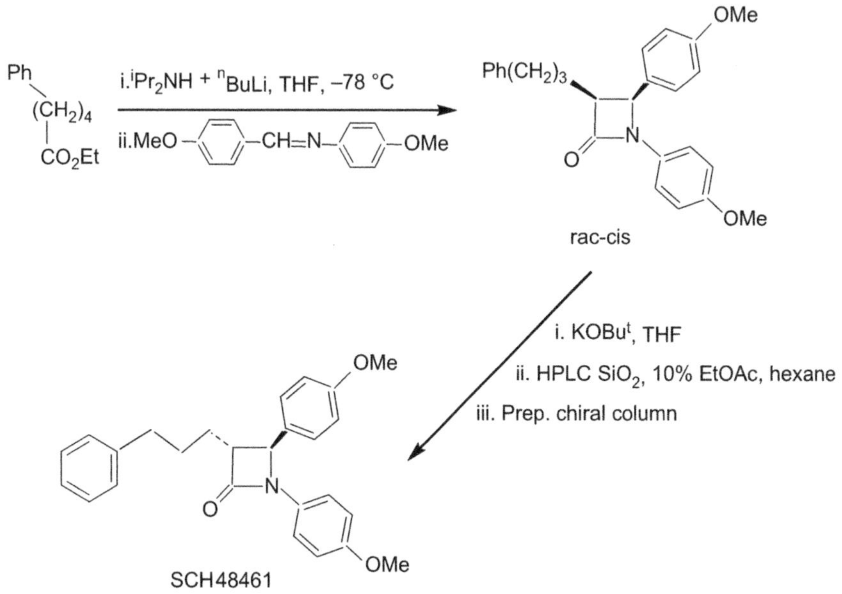

SCHEME 15.2

Synthesis of SCH 48461 (*3*).

significant effect in reducing the level of hepatic cholesteryl esters when given at the much higher dose of 50 mg/kg/day (3).

15.4 INHIBITION OF A NOVEL MECHANISM FOR CHOLESTEROL UPTAKE AND THE DISCOVERY OF SCH 48461

The differences in *in vivo* activity of the lead compounds are clearly not reflected in the *in vitro* ACAT activity, although selective metabolism *in vivo* could be a possible explanation. It was also shown that this class of compounds blocks absorption of free cholesterol, whereas ACAT inhibitors block the subsequent esterification of intracellular cholesterol. Therefore it was hypothesized that this class of inhibitors blocked an earlier step in the absorption of dietary cholesterol than do ACAT inhibitors. This suggested that these compounds acted by a novel mechanism. Further studies indicated that the SCH 48461 acts at the intestinal wall to directly inhibit cholesterol absorption through an unknown mechanism, which fundamentally challenged the passive cholesterol transport model (4).

SCH 48461 was synthesized (Scheme 15.2) by the ester enolate-imine condensation of ethyl phenylvalerate with *N*-(4-methoxybenzylidene)anisidine in the presence of lithium diisopropylamide base, followed by epimerization with KO^{t}Bu giving a 4:1 mixture of *trans*- and *cis*-diastereomeric azetidinones, respectively, which were separated chromatographically. The two enantiomers of the *trans*-racemate were separated by chiral chromatography; SCH 48461 being the (−) enantiomer with absolute 3R, 4S stereochemistry (3). A chiral synthesis was also devised (5).

SCH 48461 was demonstrated to reduce serum levels of total cholesterol and of liver cholesteryl esters in a dose-dependant manner when given orally to 7 day cholesterol-fed hamsters with an $ED_{50} = 2.2$ mg/kg/day. It was also tested in a 7 day rat model ($ED_{50} = 2$ mg/kg/day), in 21 day rhesus monkey ($ED_{50} = 0.2$ mg/kg/day) and in a dog ($ED_{50} = 0.1$ mg/kg/day) models, establishing that its action was not limited only to hamsters as a species (6). The (−) compound, SCH 48461, was selected as a drug candidate for investigation in human subjects despite a clear understanding of its mechanism of action. After appropriate safety testing and initial human studies the compound was evaluated in an 8 week human trial in which it was found to reduce serum LDL-cholesterol levels by 15% at a dose of 25 mg/day (7). Pharmacokinetic characterization indicated rapid and complex metabolism in humans.

15.5 DESIGN OF EZETIMIBE (SCH 58235)

The divergence in the structure–activity patterns between the *in vitro* (ACAT) and *in vivo* (plasma cholesterol levels) led the researchers to initiate an intensive search for the mechanism of action and to abandon the *in vitro* assay. Structure optimization using a 7 day *in vivo* test system without the aid of any *in vitro* assays complicated the programme and was a very brave decision. The biological characterization initially required >100 mg quantities of analogues, and the dosing of animals over a full week with subsequent serum and tissue analysis. A wide ranging structure–activity investigation was undertaken with this paradigm, and the hypothesis that the metabolic fate of close analogues would be similar [e.g. Refs (8,9)]. Initial follow-up structure–activity studies on SCH 48461 showed that the azetidinone ring was critical for *in vivo* activity. Neither the thioazetidinone nor the amino acid derived

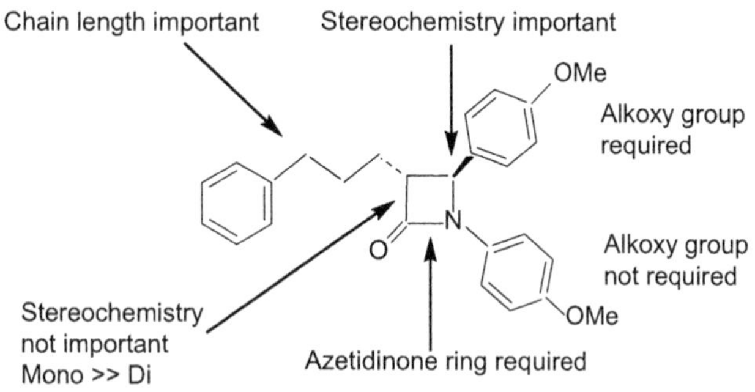

FIGURE 15.8

Summary of azetidinone structure–activity relationships.

from ring opening SCH 48461 had any significant activity, and other related lactams showed, at best, weak activity. An investigation of Structure Activity relationship (SAR) focused around the 2-azetidinone scaffold was therefore initiated. Over 150 compounds were synthesized and tested *in vivo* and the SAR results are summarized in Figure 15.8. In an effort to understand and separate effects on intrinsic potency, from potential effects on pharmacokinetics, it became very important to examine the metabolic fate of the lead compound SCH 48461.

SCH 48461 was found to be rapidly and completely metabolized in rodents and mammals. The metabolism of SCH 48461 was studied in the rat, which showed a rapid appearance of a complex metabolite mixture in the bile. Furthermore, the metabolite mixture was tested and found to have greater inhibitory activity of cholesterol absorption than did SCH 48461. An experiment was devised to determine what were active metabolites, using ^{3}H-SCH 48461 and ^{14}C cholesterol in intestinally cannulated bile-duct-diverted rats from which was isolated the glucuronide of compound **3** (Table 15.1), a phenolic metabolite of SCH 48461 (*10*). The identity of this compound was determined and tests suggested that much of the *in vivo* activity of SCH 48461 was due to the formation of 3 and that the metabolism of SCH 48461 to 3 helped to localize the compound at the intestinal wall, the putative site of action. This led the researchers to believe that a metabolite-like analogue with improved pharmacokinetic and pharmacodynamic action would provide an improved candidate for inhibiting cholesterol absorption.

Likely modes of primary metabolism of SCH 48461 were hypothesized (*11*) (shown in Figure 15.9) as:

i, ii – dealkylation of the *p*-methoxyphenyl groups at N1 and C4
iii – para-hydroxylation of the phenyl propyl group
iv – benzylic oxidation of the phenyl propyl group, possibly followed by ketone formation
v – 2-azetidinone ring opening.

From these five likely sites of metabolism, more than 40 different potential metabolites were predicted (*11*). Furthermore, the stereochemical consequences of benzylic hydroxylation and potential intramolecular cyclization/lactone formation of the benzylic hydroxyl metabolites made the preparation of all possible metabolites a mountainous task. The synthesis of the most probable metabolites

Table 15.1 Activity of SCH 48461 and Some of Its Putative Metabolites (*11*)

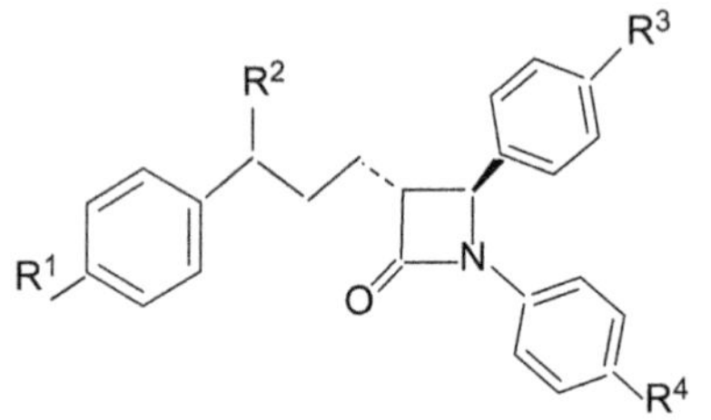

Compound	R^1	R^2	R^3	R^4	Activity*
SCH 48461	H	H	OMe	OMe	2.2
1	OH	H	OMe	OMe	−16%
2	H	H	OMe	OH	−78%
3	H	H	OH	OMe	5
4	H	OH(S)	OH	OMe	0.3
5	H	OH(R)	OH	OMe	3
6	H	=O	OH	OMe	3

*Activity is the ED_{50} mg/kg/day to reduce the level of liver-cholesterol esters, or percentage reduction at 10 mg/kg/day, in the cholesterol-fed hamster during 7 days.

was undertaken and the compounds were assayed for activity using the 7 day cholesterol-fed hamster test. Those putative metabolite series synthesized are represented by the general structure in Table 15.1. Some of the examples with their activities are shown in Table 15.1. Hydroxylation at R^1 and R^4 gave compounds (1 and 2 respectively) that were less active than SCH 48461 suggesting that such metabolites were not required for activity. Hydroxylation at R^3 (**3**), as indicated above, retains potency. Another route of metabolism involved hydroxylation of the 3-phenylpropyl side chain to produce alcohols and ketones. The chiral (S) alcohol was substantially more active than SCH 48461; the chiral (R) alcohol and the corresponding ketone were less active.

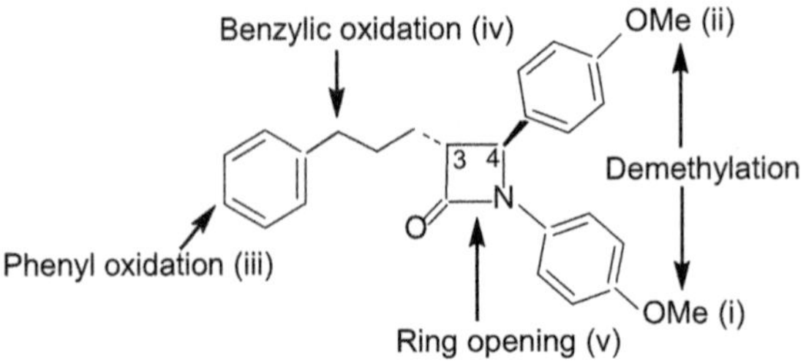

FIGURE 15.9

Putative sites of metabolism of SCH 48461.

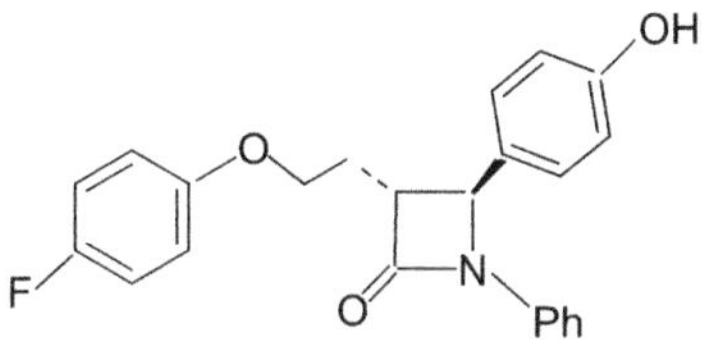

FIGURE 15.10

One example of designing a metabolically more stable analogue.

The most potent of the putative metabolites was analogue **4** (Table 15.1) having $R^1 = H$, $R^2 = OH(S)$, $R^3 = OH$, and $R^4 = OCH_3$. Previous structure–activity results had shown that R^3 should be OH or alkoxyl but that R^4 could take a wide range of substituents. The use of halogen to block sites of metabolism is well known and fluorine was selected due to its small steric demand and its deactivating effect to lessen the likelihood of P_{450}-mediated aromatic hydroxylation of a phenyl ring. The various observations on metabolism together with the results from putative metabolites gave rise to a strategy for the design of an improved analogue: provide the functionality at the sites where the metabolism increases potency (i.e. at R^2 and R^3) and block the sites where oxidative metabolism decreases potency (i.e. at R^1 and R^4). This strategy was applied to several chemical series related to SCH 48461, such as the alkoxy series with a blocked benzylic position, e.g. Figure 15.10, wherein the $(-)$ enantiomer was equipotent with $(-)$ SCH 48461 and, as desired, was metabolically more stable and had very low plasma levels (*12*).

However, the above reasoning was used most successfully by Rosenblum et al. (*11*). This led directly to the synthesis of the structure with $R^1 = F$, $R^2 = OH(S)$, $R^3 = OH$, and $R^4 = F$. The logical design of this compound, which is ezetimibe, from SCH 48461 is summarized in Figure 15.11 (*12*).

A representative synthesis of ezetimibe is shown in Scheme 15.3. It is based on an Evenan-type oxazolidinone condensation to establish the correct stereochemistry on the azetidinone ring and thereafter includes a Corey oxazaborolidine reduction to install the (S)-stereochemistry of the side chain OH group (*6*).

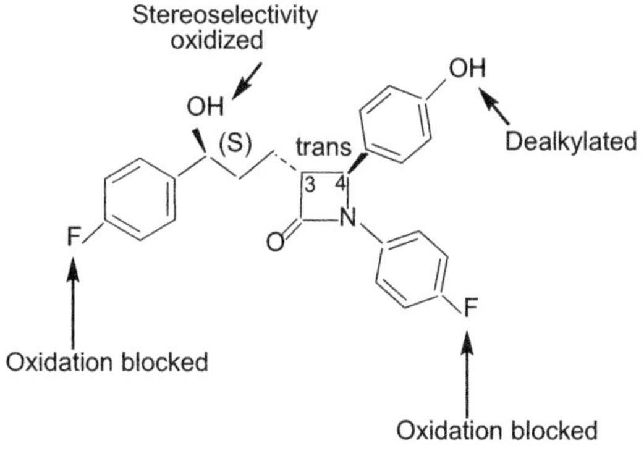

FIGURE 15.11

Design of ezetimibe (SCH 58235) (*12*).

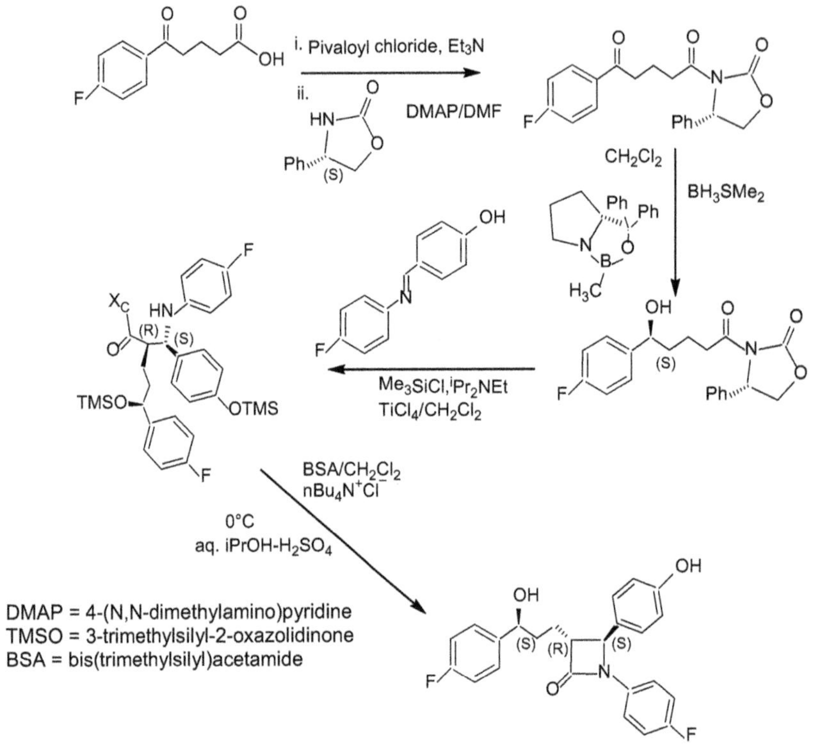

SCHEME 15.3

Representative synthesis of ezetimibe (*6*).

Ezetimibe had an ED_{50} of 0.04 mg/kg/day in the 7 day cholesterol-fed hamster, a 50-fold increase in potency relative to SCH 48461. In the rat the ED_{50} was 0.03 mg/kg/day and in the dog the ED_{50} was 0.007 mg/kg/day. The cholesterol-fed rhesus monkey showed remarkable activity, having an ED_{50} of 0.0005 mg/kg/day, 400 times more potent than SCH 48461 (*6*). Furthermore, ezetimibe showed substantially lower blood plasma drug levels of drug than did SCH 4846, the expected simplified metabolic profile and better physical properties than SCH 48461.

Ezetimibe undergoes rapid glucuronide formation in the intestine of its phenolic (at R^3) site. The glucuronide is more potent than ezetimibe itself since it localizes more avidly on the intestinal villi (*13*), and is reabsorbed and then excreted in the bile, thereby sending the drug back to the site of action. Ezetimibe and its glucuronide both undergo enterohepatic recycling, with a half-life of approximately 24 h in humans, and are excreted in the faeces (90%) and urine (10%) (*13*).

15.6 EZETIMIBE IN HUMAN STUDIES

Table 15.2 shows the results of phase III human clinical trials with ezetimibe in comparison with placebo (*6*). There is a significant reduction in total cholesterol, LDL cholesterol and triglycerides, and

Table 15.2 Ezetimibe Phase III Monotherapy Efficacy Results (6)

Treatment	Mean% Change from Baseline at Endpoint			
	LDL cholesterol	Total cholesterol	HDL cholesterol	Triglycerides
Placebo (n = 226)	+0.4	+0.8	−1.6	+5.7
Ezetimibe, 10 mg (n = 666)	−16.9*	−12.5*	+1.3*	−5.7

*Significantly different from placebo results (p < 0.01).

a small but significant increase in HDL cholesterol. Ezetimibe is available as a 10 mg tablet for a once-daily treatment and is marketed as Zetia™ (USA) or Ezetrol™ (UK and elsewhere). Zetia was approved by the Food and Drug Administration (FDA) (in the USA) in October 2002, at least 12 years after the research project started. The time from the initiation of the research programme to the discovery of ezetimibe was approximately 6 years (6). Even though ezetimibe has been proven to decrease cholesterol absorption in the intestine by a novel mechanism, and significantly decrease plasma cholesterol levels, its effect in preventing cardiovascular events and heart disease is still a matter of controversy.

Although ezetimibe was very active in cholesterol-fed animals, it was not active in animals fed a normal diet. This observation led Davis (14) to reason that a cholesterol-absorption inhibitor might upregulate hepatic HMG-CoA reductase activity, so that it could compensate for the reduced cholesterol level. To test this idea ezetimibe (at 0.007 mg/kg/day) or lovastatin (5 mg/kg/day) was administered to dogs over 14 days. At these doses the compounds had only a small effect on serum cholesterol levels but when given in combination they produced a profound reduction in serum cholesterol. Similar effects were seen in other animal species and with other statins.

This led to a study in human subjects where it was shown that the reduction in LDL cholesterol by a statin could be potentiated by the addition of 10 mg of ezetimibe. Furthermore a 40 mg dose of lovastatin produced a similar effect to 10 mg of lovastatin plus 10 mg of ezetimibe in lowering LDL-cholesterol levels. Similar results were obtained using other statins such as simvastatin, pravastatin and atorvastatin (15,16). The combination of simvastatin (10, 20, 40 or 80 mg) and 10 mg ezetimibe, marketed as Vytorin™ (USA) or Inergy™ (UK and elsewhere) was approved by the FDA in July 2004.

In June 2006, the FDA approved the use of ezetimibe in combination with fenofibrate to treat mixed hyperlipidaemia. The combination produces a significantly greater reduction in LDL cholesterol than with either drug alone and greater reductions in triglycerides than with fenofibrate alone.

15.7 IDENTIFICATION OF A NEW MECHANISM FOR CHOLESTEROL UPTAKE

The discovery of SCH 48461 and ezetimibe led to a determined search for the molecular target. The remarkable activity in animal tests suggested interaction with a fundamental mechanism for a cholesterol trafficking. The requirement for a precise stereochemical preference also suggested a specific molecular interaction at a protein site in the cholesterol absorption pathway (17). Using a genetic approach, scientists at Schering-Plough Research Institute identified Niemann–Pick C1-like 1 (NPC1L1) protein as a critical mediator of cholesterol absorption and an essential component of the ezetimibe-sensitive pathway. Ezetimibe had no effect in NPC1L1-knockout mice (*18*). This protein is abundantly expressed in the small intestine, particularly in the jejunum at the brush border membrane.

They, together with scientists at Merck, Rahway, NJ, then determined that NPC1L1 is the direct molecular target of ezetimibe. They showed that ezetimibe glucuronide binds specifically to a single site in intestinal brush border membranes. Furthermore the binding affinities of ezetimibe and key analogues to recombinant NPC1L1 were virtually identical to those observed for native enterocyte membranes. K_D values of ezetimibe glucuronide for mouse, rat, rhesus monkey and human NPC1L1 were 12,000, 540, 40 and 220 nM, respectively (*19*). For a review, see Ref. (*20*).

Niemann–Pick diseases are genetic diseases (there are four types caused by mutations in the NPC1 gene) which cause a lipid storage disorder in which harmful quantities of fatty substances or lipids accumulate in the spleen, liver, lungs, bone marrow and brain. Loss of myelin in the central nervous system is considered to be the main pathogenic factor.

15.8 CONCLUSION

Thus, this remarkable example of drug discovery started out as an attempt to make an analogue of leads which, though characterized pharmacologically, had not yet yielded a clinically useful product. A perceptive analysis with a decision to take on the challenge of using an *in vivo* assay that required 7 days of dosing paid off. It led to a serendipitous discovery of a new mechanism of drug action and the characterization of a previously unknown mechanism of cholesterol uptake. It also involved the clever design, using biochemical analysis together with medicinal chemistry and non-routine synthetic chemistry, of a very potent candidate drug. Subsequent efforts to improve the potency of ezetimibe by *in vitro* screening have identified compounds with greater receptor affinity, but not with better *in vivo* potency. This work represents a true breakthrough in the search for drugs that reduce the intake of dietary cholesterol and has made a fundamental contribution to further understanding of the intake mechanism of cholesterol into the body. This discovery also shows how cholesterol metabolism, which is one of the most studied of biological systems, is still not fully elucidated. It emphasizes how our understanding of body functions is continually developing. This research provides an outstanding example of how *analogue-based* drug research can lead to the discovery of a *pioneer drug*. It furnishes a good riposte to those critics of the pharmaceutical industry who claim that the recent decade has seen nothing new.

Acknowledgement

The author gratefully acknowledges the contribution in helpful reviewing by Dr Stuart B. Rosenblum, one of the inventors of Ezetimibe.

References

1. Sliskovic, D. R.; Picard, J. A.; Krause, B. R. ACAT Inhibitors. The Search for a Novel and Effective Treatment of Hypercholesterolemia and Atherosclerosis. *Prog. Med. Chem.* **2002**, *3*, 121–171.

2. Clader, J. W.; Berger, J. G.; Burrier, R. E.; Davis, H. R.; Domalski, M.; Dugar, S.; Kogan, T. P.; Salisbury, B.; Vaccaro, W. Substituted (1,2-Diarylethyl)amide acyl-CoA:acyltransferase Inhibitors: Effect of Polar Groups on *In vitro* and *In vivo* Activity. *J. Med. Chem.* **1995,** *38,* 1600–1607.

3. Burnett, D. A.; Caplen, M. A.; Davis, H. R., Jr.; Burrier, R. E.; Clader, J. W. 2-Azetidinones as Inhibitors of Cholesterol Absorption. *J. Med. Chem.* **1994,** *37,* 1733–1736.

4. Salisbury, B. G.; Davis, H. R.; Burrier, R.; Burnett, D. A.; Boykow, G.; Caplen, M. A.; Clemmons, A. L.; Compton, D. S.; Hoos, L. M.; McGregor, D. G.; Schnitzer-Polokoff, R.; Smith, A. A.; Weig, B. C.; Zilli, D. L.; Clader, J. W.; Sybertz, E. J. Hypocholesterolemic Activity of a Novel Inhibitor of Cholesterol Absorption SCH 48461. *Atherosclerosis* **1995,** *115,* 45–63.

5. Burnett, D. A. Asymmetric Synthesis and Absolute Stereochemistry of Cholesterol Absorption Inhibitor, SCH 48461. *Tetrahedron Lett.* **1994,** *35,* 7339–7342.

6. Clader, J. W. The Discovery of Ezetimibe: A View from Outside the Receptor. *J. Med. Chem.* **2004,** *47,* 1–9.

7. Bergman, M.; Morales, H.; Mellars, L.; Kosoglou, T.; Burrier, R.; Davis, H. R.; Sybertz, E. J.; Pollare, T. The Clinical Development of a Novel Cholesterol Absorption Inhibitor. *Proceedings of the XII International Symposium on Drugs Affecting Lipid Metabolism* **1995.** Nov. 7–10, Houston, TX, p.5.

8. Clader, J. W.; Burnett, D. A.; Caplen, M. A.; Domalski, M. S.; Dugar, S.; Vaccaro, W.; Sher, R.; Brocone, M. E.; Zhao, H.; Burrier, R. E.; Salisbury, B.; Davis, H. R., Jr. 2-Azetidone Cholesterol Absorption Inhibitors: Structure–Activity Relationships on the Heterocyclic Nucleus. *J. Med. Chem.* **1996,** *39,* 3684–3693.

9. McKittrick, B. A.; Ma, K.; Huie, K.; Yumibe, N.; Davis, H., Jr.; Clader, J. W.; Czarniecki, M. Synthesis of C3 Heteroatom-Substituted Azetidinones that Display Potent Cholesterol Absorption Inhibitory Activity. *J. Med. Chem.* **1998,** *41,* 752–759.

10. van Heek, M.; France, C. F.; Compton, D. S.; McLeod, R. L.; Yumibe, N. P.; Alton, K. B.; Sybertz, E. J.; Davis, H. R., Jr. In vivo Metabolism-Based Discovery of a Potent Cholesterol Absorption Inhibitor, SCH 58235, in the Rat and Rhesus Monkey through Identification of the Active Metabolites of SCH 48461. *J. Pharmacol. Exp. Ther.* **1997,** *283,* 157–163.

11. Rosenblum, S. B.; Huynh, T.; Afonso, A.; Davis, H. R., Jr.; Yumibe, N.; Clader, J. W.; Burnett, D. A. Discovery of 1-(4-fluorophenyl)-(3R)-[3-(4-fluorophenyl)-(3S)-hydroxypropyl]-(4S)-(4-hydroxyphenyl)-2-azetidinone (SCH 58235): A Designed, Potent, Orally Active Inhibitor of Cholesterol Absorption. *J. Med. Chem.* **1998,** *41,* 973–980.

12. Dugar, S.; Yumibe, N.; Clader, J. W.; Vizziano, M.; Huie, K.; Van Heek, M.; Compton, D. S.; Davis, H. R. Metabolism and structure activity data based drug design: discovery of (−) SCH 53079 an analog of the potent cholesterol absorption inhibitor (−) SCH 48461. *Bioorg. Med. Chem. Lett.* **1996,** *6,* 1271–1274.

13. van Heek, M.; Farley, C.; Compton, D. S.; Hoos, L.; Alton, K. B.; Sybertz, E. J.; Davis, H. R. Comparison of the Activity and Disposition of the Novel Cholesterol Absorption Inhibitor, SCH 58235, and Its Glucuronide, SCH 60663. *Br. J. Pharmacol.* **2000,** *129,* 1748–1754.

14. Davis, H. R., Jr.; Pula, K. K.; Alton, K. B.; Burrier, R. E.; Watkins, R. W. The Synergistic Hypocholesterolemic Activity of the Potent Cholesterol Absorption Inhibitor, Ezetimibe, in Combination with 3-Hydroxy-3-methylglutaryl Coenzyme A Reductase Inhibitors in Dogs. *Metab. Clin. Exp.* **2001,** *50,* 1234–1241.

15. Kosoglou, T.; Meyer, I.; Veltrie, E.; Statkevich, P.; Yang, B.; Zhu, Y.; Mellars, L.; Maxwell, S. E.; Patrick, J.; Cutler, D.; Batra, V.; Affrime, M. Pharmacodynamic Interaction between the New Selective Cholesterol Absorption Inhibitor Ezetimibe and Simvastatin. *Br. J. Clin. Pharmacol.* **2002,** *54,* 309–319.

16. Gagne, C.; Gaudet, D.; Brucker, E. Efficacy and Safety of Ezetimibe Coadministered with Atorvastatin or Simvastatin in Patients with Homozygous Familial Hypercholesterolemia. *Circulation* **2002,** *105,* 2469–2475.

17. Dugar, S.; Clader, J. W.; Chan, T.-M.; Davis, H., Jr. Substituted 2-Azaspiro[5.3]nonan-1-ones as Potent Cholesterol Absorption Inhibitors: Defining a Binding Conformation for SCH 48461. *J. Med. Chem.* **1995,** *38,* 4875–4877.

18. Altmann, S. W.; Davis, H. R., Jr.; Zhu, L. J.; Yao, X.; Hoos, L. M.; Tetzloff, G.; Iyer, S. P.; Maguire, M.; Golovko, A.; Zeng, M.; Wang, L.; Murgolo, N.; Graziano, M. P. Niemann–Pick C1 Like 1 Protein Is Critical for Intestinal Cholesterol Absorption. *Science* **2004,** *303,* 1201–1204.

19. Garcia-Calvoa, M.; Lisnock, J. M.; Bull, H. G.; Hawes, B. E.; Burnett, D. A.; Braun, M. P.; Crona, J. H.; Davis, H. R., Jr.; Dean, D. C.; Detmers, P. A.; Graziano, M. P.; Hughes, M.; MacIntyre, D. E.; Ogawa, A.; ÓNeill, K. A.; Iyer, S. A. N.; Shevell, D. E.; Smith, M. M.; Tang, Y. S.; Makarewicz, A. M.; Ujjainwalla, F.; Altmann, S. W.; Chapman, K. T.; Thornberry, N. A. The Target of Ezetimibe is Niemann–Pick C1-Like 1 (NPC1L1). *Proc. Natl. Acad. Sci. U S A* **2005,** *102,* 8132–8137.

20. Lammert, F.; Wang, D. Q. H. New Insights into the Genetic Regulation of Intestinal Cholesterol Absorption. *Gastroenterology* **2005,** *129,* 718–734.

Index

Note: Page numbers with "f" denote figures; "t" tables; "b" boxes.

A

Abciximab, 165t–168t, 192, 194t, 195
Acetylcholine receptors (AChRs), 11–13, 13t
Actimmune, 135
Acyl-coA:cholesterol acyltransferase (ACAT) inhibitors. *See* Cholesterol
Adenosine triphosphate (ATP). *See* Receptors
AIDS (acquired immunodeficiency syndrome), 356–357, 372–374, 378–379
Amino acids, 10t, 23, 59, 60t–62t
    cysteine, 59, 63f
    glycine, 59
    histidine, 59
    structure, 58–59
Amoxicillin, 89, 90f
Amprenavir, 363, 364f, 370–372, 372t
Amyloid, 151–152
Anderson-Fabry disease, 332
Anaemia, 308–310, 311f, 317, 320–321, 323–324
Angiotensin cleaving enzyme (ACE), 73, 73f
Anilinoquinazolines, 259–262, 260f, 260t, 261f
Antibody, 131, 131f, 142. *See also* Monoclonal antibody (mAb/Mab)
Antibody-dependent cellular cytotoxicity (ADCC), 288–289
Antibody–drug conjugates (ADCs), 300–301
Anticancer agent gefitinib. *See* Epidermal growth factor receptor (EGFR) tyrosine kinase
Antidrug antibodies (ADA), 220
Antigenicity, 172t, 175, 187t–188t, 198
Anti-HIV protease inhibitor. *See* Darunavir
Anti-tumour necrosis factor (anti-TNF) agents
    binding mode, 387, 388f–389f
    biopharmaceutical agents, 387, 390f
    efficacy, 392–393
    granulomas, 394–395
    IgG4 format, 389
    immunogenicity, 394
    mAbs, 387–389, 394–395
    membrane TNF, 394–395
    pegylation, 389, 391f
    soluble receptors, 391–392
    TNFR1 structure, 392f
    tolerance, 393
    septic shock to inflammatory diseases
        collagen-induced arthritis model, 386
        gram-negative shocks, 386
        haemorrhagic necrosis, 386
Aprepitant discovery, 121–123, 122t, 123f

B

Best-selling drugs, US, 227, 228t
Beta-blockers, 13–14
Bicyclic bis-tetrahydrofuran (bis-THF) ligand
    P2 ligand, 366, 366f
    sulfolane ring, 365f, 366
    sulfonamide-based isosteres, 367, 367f
    synthesis of, 376, 376f
    X-ray crystal structure, 367
Bile acid sequestrants. *See* cholesterol
Biobetter, 319
Biologics, 163, 165t, 170t, 176–191, 177f, 179t, 181t, 182t, 183t, 184t, 185f, 186t, 187t, 189t, 195–200, 230–231, 234
Biologics Price Competition and Innovation Act (BPCIA), 164
Biosimilars, 163–164, 176–177, 192, 195–196, 200, 318–319, 318t, 322, 324
Biotech, 163–164, 176–179, 177f, 178f, 179t, 184t, 196, 200
Brecanavir, 377, 378f

C

Cell-cell communication
    ACh, 7
    antagonist substance, 8–9
    blood–brain barrier, 15
    cyclic AMP, 6
    cytosolic proteins, 15
    enzymes. *See also* Enzymes
        activation, 5, 5f
        inhibitors, 15
    intercellular communication processes, 7, 8f
    ion channel, 5, 6f, 41t
    neuroeffector junction, 5, 5f
    neurojunctions, 11, 12f
    neurotransmitter. *See* Neurotransmitters
    parasympathetic nerves, 9, 10f, 10t
    presynaptic terminal glial cells, 7
    prokaryotic bacterial cell, characteristics, 3, 3f

417

Cell-cell communication (*Continued*)
  receptors. *See* Receptors
  sympathetic nervous system. *See* Sympathetic nervous
    system
  voluntary motor functions, 9, 9f
Center for Biologics Evaluation and Research (CBER), 177
Center for Drugs and Biologics (CDB), 176–177
Centres of Excellence in Drug Discovery (CEDDs), 249
Chaperone therapy, 337
Chimeric mab, 173–174, 186t
Chinese hamster ovary (CHO) cells, 287
Cholesterol absorption inhibitor, 400f. *See also* Ezetimibe;
    Hypocholesteroaemic drugs
  Acyl-coA:cholesterol acyltransferase (ACAT) inhibitors,
    403, 404f
    hypolipidaemic drugs, 406, 406f
    lead compound synthesis, 406f, 407
    SCH 48461 synthesis, 407–408, 407f
  bile acid sequestrants, 401–402, 402f
  biochemical pathways, 400
  de novo synthesis, 400
  HMG-CoA reductase inhibitors, 404–405, 404f
  lipoprotein particles, 401
  low-density lipoprotein and high-density lipoprotein
    cholesterol, 401
  Niemann–Pick diseases, 414
  saponins, 405–406, 405f
Chorus, 250
Contract research organization (CRO), 246
Crohn's disease, 387
Cyclic adenosine-3′,5′-monophosphate (cyclic AMP), 6, 35
Cyclooxygenase, 72
Cytochrome P450 (CYP) enzymes
  CYP3A4 metabolism, 107–108, 108f
  isoforms of, 107, 107t
  time-dependent inhibition, 108–109, 108f, 109t
Cytokines, 387

D

Darunavir
  antiviral activity, 372–374, 373t
  backbone binding concept, drug resistance
    protein backbone target, 367–368, 369f
    TMC-126, 370
  clinical usage, 377
  cyclic ether-derived ligands
    bicyclic bis-tetrahydrofuran ligand. *See* Bicyclic
      bis-tetrahydrofuran (bis-THF) ligand
    cyclic sulfones, 364, 365f
    P2 asparagine side chain carbonyl, 361–362, 363f
    tetrahydrofuran oxygen, 361–362
    tetrahydrofuranyl P2 ligand, 363, 364f

dimerization, 375–376
hydroxyethyl(arylsulfonamide) isosteres, 369–370, 370f
peptide bond mimetic
  natural products and analogues, 360, 361f
  platensimycin, X-ray crystal structure, 360–361,
    362f
  polar hydrogen bonding interactions, 360–361
protease inhibitor structure, 377, 378f
structure, 356–357, 357f
thermodynamic and kinetics effects, 374–375
TMC-126 and TMC-114, 372, 373f
X-ray crystal structure, 374, 375f
Diabetes. *See* Type 2 diabetes
Diethylstilbestrol, 346
Dipeptidyl peptidase-4 (DPP-4), 214, 215t
Discovery performance units (DPUs), 249
DMPK considerations. *See* Drug metabolism and
    pharmacokinetic (DMPK) considerations
Drug administration, 18
  vs. elimination, 16, 17f
Drug discovery process
  actives, definition, 84
  biological targets, 82–83
  clinical efficacy, 83–84
  fast follower approaches, 85, 94, 95f
  fragment-based drug discovery
    aminoindazole fragment hit, 91, 92f
    fragment hits growth, 91, 91f
  hits, definition, 84
  HTS process, 85–86, 85f, 86f
  IKK2 inhibitor, case study. *See* IKK2
  lead generation process, 94, 95f. *See also* Lead
    generation process
  lead optimisation, 83f, 84, 123f. *See also* Lead
    optimisation
  leads, definition, 84
  natural products, 89–90, 89f
  proof of mechanism biomarker, 84
  proof of principle, 84
  structure-based drug design, 86–88
  structure-based virtual screening. *See* Structure-based
    virtual screening
  target validation, 84
  Vardenafil from Sildenafil evolution, 94, 94f
Drug metabolism and pharmacokinetic (DMPK)
    considerations
  biliary excretion, 103
  compound levels, effect of, 103, 103f
  definitions, 98
  drug absorption process and factors, 98, 99f
  glutathione Phase II metabolism, 99, 100f
  haeme-containing P450 enzymes structure, 99, 99f

in vitro assays, 99–100, 101t
in vitro/in vivo PK screening, 98
phase I metabolism, 99, 100f
plasma concentration, 101, 102f
quantitative PK measurements, 101
rat clearance *vs.* microsomal metabolism rate,
    103, 104f
renal secretion, 103

E

EGFR tyrosine kinase. *See* Epidermal growth factor
    receptor (EGFR) tyrosine kinase
EGFRTK signal transduction cascade, 256–257,
    257f
Enzymes
    Acyl-coA:cholesterol acyltransferase (ACAT). *See*
        Cholesterol
    activation energy, 23, 23f
    allosteric regulation, 34t
        of biosynthetic pathways, 33, 34f
        of enzyme activity, 33, 33f
        of glycogen, 33–34
    angiotensin cleaving enzyme, 73, 73f
    aspartic acid protease, 357–358, 358f
    aspirin and COX, 72
    binding and catalytic site, 23, 24f
    catalysis rules, 21–22
    classification, 71
    coenzymes, 25, 25t
    cofactor, 22t, 24
    COMT, 7
    classification, 21, 22t
    denaturation, 21
    dopa decarboxylase, 27–28
    enzyme-controlled biosynthesis, 27
    enzyme replacement therapy. *See* Protein therapeutics
    enzyme–substrate complex, 22, 24
    HMG-CoA reductase inhibitors. *See* Cholesterol
    inhibition and inhibitors, drugs, 31, 31t, 71–72
        activity loss, 28
        antibacterial agent, 29
        folates, biosynthesis, 28, 30f
        irreversible inhibitors, 28
        methotrexate, 29
        uric acid, biosynthesis, 31, 31f
    isoenzymes, 20
    kinases. *See* IKK2 inhibitors; EGFR tyrosine
        kinase
    multienzyme complexes, 20–21
    noncompetitive inhibitors, 32, 32f, 32t
    properties, 21
    prosthetic groups, 24–25

reaction rates
    enzyme activity measurement, 25
    Lineweaver–Burk plot, 26–27, 26f
    Michaelis–Menten equation, 25–26
    substrate concentration, 25–26, 26f
    replacement therapies, 310
structure of, 19, 20f. *See also* Protein structure
therapeutic, 138–140, 139t
types, 19
Epidermal growth factor receptor (EGFR) tyrosine kinase.
    See also Gefitinib
    A431 human vulvar carcinoma cells, 258
    A431 xenograft, 272–273, 272f
    anilinoquinazolines
        4-chloroquinazolines, 260–262, 261f
        enzyme screening hits, 259–260, 260f
        in vitro data, 259–260, 260t
        stimulated and non-stimulated KB cell growth,
            259–260, 261f
    ATP *vs.* tyrosine residue, 257–258, 257f
    cytotoxic agents, 256
    dual-ErbB inhibitors, 277, 278f, 278t
    EGF, 256–257
    first-generation irreversible inhibitors, 277–279
    in vivo model
        A431 human vulvar squamous cell carcinoma cell
            line, 267
        candidate molecules, blood levels, 269, 269f
        in vitro and exposure data, 268t, 270, 270f
        pharmacokinetic profiles, 267, 267f
    isoflavones, 258, 259f, 259t
    monoclonal antibodies, 279
    second-generation irreversible inhibitors, 279
    structure–activity studies
        cell potency *vs.* receptor autophosphorylation,
            265–266, 266f
        6,7-dimethoxyquinazoline, 262, 263f, 263t
        electron-donating groups, 265
        heterocycle reduced EGF kinase activity, 263–264,
            264f, 264t
        isoquinoline and quinoline analogues, 262, 262f, 262t
        KB cell potency *vs.* enzyme activity, 265–266, 265f
    xenograft activity *vs.* c-fos inhibition, 272–273, 273t
Epigenetics, 230
Epoetin, 135–136, 317
    biosimilar epoetins, 318–319, 318t
Eptifibatide, 189t, 192, 194t
Erythropoiesis-stimulating agents (ESAs)
    AIDS patients treatment, 321
    cancer chemotherapy, 320–321
    chronic kidney disease, anaemia, 320
    Darbepoetin alfa, 319

Erythropoiesis-stimulating agents (ESAs) (*Continued*)
  immunogenicity, 321–322
  methoxy polyethylene glycol-epoetin beta, 319
  non-haematopoietic actions, 322–323
  non-human cell-derived products, 321
  rhEpo preparations, 316–318
Erythropoietin (Epo), 135
  bioassays, 316
  colony-forming units, 313–314
  colony-stimulating factors, 308
  Epo receptor, 315–316, 315f
  erythropoiesis, 308
  ESAs. *See* Erythropoiesis-stimulating agents (ESAs)
  expression, 313, 314f
  gene therapy, 324
  hematopoietic growth factors, 308, 309t
  myeloid haematopoietic cells, 308
  O_2 pressure, 309–310
  peginesatide structure, 323–324, 323f
  production sites and control, 310, 311f
  structure of, 310–311, 312f
ESAs. *See* Erythropoiesis-stimulating agents (ESAs)
Ezetimibe
  alkoxy series, 411, 411f
  azetidinone structure–activity relationship, 408–409, 409f
  cholesterol-fed animals, 413
  metabolism of, 409, 410f
  phase III human clinical trials, 412–413, 413t
  putative metabolites, 409, 410t
  SCH 58235 design, 411, 411f
  synthesis of, 411, 412f

F

Fc fusion proteins (FcFPs), 147–149, 148f
Fibrates, 403, 404f
Fragment antigen binding (Fab), 387
Fragment-based drug discovery (FBDD)
  aminoindazole fragment hit, 91, 92f
  fragment hits growth, 91, 91f
  inhibitors binding, 91, 92f

G

Gaucher disease, 330–332
Gefitinib. *See also* EGFR tyrosine kinase
  Iressa pan-Asia study study, 274, 275f
  Kaplan–Meier curves, 274–277, 275f, 276f
  non small cell lung cancer, 274
  structure, 271, 271f
  synthetic route, 273, 274f
Genetic biomarker, 230–231
Glucagon-like peptide-1 (GLP-1), 211–214, 213t. *See also* Incretin-based therapies

Gonadotropin, 140
G-protein-coupled receptor (GPCR), 6, 41t, 43–44, 45f, 46, 47f, 51, 212
Granulocyte-colony stimulating factor (G-CSF), 136–137
Granulocyte–macrophagecolony stimulating factor (GM-CSF), 136–137

H

Haematopoietic stem cell therapy (HSCT), 336
Hatch-Waxman Act, 176
Hepatitis C virus, 134–135
HER2. *See* Human epidermal growth factor receptors
Herceptin™. *See* Trastuzumab
hERG channels. *See* Human ether-a-go-go related gene (hERG) channels
High-throughput screening (HTS), 85–86, 85f, 86f
Hill–Langmuir equation, 38
HIV-1. *See* Human immunodeficiency virus
HIV-1 aspartic acid protease, 357–358, 358f
HMG-CoA reductase. *See* Cholesterol
Hormone replacement therapy (HRT)
  cancer, 346–348
  heart disease, 348
  oestrogen products, 344
  oestrogen therapy
    17-beta-oestradiol, 346
    diethylstilbestrol, 346
    hormone response element, 344
    Premarin®, 344–345
  osteoporosis, 348
  protocols, 352
  WHI
    breast cancer, 348–349, 351, 351t
    retro-progestogen, 349
HSCT. *See* Haematopoietic stem cell therapy (HSCT)
HTS. *See* High-throughput screening (HTS)
Human epidermal growth factor receptors (HER2), 192, 194–195, 284–286
  cell membrane-bound proteins, 284
  human mammary carcinoma, 284–285
  pertuzumab. *See* Pertuzumab
  proto-oncogenes, 284
  structure and function
    ECD shedding, 286
    features, 285
    HER2 overexpression, 286
    ligand-activated heterodimerization, 285, 285f
    phosphotyrosine sites, 286
  trastuzumab. *See* Trastuzumab
  trastuzumab emtansine. *See* Trastuzumab emtansine
  tyrosine kinases, 284

Human ether-a-go-go related gene (hERG) channels
    in heart cells, 115–118
    QT interval, 115–117
    X-ray crystal structure, 115–117, 117f
Human ether-a-go-go related gene (hERG) channel blockers
    basicity and lipophilicity reduction, 115–118, 117f, 118f
    cetirizine, 117, 118f
    cisapride, 117–118
    genotoxicity, 119, 119f
Human genome, 19
Human immunodeficiency virus/acquired immune deficiency
        syndrome (HIV/AIDS). See Anti-HIV protease
        inhibitor Darunavir
Hunter syndrome, 334
Hurler–Scheie syndrome, 334
Hurler syndrome, 334
Hydrogen bonding, 17–18, 18f
Hydrophobic bonding, 16, 17f
Hypocholesterolaemic drugs, 402
    niacin, 403, 403f
    probucol, 403, 403f
Hypoxia, 313, 314f, 321–322

I

Idiotype, 389, 394
IKK2 inhibitors, 111–113, 111f, 112f, 112t, 113t
Immunogenicity, 153, 321–322, 324
Immunoglobulins, 142
Incretin-based therapies
    ADA, 220
    DPP4, 220–221
    GLP-1, 212, 213t
    glucose control, 216–218, 217t
    incretin effect, 211–212, 212f
    injection site reactions, 220
    nausea, 220
    pharmacokinetics, 214–216, 215t, 216f
    plasma glucose excursion, 211–212
    weight loss, 218
IND (investigational new drug), 177, 181t
Indeglitazar discovery, 93, 94f
Insulin, 76, 77f, 133–134, 171, 173f
Interferons (IFNs), 134–135

L

Lapatinib, 194–195
Lead discovery, 197, 232f, 234, 239t, 240
Lead generation process
    active-to-hit workflow, 95–96, 95f
    hit-to-lead
        carboxylic acids potential, 106, 107t
        CRTh2 receptor, 105

CYP enzymes. See Cytochrome P450 (CYP) enzymes
DMPK considerations. See Drug metabolism and
        pharmacokinetic (DMPK) considerations
    lead target profile, 109–110, 110t
    lipophilicity, 96, 96t
    LLE, 109, 110f
    plasma protein binding effect, 105, 106f
    solubility, 97–98, 97f
Lead optimisation
    aprepitant, 121–123, 121f, 122t, 123f
    candidate biological target profile, 114, 114t
    cardiovascular side-effects. See Human ether-a-go-go
        related gene (hERG) channels
    common toxicities, 115, 116t
    dose-to-man prediction, 120, 121f
    genotoxicity, 119, 119f
    pharmacokinetic–pharmacodynamic study, 120
    phospholipidosis, 119–120, 120f
    safety considerations and potential liabilities, 114–115
    screening cascade and design-make-test cycle,
        113, 113f
    therapeutic ratio/safety margin, 115, 116f
Leukaemia, 134–135, 143t, 150
Ligand-gated ion channel, 42–43, 43f
    AChR activation, 46
    equilibrium fraction, 46
Ligand lipophilicity efficiency (LLE), 109, 110f
LSD (lysergic acid diethylamide), 40
Lysosomal storage disorders (LSDs)
    acid hydrolases, 329
    cell dysfunction, 329
    infantile-onset forms, 329
    inherited metabolic diseases, 328–329, 328t
    efficacy improvement
        bone targeting, 335–336
        muscle targeting, 335
        neuropathic LSDs, CNS target, 336
        target tissues, 335
    enzyme replacement therapy (ERT)
        Anderson-Fabry disease, 332
        Gaucher disease, 330–332
        mannose-6-phosphate, 330
        MPS. See Mucopolysaccharidosis (MPS)
        Pompe disease, 333
        residual 'health', 332
        synthesized lysosomal hydrolases transport,
            330, 331f
    stem cell therapies
        enzyme stabilization, 337
        HSCT, 336
        substrate reduction therapy, 337–338
    ultra-orphan diseases, 329–330

M

Maroteaux–Lamy syndrome, 334–335
Medroxyprogesterone acetate (Provera™), 345, 348, 350t
Monoclonal antibody (mAb), 142–146, 143t, 173–174, 177–179, 387–389, 394–395
   antibody–antigen binding fragment, 146
   constant domain amino acid sequence, 142
   de novo mAb discovery, 146
   engineered antibody fragment therapeutics, 149–151
   Fc engineering, 147
   FcFPs, 147–149, 148f
   heavy and light chains, 171
   humanization technology, 146–147
   IgG class, 142–145
   marketed antibody therapeutics, 142, 143t
   NABPs, 151–152, 152f
   N297 glycosidation, 145–146
   structure, 145–146, 146f
Mucopolysaccharidosis (MPS)
   clinical disorders, 333–334
   MPS I, 334
   MPS II, 334
   MPS VI, 334–335

N

NDA (new drug application), 176
Natural killer cell (NK), 142–145
Neurotransmitters, 4–5, 7–15, 8f, 9f, 10f, 10t, 11f, 12f, 13t, 14t, 18, 36, 41t, 42, 44
Niemann-Pick, 328t, 414
Non-antibody binding proteins (NABPs), 151–152, 152f
Non-small-cell lung cancer (NSCLC), 274

O

Outsourcing, 246–248

P

Pan assay interference compounds (PAINS), 85–86, 86f
Paralogue, 233–234
Paratope, 389, 390f, 391f
Parasympathetic nervous system
   activation of, 9, 10t
   organ control, 9, 10f
Peptide drugs, 186–191, 187t, 189t
   approved drugs, 188–191, 189t
   native and nonnative structure, 186
   properties of, 186, 187t
Peptide therapeutics, 140–141
Perjeta™. *See* Pertuzumab
Pertuzumab

   action and preclinical activity, 297–298, 297f, 298f
   CDR sequences, 297
   pathological complete response rate, 300
   pharmacokinetic behaviour, 299
   progression-free survival, 299, 300f
   synergistic antitumour effect, 299
   taxane-based chemotherapy, 299
Pharmacokinetic–pharmacodynamic study, 120
   dose-to-man prediction, 120, 121f
Phospholipidosis, 119–120, 120f
Pipeline(for new drug medicines), 229, 232–234, 233f, 236f, 237f, 240–243, 248–249, 249f
Plasminogen activator, 138, 154
Polyethylene glycol (PEG), 387–389, 390f, 391f
Pompe disease, 333
Portfolio management, 244–245
PPAR-alpha (peroxisome proliferated receptor), 403
Premarin, 344–346
Prodrug, 16
Progesterone, 344, 347–349
Proof of concept (POC), 235, 250
Proof of mechanism biomarker, 84
Protein structure
   enzymes. *See* Enzymes
   folds and structural bioinformatics, 69–71
   α-helix, 64, 66f
   ion-pair network, 67, 69f
   peptide bond formation, 63, 64f
   planar peptide bond, 63–64, 65f
   Protein Data Bank, 68–69
   Ramachandran plot, 65, 68f
   receptors
      G-protein-linked receptors, 74, 76f
      kinase-linked receptors, 74–75, 77f
   β-sheet, 64–65, 67f
   types, 66, 70f
Protein therapeutics
   anthrax bacterium, 129
   biologics sales, 128–130, 129f, 130f
   biosimilars, 157
   catumaxomab, 156–157
   cell banking, 154–155
   cell-based expression systems, 153–154
   enzyme replacement therapy, 130f, 138–139, 139t
   high-quality recombinant protein, 154
   immunogenicity, 153
   microbial fermentation, 129
   monoclonal antibody. *See* Monoclonal antibody (mAb)
   regulatory and enzymatic activity, 133
   replacement therapies
      Cerezyme®, 138–139
      coagulation and fibrinolytic regulation, 138

enzyme replacement therapy, 138–139, 139t
EPO, 135
G-CSF and GM-CSF, 136–137
growth factors, 137, 137t
IFNs, 134–135
insulin, 133–134
Myozyme®, 139–140
peptide therapeutics, 140–141
Replagal®, 139–140
VPRIV®, 138–139
vs. small molecule drugs
aspirin, molecular structure, 131, 131f
characteristics, 131–132, 132t
functional classification, 133
immunoglobulin-based therapeutics, 132–133
subcutaneous administration, 155
therapeutic molecules, definition, 128

R

Receptors
AChRs, 11–13, 13t, 42–43, 43f, 46. *See also* nicotinic
receptor
adenosine, 41t
adrenaline and noradrenaline receptors, 4, 7, 12f, 13,
14t, 15, 18, 33–34, 41t, 46, 89f
β-adrenoceptor structure, 43–44, 45f
agonist-binding reaction, 53
agonist efficacy, 51
ATP, 41t, 259–260, 266–267, 271, 277
bioassay, 35–37, 36f
constitutive activity, 53
desensitization, 49–51
dopamine, 4, 15, 27, 27f, 41t
drug–receptor interactions, 37–39, 39f
enkephalin receptor, 14
equilibrium fraction, 46
GABA, 41t, 42–43
Glutamate, 41t
GPCR. *See* G-protein coupled receptor
HER2. *See* Human epidermal growth factor receptors
histamine receptors, 7, 15, 18, 41t, 44
kinase-linked receptors, 74–75, 77f
nicotinic receptor, 74, 75f. *See also* AChRs
partial agonism, 51
protein receptors, 74–75
radioligand-binding studies, 39–40
Schild equation. *See* Schild equation
Serotonin (5-HT), 4, 7, 15, 37, 40, 41t
SSRI (selective serotonin reuptake inhibitor), 7
types, 41t, 42
Recombinant proteins, 163–169, 175–177, 179,
183, 186t, 187t, 189t, 197

Regulator/effector molecule, 32–33
Research and development strategies
attrition-based pipeline, 235–237, 236f, 237f
big *vs.* small companies, 242–243
cost reduction
Centres of Excellence in Drug Discovery, 249
discovery performance units, 249
line departments, 245
open innovation, 250
process improvement, 245
proof of concept, 250
vertical integration *vs.* disintegration. *See* Decentralization
decentralization
in-house resource location, 245–246
outsourcing and partnering, 246–248
virtual discovery and development, 246, 246f,
248
virtual pipeline, 248–249, 249f
failed drug cost, 227
hypothetical cash flow *vs.* time plot, 227, 228f
in-house
vs. in-license, 240–241, 241f
vs. out-license, 241–242
lead discovery, 234
lead optimization, 234–235
preclinical evaluation, 235
probability of success, 230
project management, 244
proof of concept, 235
Registration and Launch, 235
resource
line departments, 238, 239t
nonaligned resource, 240
therapy areas, 240
revenue opportunities, 230
stage-gate organization, 232–234, 233f
standard operating procedures, 243
target discovery, 234
Retro-progestogen, 349
Ritonavir, 372, 372t

S

Saquinavir (SQV), 358–360, 359f, 372t
Scatchard plot, 39–40
SCH 48461. *See* Ezetimibe
Scheie syndrome, 334
Schild equation
dose ratio, 48
nicotinic antagonist, 48, 50f
open ion channels, 49
for tubocurarine blockade, 49, 50f
Sinoatrial node (SA), 46

Small-molecule drugs (SMDs)
  acquisitions and partnerships, 164, 170t
  antithrombotic agents
    GPIIb/IIIa antagonists, 192, 193f, 194t
    Her2 inhibitors, 194–195
  vs. biologics
    biologic drug approvals, 177, 178f, 179t
    biotherapies, 177, 178f
    Center for Drugs and Biologics, 176–177
    Center for Drugs Evaluation and Research, 177
    clinical stages, 179
    clinical success rates, 180, 181t
    drug lifetime, 176, 177f
    Mabs. *See* Monoclonal antibody (mAb)
    mean success rates, 185–186, 186t
    New drug application, 176
    R&D cost, 183–185, 184t, 185f
    resource-constrained environment, 200
    stage-related clinical cycle times, 180–183,
      182t, 183t
    stage-related differences, 198, 199t
    stage-related success rates, 180, 182t
    World Health Organization therapy area classes,
      198, 199t
  Biologics Price Competition and Innovation Act,
    164
  vs. biomolecular drugs
    active metabolite, 174
    antigenicity and hypersensitivity, 175
    aspirin *vs.* antibody, 169, 171f
    clearance mechanisms, 175
    clinical phases, 198
    cytochrome P450s, 174
    discovery and development stages, 196, 197f
    distribution, 174
    dosing regimen, 174
    drug–drug interactions, 175
    lead discovery, 197
    lead optimization, 197–198
    manufacture and supply, 191
    natural sources, 164–169
    papain and pepsin, 173
    preclinical evaluation, 198
    pricing, 191–192
    properties of, 169, 172t
    serum half-life, 174
    species reactivity, 174–175
    target discovery, 197
    variable and constant regions, antibody, 171, 173f
  biosimilar biomolecules *vs.* generic small molecules,
    195–196
  biotechnology company, 163–164, 165t

  early biotech approvals, 164, 169t
  peptide drugs. *See* Peptide drugs
SQV. *See* Saquinavir (SQV)
Stage-gate organization, 232–234, 233f
Standard operating procedure (SOP), 243
Statins, 404–405, 404f
Structure-based virtual screening
  Chk-1 kinase,, 86–87, 87f
  CXCR4 antagonists, 88, 88f
  ligand-based VS, 87
  morphine and methadone, 87, 87f
  pharmacophore model, 87–88, 88f
  targeted library design, 88
  virtual screening, definition, 86
Sulphonylurea drugs, 210, 211t, 221–222
Sympathetic nervous system
  activation of, 9, 10t
  fight/flight, 9, 11f
  organs control, 9, 10f

T
Thrombopoietin receptor, 175
Tipranavir, 375–376
Tirofiban, 192, 193f, 194t
Tissue nonspecific alkaline phosphatase (TNSALP), 335–336
Transporter proteins, 82–83, 104, 105f
Trastuzumab
  ADCC, 288–289
  antitumour activity, 288
  chemotherapy administration, 294, 295t
  clinical development, 290
  endocytosis, 289
  immune effector cells, 288–289
  metastatic breast cancer, 290–291, 292t
  metastatic gastric cancer, 294–296
  nonclinical pharmacology, 289–290
  pharmacokinetic profile, 290
  and quality control, 287
  resistance and treatment, 291–293
  subcutaneous formulation, 296
  tolerability and safety, 293–294
  Western blotting, 288
Trastuzumab emtansine
  ADC, 300–301
  characteristics of, 301–302, 301f
  objective response rate, 302
  phase III clinical trial, 302
Tumour necrosis factor (TNF), 147–148, 386–387, 388f,
    390f, 391–392, 392f
Type 2 diabetes. *See also* Incretin-based therapies
  antihyperglycaemic drug therapy, 207b–208b
  cardiovascular risk, 219–220

diagnosis of, 207b–208b
disease modification, 218–219
history of, 209–210, 209f
hyperglycaemia, 206
pharmacotherapy
   β-cell function, 211
   PPAR-gamma transcription factor, 210
   treatment hierarchy, 210, 211t

V
Vertical integration, 245–246
Virtual screening, 86–88

W
Women's Health Initiative (WHI)
   breast cancer, 348–349, 351, 351t

Colour Plate

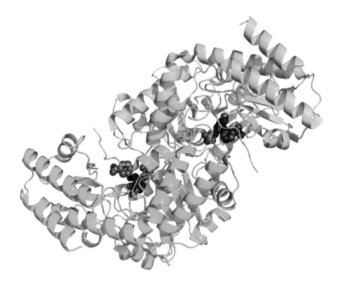

FIGURE 1.15

Structure of an enzyme from a microorganism (Sulfolobus solfatarius).

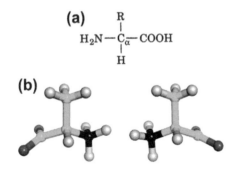

FIGURE 2.1

(a) The basic structure of an amino acid; (b) the L and D forms of the amino acid alanine.

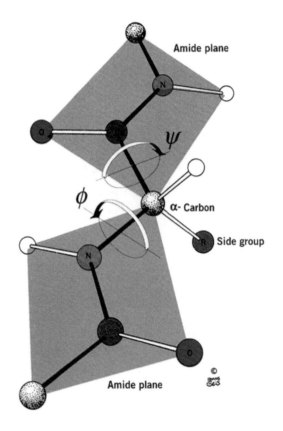

FIGURE 2.6

Figure showing the planar peptide bond.

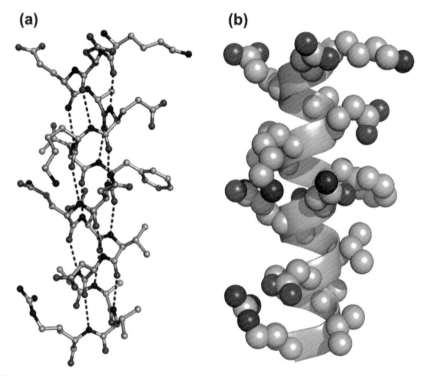

FIGURE 2.7

The α-helix with different R groups decorating the outside. (a) Representation in stick mode showing the H-bond network between the carboxyl and amino groups at every fourth residue along the chain. (b) Representation in ribbon for the C α backbone and space filling for the side chains.

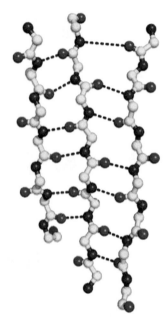

FIGURE 2.8

The β-sheet taken from a sample protein structure. The side chains have been removed. The H-bonding between the carboxyl and amino groups of the different polypeptide chains is shown in black dashed line. The sheet on the right is anti-parallel and the sheets on the left are parallel.

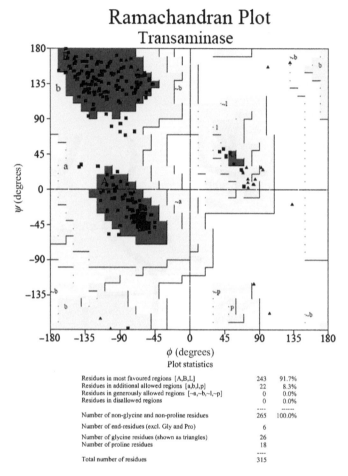

Ramachandran Plot
Transaminase

Plot statistics

Residues in most favoured regions [A,B,L]	243	91.7%
Residues in additional allowed regions [a,b,l,p]	22	8.3%
Residues in generously allowed regions [~a,~b,~l,~p]	0	0.0%
Residues in disallowed regions	0	0.0%
	----	------
Number of non-glycine and non-proline residues	265	100.0%
Number of end-residues (excl. Gly and Pro)	6	
Number of glycine residues (shown as triangles)	26	
Number of proline residues	18	

Total number of residues	315	

Based on an analysis of 118 structures of resolution of at least 2.0 Å
and R-factor no greater than 20%, a good quality model would be expected
to have over 90% in the most favoured regions.

FIGURE 2.9

An example of a Ramachandran plot produced from PROCHECK (*9*). The red region on the top left is for parallel and anti-parallel β-sheets, the lower left red region for α-helices and the small red region on the right for left-handed helices which are not found in globular proteins; however, some amino acids with important catalytic function can fall into this region. The number of amino acids falling in disallowed regions is indicative of the quality of the crystal structure.

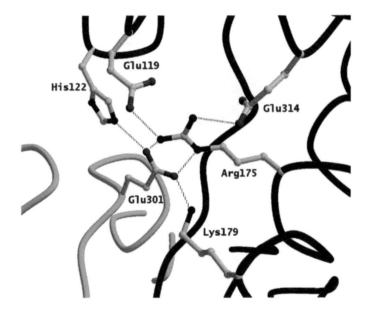

FIGURE 2.10

An ion-pair network on the interface between two protein subunits A (green) and subunit B (blue) of a thermophilic alcohol dehydrogenase enzyme (*24*).

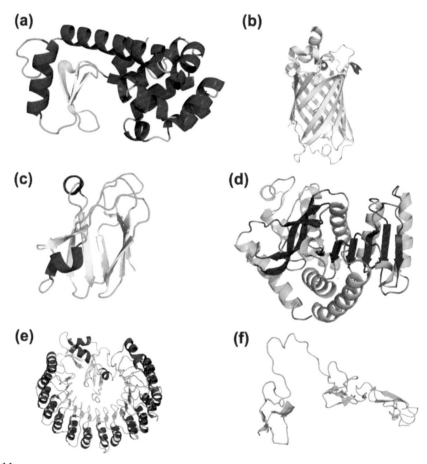

FIGURE 2.11

A figure showing examples of the different types of protein fold as described. (a) Mainly α-helical, lysozyme, (b) and (c) mainly β-sheet, yellow fluorescent protein and plastocyanin (d) αβ-protein, ribonuclease inhibitor (e) α + β protein, lactate dehydrogenase and (f) unstructured proteins with disulfide bridges, EFG domain.

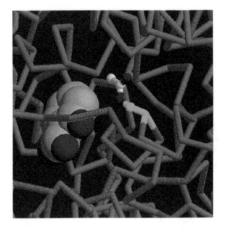

FIGURE 2.12

The aspirin bound to the serine in the active site of the cyclooxygenase.

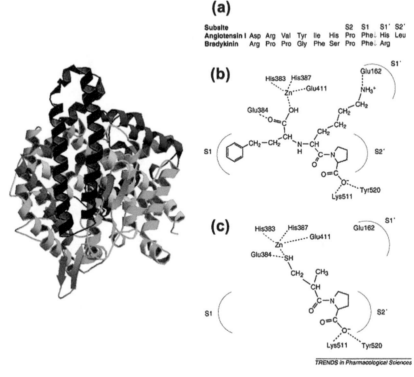

(a)

Subsite								S2	S1	S1'	S2'
Angiotensin I	Asp	Arg	Val	Tyr	Ile	His		Pro	Phe↓	His	Leu
Bradykinin		Arg	Pro	Pro	Gly	Phe	Ser	Pro	Phe↓	Arg	

(b)

(c)

TRENDS in Pharmacological Sciences

FIGURE 2.13

The human ACE enzyme and the active site complexes with inhibitors taken from Ref. (*21*). (a) The sites of cleavage by ACE in the substrates angiotensin I and bradykinin. (b) The principle interactions between the ACE inhibitor lisinopril and the active site of testicular angiotensin cleaving enzyme (tACE). Non-covalent interactions between enzyme groups and the inhibitor are indicated by broken lines and the sub-sites corresponding to those in (a) are labelled. (c) Projected interactions between captopril and tACE, based on the structure of the ACE–captopril complex with ACE sub-sites indicated.

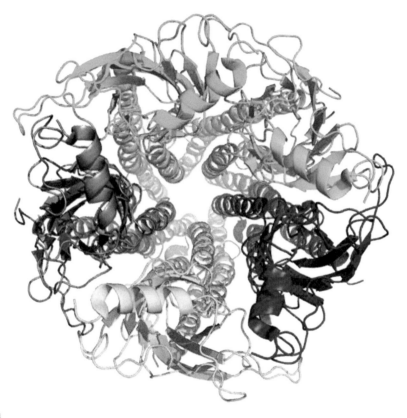

FIGURE 2.14

End on view of the structure of the nicotinic acetylcholine receptor PDB 2BG9 showing the five helical poly-peptides that pass through the membrane in different colours.

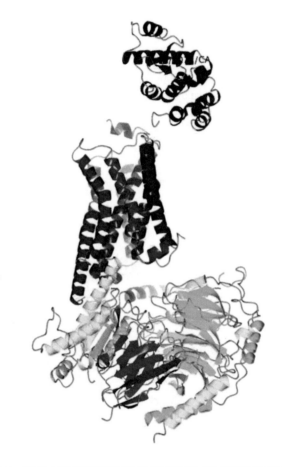

FIGURE 2.15

Structure of the b2 adrenergic receptor–G-protein complex. PDB 3SN6. Figure taken from Ref. (*22*). The structure gives a detailed view of transmembrane signalling. The ternary complex made up of the agonist, the receptor and the G-protein undergoes conformational changes to enhance the affinity of the agonist for the receptor and to favour the hydrolysis of guanosine triphosphate to guanosine diphosphate by the G-protein. At the top shown in purple is an easily crystallizable protein lysozyme which is not a part of this complex but facilitates the crystallization. The agonist binds to the transmembrane helices which contact the G - protein which is a trimer. The protein in red is called a nanobody which is present to aid the crystallization.

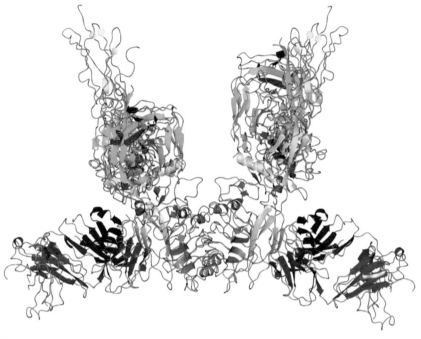

FIGURE 2.16

Part of insulin receptor which binds the insulin protein PDB 3LOH.

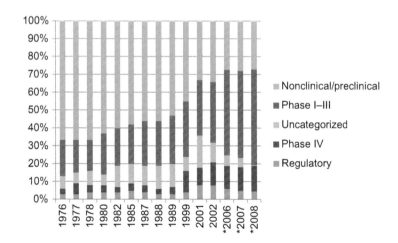

FIGURE 5.7

Pharmaceutical industry percentage allocation of R&D expenditures from 1976 to 2008. Data for 1976–2002 is provided by PhRMA and analyzed by Cohen (*36*). Nonclinical/preclinical is the term by PhRMA for discovery and preclinical research.

*Data supplied from PhRMA profiles for 2008–2010 (34,39,40).

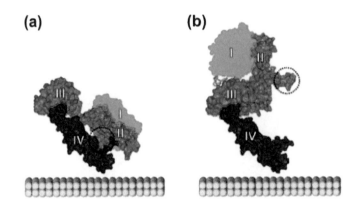

FIGURE 9.1

Conformation of inactive (a) and EGF (orange-coloured) activated (b) EGFR extracellular domain.

Source: Illustrations taken from Ref. (22).

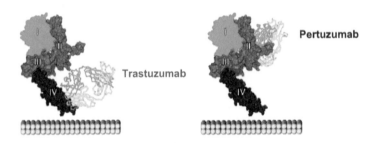

FIGURE 9.2

Trastuzumab and pertuzumab bind to distinct epitopes on HER2 extracellular domain.

Source: Illustrations taken from Ref. (22), according to data from Ref. (23).

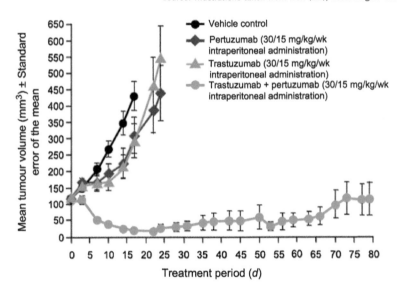

FIGURE 9.3

Synergistic efficacy of pertuzumab + trastuzumab combination therapy in a HER2-positive breast cancer xenograft model.

Source: From Ref. (14).

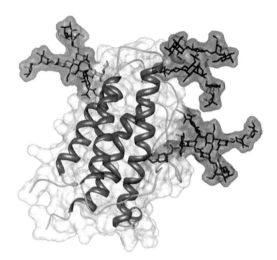

FIGURE 10.2 Predicted structure of human Epo.

The three complex *N*-linked glycans are shown in purple and the single *O*-linked glycan in pink.

Source: The figure is reproduced with permission from the Woods Group using Chimera. [Woods Group. (2005–2012) GLYCAM Web. Complex Carbohydrate Research Center, University of Georgia, Athens, GA, USA (http://www.glycam.org)].

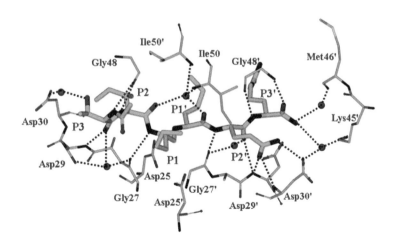

FIGURE 13.2

X-ray crystal structure of the HIV-1 protease in complex with a substrate analogue (green) showing closed conformation flaps.

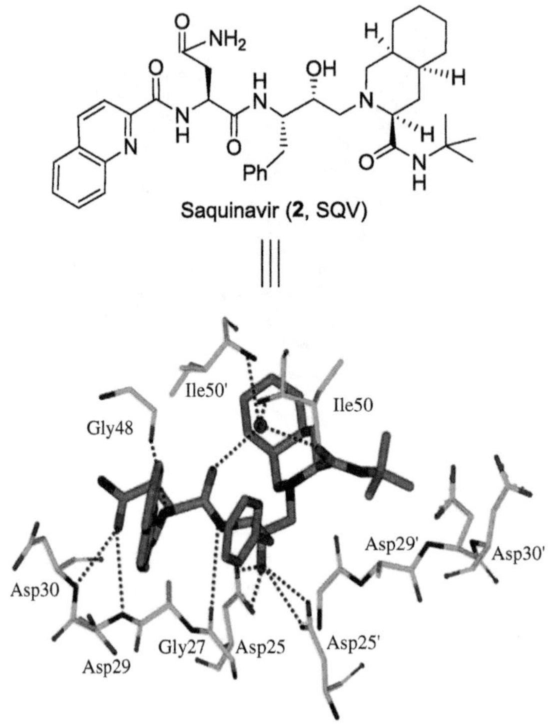

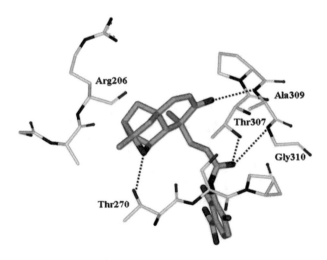

FIGURE 13.3

Saquinavir (2) and X-ray structure of SQV–HIV-1 protease complex.

FIGURE 13.5

X-ray crystal structure of platensimycin (5) bound to bacterial FabF enzyme.

Asp 30
Asp 25
Asp 25'
Asp 29
Gly 48
8 IC_{50} = 0.05 nM
CIC_{95} = 8 nM
11 IC_{50} = 132 nM
CIC_{95} = 800 nM
9 X = O, IC_{50} = 5.4 nM
CIC_{95} = 100 nM
10 X = CH_2, IC_{50} = 2.6 nM
12 IC_{50} = 694 nM

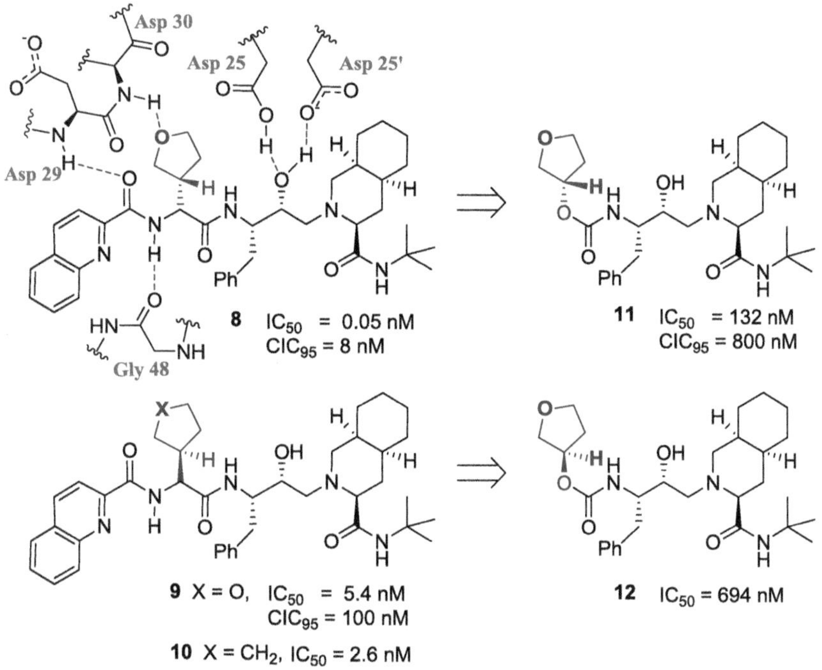

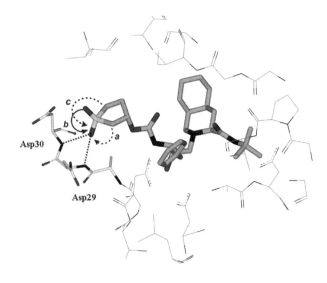

FIGURE 13.6

Introducing stereochemically defined cyclic ether-based ligands into PIs.

c
b
a
Asp30
Asp29

FIGURE 13.9

Design strategy based on modelling of 3-(*S*)-sulfolane-based inhibitor **16**.

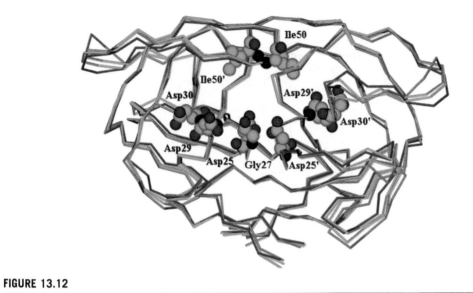

FIGURE 13.12

Superimposition of X-ray crystal structures of mutant proteases' backbones.

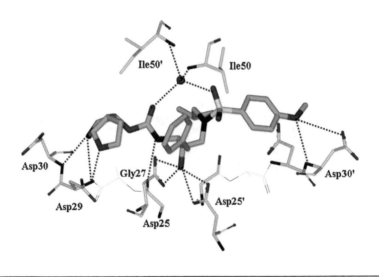

FIGURE 13.14

X-ray crystal structure of inhibitor **28**–HIV-1 protease complex.

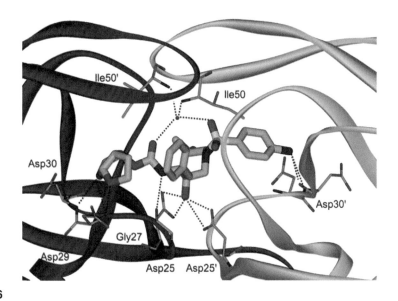

FIGURE 13.16

X-ray structure of DRV-bound HIV-1 protease complex.

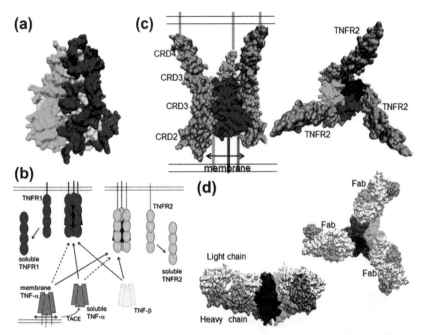

FIGURE 14.1 Mode of binding of TNF-α and TNF-β to their receptors and TNF-α neutralization by monoclonal antibodies.

(a) Tridimensional structure of TNF-α drawn from pdb 1tnf, showing the surfaces of the three trimers in green, red and blue (behind). TNF-β, which has the same trimeric structure as TNF-α, shares with it a limited number of residues in surface; these residues are shown in yellow. This limited surface homology explains why anti-TNF-α mAbs do not bind to TNF-β, and vice versa. (b) Although different, TNF-α (in blue) and TNF-β (in yellow) bind to the same receptors, TNFR1 (in red) and TNFR2 (in orange). TNFR1 and TNFR2 have four cysteine-rich domains (CRDs); TNFR2 (p75) has a higher molecular weight than TNFR1 (p55), mainly due to the presence of glycans in the membrane proximal part of the receptor. Binding of TNF-α or TNF-β to TNFR1 or TNFR2 leads to the trimerization of the receptor, triggering the signalling cascade. TNF-α also differs from TNF-β in that it is a membrane protein secondarily cleaved by the metalloprotease TACE to generate soluble TNF-α, while TNF-β is secreted as a soluble cytokine. Both membrane and soluble TNF-α bind to TNFR1 and TNFR2, although with subtle differences (dotted lines indicate a slightly lower affinity). TNFR1 and TNFR2 can also be cleaved by metalloproteases (or be subjected to alternative spicing), generating soluble receptors that maintain affinity for TNF-α and TNF-β. (c) Tridimensional structure of TNF-α bound to three truncated forms of TNFR2 (limited to the four CRDs), drawn from pdb 3alq. The three monomers of TNF-α are represented with the same colours as in (a); TNFR2 is in orange. The left and right figures are distinguished by a 90° rotation, showing the cytokine laterally and from the top, respectively. CRD2 and CRD3 of TNFR2 bind laterally to two monomers of TNF-α, in a furrow which parallels the main axis of the trimer. This mode of binding, which characterizes the TNF-TNFR family, is at the origin of the receptor trimerization. (d) Representation of what could be the binding of anti-TNF Fab fragments to TNF-α. In the absence of available tridimensional structures of anti-TNF-α mAbs complexed with TNF-α, this representation is that of the binding of Fab fragments of ruplizumab, a neutralizing mAb binding to CD40L, another member of the TNF-α family (pdb 1i9r). VH and CH1 domains from the ruplizumab heavy chain are shown in dark grey, and the VL and CL domains of the ruplizumab light chain are in light grey. The left and right figures are distinguished by a 90° rotation on the Fab axis, showing the cytokine from the top and laterally, respectively. In the present example, the main axis of the Fab is orthogonal to the main axis of the cytokine. The epitope is straddling two monomers and the trimeric symmetry allows the binding of three Fabs to a single cytokine (3:1 ratio). Taking into consideration the mode of binding of TNF-α to its receptors (c), it is obvious that the binding of anti-TNF-α mAbs like that depicted here will prevent the binding of TNF-α to its receptor, which characterizes the neutralization function.

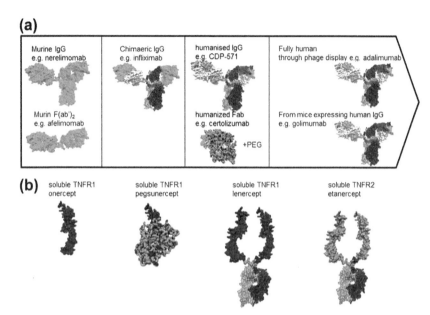

FIGURE 14.2 Biopharmaceutical agents developed to neutralize TNF-α.

(a) Technological improvements in the design of monoclonal antibodies against TNF-α. Starting from murine mAbs (green) produced by the hybridoma technology, full-length IgG such as nerelimomab or F(ab')$_2$ fragments such as afelimomab were developed and tested in humans. The next technological advance was the development of a chimeric mAb through recombining cDNAs coding for VH and VL domains deriving from an anti-TNF-α murine mAb (green) with cDNAs coding for a human heavy (γ) chain and a kappa (κ) light chain, respectively (the κ-chains are shown in light grey, one γ-chain is in dark grey and the other in black). Infliximab was generated with a γ1-heavy chain and is thus is an IgG1 subclass protein. The next step was the development of humanized mAbs, always starting from a murine mAb with a given antigenic specificity, by using refinements of DNA recombination technologies. CDRs from the murine mAbs (in green) are grafted onto frameworks derived from human VH and VL domains. Development of the CDP-571 was discontinued due to lack of efficacy, may be because it was generated as an IgG4 subclass protein (γ4 heavy chain); certolizumab, was developed as an IgG1 Fab fragment complexed with polyethylene glycol chains (in blue and red; see details in Figure 14.3). The last step was the development of fully human mAb, produced directly from human immunoglobulin genes, expressed either in phage libraries (top) or in transgenic mice (bottom). Note that the paratope (in blue) remains the product of both recombinations between human genes and hypermutations (artificial in phage display and natural in transgenic mice), and will have a unique structure, i.e. non-self. The two mAbs currently approved are both IgG1. (b) Initial grouping in the development of anti-TNF soluble receptors. Initially TNFR1 (in red) was developed as a soluble single-chain receptor (onercept), as a single chain complexed to polyethylene glycol chains (schematized in blue and red) (pegsunercept), or as a dimeric receptor fused with an IgG1 Fc fragment. TNFR2 was only developed as a fusion protein with human IgG1 Fc fragment, and is the only soluble TNFR approved (etanercept); its detailed structure is shown in Figure 14.4.

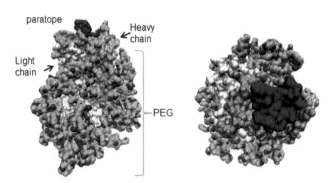

FIGURE 14.3 Model of pegylated Fab.

The Fab fragment of an anti-HIV-1 gp120 IgG1κmAb (clone b12) was used as a model mAb. Its three-dimensional structure (pdb #1hzh) was drawn using surface representation software (VMD software, University of Illinois), with the light and heavy chains in light and dark grey, respectively. The three CDRs from the light chain and the three CDRs from the heavy chain are shown in orange and red, respectively, and together form the paratope (antigen-binding site), better seen on the right. It is known that two 20 kDa molecules of polyethylene glycol (PEG) are bound to the C-terminal part of the heavy chain. The exact conformation of the two PEG chains is still debated. Starting from small PEG chains co-crystallized with proteins (pdb 1to2 and 3b6j), PEG chains of the correct molecular weight were built by juxtaposing existing structures, and were freely represented covering the Fab. Carbon and oxygen atoms of the PEG are shown in blue and red, respectively. In this ideal representation, PEG seems to protect the Fab from degradation and capture by antigen-presenting cells whilst not occluding the antigen-binding site of the mAb. However, PEG chains are very mobile structures and could adopt many different configurations. Precisely how they protect the Fab is not known.

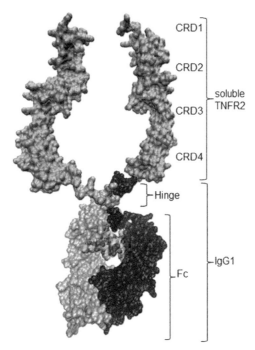

FIGURE 14.4 Model of etanercept.

Although etanercept is TNFR2-Fc, this representation was made by juxtaposing the structure of TNFR1 (pdb #1ext) and the hinge and the Fc portion of an anti-HIV-1 gp120 IgG1 mAb (clone b12) (pdb #1hzh). Both are drawn as surface representations (VMD software, University of Illinois), in orange for TNFR, and dark grey and black for the two γ1-heavy chains. The N-glycans of the Fc are shown in red and blue. TNFR1 was preferred to TNFR2 because TNFR2 (pdb #3alq) is only known in association with TNF-α (Figure 14.1(c)) while TNFR1 alone has been crystallized. The four CRDs of TFR1 and TNFR2 are very similar in their structure.

Printed and bound by CPI Group (UK) Ltd, Croydon, CR0 4YY

15/10/2024

01774737-0001